TENSOR CALCULUS AND ANALYTICAL DYNAMICS

Library of Engineering Mathematics

Series Editor
Alan Jeffrey, University of Newcastle upon Tyne and University of Delaware

Linear Algebra and Ordinary Differential Equations
A. Jeffrey

Nonlinear Ordinary Differential Equations
R. Grimshaw (University of New South Wales)

Complex Analysis and Applications
A. Jeffrey

Perturbation Methods for Engineers and Scientists
A. W. Bush (University of Teesside)

Theory of Stability of Continuous Elastic Structures
M. Como and A. Grimaldi (University of Rome T.V.)

Qualitative Estimates for Partial Differential Equations: An Introduction
J.N. Flavin (University College Galway) and
S. Rionero (University of Naples, Federico II)

Numerical Methods for Differential Equations
J.R. Dormand (University of Teesside)

Tensor Calculus and Analytical Dynamics
John G. Papastavridis (Georgia Institute of Technology)

TENSOR CALCULUS AND ANALYTICAL DYNAMICS

A Classical Introduction to Holonomic and Nonholonomic Tensor Calculus; and Its Principal Applications to the Lagrangean Dynamics of Constrained Mechanical Systems.

For Engineers, Physicists, and Mathematicians

JOHN G. PAPASTAVRIDIS, PH.D

CRC Press
Boca Raton London New York Washington, D.C.

Library of Congress Cataloging-in-Publication Data

Papastavridis, J. G. (John G.),
Tensor calculus and analytical dynamics / John G. Papastavridis.
p. cm. -- (Library of engineering mathematics)
Includes bibliographical references and indexes.
ISBN 0-8493-8514-8 (alk. paper)
1. Calculus of tensors. 2. Dynamics. I. Title. II. Series.
QA433.P36 1998 98-29656
515'.63--dc21 CIP

International Standard Book Number 0-8493-8514-8
Library of Congress Card Number 98-29656
Printed in the United States of America 1 2 3 4 5 6 7 8 9 0
Printed on acid-free paper

To the distant but live and loving memory of
my maternal Grandfather,

IOANNIS E. MITROPOULOS
(ΙΩΑΝΝΗ ΗΛΙΑ ΜΗΤΡΟΠΟΥΛΟΥ, 1883–1956)

A noble and altruistic man;
also an industrialist and entrepreneur.

...with absolute confidence in the great things and with inexhaustible distrust of all details.

P. Frank (*Einstein*, 1947, on Einstein's "manner of confronting the world")

Science and art are not so different as they appear. The laws in the realms of truth and beauty are laid down by the masters, who create eternal works.

M. Born (*Natural Philosophy of Cause and Chance,* 1949, p.7)

...if the history of the exact sciences teaches anything it is that emphasis on extreme rigor often engenders sterility, and that the successful pioneer depends more on brilliant hunches than on the results of existence theorems....

H. Margenau and G. M. Murphy
(*The Mathematics of Physics and Chemistry,* 1943, p. iv)

Progress should be aimed at some specified goal and can only be judged in terms of that goal. Progress should not automatically imply new in place of the old, if the old is actually satisfactory. Nor does criticizing the new automatically mean that the critic wants to return to the past. Rather it means that *the new does not live up to the existing expectations and that the new things need to be better than the old before they should replace what is good from the past*....Criticism of the new may simply imply a notion of conservation [italics added for emphasis]. [In that sense, *this* book is conservationist.]

M. Shallis (*The Silicon Idol,* The Micro Revolution
and Its Social Implications, 1984, p. 4)

Author

John G. Papastavridis, a native of Athens, Greece, and a graduate of its National Technical University (E.M.P.) (B.Sc. in Civil Engineering (1970) and Purdue University (M.Sc., 1972, and Ph.D., 1976, in Continuum Mechanics and Physics), has been teaching Applied Mechanics at the Georgia Institute of Technology, first in its School of Engineering Science and Mechanics (1979 to 1986) and then in its School of Mechanical Engineering (1986 to present). Professor Papastavridis' research interests are in the areas of Analytical and Structural Mechanics (including Stability, Nonlinear Oscillations, and Variational Principles), Mathematical Methods in Engineering (such as Calculus of Variations and Tensor Analysis); also History of Mechanics, Physics, and Technology. Professor Papastavridis has taught mechanics to several thousands of engineering students; has authored over 40 research (archival) papers, and reviewed many more, for several prestigious journals. Also, he has received a number of professional awards, and was the invited speaker of the Midwest Mechanics Seminar Tours for 1993–4.

Currently (Fall 1998), Professor Papastavridis is nearing preparation of an extensive treatise on *Analytical Mechanics,* at the postgraduate/professional level (to be published by Oxford University Press); and of a smaller compendium/handbook on *Elementary Mechanics* at the senior undergraduate/professional level.

Preface

SCOPE, BACKGROUND, CONTENTS

This monograph constitutes a concise but comprehensive introduction to classical tensor calculus, in both holonomic and non- (or an-) holonomic coordinates, and its principal applications to the Lagrangean analytical dynamics of discrete and constrained systems. It is a book written by a theoretically minded mechanician (or mechanicist), for graduate students, professors, and researchers in the areas of theoretical and applied mechanics, engineering science, and mechanical/aerospace/structural (and even electrical) engineering, and also physics and applied mathematics. The latter are particularly welcome in view of the current wide and unhealthy chasm between them and the rest of us.

The sole technical prerequisites here are a solid working knowledge of (1) conventional infinitesimal calculus, including linear algebra and vector calculus (say, on the level of the well-known Schaum's outline series); (2) intermediate-level theoretical mechanics (say, on the level of Synge/Griffith's *Principles of Mechanics,* or Rosenberg's *Analytical Dynamics*); and (3) most importantly, a mature learning outlook.

Specifically, the book covers, in sequence, the following:

Part I: Tensor Calculus

Chapter 1: Introduction and Background (brief history, algebraic preliminaries, fundamental differential–geometric concepts)
Chapter 2: Tensor Algebra
Chapter 3: Tensor Analysis (including nonholonomic tensor calculus)

Part II: Analytical Dynamics

Chapter 4: Introduction to Analytical Dynamics
Chapter 5: Particle on a Curve and on a Surface (in both ordinary Euclidean and Riemannian spaces, including problems of embedding and equivalence)
Chapter 6: Lagrangean Mechanics: Kinematics (constraints, nonholonomic or quasi-variables, transitivity equations, Frobenius' theorem)
Chapter 7: Lagrangean Mechanics: Kinetics (fundamental principles and equations of motion; in both holonomic and nonholonomic variables, with or without constraint reactions, and theorems of power)

The book also contains a fair number of completely solved Examples and Problems with their Answers and/or Hints (for example, the motion of a sled, a rolling sphere, and a coin, (Section 7.7), and the problem of stability of orbits in configuration space, (Section 5.7)) strategically placed throughout the text.

The exposition ends with a far from complete, but relatively detailed and rounded list of References/Bibliography; and a Name/Subject Index.

No single volume can even pretend to cover satisfactorily all aspects of this vast and fascinating subject, much more so when, as is the case here, one must observe inflexible space limits; hence, a certain selection has been necessary and inescapable. I decided to stress what I consider to be the *fundamental minimum core* material of the geometry of Lagrangean dynamics, a fair portion of which is virtually unavailable in a readable book form, in print, and in English,* and have omitted the following important topics: canonical, or Hamiltonian mechanics in nonholonomic variables; variational principles (of Jourdain, Gauss; Maupertuis/Euler/Lagrange, and Hamilton); impulsive motion; group theoretic and invariance aspects (Poincaré/Chetaev theory, equations of Killing, theory of Noether); spinors and quaternions, and their applications to rigid-body kinematics — perhaps another volume in the future.**

However, this modest book *does* promise to its faithful readers that, with its help, they will be able to move quickly and confidently into any particular geometry-based area of theoretical dynamics, in either classical or modern form, and make their creative contributions there. It is in this sense, of Mach's *Denkökonomie,* that this text is a most effective investment of the readers' precious time and effort.

RAISON D'ETRE, AND SOME EDUCATIONAL PHILOSOPHY

The customary words of explanation, or apology, for writing "another" book on tensors and dynamics are now in order. The main theme of this work is formal structure (not formalisms) and basic geometrical/physical ideas behind the most general equations of motion of mechanical systems under linear velocity constraints (or first-order kinematical constraints). It is a book for people who place theories, ideas, and knowledge (and not just "information") above all else — irrespectively of their age or short-term profitability, expediency, or the *ad nauseam* parroted "computational efficiency" — and do not apologize for it.

The existing books on our subject, in print and in English, seem to this author to be too formalistic and, as such, of little use to engineers and other willing and able nonmathematicians; let us say of the sledgehammer-to-crack-a-nut-type, i.e., a lot of expensive theory and a few contrived and/or trivial examples. As such, this book is virtually unique in the entire contemporary English literature, and I believe *does* meet real and long overdue needs of students and teachers of theoretical mechanics.

---

* A notable exception being the masterly but brief (and not too readable by nonmathematicians) account in Schouten (1954b).

** Some of these topics and much more, are treated *nontensorially* in our extensive treatise *Analytical Mechanics* (to appear, we hope, before 2000).

On Notation, Rigor, etc.

In order to make the exposition accessible to as many interested readers as possible (and without succumbing to a lowest common denominator mentality), I have chosen, after considerable mind and soul searching and discussions with colleagues and students, not the direct (or symbolic, or matrix, or dyadic, or coordinate-free, or invariant) notation, but the not so popular today, although paedagogically superior, classical *component* or *indicial* notation — *the notation chosen by the most famous practitioners of geometry, mechanics, and physics of this century,* e.g., (chronologically), Ricci, Levi-Civita, Eisenhart, Veblen, Einstein, Eddington, Weyl, Synge, Schouten, Struik, Vagner (or Wagner), Vranceanu, Wundheiler, Horák, Dobronravov, Lur'e et al., and a host of other contemporary creative researchers ranging from theoretical physics (e.g., Kilmister, McCauley, Pirani, Schmutzer, Stephani) to applied dynamics, robotics, etc. (e.g., Korenev, Maißer).

The indicial notation "kills two birds with one stone": it combines both *coordinate generality* (form invariance) and *coordinate specificity* (i.e., one knows exactly what to do in a specific situation). Frequently, it is the *only* notation, e.g., Christoffel symbols, Ricci–Boltzmann–Hamel coefficients. The sole requirement here, perhaps a little painful and/or frustrating, to the beginner, but really quite economical in the long run, is the correct vertical and (sometimes) horizontal positioning and reading of the indices — *the indices carry the game,* like the wings the airplane!

Other contemporary mathematicians and theoretical physicists agree.*

Nevertheless, since no single notation is *uniformly* better than all the others, a certain blending of tensors with the good old-fashioned vectors has been carried out, whenever needed — different tools for different jobs, but with an overarching geometrical theme of simplicity.

---

* Here are some relevant opinions:

a. By a *mathematician/educator:* "After many hours in the classroom it is the author's opinion that the tensor component approach (replete with subscripts and superscripts) is the correct one to use for beginners, even though it may require some painful initial adjustments. Although the more sophisticated, noncomponent approach is necessary for modern applications of the subject, it is believed that a student will appreciate and have an immensely deeper understanding of this sophisticated approach to tensors after a mastery of the component approach. In fact, noncomponent advocates frequently surrender to the introduction of components after all; *some proofs and important tensor results just do not lend themselves to a completely component-free treatment*." (Kay, 1988, Preface), italics added for emphasis).

b. By a noted *geometer*: "...It is a fact that many new results in differential geometry have been first discovered by 'formal suffix manipulation', and only subsequently have they been reobtained in an invariant, though probably more significant manner. The reader is advised to acquire skill both in using tensors without reference to a basis, and also in suffix manipulation when working with components. These two methods should be regarded as complementary to one another." He adds that even the great geometer E. Cartan, one of the earliest and authoritative proponents of nonindicial notations, ...did not hesitate to use tensor components when more convenient to do so..." (Willmore, 1959, p.187).

c. By a seasoned *theoretical physicist*: "We use coordinates explicitly in the physics tradition rather than straining to achieve the most abstract (and thereby also most unreadable and least useful) presentation possible." and "[Arnold's]...treatment of Hamiltonian systems relies on Cartan's formalism of exterior differential forms, requiring a degree of mathematical preparation that is neither typical nor necessary for physics and engineering graduate students." (McCauley, 1997, p. xiii).

The currently popular and religiously promoted *matrix* notation, a side effect of single-minded and rabid...computeritis (as if accounting were the highest form of human activity) seems to me very contrived and *after-the-factish,* definitely not a daily working tool for the exploration and mastery of new and rugged terrrain. In that notation, even very simple operations require special local explanations; for example, how should one write the dot product of two ordinary vectors **a** and **b**:

$$\mathbf{a} \cdot \mathbf{b}, \text{ or } \mathbf{a}^T \cdot \mathbf{b}, \text{ or } \mathbf{a} \cdot \mathbf{b}^T, \text{ or } \mathbf{b}^T \cdot \mathbf{a}, \text{ or } \mathbf{b} \cdot \mathbf{a}^T, \quad \text{where } (\ldots)^T \equiv \text{Transpose of } (\ldots),$$

or something else? Whatever cosmetic or aesthetic advantages that notation may have, they are far outweighed by the merciless straightjacket of *noncommutativity* of matrix multiplication; whereas *indicial notation is commutative,* or orderless. Also, the matrix notation might convey the false impression that tensors and matrices are the same thing. At best, tensors and matrices are, as Willmore puts it, *complementary,* and along with vectors should all be used. Similar comments hold for the other currently fashionable, mostly in some applied mathematician circles, geometrical tool called *exterior,* or *Cartanian,* calculus (see Remarks in Section 1.1).

For similar reasons of inclusiveness and intellectual efficiency, I have avoided the language and notation of modern applied mathematics ("epsilonics"); i.e., I have not presented the material in its most "elegant," or "compact," or fashionable form. The libraries are full of such rigorous/formalistic books that are almost never read. In this sense, I have sought to restore and hand down to the average analytically minded engineer, physicist, or applied mathematician the best of the classical tradition, in its most economical fashion.

ON THE GENESIS OF THIS BOOK

During the winter of 1987–88, I began writing a textbook on Analytical Mechanics, comparable in size and scope to the well-known classics of G. Hamel (*Theoretische Mechanik,* 1949) and A. I. Lur'e (*Analiticheskaia Mechanika,* 1961; French transl. 1968). I called its final chapter *"The Tensor Calculus of Nonholonomic Systems, and the Covariant Formulation of Classical Analytical Dynamics."* However, over the next few years, that project grew into a treatise of pretty hefty proportions, and so, following the advice of friends and editors, I decided to form *two* books: a large one on Analytical Mechanics, *free of tensors;* and a second of more modest size, based exclusively on that last tensorial chapter. This is the second, smaller, book; the bigger one has already been completed, and will, I hope, become public sometime during 2000, by Oxford University Press. May both these works make many and loyal friends!

John G. Papastavridis
Fall 1998,
Atlanta, Georgia

Acknowledgments

For several reasons that cannot be elaborated here, these are not the best of times for writing "another" book on applied mathematics and mechanics. The material (monetary) benefits are virtually negligible, if not outright negative; and today's "research" universities do not actively encourage such scholarly activities. Under such adverse external conditions, authors must seek help and inspiration elsewhere. In my case, these came from the following:

WRITINGS

- The pioneering and beautifully written article (1926–27) and booklet (1936) by **J. L. Synge**;
- The lucid and lively article (1948) and books (1970, 1976), by **V. V. Dobronravov**, virtually unknown in the West;
- The tensorial parts and appendix of the monumental monograph on analytical mechanics by **A. I. Lur'e** (1961, French transl. 1968);
- A collection of dense, clear, and eye-opening articles (early 1980s) by Dr. **Peter Maißer**, professor of mechatronics at the Technical University of Chemnitz-Zwickau (Germany); in a way, the continuation of Synge (1936);
- Certain unique articles in the massive RAAG memoirs (4 volumes, 1955–1968) by **K. Kondo** and his co-workers (University of Tokyo, Japan).

To all these authors, and several others to be identified later in the text, I am truly grateful; without their works, this book would have been a poorer one. The most appropriate of the classical and modern literature has found its way into the text and examples/problems, properly referenced; failure to acknowledge an author's contribution is *not* intentional, just oversight.

PEOPLE

- Ms. **Katharine L. Calhoun** and Ms. **Martha L. Saghini**, librarians at the Georgia Tech Library, for their most expert and capable help in locating and delivering to me hundreds of critically needed and hard-to-find books and papers, from all over the world; and Ms. **Sally M. Sellers**, for her skillful and creative drawing, by computer, of the work's two dozen figures.
- Professors **Sergei A. Zegzhda** and **Mikhail P. Yushkov**, of the Mathematics and Mechanics faculty of St. Petersbourg University (Russia), for

making available to me their excellent work on Theoretical Mechanics (coauthored with the late Professor **Nikolai N. Polia'hov**), where the principles of dynamics are presented in a readable tensorial form.

- Dr. **Hartmut Bremer** (O. Univ.-Prof. Dr.-Ing. habil.), professor of mechatronics at the University of Linz (Austria; formerly, professor of mechanics at the Technical University of Munich, Germany) for his mature and thoughtful comments on an earlier draft of the book; and for invaluable quanta of knowledge and tips on some of the finer points of theoretical mechanics.
- Dr. **Leon Y. Bahar**, professor of mechanics at Drexel University (Philadelphia), for his ever wise and mentorlike contributions to my dynamics education. Leon has been guiding and channeling the flights of fancy of my applied mathematics imagination to engineeringly relevant and readable goals; and much more.
- My family: wife **Kim Ann**, daughter **Julia Konstantina**, mother **Konstantina**, and brother **Stavros** (subjectively but fairly, one of the brightest mathematicians of contemporary Greece), for their continuous and priceless moral and material support.

Last, this volume could not have been written without: (1) the institution of *academic tenure* (much maligned and curtailed recently by reactionary ideologues, demagogues, and ignoramuses) and (2) the (alas, fast disappearing) policy, of most university libraries, of *open, direct, and free access to books and journals.* Regrettably, and in spite of high-tech millenarian promises, the next generation of scholarly authors will not be as lucky as I have been, in both these areas!

John G. Papastavridis

Table of Contents

Chapter 3
Tensor Analysis

Summary of Conventions, Notations; and Basic Formulae

NUMBERING OF EQUATIONS, EXAMPLES, AND PROBLEMS

- Chapters are divided into sections; e.g., Section 3.4 means Section 4 of Chapter 3. Equations are numbered *consecutively within each section*; e.g., Equation 3.4.2 means the second equation of Section 4, of Chapter 3. Related equations are indicated, further, by letters, e.g., Equation 3.4.2a follows Equation 3.4.2 and elaborates it; or, such equations may start with a lettered one, e.g., Equation 3.4.3a, while no Equation 3.4.3 exists.
- Examples and Problems *are placed anywhere within a section*, and are numbered consecutively within it; e.g., Example 5.7.2 (Problem 5.7.3) means the second example (third problem) of Section 7 of Chapter 5. Within Examples/Problems, equations are numbered consecutively *alphabetically.*

RANGE OF INDICES

	Capital	Small
Italic	Nonholonomic: $T^{AB}{}_{C}$ $1, \ldots, n$	Holonomic: $T^{ab}{}_{c}$ $1, \ldots, n$
Greek	Nonholonomic: $T^{\Theta\Psi}{}_{\Omega}$ $1, \ldots, n+1$	Holonomic: $T^{\theta\psi}{}_{\omega}$ $1, \ldots, n+1$

BASIC FORMULAE

TENSOR CALCULUS (CHAPTERS 1 TO 3; IN APPROXIMATE ORDER OF APPEARANCE)*

Chapter 1: Introduction and Background

Summation convention (in a product) over an index that appears twice, once up, once down.

---

* See also related collections of basic formulae in Kay (1988), Kreyszig (1959, pp. 328ff).

$(\ldots)_{,k} \equiv (\ldots)/\partial q^k$	Comma notation for partial derivatives; relative to q^k, or some other well-understood coordinates, or variables (see below).
$q^k, q^{k'}, q^{k''}, \ldots$	Systems of general curvilinear coordinates, Lagrangean (i.e., holonomic, or global) system coordinates.
$y^k = y_k, y^{k'}, \ldots$	Systems of rectangular Cartesian coordinates.
$z^k, z^{k'}, \ldots$	Systems of general affine coordinates (rectilinear nonorthogonal).
$x^k, x^{k'}, \ldots$	Systems of general variables, curvilinear coordinates (no such thing as x_k).
$\lvert A \rvert \equiv \mathrm{Det}(A)$	Determinant of A.
$J = \lvert \partial q'/\partial q \rvert \neq 0, \infty$	Jacobian of a coordinate transformation (CT) $q \to q' = q'(q)$. If $J > 0$, CT is proper (oriented manifold); if $J < 0$, CT is improper.

Chapter 2: Tensor Algebra

Point spaces (in order of decreasing generality):

V_n	General space (possibly discrete, i.e., nondifferentiable).
X_n	General differentiable manifold, with curvilinear coordinates q^k and invertible + continuous + differentiable transformations $q \to q'$.
L_n	An X_n with linear, or affine, connection (in general with torsion); but no metric.
R_n	Riemannian (metric) manifold: a manifold with affine connection and (usually) positive definite metric; an L_n with (a real C^2-differentiable + symmetric + positive definite) metric; manifold of greatest interest in physical applications.
E_n	Euclidean space: metric, linear, and flat manifold; a T_n + metric (can introduce global orthogonal rectilinear coordinates; i.e., $g_{kl} \to \delta_{kl}$: rectangular Cartesian coordinates); i.e., a flat R_n.
Invariant	A quantity that remains unchanged by coordinate transformations: $T \to T'(q^{k'}) = T(q^k)$ (see "relative scalar" below).
$\delta^k{}_b = \delta_b{}^k$	Delta of Kronecker.
$\varepsilon_{ijk} = \varepsilon^{ijk}$	Permutation symbol of Levi-Civita (in three dimensions).
$g_{kb} = g_{bk}$	Metric, or fundamental, tensor (covariant components).
$ds^2 = g_{kb}\, dq^k\, dq^b$	Line element (squared); or metric form, or fundamental form (of that space).

$g = \mathrm{Det}(g_{kb}) \equiv \lvert g_{kb}\rvert > 0$	(assumed)
$g_{kb}\, g^{kh} = \delta^h{}_b$	Definition of conjugate fundamental/metric tensor $g^{kh} = g^{hk};\ g^k{}_b = \delta^k{}_b$

$$(dg^{kh} = -g^{ki}\, g^{hj}\, dg_{ij},\quad dg_{kb} = -g_{ki}\, g_{bj}\, dg^{ij},\quad dg/g = g^{kl}\, dg_{kl} = -g_{kl}\, dg^{kl}).$$

$e_{ijk} \equiv \sqrt{g}\,\varepsilon_{ijk}$	Covariant permutation tensor (in three dimensions).
$e^{ijk} \equiv \left(1/\sqrt{g}\right)\varepsilon^{ijk}$	Contravariant permutation tensor (in three dimensions).
$\mathbf{e}_k \equiv \partial\mathbf{R}/\partial q^k$	Holonomic (natural) covariant basis; $\mathbf{R}$: position vector.
$\mathbf{e}^k \equiv \partial q^k/\partial\mathbf{R}$	Holonomic contravariant (gradient) basis; dual or reciprocal to $\mathbf{e}_k$: $\mathbf{e}_k \cdot \mathbf{e}^b = \delta^b{}_k$.
$v_k = \mathbf{v} \cdot \mathbf{e}_k$	Covariant components of vector $\mathbf{v} = v_k\, \mathbf{e}^k = v^k\, \mathbf{e}_k$.
$v^k = \mathbf{v} \cdot \mathbf{e}^k$	Contravariant components of vector $\mathbf{v}$.
$g_{kb}\, v^k v^b = g^{kb}\, v_k v_b = v^k v_b$	Magnitude, or length, of vector $\mathbf{v}$.
$v_k = g_{kb}\, v^b,\ v^k = g^{kb}\, v_b$	Lowering and raising of indices of a vector.
$T_{k'} = (\partial q^k/\partial q^{k'})\, T_k \equiv A^k{}_{k'}\, T_k$	Covariant (absolute) first-rank tensor, or vector.
$T^{k'} = (\partial q^{k'}/\partial q^k)\, T^k \equiv A^{k'}{}_k\, T^{k'}$	Contravariant (absolute) first-rank tensor, or vector.
$T_{k'b'} = A^k{}_{k'}\, A^b{}_{b'}\, T_{kb}$	Covariant second-rank (absolute) tensor.
$T^{k'b'} = A^{k'}{}_k\, A^{b'}{}_b\, T^{kb}$	Contravariant second-rank (absolute) tensor.
$T_{k'}{}^{b'} = A^k{}_{k'}\, A^{b'}{}_b\, T_k{}^b$	Mixed second-rank (absolute) tensor.
$T^{b'}{}_{k'} = A^{b'}{}_b\, A^k{}_{k'}\, T^b{}_k$	Mixed second-rank (absolute) tensor.

Direct, or dyadic, representation of second-rank tensor $\mathbf{T}$:

$\mathbf{e}^l \otimes \mathbf{e}^k$: tensor product of $\mathbf{e}^l$, $\mathbf{e}^k$, etc. (their order matters),

$\mathbf{T} = T_{lk}\, \mathbf{e}^l \otimes \mathbf{e}^k = T^{lk}\, \mathbf{e}_l \otimes \mathbf{e}_k = T^l{}_k\, \mathbf{e}_l \otimes \mathbf{e}^k = T_l{}^k\, \mathbf{e}^l \otimes \mathbf{e}_k$

$[T^{kl}\, dg_{kl} = -T_{kl}\, dg^{kl}$; but for $A_{kl} \neq g_{kl}$: $T^{kl}\, dA_{kl} = T_{kl}\, dA^{kl}]$.

Symmetry of components of second-rank (absolute) tensor:

$T_{kb} = T_{bk},\ T^{kb} = T^{bk},\ T_k{}^b = T^b{}_k$.

Relative scalar (rank zero) $T' = \lvert\partial q/\partial q'\rvert^w\, T$; e.g., $\mathrm{Det}(g_{k'l'}) = \lvert\partial q/\partial q'\rvert^2\, \mathrm{Det}(g_{kl})$ (If $w = 0$, then T is an absolute scalar, or invariant; i.e., not all scalars are invariants. Form invariance, however, means that: $T(q) = T(q')$, in addition to $T(q) = T'(q')$).

$T_{k'} = \lvert\partial q/\partial q'\rvert^w\, A^k{}_{k'}\, T_k$	Covariant relative first-rank tensor, or vector, of weight w.
$T^{k'} = \lvert\partial q/\partial q'\rvert^w\, A^{k'}{}_k\, T^{k'}$	Contravariant relative first-rank tensor, or vector, of weight w.
$T_{k'b'} = \lvert\partial q/\partial q'\rvert^w\, A^k{}_{k'}\, A^b{}_{b'}\, T_{kb}$	Covariant second-rank relative tensor. ($w = 0$: absolute such tensor, etc.; $w = +1$: densities; $w = -1$: capacities).

Chapter 3: Tensor Analysis

$2\Gamma_{k,ij} \equiv g_{ki,j} + g_{kj,i} - g_{ij,k}$	First-kind holonomic Christoffels; $\Gamma_{k,ij} = \Gamma_{k,ji}$.
$\Gamma^b{}_{ij} \equiv g^{bk}\, \Gamma_{k,ij}$	Second-kind holonomic Christoffels; $\Gamma^k{}_{ij} = \Gamma^k{}_{ji}$

$$\left[\Gamma^{b}{}_{sb} = (1/2)g^{kl}g_{kl,s} = (1/2g)g_{,s} = \left(1/\sqrt{g}\right)\left(\partial\sqrt{g}/\partial q^{s}\right) = \left(\ln\sqrt{g}\right)_{,s}\right].$$

$\mathbf{e}_{k,b} = \mathbf{e}_{b,k} = \Gamma_{i,kb}\,\mathbf{e}^{i} = \Gamma^{i}{}_{kb}\,\mathbf{e}_{i}$	Gradients of covariant holonomic basis vectors.
$\mathbf{e}^{k}{}_{,b} = -\Gamma^{k}{}_{ib}\,\mathbf{e}^{i}\ (\neq \mathbf{e}^{b}{}_{,k})$	Gradients of contravariant holonomic basis vectors.
$T_{k}\vert_{r} \equiv T_{k,r} - \Gamma^{b}{}_{kr}\,T_{b}$	Covariant derivative of absolute covariant vector T_k.
$T^{k}\vert_{r} \equiv T^{k}{}_{,r} + \Gamma^{k}{}_{br}\,T^{b}$	Covariant derivative of absolute vector T^{k}.
$T_{kb}\vert_{r} \equiv T_{kb,r} - \Gamma^{s}{}_{kr}\,T_{sb} - \Gamma^{s}{}_{br}\,T_{ks}$	Covariant derivative of absolute tensor T_{kb}.
$T^{kb}\vert_{r} \equiv T^{kb}{}_{,r} + \Gamma^{k}{}_{sr}\,T^{sb} + \Gamma^{b}{}_{sr}\,T^{ks}$	Covariant derivative of absolute tensor T^{kb}.
$T^{k}{}_{b}\vert_{r} \equiv T^{k}{}_{b,r} + \Gamma^{k}{}_{sr}\,T^{s}{}_{b} - \Gamma^{s}{}_{br}\,T^{k}{}_{s}$	Covariant derivative of absolute tensor $T^{k}{}_{b}$.
$T_{b}{}^{k}\vert_{r} \equiv T_{b}{}^{k}{}_{,r} + \Gamma^{k}{}_{sr}\,T_{b}{}^{s} - \Gamma^{s}{}_{br}\,T_{s}{}^{k}$	Covariant derivative of absolute tensor $T_{b}{}^{k}$.
$T\rceil_{r} \equiv T_{,r} - w\,\Gamma^{k}{}_{kr}\,T$, etc.	Covariant derivative of relative scalar T (i.e., no weight change).

Transformation of covariant derivative of absolute vector T_k. Generally, covariant derivatives of absolute (relative) tensors are absolute (relative) tensors of the kind indicated by their indices: $T_{k'}\vert_{r'} = A^{k}{}_{k'}\,A^{r}{}_{r'}\,T_{k}\vert_{r}$, etc.

$DT^{\cdots}{}_{\cdots} \equiv (T^{\cdots}{}_{\cdots}\vert_{k})\,dq^{k}$	Absolute differential of $T^{\cdots}{}_{\cdots}$.
$DT^{\cdots}{}_{\cdots}/Du \equiv (T^{\cdots}{}_{\cdots}\vert_{k})\,(dq^{k}/du)$	Absolute derivative of $T^{\cdots}{}_{\cdots}$ along a curve $q^{k}(u)$.

Ricci's lemma: $g_{kb}\vert_{s} = 0,\ g^{kb}\vert_{s} = 0,\ g^{k}{}_{b}\vert_{s} = \delta^{k}{}_{b}\vert_{s} = 0.$

Differential of n-Vector $\mathbf{V}(t, q)$:

$$d\mathbf{V} = [dV^{k} + \Gamma^{k}{}_{sl}\,V^{s}(dq^{l}/dt)]\mathbf{e}_{k} = [dV_{k} - \Gamma^{s}{}_{kl}\,V_{s}(dq^{l}/dt)]\mathbf{e}^{k} \equiv DV_{k}\,\mathbf{e}^{k} \equiv DV^{k}\,\mathbf{e}_{k}$$

Particle acceleration (or material derivative) of $\mathbf{V}$:

$$(d\mathbf{V}/dt)^{k} \to DV^{k}/Dt \equiv \partial V^{k}/\partial t + V^{k}\vert_{l}\,V^{l}$$

$d/dt(\ldots)$–derivative of $\mathbf{T} = T_{kl}\,\mathbf{e}^{k} \otimes \mathbf{e}^{l}$:

$$(d\mathbf{T}/dt)^{kl} = dT^{kl}/dt + (\Gamma^{k}{}_{rs}\,T^{rl} + \Gamma^{l}{}_{rs}\,T^{kr})(dq^{s}/dt).$$

Kinematics of an orthonormal triad; tensor and vector of angular velocity of $\{\mathbf{e}_k\}$:

$$d\mathbf{e}_{k} = d\phi_{kl}\,\mathbf{e}^{l} = d\phi_{kl}\,\mathbf{e}_{l},$$

$$d\phi_{kl} \equiv d\mathbf{e}_{k}\cdot\mathbf{e}_{l} = -d\phi_{lk} = \Gamma_{l,ks}\,dq^{s} \qquad (\Gamma_{l,ks} + \Gamma_{k,ls} = 0,\ \Gamma^{l}{}_{ks} + \Gamma^{k}{}_{ls} = 0),$$

$$d\phi_{ks}/dt \equiv \omega_{ks} = (d\mathbf{e}_{k}/dt)\cdot\mathbf{e}_{s} = \Gamma_{s,kl}(dq^{l}/dt) = -\omega_{sk} = -(d\mathbf{e}_{s}/dt)\cdot\mathbf{e}_{k}$$

$$= -\Gamma_{k,sl}(dq^{l}/dt),$$

$$\omega_{sk} \equiv e_{skh}\, \omega^h \Leftrightarrow \omega^h = (1/2)\, e^{hsk}\, \omega_{sk} \text{ (axial vector of } \omega_{ks});$$

$$\omega = (\omega_1, \omega_2, \omega_3):$$

$$\omega_1 = \omega_{23} = -\omega_{32} = (d\mathbf{e}_2/dt) \cdot \mathbf{e}_3 = -(d\mathbf{e}_3/dt) \cdot \mathbf{e}_2,$$

$$\omega_2 = \omega_{31} = -\omega_{13} = (d\mathbf{e}_3/dt) \cdot \mathbf{e}_1 = -(d\mathbf{e}_1/dt) \cdot \mathbf{e}_3,$$

$$\omega_3 = \omega_{12} = -\omega_{21} = (d\mathbf{e}_1/dt) \cdot \mathbf{e}_2 = -(d\mathbf{e}_2/dt) \cdot \mathbf{e}_1.$$

Moving axes theorem:

$$(d\mathbf{V}/dt)_k = dV_k/dt + (\omega \times \mathbf{V})_k,$$

$$= dV_k/dt + \omega_{sk}\, V^s = dV_k/dt + (e_{skh}\, \omega^h)V^s \; [\neq dV_k/dt]$$

Gradient of an invariant T:

$$(\text{grad } T)^k = T|_k = T_{,k}.$$

Divergence of a contravariant vector T^k:

$$\text{div } \mathbf{T} = T^k|_k = \left(1/\sqrt{g}\right)\left(\sqrt{g}T^k\right)_{,k}.$$

Divergence of a second-rank tensor $T^k{}_s$:

$$T^k{}_s|_k = \left(1/\sqrt{g}\right)\left(\sqrt{g}T^k{}_s\right)_{,k} - \Gamma^b{}_{sk}\, T^k{}_b, \quad \text{for general tensors,}$$

$$T^k{}_s|_k = \left(1/\sqrt{g}\right)\left(\sqrt{g}T^k{}_s\right)_{,k} - (1/2)(\partial g_{ab}/\partial q^s)T^{ab},$$

$$= \left(1/\sqrt{g}\right)\left(\sqrt{g}T^k{}_s\right)_{,k} + (1/2)(\partial g^{ab}/\partial q^s)T_{ab}, \quad \text{for symmetric tensors,}$$

$$T^{ks}|_k = \left(1/\sqrt{g}\right)\left(\sqrt{g}T^{ks}\right)_{,s}, \quad \text{for antisymmetric tensors.}$$

Laplacian operator:

$$\nabla^2(\ldots) \equiv \text{div grad}(\ldots) = g^{kl}(\ldots)|_{kl}.$$

Laplacian operator applied to invariant T:

$$\nabla^2 T \equiv g^{kl}(T|_{kl}) = \left(1/\sqrt{g}\right)\left(\sqrt{g}g^{kl}T_{,l}\right)_{,k}.$$

Curl(ing), or rot(ation) vector, of a three-dimensional covariant vector T_k (assuming symmetric Christoffels):

$$(\text{curl}\,\mathbf{T})^k = e^{klr}\, T_r|_l = \left(\varepsilon^{klr}/\sqrt{g}\right) T_{r,l} = \left(\varepsilon^{klr}/2\sqrt{g}\right)(T_r|_l - T_l|_r)$$

$$= \left(\varepsilon^{klr}/2\sqrt{g}\right)(T_{r,l} - T_{l,r}).$$

Curl tensor of n-Vector $\mathbf{V}$ (assuming symmetric holonomic Christoffels):

$$\text{Curl}\,\mathbf{V} = \nabla\otimes\mathbf{V} - \mathbf{V}\otimes\nabla \equiv [(\ldots)_{,k}\,\mathbf{e}^k] \otimes (V_l\,\mathbf{e}^l) - (V_l\,\mathbf{e}^l) \otimes [(\ldots)_{,k}\,\mathbf{e}^k]$$

$$= (V_{k,l} - V_{l,k})\,\mathbf{e}^l \otimes \mathbf{e}^k \equiv (\partial_l V_k - \partial_k V_l)\,\mathbf{e}^l \otimes \mathbf{e}^k \equiv V_{lk}\,\mathbf{e}^l \otimes \mathbf{e}^k,$$

$$V_{lk} = (\text{Curl}\,\mathbf{V})_{lk} = \mathbf{e}_l \cdot \text{Curl}\,\mathbf{V} \cdot \mathbf{e}_k.$$

Riemann–Christoffel (R–C) curvature tensor:

$$R^l{}_{krs} \equiv \Gamma^l{}_{ks,r} - \Gamma^l{}_{kr,s} + \Gamma^b{}_{ks}\,\Gamma^l{}_{br} - \Gamma^b{}_{kr}\,\Gamma^l{}_{bs}.$$

Covariant R–C [in a torsionless manifold, the number of independent R–C components is $n^2(n^2 - 1)/12$, while, if the manifold has torsion, that number is $n^3(n - 1)/2$]:

$$R_{lkrs} \equiv g_{lb}\,R^b{}_{krs} \Leftrightarrow R^b{}_{krs} = g^{bl}\,R_{lkrs}.$$

Noncommutativity of mixed covariant derivatives of vector V_k, V^k

$$V_k|_{rs} - V_k|_{sr} = R^l{}_{krs}\,V_l = R_{lkrs}\,V^l, \qquad V^k|_{rs} - V^k|_{sr} = -R^k{}_{lrs}\,V^l.$$

In a general manifold with affinities $\Lambda^k{}_{rs}$ and torsion $S^k{}_{rs} = \Lambda^k{}_{[rs]}$, where $R^l{}_{kbc}$ is given by an RC-like Formula, but with the Γs replaced by Λs:

$$V_k|_{bc} - V_k|_{cb} = R^l{}_{kbc}\,V_l + 2S^l{}_{cb}\,V_k|_l.$$

Covariant R–C in terms of first-kind Christoffels:

$$R_{lkrs} = g_{lp}\,R^b{}_{krs} = (1/2)(g_{ls,kr} + g_{kr,ls} - g_{lr,ks} - g_{ks,lr}) + g_{bm}(\Gamma^b{}_{kr}\,\Gamma^m{}_{ls} - \Gamma^m{}_{ks}\,\Gamma^b{}_{lr})$$

Antisymmetry in first pair:	$R_{lkrs} = -R_{klrs}$	[$\Rightarrow R_{kkrs} = 0$ (no sum on k)]
Antisymmetry in second pair:	$R_{lkrs} = -R_{lksr}$	[$\Rightarrow R_{lkss} = 0$ (no sum on s)]
	$R^l{}_{krs} = -R^l{}_{ksr}$	[$\Rightarrow R^l{}_{kss} = 0$ (no sum on s)]
Block symmetry:	$R_{lkrs} = R_{rslk}$	
Bianchi identities:	$R_{lkrs} + R_{lrsk} + R_{lskr} = 0$	[$\Rightarrow R^l{}_{krs} + R^l{}_{rsk} + R^l{}_{skr} = 0$]

[Also cyclically in k, r, s: $k \to r \to s$, $r \to s \to k$, $s \to k \to r$; e.g., $R_{klrs} + R_{krsl} + R_{kslr} = 0$, etc.; i.e., given a group of four indices, there exists only one independent Bianchi identity. R–C components with *more than two* equal indices vanish.]

Nonholonomic, or anholonomic contravariant bases

$$\mathbf{e}^K = A^K{}_k\, \mathbf{e}^k \; (\Leftrightarrow \mathbf{e}^k = A^k{}_K\, \mathbf{e}^K), \quad \mathbf{e}_K = A_K{}^k\, \mathbf{e}_k \; (\Leftrightarrow \mathbf{e}_k = A_k{}^K\, \mathbf{e}_K),$$

$$A^k{}_K\, A^K{}_l = \delta^k{}_l \; (\Leftrightarrow A^K{}_k\, A^k{}_L = \delta^K{}_L), \quad A^K{}_L = \mathbf{e}^K \cdot \mathbf{e}_L = \delta^K{}_L,$$

$A^K{}_k = \mathbf{e}^K \cdot \mathbf{e}_k = (\mathbf{e}^K)_k$	Covariant components of $\mathbf{e}^K$ along $\mathbf{e}_k$
$= (\mathbf{e}_k)^K$	Contravariant components of $\mathbf{e}_k$ along $\mathbf{e}^K$
$A^k{}_K = \mathbf{e}^k \cdot \mathbf{e}_K = (\mathbf{e}_K)^k$	Contravariant components of $\mathbf{e}_K$ along $\mathbf{e}^k$
$= (\mathbf{e}^k)_K$:	Covariant components of $\mathbf{e}^k$ along $\mathbf{e}_K$.

Nonintegrability conditions:

$$\partial A^K{}_k/\partial q^s - \partial A^K{}_s/\partial q^k \neq 0 \quad \text{(for at least one value of } K, k, s).$$

Local Pfaffian transformations:

$$dq^K = A^K{}_k\, dq^k \Leftrightarrow dq^k = A^k{}_K\, dq^K, \quad q^K \equiv \theta^K \text{ and } dq^K \equiv d\theta^K,$$

$$d\theta^K = A^K{}_k\, dq^k \Leftrightarrow dq^k = A^k{}_K\, d\theta^K, \quad \theta^K\text{: quasi-coordinates}$$

[If $d\theta^K$: Cartesian, then

$$A^K{}_k \equiv \partial(d\theta^K)/\partial(dq^k) = \cos(d\theta^K, dq^k) = \cos(dq^k, d\theta^K) = \partial(dq^k)/\partial(d\theta^K) \equiv A^k{}_K].$$

Elementary displacement vector: $d\mathbf{R} = \mathbf{e}_k\, dq^k = \mathbf{e}_K\, d\theta^K$.
Elementary arc-length (line element): $(ds)^2 = g_{kl}\, dq^k\, dq^l = g_{KL}\, d\theta^K\, d\theta^L$.
Nonholonomic metric tensors:

$$g_{KL} = \mathbf{e}_K \cdot \mathbf{e}_L \; (= g_{LK}), \quad g^{KL} = \mathbf{e}^K \cdot \mathbf{e}^L \; (= g^{LK}),$$

$$g_{KL} = A^k{}_K\, A^l{}_L\, g_{kl} \Leftrightarrow g_{kl} = A^K{}_k\, A^L{}_l\, g_{KL},$$

$$g^{KL} = A^K{}_k\, A^L{}_l\, g^{kl} \Leftrightarrow g^{kl} = A^k{}_K\, A^l{}_L\, g^{KL},$$

$$g_{KL}\, g^{LR} = \delta^R{}_K \; (= g^R{}_K), \quad g_{KL}\, g^{KR} = \delta^R{}_L \; (= g^R{}_L).$$

Nonholonomic R_n-vectors and their components:

$$\mathbf{V} = V_k\, \mathbf{e}^k = V^k\, \mathbf{e}_k = V_K\, \mathbf{e}^K = V^K\, \mathbf{e}_K,$$

$$V_k \equiv \mathbf{V} \cdot \mathbf{e}_k, \quad V^k \equiv \mathbf{V} \cdot \mathbf{e}^k, \quad V_K \equiv \mathbf{V} \cdot \mathbf{e}_K, \quad V^K \equiv \mathbf{V} \cdot \mathbf{e}^K,$$

$$V_K = A^k{}_K \, V_k = A_{Kk} \, V^k \Leftrightarrow V_k = A^K{}_k \, V_K, \quad V^k = A^{kK} \, V_K,$$

$$V^K = A^K{}_k \, V^k = A^{Kk} \, V_k \Leftrightarrow V_k = A_{kK} \, V^K, \quad V^k = A^k{}_K \, V^K,$$

$$V_K = A^k{}_K \, V_k = A^k{}_K \, (g_{kl} \, V^l) = A^k{}_K \, g_{kl} \, (A^l{}_L \, V^L) = g_{KL} \, V^L,$$

$$V^K = \ldots = g^{KL} \, V_L.$$

Mixed third-rank (absolute) tensor $T^{kl}{}_m$, relative to the NH bases $\{P, \mathbf{e}_K\}$ and $\{P, \mathbf{e}^K\}$:

$$T^{KL}{}_M = A^K{}_k \, A^L{}_l \, A^m{}_M \, T^{kl}{}_m \Leftrightarrow T^{kl}{}_m = A^k{}_K \, A^l{}_L \, A^M{}_m \, T^{KL}{}_M.$$

Useful relations (obtained by differentiating: $A^k{}_K \, A^K{}_l = \delta^k{}_l \Leftrightarrow A^K{}_k \, A^k{}_L = \delta^K{}_L$):

$$A^k{}_{L,s} = -A^k{}_K \, A^l{}_L \, A^K{}_{l,s}, \qquad A^K{}_{l,s} = -A^L{}_l \, A^K{}_k \, A^k{}_{L,s}.$$

Symbolic "quasi-chain rule" (since the θ^K do not exist):

$$\partial(\ldots)/\partial\theta^K \equiv [\partial(\ldots)\,/\partial q^k][\partial(dq^k)/\partial(d\theta^K)] \equiv A^k{}_K[\partial(\ldots)/\partial q^k],$$

$$(\ldots)_{,K} \equiv A^k{}_K(\ldots)_{,k},$$

$$\partial(\ldots)/\partial q^k \equiv [\partial(\ldots)/\partial\theta^K][\partial(d\theta^K)/\partial(dq^k)] \equiv A^K{}_k[\partial(\ldots)/\partial q^k],$$

$$(\ldots)_{,k} \equiv A^K{}_k(\ldots)_{,K},$$

$$d(\ldots) = [\partial(\ldots)/\partial q^k]dq^k = [\partial(\ldots)/\partial q^k](A^k{}_K \, d\theta^K) = [\partial(\ldots)/\partial\theta^K] \, d\theta^K.$$

Transformation of $V_{[k,l]}$, $V_{[K,L]}$:

$$V_{[K,L]} = A^k{}_K \, A^l{}_L \, V_{[k,l]} - \Omega^R{}_{KL} \, V_R, \qquad V_{[k,l]} = A^K{}_k \, A^L{}_l[V_{[K,L]} + \Omega^R{}_{KL} \, V_R].$$

Object of nonholonomicity, or anholonomicity (AO) [$n^2(n-1)/2$ components]

$$\Omega^R{}_{KL} \equiv (1/2)(A^R{}_{k,l} - A^R{}_{l,k})A^k{}_K \, A^l{}_L \equiv A^k{}_K \, A^l{}_L \, A^R{}_{[k,l]}$$

$$\equiv (1/2)(A^R{}_{k,L} \, A^k{}_K - A^R{}_{k,K} \, A^k{}_L) \equiv A^R{}_{k,[L} \, A^k{}_{K]},$$

$$\Omega^R{}_{KL} = A^k{}_K \, A^l{}_L \, \Omega^R{}_{kl} \; (\Leftrightarrow \Omega^R{}_{kl} = A^K{}_k \, A^L{}_l \, \Omega^R{}_{KL}),$$

$$\Omega^R{}_{kl} \equiv A^R{}_{[k,l]},$$

$$\Omega^R{}_{KL} = -\Omega^R{}_{LK} \; (\Omega^R{}_{kl} = -\Omega^R{}_{kl}).$$

Hamel (–Volterra) transitivity coefficients:

$$\gamma^R{}_{KL} \equiv 2\Omega^R{}_{KL} = A^k{}_K\, A^l{}_L (A^R{}_{k,l} - A^R{}_{l,k}) = A^R{}_{k,l}(A^k{}_K\, A^l{}_L - A^l{}_K\, A^k{}_L)$$

$$= A^R{}_l (A^l{}_{L,k}\, A^k{}_K - A^l{}_{K,k}\, A^k{}_L) \equiv A^R{}_r (A^r{}_{L,K} - A^r{}_{K,L}).$$

Fundamental noncommutativity equation for NH mixed (symbolic!) partial derivatives:

$$\partial/\partial\theta^L(\partial f/\partial\theta^K) - \partial/\partial\theta^K(\partial f/\partial\theta^L) = 2\Omega^R{}_{LK}(\partial f/\partial\theta^R) \equiv \gamma^R{}_{LK}(\partial f/\partial\theta^R).$$

Additional useful Ω, γ identities:

$$\Omega^R{}_{K[L,S]} = 2\;\Omega^M{}_{[LS}\;\Omega^R{}_{M]K},$$

$$\gamma^R{}_{KL,S} - \gamma^R{}_{KS,L} = 2(\gamma^M{}_{LS}\,\gamma^R{}_{MK} - \gamma^M{}_{MS}\,\gamma^R{}_{LK}),$$

$$\gamma^R{}_{SM}\,\gamma^M{}_{KL} + \gamma^R{}_{LM}\,\gamma^M{}_{SK} + \gamma^R{}_{KM}\,\gamma^M{}_{LS} + \gamma^R{}_{KL,S} + \gamma^R{}_{SK,L} + \gamma^R{}_{LS,K} = 0.$$

Transitivity equations:

$$d_2(d_1\theta^R) - d_1(d_2\theta^R) = \gamma^R{}_{KL}\, d_1\theta^K\, d_2\theta^L + A^R{}_r[d_2(d_1q^r) - d_1(d_2q^r)],$$

$$d_2(d_1q^r) - d_1(d_2q^r) = A^r{}_R\{[d_2(d_1\theta^R) - d_1(d_2\theta^R)] - \gamma^R{}_{KL}\, d_1\theta^K\, d_2\theta^L\}.$$

Transformation of the γ coefficients:

$$\gamma^{R'}{}_{K'L'} = A^{R'}{}_R\, A^K{}_{K'}\, A^L{}_{L'}\, \gamma^R{}_{KL} + A^K{}_{K'}\, A^L{}_{L'}(A^{R'}{}_{K,L} - A^{R'}{}_{L,K}).$$

Transformation of the Ω coefficients:

$$\Omega^{R'}{}_{K'L'} = A^{R'}{}_R\, A^K{}_{K'}\, A^L{}_{L'}\, \Omega^R{}_{KL} + A^K{}_{K'}\, A^L{}_{L'}\, A^{R'}{}_{[K,L]}$$

$$= A^{R'}{}_R\, A^K{}_{K'}\, A^L{}_{L'}\, \Omega^R{}_{KL} - A^{R'}{}_K\, A^K{}_{[K',L']}$$

$$= A^{R'}{}_R\, A^K{}_{K'}\, A^L{}_{L'}\, \Omega^R{}_{KL} + A^L{}_{[L'}\, A^K{}_{K']}\, A^{R'}{}_{K,L}.$$

Second transitivity coefficients: $\gamma_{S,KL} \equiv g_{SR}\,\gamma^R{}_{KL} \leftrightarrow \gamma^R{}_{KL} = g^{RS}\,\gamma_{S,KL}$

Nonholonomic affinities and Christoffels:

$$\mathbf{e}_{K,L} = \Lambda^R{}_{KL}\,\mathbf{e}_R = \Lambda_{R,KL}\,\mathbf{e}^R, \quad \mathbf{e}^R{}_{,L} = -\Lambda^R{}_{KL}\,\mathbf{e}^K,$$

$$\mathbf{e}_{K,k} = A^L{}_k\,\mathbf{e}_{K,L} = A^L{}_k(\Lambda^R{}_{KL}\,\mathbf{e}_R) = A^L{}_k(\Lambda_{R,KL}\,\mathbf{e}^R),$$

$$\Lambda^R{}_{KL} = \mathbf{e}_{K,L} \cdot \mathbf{e}^R \equiv A^R{}_K|_L \equiv A^R{}_k\, A^k{}_K|_L \equiv A^R{}_k\, A^l{}_L\, A^k{}_K|_l,$$

$$\Lambda_{R,KL} = \mathbf{e}_{K,L} \cdot \mathbf{e}_R \equiv A_{KR}|_L \equiv A^k{}_R\, A_{Kk}|_L \equiv A^k{}_R\, A^l{}_L\, A_{Kk}|_l,$$

$$\Lambda^R{}_{KL} = g^{RS}\, \Lambda_{S,KL} \leftrightarrow \Lambda_{S,KL} = g_{SR}\, \Lambda^R{}_{KL},$$

$$\Lambda^R{}_{KL} = \mathbf{e}_{K,L} \cdot \mathbf{e}^R = -\mathbf{e}^R{}_{,L} \cdot \mathbf{e}_K,$$

$$\Lambda_{R,KL} = \mathbf{e}_{K,L} \cdot \mathbf{e}_R = (\mathbf{e}_K \cdot \mathbf{e}_R)_{,L} - \mathbf{e}_{R,L} \cdot \mathbf{e}_K = g_{KR,L} - \Lambda_{K,RL},$$

$$\gamma^R{}_{KL} = \Lambda^R{}_{LK} - \Lambda^R{}_{KL} \equiv 2\Lambda^R{}_{[L,K]} = -2\Lambda^R{}_{[KL]},$$

$$\gamma_{R,KL} = \Lambda_{R,LK} - \Lambda_{R,KL} \equiv 2\Lambda_{R,[LK]} = -2\Lambda_{R,[KL]}.$$

Integrability/transitivity equations for $\gamma^R{}_{LK} = -\gamma^R{}_{KL}$ (in an R_n, Λs $\to$ Γs):

$$(\mathbf{e}_{K,L} - \mathbf{e}_{L,K}) \cdot \mathbf{e}^R = \Gamma^R{}_{KL} - \Gamma^R{}_{LK} = \gamma^R{}_{LK},$$

$$\partial \mathbf{e}_K / \partial \theta^L - \partial \mathbf{e}_L / \partial \theta^K = \gamma^R{}_{LK}\, \mathbf{e}_R = \gamma_{R,LK}\, \mathbf{e}^R.$$

Transformation of second-kind nonholonomic Christoffels:

$$\Lambda^R{}_{KL} = A^R{}_r\, A^k{}_K\, A^l{}_L\, \Lambda^r{}_{kl} + A^R{}_r\, A^r{}_{K,L} = A^R{}_r\, A^k{}_K\, A^l{}_L\, \Lambda^r{}_{kl} + A^R{}_r\, A^l{}_L\, A^r{}_{K,l}$$

$$= A^R{}_r\, A^k{}_K\, A^l{}_L\, \Lambda^r{}_{kl} - A^k{}_K\, A^l{}_L\, A^R{}_{k,l} = A^R{}_r\, A^k{}_K\, A^l{}_L\, \Lambda^r{}_{kl} - A^k{}_K\, A^R{}_{k,L}.$$

Second-kind nonholonomic Λs, Ss, and γs:

$$\Lambda^R{}_{[KL]} = S^R{}_{KL} + (1/2)\gamma^R{}_{LK} = S^R{}_{KL} + \Omega^R{}_{LK}.$$

Transformation of first-kind nonholonomic Christoffels:

$$\Lambda_{R,KL} = A^r{}_R\, A^k{}_K\, A^l{}_L\, \Lambda_{r,kl} + A_{Rr}\, A^r{}_{K,L} = A^r{}_R\, A^k{}_K\, A^l{}_L\, \Lambda_{r,kl} + A_{Rr}\, A^l{}_L\, A^r{}_{K,l}$$

$$= A^r{}_R\, A^k{}_K\, A^l{}_L\, \Lambda_{r,kl} - A^k{}_K\, A^l{}_L\, A_{Rk,l} = A^r{}_R\, A^k{}_K\, A^l{}_L\, \Lambda_{r,kl} - A^k{}_K\, A_{Rk,L}$$

First-kind nonholonomic Λs, Ss, and γs:

$$\Lambda_{R,[KL]} = A^r{}_R\, A^k{}_K\, A^l{}_L\, S_{r,kl} + (1/2)\gamma_{R,LK} = S_{R,KL} + \Omega_{R,LK}.$$

Integrability/transitivity equations for $\gamma_{R,LK} = -\gamma_{R,KL}$:

$$\mathbf{e}_{K,L} - \mathbf{e}_{L,K} = \gamma_{R,LK}\, \mathbf{e}^R \quad \text{(in a torsionless manifold).}$$

Nonholonomic Christoffel-like symbols of the first kind:

$$C_{R,KL} = C_{R,LK} \equiv (1/2)(g_{RL,K} + g_{KR,L} - g_{KL,R}) \ (\neq \Gamma_{R,KL}, \text{ of an } R_n).$$

Relations among the nonholonomic Γs and Cs (in an R_n):

$$\Gamma_{R,KL} = C_{R,KL} + (1/2)(\gamma_{K,RL} + \gamma_{L,RK} - \gamma_{R,KL}) = A^r{}_R\, A^k{}_K\, A^l{}_L\, \Gamma_{r,kl} + g_{kr}\, A^r{}_R\, A^k{}_{K,L},$$

$$C_{R,KL} = \Gamma_{R,LK} + (\Gamma_{L,[RK]} + \Gamma_{K,[RL]} - \Gamma_{R,[LK]})$$

$$= \Gamma_{R,LK} + (1/2)(\gamma_{L,KR} + \gamma_{K,LR} - \gamma_{R,KL}),$$

$$\Gamma_{R,(KL)} = C_{R,KL} + (1/2)(\gamma_{K,RL} + \gamma_{L,RK}).$$

Nonholonomic covariant derivatives of V_K, V^K, $T^K{}_L$:

$$V^K|_L \equiv A^K{}_k\, A^l{}_L\, V^k|_l = V^K{}_{,L} + \Lambda^K{}_{RL}\, V^R,$$

$$V_K|_L \equiv A^k{}_K\, A^l{}_L\, V_k|_l \equiv V_{K,L} - \Lambda^R{}_{KL}\, V_R,$$

$$T^K{}_L|_R \equiv T^K{}_{L,R} + \Lambda^K{}_{SR}\, T^S{}_L - \Lambda^S{}_{LR}\, T^K{}_S = A^K{}_k\, A^l{}_L\, A^r{}_R\, T^k{}_l|_r,$$

$$d\mathbf{V} = (\partial\mathbf{V}/\partial\theta^L)d\theta^L = (V^K|_L\, \mathbf{e}_K)d\theta^L \equiv DV^K\, \mathbf{e}_K = (V_K|_L\, \mathbf{e}^K)d\theta^L \equiv DV_K\, \mathbf{e}^K.$$

Parallel transport conditions for V^K and V_K:

$$d\mathbf{V} = \mathbf{0} \Rightarrow DV^K = 0 \Rightarrow d{*}V^K = -\Lambda^K{}_{RL}\, V^R\, d\theta^L,$$

$$DV_K = 0 \Rightarrow d{*}V_K = \Lambda^R{}_{KL}\, V_R\, d\theta^L.$$

Nonholonomic counterpart of Ricci's theorem: $g_{KL}|_R = 0$.
Kinematical interpretation of the Christoffels ($\mathbf{e}_{K,L} = \Lambda^R{}_{KL}\, \mathbf{e}_R$, $\mathbf{e}^K{}_{,L} = -\Lambda^K{}_{RL}\, \mathbf{e}^R$):

$$d\mathbf{e}_K = \mathbf{e}_{K,L}\, d\theta^L = (\Lambda^R{}_{KL}\, \mathbf{e}_R)d\theta^L, \quad d\phi_{KR} \equiv d\mathbf{e}_K \cdot \mathbf{e}_R = -d\mathbf{e}_R \cdot \mathbf{e}_K \equiv -d\phi_{RK}.$$

Ricci coefficients of rotation (of orthonormal basis in rigid-body motion):

$$\Lambda_{R,KL} \Rightarrow \lambda_{R,KL} = \mathbf{e}_{K,L} \cdot \mathbf{e}_R = -\lambda_{K,RL} = -\mathbf{e}_{R,L} \cdot \mathbf{e}_K.$$

Hamel/Ricci coefficients in an orthonormal basis: $2\lambda_{R,[KL]} = \gamma_{R,LK}$.
Ricci coefficients in terms of the Hamel coefficients: $2\lambda_{R,KL} = \gamma_{K,RL} + \gamma_{L,RK} - \gamma_{R,KL}$.
$A_{Kr}|_s$ in terms of the Ricci coefficients: $A_{Kr}|_l = A^R{}_r\, A^L{}_l\, \lambda_{R,KL}$.

Anholonomicity object in three–dimensional space:

$$(\mathrm{curl}\ \mathbf{e}^R)^b = e^{blk}\ A^R{}_{k,l} = (e^{blk}/2)(A^R{}_{k,l} - A^R{}_{l,k}) = e^{blk}\ \Omega^R{}_{kl} \equiv (e^{blk}/2)\gamma^R{}_{kl}.$$

Rotation/curling of the NH contravariant three-dimensional vectors $\mathbf{e}^R$:

$$\mathrm{curl}\ \mathbf{e}^R = e^{BLK}\ \Omega^R{}_{KL}\ \mathbf{e}_B = -(1/2)e^{BKL}\ \gamma^R{}_{KL}\ \mathbf{e}_B,$$

$$\gamma^R{}_{KL} = -e_{BKL}(\mathrm{curl}\ \mathbf{e}^R)^B = e_{BLK}(\mathrm{curl}\ \mathbf{e}^R)^B, \quad \gamma^R{}_{kl} \equiv 2\Omega^R{}_{kl} = e_{blk}(\mathrm{curl}\ \mathbf{e}^R)^b,$$

$$\mathbf{e}^B \cdot \mathrm{curl}\ \mathbf{e}^R = (1/2)e^{BLK}\ \gamma^R{}_{KL};$$

Curl (...) of n-vector in nonholonomic coordinates:

$$\mathrm{Curl}^*\mathbf{V} \equiv [(\ldots)_{,K}\ \mathbf{e}^K] \otimes (V_L\ \mathbf{e}^L) - (V_L\ \mathbf{e}^L) \otimes [(\ldots)_{,K}\ \mathbf{e}^K]$$

$$= (V_{K,L} - V_{L,K})\mathbf{e}^L \otimes \mathbf{e}^K \equiv V_{LK}\ \mathbf{e}^L \otimes \mathbf{e}^K$$

$$= [A^k{}_K\ A^l{}_L(V_{k,l} - V_{l,k}) + \gamma^R{}_{LK}\ V_R]\ \mathbf{e}^L \otimes \mathbf{e}^K,$$

$$V_{LK} = A^k{}_K\ A^l{}_L\ V_{lk} + \gamma^R{}_{LK}\ V_R,$$

$$\gamma^R{}_{LK} = \mathbf{e}_K \cdot \mathrm{Curl}\ \mathbf{e}^R \cdot \mathbf{e}_L = A^k{}_K\ (\mathrm{Curl}\ \mathbf{e}^R)_{kl}\ A^l{}_L.$$

Analytical Dynamics (Chapters 4 to 7; in approximate order of appearance)

Chapter 5: Particle on a Curve and on a Surface

Curves

Equations of Frenet–Serret (τ: tangent, ν: normal, β: binormal):

$$d\tau/ds = (0)\tau + (\kappa_{(\nu)})\nu + (0)\beta \qquad \text{or} \qquad D\tau^l/Ds = \kappa_{(\nu)}\ \nu^l,$$

$$d\nu/ds = (-\kappa_{(\nu)})\tau + (0)\nu + (\kappa_{(\beta)})\beta \qquad \text{or} \qquad D\nu^l/Ds = -\kappa_{(\nu)}\ \tau^l + \kappa_{(\beta)}\ \beta^l,$$

$$d\beta/ds = (0)\tau + (-\kappa_{(\beta)})\nu + (0)\beta \qquad \text{or} \qquad D\beta^l/Ds = -\kappa_{(\beta)}\ \nu^l.$$

Darboux vector: $\mathbf{D} \equiv \kappa_{(\beta)}\ \tau + \kappa_{(\nu)}\ \beta \equiv \mathbf{D}_\tau + \mathbf{D}_\beta$,
Total local curvature: $|\mathbf{D}| = [\kappa_{(\nu)}{}^2 + \kappa_{(\beta)}{}^2]^{1/2}$,

$$\kappa_{(\nu)} = \mathbf{D} \cdot \beta = \mathbf{D}_\beta \cdot \beta, \qquad \kappa_{(\beta)} = \mathbf{D} \cdot \tau = \mathbf{D}_\tau \cdot \tau.$$

Frenet–Serret Equations in terms of the Darboux vector **D**:

$$d\tau/ds = \mathbf{D}\times\tau = \kappa_{(\nu)}\,\nu,$$

$$d\nu/ds = \mathbf{D}\times\nu = -\kappa_{(\nu)}\,\tau + \kappa_{(\beta)}\,\beta,$$

$$d\beta/ds = \mathbf{D}\times\beta = \ldots = -\kappa_{(\beta)}\,\nu.$$

Surfaces

$\mathbf{e}_\alpha \equiv \partial\mathbf{r}(q^\beta)/\partial q^\alpha$ Surface covariant basis vectors; q^β: surface (Gaussian) coordinates. Ordinary surfaces: $\alpha, \beta, \gamma, \ldots = 1, 2$.

Gauss–Weingarten (or structure) equations:

$$\mathbf{e}_{\alpha,\beta} = \mathbf{e}_{\beta,\alpha} = A^\gamma{}_{\alpha\beta}\,\mathbf{e}_\gamma + b_{\alpha\beta}\,\mathbf{n}, \quad \mathbf{e}^\alpha{}_{,\beta} = -A^\alpha{}_{\beta\gamma}\,\mathbf{e}^\gamma + b^\alpha{}_\beta\,\mathbf{n},$$

$$\mathbf{e}^\alpha{}_{,\beta}\cdot\mathbf{e}_\gamma = -\mathbf{e}^\alpha\cdot\mathbf{e}_{\gamma,\beta} = -A^\alpha{}_{\beta\gamma}.$$

First-kind surface Christoffels:

$$A_{\gamma,\alpha\beta} = A_{\gamma,\beta\alpha} \equiv \mathbf{e}_{\alpha,\beta}\cdot\mathbf{e}_\gamma = \mathbf{e}_{\beta,\alpha}\cdot\mathbf{e}_\gamma$$

$$= (1/2)(g_{\gamma\beta,\alpha} + g_{\alpha\gamma,\beta} - g_{\alpha\beta,\gamma}) = A^\delta{}_{\alpha\beta}\,g_{\delta\gamma}.$$

Second-kind surface Christoffels:

$$A^\delta{}_{\alpha\beta} = A^\delta{}_{\beta\alpha} \equiv \mathbf{e}_{\alpha,\beta}\cdot\mathbf{e}^\delta = (g^{\delta\gamma}/2)(g_{\gamma\beta,\alpha} + g_{\alpha\gamma,\beta} - g_{\alpha\beta,\gamma}) = A_{\gamma,\alpha\beta}\,g^{\delta\gamma}.$$

Metric–Christoffels relations $\quad A_{\gamma,\beta\alpha} + A_{\beta,\gamma\alpha} = g_{\beta\gamma,\alpha}$.

Normal component of surface gradients:

$$b_{\alpha\beta} = \mathbf{e}_{\alpha,\beta}\cdot\mathbf{n} = \mathbf{e}_{\beta,\alpha}\cdot\mathbf{n} = -\,\mathbf{e}_\alpha\cdot\mathbf{n}_{,\beta} = -\,\mathbf{e}_\beta\cdot\mathbf{n}_{,\alpha}$$

$$= (\mathbf{e}_{\alpha,\beta}, \mathbf{e}_1, \mathbf{e}_2)\,/\,[\mathrm{Det}(g_{\alpha\beta})]^{1/2},$$

$$b^\alpha{}_\beta \equiv \mathbf{e}^\alpha{}_{,\beta}\cdot\mathbf{n}.$$

Gauss–Weingarten (or structure) equations:

$$\mathbf{n}_{,\alpha} = -b_{\alpha\beta}\,\mathbf{e}^\beta = -b_{\alpha\beta}(g^{\beta\gamma}\,\mathbf{e}_\gamma) = -b^\gamma{}_\alpha\,\mathbf{e}_\gamma.$$

Fundamental quadratic surface forms:

$$i \equiv d\mathbf{r}\cdot d\mathbf{r} = g_{\alpha\beta}\,dq^\alpha\,dq^\beta, \qquad ii \equiv d\mathbf{r}\cdot d\mathbf{n} = -b_{\alpha\beta}\,dq^\alpha\,dq^\beta.$$

Differential of surface vector: $\mathbf{V} = \mathbf{V}(q) = V^\alpha \mathbf{e}_\alpha$,

$$d\mathbf{V} = (\partial\mathbf{V}/\partial q^\alpha)dq^\alpha = dq^\alpha(V^\beta \mathbf{e}_\beta)_{,\alpha} = dq^\alpha(V^\beta{}_{,\alpha}\mathbf{e}_\beta + V^\beta \mathbf{e}_{\beta,\alpha})$$

$$= dq^\alpha[V^\beta{}_{,\alpha}\mathbf{e}_\beta + V^\beta(A^\gamma{}_{\beta\alpha}\mathbf{e}_\gamma + b_{\alpha\beta}\mathbf{n})]$$

$$= (V^\alpha{}_{,\beta} + A^\alpha{}_{\gamma\beta}V^\gamma)\mathbf{e}_\alpha\, dq^\beta + (V^\alpha b_{\beta\alpha}\mathbf{n})dq^\beta$$

$$\equiv (V^\alpha|_\beta\, \mathbf{e}_\alpha + V^\alpha b_{\alpha\beta}\mathbf{n})dq^\beta \equiv d\mathbf{V}_{(t)} + d\mathbf{V}_{(n)},$$

Tangential (to surface) part of $d\mathbf{V}$: $d\mathbf{V}_{(t)} \equiv (V^\alpha|_\beta\, dq^\beta)\mathbf{e}_\alpha \equiv DV^\alpha\, \mathbf{e}_\alpha$,
Normal (to surface) part of $d\mathbf{V}$: $d\mathbf{V}_{(n)} \equiv (V^\alpha b_{\alpha\beta}\, dq^\beta)\mathbf{n}$.

Curve on a surface

$$\mathbf{V} \to \mathbf{V}[q^\alpha(s)] = \mathbf{V}(s), \quad dq^\alpha \to [\partial q^\alpha(s)/\partial s]ds.$$

$\mathbf{V}$ is parallel transported along a surface curve ($\mathbf{V}$ is constant):

$$d\mathbf{V}_{(t)} = \mathbf{0} \to DV^\alpha = 0, \quad \text{i.e.,}\ V^\alpha|_\beta = 0.$$

Specialization: $\mathbf{V} \to \tau$:

$d\mathbf{V}_{(t)}/ds \to d\tau_{(t)}/ds = [(d^2q^\alpha/ds^2) + A^\alpha{}_{\beta\gamma}(dq^\beta/ds)\,(dq^\gamma/ds)]\ \mathbf{e}_\alpha \equiv \mathbf{k}_{(g)}$:
geodesic curvature of curve.

$d\mathbf{V}_{(n)}/ds \to d\tau_{(n)}/ds = [b_{\alpha\beta}(dq^\alpha/ds)\,(dq^\beta/ds)]\ \mathbf{n} \equiv \mathbf{k}_{(n)}$:
normal curvature of curve

$$[-ii/i = -(d\mathbf{r}/ds)\cdot(d\mathbf{n}/ds) = k_{(n)} = \mathbf{k}_{(n)}\cdot\mathbf{n}].$$

Meusnier's theorem: $k_{(n)} = \kappa_{(\nu)}\cos\theta$, $\theta = \text{angle}(\nu, \mathbf{n})$.
Geodesic lines of surface: $k_{(g)} = 0$.
Equations of geodetic lines: $(d^2q^\alpha/ds^2) + A^\alpha{}_{\beta\gamma}(dq^\beta/ds)\,(dq^\gamma/ds) = 0$.
Extension to general nonsurface vectors: $\mathbf{V} = V^\alpha\mathbf{e}_\alpha + V^n\,\mathbf{n} = V_\alpha\mathbf{e}^\alpha + V_n\,\mathbf{n}$.

$$\partial\mathbf{V}/\partial q^\alpha \equiv \mathbf{V}_{,\alpha} = (V_\beta|_\alpha - b_{\alpha\beta}V_n)\mathbf{e}^\beta + (V_{n,\alpha} + b^\beta{}_\alpha V_\beta)\mathbf{n}$$

$$= (V_\beta|_\alpha - b_{\alpha\beta}V^n)\mathbf{e}^\beta + (V^n{}_{,\alpha} + b_{\beta\alpha}V^\beta)\mathbf{n},$$

$$\partial\mathbf{V}/\partial n \equiv \mathbf{V}_{,n} = V_{\alpha,n}\,\mathbf{e}^\alpha + V_{n,n}\,\mathbf{n} = V^\alpha{}_{,n}\,\mathbf{e}_\alpha + V^n{}_{,n}\,\mathbf{n},$$

$$V_\beta|_\alpha \equiv V_{\beta,\alpha} - A^\gamma{}_{\beta\alpha}V_\gamma, \quad V^\beta|_\alpha \equiv V^\beta{}_{,\alpha} + A^\beta{}_{\alpha\gamma}V^\gamma.$$

Acceleration of a particle on a surface (with $v^\alpha \equiv dq^\alpha/dt$):

$$\mathbf{a} \equiv d\mathbf{v}/dt = d(v^{\alpha}\, \mathbf{e}_{\alpha})/dt = (dv^{\gamma}/dt + A^{\gamma}{}_{\alpha\beta}\, v^{\alpha}\, v^{\beta})\mathbf{e}_{\gamma} + (b_{\alpha\beta}\, v^{\alpha}\, v^{\beta})\mathbf{n}.$$

Equation of motion in configuration space ($V^k \equiv dq^k/dt$, Latin indices: 1, …, n):

$$A^k \equiv DV^k/Dt \equiv V^l\, V^k|_l \equiv (dq^l/dt)\left[(dq^k/dt)|_l\right]$$

$$\equiv d^2q^k/dt^2 + \Gamma^k{}_{rs}(dq^r/dt)(dq^s/dt) = Q^k.$$

Q^k: system force (see Chapter 7)

Linearized equation of isochronous perturbation $\eta^k(t)$ from a fundamental trajectory:

Contravariant form: $D^2\eta^k/Dt^2 + (R^k{}_{rlb}\, V^r\, V^b)\eta^l = Q^k|_l\, \eta^l,$

Covariant form: $g_{sk}(D^2\eta^k/Dt^2) + (R_{srlb}\, V^r\, V^b)\eta^l = Q_s|_l\, \eta^l.$

Linearized energy perturbation equation (Π: potential):

$$(g_{kl}V^k)D\eta^l/Dt + \Pi_{,k}\, \eta^k = V_l(D\eta^l/Dt) + \Pi_{,l}\, \eta^l = \Delta E,$$

Linearized equation of normal (nonisochronous) perturbation $n^k(t)$:

$$D^2n^k/Dt^2 + (d^2\mu/dt^2)V^k + 2(d\mu/dt)Q^k + R^k{}_{rlb}\, V^r\, n^l\, V^b = Q^k|_l\, n^l,$$

$$\eta^k = n^k + \mu\, V^k, \quad \mu \equiv \eta^k\, V_k/V^2.$$

Chapter 6: Lagrangean Kinematics

q^k ($k = 1, \ldots, n$) Lagrangean/system coordinates.

Position of typical system particle P, in terms of Lagrangean coordinates:

$$\mathbf{r} = \mathbf{r}(q^k,\, t) \equiv \mathbf{r}(q^{\alpha}).$$

Velocity of P:

$$\mathbf{v} \equiv d\mathbf{r}/dt \equiv v^k\, \mathbf{e}_k + v^{n+1}\, \mathbf{e}_{n+1} \equiv v^k\, \mathbf{e}_k + \mathbf{e}_0 \equiv v^{\alpha}\, \mathbf{e}_{\alpha}.$$

Holonomic basis vectors, for a particle:

$$\mathbf{e}_k \equiv \partial\mathbf{r}/\partial q^k,\ \mathbf{e}_{n+1} \equiv \partial\mathbf{r}/\partial q_{n+1} \equiv \partial\mathbf{r}/\partial t \equiv \mathbf{e}_0.$$

Contravariant, holonomic, velocity components of a particle:

$$v^k \equiv dq^k/dt, \quad v^{n+1} \equiv v^0 \equiv dq^{n+1}/dt \equiv dt/dt = 1.$$

Kinematically admissible, or possible, displacement of P:

$$d\mathbf{r} = \mathbf{e}_k\, dq^k + \mathbf{e}_0\, dt \equiv \mathbf{e}_\alpha\, dq^\alpha.$$

Virtual displacement of P:

$$\delta\mathbf{r} \equiv (d\mathbf{r})|_{dt \to \delta t = 0} = \mathbf{e}_k\, \delta q^k.$$

Acceleration of P:

$$\mathbf{a} \equiv d\mathbf{v}/dt = (dv^k/dt)\mathbf{e}_k + v^k\, v^l\, \mathbf{e}_{k,l} + 2v^k\, \mathbf{e}_{k,0} + \mathbf{e}_{0,0}.$$

Position of figurative system particle P: $\mathbf{R}$.
Velocity of figurative system particle P:

$$\mathbf{V} \equiv d\mathbf{R}/dt = V^k\, \mathbf{E}_k + \mathbf{E}_0 \equiv V^\alpha\, \mathbf{E}_\alpha,$$

$$V^k \equiv dq^k/dt (= v^k), \quad V^{n+1} \equiv V^0 \equiv dq^{n+1}/dt \equiv dt/dt = 1 (= v^{n+1} = v^0).$$

Kinematically admissible displacement of P:

$$d\mathbf{R} \equiv d\mathbf{q} = \mathbf{E}_k\, dq^k + \mathbf{E}_0\, dt \equiv \mathbf{E}_\alpha\, dq^\alpha.$$

Virtual displacement of P:

$$\delta\mathbf{R} \equiv \delta\mathbf{q} = (d\mathbf{R})|_{dt \to \delta t = 0} = \mathbf{E}_k\, \delta q^k.$$

Acceleration of P:

$$\mathbf{A} \equiv d\mathbf{V}/dt = (dV^k/dt)\mathbf{E}_k + V^k\, V^l\, \mathbf{E}_{k,l} + 2V^k\, \mathbf{E}_{k,0} + \mathbf{E}_{0,0}.$$

Fundamental kinematical identities:

$$\partial\mathbf{r}/\partial q^k = \partial\mathbf{v}/\partial v^k = \partial\mathbf{a}/\partial(dv^k/dt) = \ldots = \mathbf{e}_k,$$

$$\partial\mathbf{R}/\partial q^k = \partial\mathbf{V}/\partial V^k = \partial\mathbf{A}/\partial(dV^k/dt) = \ldots = \mathbf{E}_k,$$

$$\partial\mathbf{v}/\partial q^k = \partial/\partial q^k(d\mathbf{r}/dt) = d/dt(\partial\mathbf{r}/\partial q^k) = d/dt(\partial\mathbf{v}/\partial v^k) = d\mathbf{e}_k/dt,$$

$$\partial\mathbf{V}/\partial q^k = \partial/\partial q^k(d\mathbf{R}/dt) = d/dt(\partial\mathbf{R}/\partial q^k) = d/dt(\partial\mathbf{V}/\partial V^k) = d\mathbf{E}_k/dt.$$

Basic integrability conditions:

$$\mathbf{e}_{k,l} = \mathbf{e}_{l,k}, \quad \mathbf{e}_{k,0} = \mathbf{e}_{0,k},$$

$$E_k(\mathbf{v}) \equiv d/dt(\partial\mathbf{v}/\partial v^k) - \partial\mathbf{v}/\partial q^k = \mathbf{0},$$

$$\mathbf{E}_{k,l} = \mathbf{E}_{l,k}, \quad \mathbf{E}_{k,0} = \mathbf{E}_{0,k},$$

$$E_k(\mathbf{V}) \equiv d/dt(\partial \mathbf{V}/\partial V^k) - \partial \mathbf{V}/\partial q^k = \mathbf{0}.$$

Euler–Lagrange operator in holonomic variables:

$$E_k(\ldots) \equiv d/dt[\partial(\ldots)/\partial v^k] - \partial(\ldots)/\partial q^k.$$

Nonholonomic system velocities ($V^{n+1} \equiv dt/dt = 1$, $A^L{}_{n+1} \equiv A^L$):

$$\omega^L \equiv d\theta^L/dt \equiv A^L{}_\lambda(dq^\lambda/dt) \equiv A^L{}_\lambda V^\lambda = A^L{}_l V^l + A^L.$$

Kinematically admissible system displacements:

$$d\theta^L = A^L{}_\lambda \, dq^\lambda = A^L{}_l \, dq^l + A^L{}_{n+1} \, dt$$

Virtual system displacements: $\delta\theta^L = A^L{}_l \, \delta q^l$.
Time constraint: $\delta t = 0$.
Compatibility conditions among coordinate transformation coefficients:

$$A^\Lambda{}_\lambda A^\lambda{}_\Gamma = \delta^\Lambda{}_\Gamma, \quad A^\Lambda{}_\lambda A^\beta{}_\Lambda = \delta^\beta{}_\lambda,$$

$$A^L{}_l A^l{}_{N+1} = -A^L{}_{N+1}, \quad A^L{}_l A^k{}_L = \delta^k{}_l, \quad A^l{}_L A^L{}_{n+1} = -A^l{}_{n+1}.$$

Temporal quasi-derivatives:

$$\partial(\ldots)/\partial\theta^{N+1} \equiv [\partial(\ldots)/\partial q^\lambda](\partial q^\lambda/\partial\theta^{N+1}) = A^l{}_{N+1} [\partial(\ldots)/\partial q^l] + \partial(\ldots)/\partial t,$$

$$\partial(\ldots)/\partial q^{n+1} \equiv \partial(\ldots)/\partial t \equiv [\partial(\ldots)/\partial\theta^\Lambda](\partial\theta^\Lambda/\partial q^{n+1})$$

$$= A^L{}_{n+1} [\partial(\ldots)/\partial\theta^L] + \partial(\ldots)/\partial\theta^{N+1}.$$

Nonholonomic variable notation:

$$f = f(t, q, dq/dt) = f[t, q, dq(t, q, \omega)/dt] = f^*(t, q, \omega) = f^*.$$

Particle velocity in nonholonomic variables:

$$\mathbf{v}^* = v^\lambda \mathbf{e}_\lambda = \omega^\Lambda \mathbf{e}_\Lambda = \omega^L \mathbf{e}_L + \mathbf{e}_{N+1}.$$

Nonholonomic basis vectors, for a particle:

$$\mathbf{e}_L = A^l{}_L \mathbf{e}_l \Leftrightarrow \mathbf{e}_l = \ldots = A^L{}_l \mathbf{e}_L,$$

$$\mathbf{e}_{N+1} = -A^L{}_{N+1} \mathbf{e}_L + \delta^{n+1}{}_{N+1} \mathbf{e}_{n+1}, \quad \mathbf{e}_{n+1} = -A^l{}_{n+1} \mathbf{e}_l + \delta^{N+1}{}_{n+1} \mathbf{e}_{N+1}.$$

Particle acceleration in nonholonomic variables:

$$\mathbf{a}^* = (d\omega^L/dt)\ \mathbf{e}_L + \ldots.$$

Kinematically admissible particle displacement in nonholonomic variables:

$$d\mathbf{r}^* = d\theta^L\ \mathbf{e}_L + dt\ \mathbf{e}_{N+1}.$$

Virtual particle displacement in nonholonomic variables:

$$\delta\mathbf{r}^* = \delta\theta^L\ \mathbf{e}_L.$$

Transitivity equations [with $(\ldots)^{\cdot} \equiv d/dt(\ldots)$]:

$$(\delta\mathbf{r})^{\cdot} - \delta\mathbf{v} = [(\delta\theta^K)^{\cdot} - \delta\omega^K]\ \mathbf{e}_K - (\gamma^L{}_{K\Sigma}\ \omega^\Sigma\ \delta\theta^K)\ \mathbf{e}_L,$$

$$d(\delta\theta^K) - \delta(d\theta^K) = \gamma^K{}_{L\Lambda}\ d\theta^\Lambda\ \delta\theta^L + A^K{}_k\ [(d(\delta q^k) - \delta(dq^k)]$$

$$\equiv (\gamma^K{}_{LR}\ d\theta^R + \gamma^K{}_L\ dt)\ \delta\theta^L + A^K{}_k\ [(d(\delta q^k) - \delta(dq^k)],$$

$$(\delta\theta^K)^{\cdot} - \delta\omega^K = \gamma^K{}_{L\Lambda}\ \omega^\Lambda\ \delta\theta^L + A^K{}_k\ [(\delta q^k)^{\cdot} - \delta(dq^k/dt)]$$

$$\equiv h^K{}_L\ \delta\theta^L + A^K{}_k\ [(\delta q^k)^{\cdot} - \delta V^k]$$

$$[h^K{}_L \equiv \gamma^K{}_{LR}\ \omega^R + \gamma^K{}_{L,N+1}\ \omega^{N+1}$$

$$\equiv \gamma^K{}_{LR}\ \omega^R + \gamma^K{}_L\text{: two-index Hamel coefficients}];$$

$$d(\delta t) - \delta(dt) = \gamma^{N+1}{}_{\Lambda\Psi}\ d\theta^\Psi\ \delta\theta^\Lambda = \gamma^{N+1}{}_{L\Psi}\ d\theta^\Psi\ \delta\theta^L = 0$$

$$\Rightarrow \gamma^{N+1}{}_{\Lambda\Psi} = 0 \Rightarrow \gamma^{N+1}{}_{L\Psi} = 0,\ \gamma^K{}_{L,N+1} \equiv \gamma^K{}_L$$

$$\equiv (\partial A^K{}_l/\partial q^s - \partial A^K{}_s/\partial q^l)\ A^l{}_L\ A^s{}_{N+1} + (\partial A^K{}_l/\partial t - \partial A^K{}_{n+1}/\partial q^l)\ A^l{}_L.$$

Inverted transitivity equations:

$$d(\delta q^k) - \delta(dq^k) = A^k{}_K\{[d(\delta\theta^K) - \delta(d\theta^K)] - \gamma^K{}_{L\Lambda}\ d\theta^\Lambda\ \delta\theta^L\},$$

$$(\delta q^k)^{\cdot} - \delta V^k = A^k{}_K\ \{[(\delta\theta^K)^{\cdot} - \delta\omega^K] - \gamma^K{}_{L\Lambda}\ \omega^\Lambda\ \delta\theta^L\}$$

$$= A^k{}_K\{[(\delta\theta^K)^{\cdot} - \delta\omega^K] - (\gamma^K{}_{LR}\ \omega^R + \gamma^K{}_L)\delta\theta^L\}$$

$$= A^k{}_K\{[(\delta\theta^K)^{\cdot} - \delta\omega^K] - h^K{}_L\ \delta\theta^L\}.$$

Fundamental kinematical identities:

$$\partial \mathbf{r}/\partial\theta^K = \partial \mathbf{v}^*/\partial\omega^K = \partial \mathbf{a}^*/\partial(d\omega^K/dt) = \ldots = \mathbf{e}_K,$$

$$\partial \mathbf{R}^*/\partial\theta^K = \partial \mathbf{V}^*/\partial\omega^K = \partial \mathbf{A}^*/\partial(d\omega^K/dt) = \ldots = \mathbf{E}_K.$$

Euler–Lagrange operator in nonholonomic variables:

$$E_K(\ldots) \equiv d/dt[\partial(\ldots)/\partial\omega^K] - \partial(\ldots)/\partial\theta^K.$$

Nonholonomic deviation vector:

$$\mathbf{g}_K \equiv E_K(\mathbf{v}^*) \equiv d\mathbf{e}_K/dt - \partial\mathbf{v}^*/\partial\theta^K = E_K(v^k)\mathbf{e}_k \equiv W^k{}_K\,\mathbf{e}_k$$

$$= E_K(v^k)(A^L{}_k\,\mathbf{e}_L) \equiv -H^L{}_K\,\mathbf{e}_L$$

$$= (\partial\mathbf{e}_K/\partial\theta^\Delta - \partial\mathbf{e}_\Delta/\partial\theta^K)\,\omega^\Delta = -(\gamma^L{}_{K\Delta}\,\omega^\Delta)\mathbf{e}_L = -h^L{}_K\,\mathbf{e}_L,$$

$$W^k{}_K \equiv E_K(v^k) = -(\partial v^k/\partial\omega^L)H^L{}_K \equiv -A^k{}_L\,H^L{}_K,$$

$$H^L{}_K \equiv -(\partial\omega^L/\partial v^k)W^k{}_K \equiv -A^L{}_k\,W^k{}_K.$$

Constraints ("equilibrium" choice of quasi-velocities):

$[D, D', D''; d, d', d'' = 1, \ldots, m; I, I', I'', \ldots; i, i', i'', \ldots = m+1, \ldots, n]$

$$\omega^D \equiv A^D{}_k\,V^k + A^D{}_{n+1} = 0, \qquad \omega^I \equiv A^I{}_k\,V^k + A^I{}_{n+1} \neq 0;$$

$$V^k = A^k{}_I\,\omega^I + A^{n+1}, \qquad V^{n+1} = \omega^{N+1} = 1;$$

$$d\theta^D \equiv A^D{}_k\,dq^k + A^D{}_{n+1}\,dt = A^D{}_\beta\,dq^\beta = 0 \quad (\text{since } dq^{n+1} = dt),$$

$$d\theta^I \equiv A^I{}_k\,dq^k + A^I{}_{n+1}\,dt = A^I{}_\beta\,dq^\beta \neq 0;$$

$$dq^k = A^k{}_K\,d\theta^K + A^{n+1}\,dt = A^k{}_I\,d\theta^I + A^{n+1}\,dt,$$

$$dq^{n+1} = A^{n+1}{}_K\,d\theta^K + A^{n+1}{}_{N+1}\,dt = (0)d\theta^K + (\delta^{n+1}{}_{N+1})dt = dt \;(= d\theta^{N+1})$$

$$\delta\theta^D \equiv A^D{}_k\,\delta q^k + A^D{}_{n+1}\,\delta t = A^D{}_k\,\delta q^k = 0 \quad (= A^D{}_\beta\,\delta q^\beta,\ \delta q^{n+1} = \delta t = 0),$$

$$\delta\theta^I \equiv A^I{}_k\,\delta q^k + A^I{}_{n+1}\,\delta t = A^I{}_k\,\delta q^k \neq 0 \quad (= A^I{}_\beta\,\delta q^\beta;\ I = m+1, \ldots, n);$$

$$\delta q^k = A^k{}_K\,\delta\theta^K + A^{n+1}\,\delta t = A^k{}_I\,\delta\theta^I,$$

$$\delta q^{n+1} = A^{n+1}{}_K\,\delta\theta^K + A^{n+1}{}_{N+1}\,\delta t = (0)\delta\theta^K + (\delta^{n+1}{}_{N+1})(0) = \delta\theta^{N+1} = 0.$$

Constrained transitivity equations:

$$d(\delta\theta^D) - \delta(d\theta^D) = \gamma^D{}_{II'}\, d\theta^{I'}\, \delta\theta^I + \gamma^D{}_I\, dt\, \delta\theta^I,$$

$$d(\delta\theta^I) - \delta(d\theta^I) = \gamma^I{}_{I'I''}\, d\theta^{I''}\, \delta\theta^{I'} + \gamma^I{}_{I'}\, dt\, \delta\theta^{I'};$$

$$(\delta\theta^D)^{\cdot} - \delta\omega^D = \gamma^D{}_{II'}\, \omega^{I'}\, \delta\theta^I + \gamma^D{}_I\, \delta\theta^I \equiv h^D{}_I\, \delta\theta^I,$$

$$(\delta\theta^I)^{\cdot} - \delta\omega^I = \gamma^I{}_{I'I''}\, \omega^{I''}\, \delta\theta^{I'} + \gamma^I{}_{I'}\, \delta\theta^{I'} \equiv h^I{}_{I'}\, \delta\theta^{I'}$$

[$\Rightarrow$ In general, $d(\delta\theta^D) - \delta(d\theta^D) \neq 0$; we cannot have both $d(\delta\theta^D) = 0$ and $\delta(d\theta^D) = 0$ (even though $d\theta^D = 0$ and $\delta\theta^D = 0$); it is either the one or the other.]

Theorem of Frobenius: the necessary and sufficient conditions for the holonomicity of the system of Pfaffian constraints; geometrical interpretation of these conditions:

$$d\theta^D \equiv A^D{}_\lambda\, dq^\lambda = A^D{}_k\, dq^k + A^D{}_{n+1}\, dq^{n+1} \equiv A^D{}_k\, dq^k + A^D\, dt = 0,$$

$$\delta\theta^D \equiv A^D{}_\lambda\, \delta q^\lambda = A^D{}_k\, \delta q^k + A^D{}_{n+1}\, \delta q^{n+1} = A^D{}_k\, \delta q^k = 0,$$

are

$$\gamma^D{}_{II'} = 0,\ \gamma^D{}_I = 0\ [D = 1, \ldots, m;\ I, I' = m + 1, \ldots, n];$$

$$(\text{Curl}\ \mathbf{E}^D)_{hk} = (\partial A^D{}_k/\partial q^h - \partial A^D{}_h/\partial q^k) \equiv (A^D{}_{k,h} - A^D{}_{h,k}) \equiv (\gamma^D{}_{kh}),$$

$$\gamma^D{}_{II'} = A^h{}_{I'}\, (\partial A^D{}_k/\partial q^h - \partial A^D{}_h/\partial q^k)\, A^k{}_I = A^h{}_{I'}\, \gamma^D{}_{kh}\, A^k{}_I$$

$$= (\mathbf{E}_{I'})^h (\text{Curl}\ \mathbf{E}^D)_{hk} (\mathbf{E}_I)^k = \mathbf{E}_{I'} \cdot \text{Curl}\ \mathbf{E}^D \cdot \mathbf{E}_I = \mathbf{E}^D \cdot \mathbf{E}_{II'} = 0,$$

$$\mathbf{E}_{II'} = -\mathbf{E}_{I'I} \equiv \partial\mathbf{E}_{I'}/\partial\theta^I - \partial\mathbf{E}_I/\partial\theta^{I'} \quad (\text{since } \mathbf{E}_{k,l} = \mathbf{E}_{l,k})$$

$$[\partial\mathbf{E}_K/\partial\theta^\Delta - \partial\mathbf{E}_\Delta/\partial\theta^K = \gamma^\Psi{}_{\Delta K}\, \mathbf{E}_\Psi \Rightarrow \gamma^\Psi{}_{\Delta K} = (\partial\mathbf{E}_K/\partial\theta^\Delta - \partial\mathbf{E}_\Delta/\partial\theta^K) \cdot \mathbf{E}^\Psi].$$

Chapter 7: Lagrangean Kinetics

Kinetic energy:

$$2T \equiv S\, dm\ \mathbf{v} \cdot \mathbf{v} = 2(T_{(2)} + T_{(1)} + T_{(0)}) \quad \text{(holonomic variables)},$$

$$2T_{(2)} \equiv M_{kl}\, V^k\, V^l, \quad T_{(1)} \equiv M_{k,n+1}\, V^k\, V^{n+1} \equiv M_k\, V^k,$$

$$2T_{(0)} = M_{n+1,n+1} V^{n+1}\, V^{n+1} \equiv M_0; \quad M_{\alpha\beta} = M_{\beta\alpha} = S\, dm\ \mathbf{e}_\alpha \cdot \mathbf{e}_\beta = \mathbf{E}_\alpha \cdot \mathbf{E}_\beta$$

$$2T^* = M_{\Gamma\Delta}\,\omega^\Gamma\,\omega^\Delta = 2(T^*_{(2)} + T^*_{(1)} + T^*_{(0)}) \quad \text{(nonholonomic variables)},$$

$$2T^*_{(2)} \equiv M_{KL}\,\omega^K\,\omega^L, \quad T^*_{(1)} \equiv M_K\,\omega^K, \quad 2T^*_{(0)} \equiv M^*_0,$$

$$M_{KL} = M_{LK} \equiv S\,dm\;\mathbf{e}_K \cdot \mathbf{e}_L = \mathbf{E}_K \cdot \mathbf{E}_L \quad \text{(inertia coefficients)}.$$

Constrained form of: $T \quad 2T^*_o(t, q, \omega_I) = 2T^*_o = M_{II'}\,\omega^I\,\omega^{I'} + 2M_I\,\omega^I + M^*_0$.
Kinematical line element:

$$(dS)^2 = 2T(dt)^2 = M_{\gamma\delta}\,dq^\gamma\,dq^\delta \equiv M_{kl}\,dq^k\,dq^l + 2M_k\,dq\,dt + M_0(dt)^2.$$

Gibbs–Appell function, or Appellian:

$$2S \equiv S\,dm\;\mathbf{a}\cdot\mathbf{a} = S\,dm\;a^2 = S\,dm\;\mathbf{a}^*\cdot\mathbf{a}^* \equiv 2S^*.$$

Kinematico-inertial identities:

$$\partial S/\partial(dV^k/dt) = S\,dm\;\mathbf{a}\cdot[\partial\mathbf{a}/\partial(dV^k/dt)] = S\,dm\;\mathbf{a}\cdot\mathbf{e}_k = I_k,$$

$$\partial S^*/\partial(d\omega^K/dt) = S\,dm\;\mathbf{a}^*\cdot[\partial\mathbf{a}^*/\partial(d\omega^K/dt)] = S\,dm\;\mathbf{a}\cdot\mathbf{e}_K = I_K.$$

System acceleration:

$$\mathbf{A} \equiv d\mathbf{V}/dt$$

$$= (\partial\mathbf{R}/\partial q^\alpha)(dV^\alpha/dt) + (\partial^2\mathbf{R}/\partial q^\alpha\partial q^\beta)V^\alpha\,V^\beta \equiv (dV^\alpha/dt)\mathbf{E}_\alpha + V^\alpha(d\mathbf{E}_\alpha/dt)$$

$$= (\partial\mathbf{R}/\partial q^k)(dV^k/dt) + (\partial^2\mathbf{R}/\partial q^k\partial q^l)V^k\,V^l + 2(\partial^2\mathbf{R}/\partial q^k\partial t)V^k + \partial^2\mathbf{R}/\partial t^2$$

$$= (DV^\beta/Dt)\mathbf{E}_\beta \equiv A^\beta\,\mathbf{E}_\beta, \quad (V^k \equiv dq^k/dt),$$

$$A^\beta \equiv DV^\beta/Dt \equiv V^\beta|_\delta\,V^\delta \equiv (V^\beta{}_{,\delta} + \Gamma^\beta{}_{\delta\zeta}\,V^\zeta)V^\delta = dV^\beta/dt + \Gamma^\beta{}_{\delta\zeta}\,V^\delta\,V^\zeta,$$

$$A_\beta = A^\delta\,M_{\beta\delta} = DV_\beta/Dt = dV_\beta/dt - \Gamma_{\delta,\beta\zeta}\,V^\delta\,V^\zeta,$$

$$A_k = M_{kb}\,(dV^b/dt) + \Gamma_{k,bh}\,V^b\,V^h + 2\Gamma_{k,b0}\,V^b + \Gamma_{k,00},$$

$$A^k = dV^k/dt + \Gamma^k{}_{bh}\,V^b\,V^h + 2\Gamma^k{}_{b0}\,V^b + \Gamma^k{}_{00}.$$

Christoffel symbols of first kind:

$$2\Gamma_{\alpha,\beta\delta} \equiv (\partial M_{\alpha\beta}/\partial q^\delta + \partial M_{\alpha\delta}/\partial q^\beta - \partial M_{\beta\delta}/\partial q^\alpha)$$

$$= \mathcal{S}\, dm\ \mathbf{e}_\alpha \cdot (\partial \mathbf{e}_\beta/\partial q^\delta) = \mathbf{E}_\alpha \cdot (\partial \mathbf{E}_\beta/\partial q^\delta).$$

Christoffel symbols of second kind:

$$\Gamma^\alpha{}_{\beta\delta} \equiv M^{\alpha\zeta}\, \Gamma_{\zeta,\beta\delta},$$

$$\Gamma^k{}_{bh} \equiv M^{kd}\, \Gamma_{d,bh}, \quad \Gamma^k{}_{b,n+1} \equiv M^{kd}\, \Gamma_{d;b,n+1}, \quad \Gamma^k{}_{n+1,n+1} \equiv M^{kd}\, \Gamma_{d;n+1,n+1},$$

$$2\Gamma_{d;b,n+1} \equiv \partial M_{d,n+1}/\partial q^b + \partial M_{db}/\partial q^{n+1} - \partial M_{b,n+1}/\partial q^d$$

$$\text{[or, simply: } 2\Gamma_{d,b0} = (\partial M_d/\partial q^b - \partial M_b/\partial q^d) + \partial M_{db}/\partial t],$$

$$\Gamma_{d,00} = \partial M_d/\partial t - (1/2)\partial M_0/\partial q^d, \quad \Gamma_{0,00} = (1/2)\partial M_0/\partial t, \quad \Gamma_{0,b0} = (1/2)\partial M_0/\partial q^b.$$

System inertia "force":

$$I_k \equiv \mathcal{S}\, dm\ \mathbf{a} \cdot \mathbf{e}_k = d/dt(\partial T/\partial V^k) - \partial T/\partial q^k$$

$$= E_k(T) \quad \text{(holonomic);}$$

$$I_K \equiv \mathcal{S}\, dm\ \mathbf{a}^* \cdot \mathbf{e}_K = d/dt(\partial T^*/\partial \omega^K) - \partial T^*/\partial \theta^K - \Gamma_K$$

$$\equiv E_K(T^*) - \Gamma_K \text{ (nonholonomic),}$$

$$-\Gamma_K \equiv -\mathcal{S}\, dm\ \mathbf{v}^* \cdot [d/dt(\partial \mathbf{v}^*/\partial \omega^K) - \partial \mathbf{v}^*/\partial \theta^K]$$

$$= \gamma^L{}_{K\Lambda}(\partial T^*/\partial \omega^L)\, \omega^\Lambda \equiv h^L{}_K\, P_L,$$

$$E_k(T) = A^K{}_k\, E_K(T^*) + E_k(\omega^K)(\partial T^*/\partial \omega^K),$$

$$E_k(\omega^K) \equiv d/dt(\partial \omega^K/\partial V^k) - (\partial \omega^K/\partial q^k) = A^L{}_k\, \gamma^K{}_{L\Lambda}\, \omega^\Lambda \equiv A^L{}_k\, h^K{}_L,$$

$$E_k(\omega^K)(\partial T^*/\partial \omega^K) = A^K{}_k[\gamma^L{}_{K\Lambda}(\partial T^*/\partial \omega^L)\omega^\Lambda] \equiv -A^K{}_k\, \Gamma_K,$$

$$E_k(T) = A^K{}_k[E_K(T^*) - \Gamma_K] \Leftrightarrow E_K(T^*) - \Gamma_K = A^k{}_K\, E_k(T),$$

$$I_k = A^K{}_k\, I_K \Leftrightarrow I_K = A^k{}_K\, I_k.$$

D'Alembert's physical decomposition:

$$d\mathbf{f}_{\text{total force}} = d\mathbf{F}_{\text{impressed force}} + d\mathbf{R}_{\text{constraint reaction}}.$$

System forces:

$$Q_\gamma \equiv S\, d\mathbf{F} \cdot \mathbf{e}_\gamma, \quad \Lambda_\gamma \equiv S\, d\mathbf{R} \cdot \mathbf{e}_\gamma,$$

$$Q_\Gamma \equiv S\, d\mathbf{F} \cdot \mathbf{e}_\Gamma, \quad \Lambda_\Gamma \equiv S\, d\mathbf{R} \cdot \mathbf{e}_\Gamma,$$

$$Q_\gamma = A^\Gamma{}_\gamma Q_\Gamma \Leftrightarrow Q_\Gamma = A^\gamma{}_\Gamma Q_\gamma, \quad \Lambda_\gamma = A^\Gamma{}_\gamma \Lambda_\Gamma \Leftrightarrow \Lambda_\Gamma = A^\gamma{}_\Gamma \Lambda_\gamma.$$

Newton–Euler law, in d'Alembert form: $dm\, \mathbf{a} = d\mathbf{F} + d\mathbf{R}$.

Principle of d'Alembert: $-\delta' W_{(r)} \equiv S(-d\mathbf{R}) \cdot \delta\mathbf{r} = \Lambda_k\, \delta q^k = \Lambda_K\, \delta\theta^K = 0,$

Principle of Lagrange: $\delta I = \delta' W,$

$$\delta I \equiv S\, dm\, \mathbf{a} \cdot \delta\mathbf{r} = I_k\, \delta q^k = I^k\, \delta q_k = I_K\, \delta\theta^K = I^K\, \delta\theta_K,$$

$$\delta' W \equiv S\, d\mathbf{F} \cdot \delta\mathbf{r} = Q_k\, \delta q^k = Q^k\, \delta q_k = Q_K\, \delta\theta^K = Q^K\, \delta\theta_K.$$

Equations of Lagrange: $I_k \equiv E_k(T) \equiv d/dt(\partial T/\partial V^k) - \partial T/\partial q^k = Q_k$.
Equations of Routh–Voss: $E_k(T) \equiv d/dt(\partial T/\partial V^k) - \partial T/\partial q^k = Q_k + \lambda_D\, A^D{}_k$.
General kinetic equations: $I_I = Q_I \quad [\delta\theta^I \neq 0 \Rightarrow \Lambda_I = 0]$.
General kinetostatic equations: $I_D = Q_D + \Lambda_D \quad [\delta\theta^D = 0 \Rightarrow \Lambda_D \neq 0]$.
Equations of Maggi (kinetic): $A^k{}_I\, E_k(T) = A^k{}_I\, Q_k\ (= Q_I), \quad \Lambda_I = 0$.
Equations of Maggi (kinetostatic): $A^k{}_D\, E_k(T) = A^k{}_D\, Q_k + \Lambda_D\ (= Q_D + \Lambda_D)$.
Equations of Hamel (kinetic): $d/dt(\partial T^*/\partial\omega^I) - \partial T^*/\partial\theta^I - \Gamma_I = Q_I$.
Equations of Hamel (kinetostatic): $d/dt(\partial T^*/\partial\omega^D) - \partial T^*/\partial\theta^D - \Gamma_D = Q_D + \Lambda_D$.
Explicit form of kinetic Hamel equations:

$$d\omega^K/dt + \Gamma^K{}_{(\Lambda\Psi)}\, \omega^\Lambda\, \omega^\Psi = d\omega^K/dt + \Gamma^K{}_{(LR)}\, \omega^L\, \omega^R + 2\Gamma^K{}_{(L,N+1)}\, \omega^L + \Gamma^K{}_{N+1,N+1} = Q^K.$$

Special form of constraints: $V^d = B^d{}_i(t, q)\, V^i + B^d{}_{n+1}(t, q) \equiv B^d{}_i(t, q)\, V^i + B^d(t, q)$.
Equations of Hadamard–Chaplygin: $E_i(T) + B^d{}_i\, E_d(T) = Q_i + B^d{}_i\, Q_d$.
Equations of Voronets:

$$d/dt(\partial T_o/\partial V^i) - \partial T_o/\partial q^i - B^d{}_i(\partial T_o/\partial q^d) - W^d{}_{ii'}(\partial T/\partial V^d)_o\, V^{i'}$$

$$\pm\, W^d{}_i(\partial T/\partial V^d)_o = Q_i + B^d{}_i\, Q_d,$$

$$T = T(t, q, V^d, V^i) = T[t, q, V^d(t, q, V^i), V^i] = T_o(t, q, V^i) = T_{o(constrained\ kin.energy)},$$

$$\gamma^D{}_{II'} \Rightarrow -W^d{}_{ii'} = [\partial B^d{}_{i'}/\partial q^i + B^{d'}{}_i(\partial B^d{}_{i'}/\partial q^{d'})] - [\partial B^d{}_i/\partial q^{i'} + B^{d'}{}_{i'}(\partial B^d{}_i/\partial q^{d'})]$$

$$= (\partial B^d{}_{i'}/\partial q^i - \partial B^d{}_i/\partial q^{i'}) + [B^{d'}{}_i(\partial B^d{}_{i'}/\partial q^{d'}) - B^{d'}{}_{i'}(\partial B^d{}_i/\partial q^{d'})],$$

$$\gamma^D{}_{I,N+1} \equiv \gamma^D{}_I \Rightarrow -W^d{}_{i,n+1} \equiv -W^d{}_i$$

$$\equiv [\partial B^d/\partial q^i + B^{d'}{}_i(\partial B^d/\partial q^{d'})] - [\partial B^d{}_i/\partial t + B^{d'}(\partial B^d{}_i/\partial q^{d'})]$$

$$= (\partial B^d/\partial q^i - \partial B^d{}_i/\partial t) + [B^{d'}{}_i(\partial B^d/\partial q^{d'}) - B^{d'}(\partial B^d{}_i/\partial q^{d'})].$$

Equations of Chaplygin $[B^d{}_i = B^d{}_i(q^{i'}), \quad B^d = 0, \quad T = T_o(t, q^i, V^i) = T_o]$:

$$d/dt(\partial T_o/\partial V^i) - \partial T_o/\partial q^i + (\partial B^d{}_{i'}/\partial q^i - \partial B^d{}_i/\partial q^{i'})(\partial T/\partial V^d)_o\, V^{i'}$$

$$\equiv E_i(T_o) - T^d{}_{ii'}(\partial T/\partial V^d)_o\, V^{i'} = Q_i + B^d{}_i\, Q_d,$$

$$-W^d{}_{ii'} \Rightarrow -T^d{}_{ii'} \equiv \partial B^d{}_{i'}/\partial q^i - \partial B^d{}_i/\partial q^{i'}.$$

Equations of Jacobi–Synge: $E_k(T) = Q_k - [M_{D'D}(A^D{}_\beta|_\lambda V^\beta V^\lambda + A^D{}_k Q^k)]A^{D'}{}_k$.
Central equation of mechanics: $\delta T + \delta'\Gamma + \delta' W = d(\delta P)/dt,$

$$\delta'\Gamma \equiv \int dm\, \mathbf{v}\cdot[(\delta\mathbf{r})^{\cdot} - \delta\mathbf{v}] = P_K\,[(\delta\theta^K)^{\cdot} - \delta\omega^K] - (\gamma^L{}_{K\Sigma}\, P_L\, \omega^\Sigma)\,\delta\theta^K,$$

$$\delta P \equiv \int dm\, \mathbf{v}\cdot\delta\mathbf{r} = P_K\,\delta\theta^K = P_k\,\delta q^k \equiv p_k\,\delta q^k.$$

Central equation in quasi-variables:

$$(dP_K/dt)\delta\theta^K - (\partial T^*/\partial\theta^K)\delta\theta^K + P_K[(\delta\theta^K)^{\cdot} - \delta\omega^K] = Q_K\,\delta\theta^K,$$

$$[(\delta q^k)^{\cdot} - \delta V^k = 0 \Rightarrow (\delta\theta^K)^{\cdot} - \delta\omega^K = \gamma^K{}_{L\Sigma}\,\omega^\Sigma\,\delta\theta^L].$$

Kinematico-inertial identity [for any $f = f(t, q, V)$]:

$$E_k(f)V^k \equiv [d/dt(\partial f/\partial V^k) - \partial f/\partial q^k]V^k$$

$$= d/dt[(\partial f/\partial V^k)\, V^k - f] + \partial f/\partial t \equiv dh(f)/dt + \partial f/\partial t.$$

Power equation in Holonomic variable:

$$dh(T)/dt = -\partial T/\partial t + Q_k\, V^k - \lambda_D\, A^D.$$

Generalized energy function (holonomic variables):

$$h \equiv h(L) = h(t, q, V) = (\partial L/\partial V^k)V^k - L$$

$$= L_{(2)} - L_{(0)} = T_{(2)} + (\Pi - T_{(0)}),$$

$$L \equiv T(t, q, V) - \Pi(t, q) = T_{(2)} + T_{(1)} + T_{(0)} - \Pi$$

$$= T_{(2)} + T_{(1)} + (T_{(0)} - \Pi) = L_{(2)} + L_{(1)} + L_{(0)}.$$

Jacobi–Painlevé integral (under certain conditions):

$$h = T_{(2)} + (\Pi - T_{(0)}) = \text{constant}.$$

Kinematico–inertial identity [for any $L^*(t, q, \omega)$]:

$$E_I(L^*)\omega^I = dh^*/dt + \partial L^*/\partial t + (\partial L^*/\partial q^k)A^k.$$

Generalized energy function (nonholonomic variables):

$$h^* \equiv (\partial L^*/\partial \omega^I)\omega^I - L^* = h^*(t, q, \omega) = T^*_{(2)} + (\Pi^* - T^*_{(0)}).$$

Power equation in nonholonomic variables: $dh^*/dt = -\partial L^*/\partial t + Q_I\, \omega^I + R,$

$R \equiv R' + R''$: rheonomic nonholonomic power:

$$R' \equiv -(\partial L^*/\partial q^k)A^k, \quad R'' \equiv -\gamma^K{}_I\, \omega^I(\partial T^*/\partial \omega^K).$$

Jacobi–Painlevé integral (under certain conditions):

$$h^* = T^*_{(2)} + (\Pi^* - T^*_{(0)}) = \text{constant}.$$

Part I

Tensor Calculus

1 Introduction and Background

Tensor Calculus is one of the more satisfactory chapters of mathematics.

C. Lanczos (1970, p. 99)

Perhaps the best evidence of the remarkable effectiveness of the tensor apparatus in the study of Nature is in the fact that it was possible to include, between the covers of one small volume [335 pages] a large amount of material that is of interest to mathematicians, physicists and engineers."

Sokolnikoff (1951, p. vi)

Sans l'invention des coordonnées rectilignes, l'algèbre en serait peut—etre encore au point où Diophante et ses commentateurs l'ont laissaient, et nous n'aurions, ni le Calcul infinitésimal, ni la Mécanique analytique. Sans l'introduction des coordonnées sphériques, la Mécanique céleste était absolument impossible....Alors viendra nécessairement le règne des coordonnées curvilignes quelconques, qui pourront seules aborder les nouvelles questions dans toutes leur généralite. Oui, cette époque definitive arrivera, mais bien tard: ceux qui, les premiers, ont signalé ces nouveaux instruments, n'existeront plus et seront complètement oubliés, à moins que quelque géomètre archéologue ne ressuscite leurs noms. Eh! qu'importe, d'ailleurs, si la science a marché!

G. Lame (*Lécons sur les Coordonnées Curvilignes et Leurs Diverses Applications*, Paris, 1859, pp. 367–368)

1.1 SOME HISTORY

1.1.1 Aims of Tensor Calculus

Tensor calculus (TC) is a branch of geometry that allows us to formulate geometrical and physical theorems (usually as differential equations) in terms of general, i.e., curvilinear, *coordinates* and *components* of the pertinent quantities, that are independent, or *form invariant*, of the particular system of coordinates used for their descriptions — hence its older name: *absolute differential calculus*. Thus, TC combines the best of both worlds, i.e., (1) *specificity* and *concreteness* for the analytical operations involved and, simultaneously, (2) *generality* and *invariance* of description. It formulates equations valid for a family, or *group*, of coordinates that are obtained from each other by well-defined transformations, of various degrees of generality. This makes TC the ideal tool in several areas of *mathematics* (e.g., differential geometry), *discrete* and *continuum mechanics* (e.g., ***analytical dynamics***, elasticity, shell theory, dislocations), and *theoretical physics* (field theories, e.g., electrodynamics, relativity, cosmology).*

1.1.2 Tensors and Geometry

Classical TC (in *holonomic* coordinates**) was initiated and systematized during the second half of the 19th century, and was brought to definitive form during the first two decades of this century, by such first-rate mathematicians, mathematical physicists, and mechanicians as: Riemann (1854), Beltrami (1865, 1868, 1869), Christoffel (1869, 1870), Lipschitz (1869, 1870, 1873, 1874, 1876, 1882), Voss (1880), Ricci (1884, 1887, 1888, 1889, 1892, 1893, 1894, 1895, 1897, 1898, 1901, 1902, 1903, 1904, 1912), Levi-Civita (1899, 1901, 1917) et al.***

Schematically, the line of development, the "idea tree", is:

(Gauss →) Riemann → Christoffel → Bianchi →

Ricci (1887–96) → Levi-Civita → ...

Additional important work was done by Weyl (1918), and Veblen and Whitehead (1920s), but their contributions will not be needed here.

TC represents the next "natural" step in the long and fruitful process of the *algebrization of geometry.* Schematically, geometry went through the following evolution:

Practical geometry (ancient builders and carpenters, e.g., Babylonians, Egyptians)
Theoretical geometry (ancient Greeks, e.g., Pythagoras, Hippocrates, Archytas, Theaetetus, Eudoxus, Euclid, Archimedes, Appolonius)
Analytic geometry (Descartes, 17th century)
Vector geometry/calculus (Euler, 18th century; fully developed in 19th century)
Differential geometry $\xrightarrow{\text{(curvilinear coordinates)}}$ *Tensor* calculus, theory of Invariants (mid to late 19th century, early 20th century)

Each of these geometrical methods grew out of the limitations of its predecessor. Thus in analytic geometry one has to choose a set of (usually rectangular Cartesian) axes,

* Tensorial methods have also been pioneered, since the 1930s, by G. Kron, in a somewhat unorthodox and controversial fashion, in electrical machine theory, and have been extended and elaborated later by K. Kondo and his co-workers in many other areas of engineering science, such as solid mechanics (plasticity, dislocations), electrical engineering (rotating machines, diacoptics, etc.); see RAAG *Memoirs*.

** These concepts are elaborated later in this and subsequent chapters.

*** For details, see, e.g., (alphabetically): Berwald (1923, publ. 1927), Kline (1972), Reich (1994), Weatherburn (1938, 1963). TC was formally launched, as a relatively complete subject, with the famous memoir of 1901 by the distinguished Italian mathematicians G. C. Ricci and (his student) T. Levi-Civita: "Méthodes de Calcul Différentiel Absolu et leurs Applications," in the prestigious German journal *Mathematische Annalen* (v. 54, pp. 125–201).

for a particular problem, and there are no clear rules as to how to select the "best" axes for it. On the other hand, clearly, such axes are superfluous: the general propositions of Geometry, Mechanics, and Physics have an *intrinsic* validity, i.e., on the level of the equations of motion (or field and constitutive equations) such propositions are independent of the particular coordinates used for their study. (This is not to be confused with the much more serious problem of invariance, or *objectivity*, of the laws of nature relative to arbitrary *frame of reference*, or *kinematical*, transformations (see Chapter 4) a problem that led, eventually, to the General Theory of Relativity.) To overcome these drawbacks, natural scientists tried, first, to deal *directly* with the geometrical and physical quantities involved, by corresponding to them simple geometrical entities from ordinary three-dimensional Euclidean geometry, with well-defined properties — this led to vector calculus (VC). But even VC, despite its remarkable simplicity and success, did not answer the problem completely. Vectors are an excellent tool for studying classical mechanics and physics in the familiar Euclidean three-dimensional physical space (E_3). However, when one has to go to Euclidean spaces of higher (than three) dimensionality, or take the more complex but, as we will see in analytical dynamics (AD) *unavoidable*, route to *non-Euclidean* spaces (i.e., even for classical *nonrelativistic* mechanics), one exceeds the limits of ordinary vectors. Ingenious attempts to generalize VC to meet such demands resulted in extremely cumbersome, unmemorable, and frequently after-the-fact notations — pretty much like the Ptolemaic epicyclical system vs. the (Aristarchian →) Copernican heliocentric one; or the ether theory of the 19th century vs. relativity of this century. Overcoming the limitations of vectors led to a new, more general, powerful, and yet simple tool: the *absolute differential calculus*, or *tensor calculus*. The earlier analogy is more than symbolic: *tensors became the* sine qua non *tool of Relativity*; pretty much like ordinary infinitesimal calculus became the corresponding tool of classical dynamics of Newton–Euler, and matrices/operators that of quantum mechanics.

This new geometrical calculus sought to avoid attachment to any particular set of axes, or system of coordinates, and, instead, wrote down equations in forms valid for any such coordinates.* Similarly, later in this century, out of the limitations of TC grew more general and complex quantities, such as *holors*, *spinors* (and *bispinors*), *twistors*, etc. However, a solid knowledge of classical tensors (this book) is time and effort well spent, no matter what the future holds, because one can then go to the earlier, more-advanced concepts with modest effort. Perhaps TC vs. spinors, twistors, etc. can be likened to the situation of classical vs. quantum mechanics; one needs to learn the former before attempting to understand the latter.

1.1.3 Tensors and Physics

The development of the tensor concepts, along with some of their algebra (but not their differential calculus), as a direct result of classical physics problems, is due to the famous Göttingen physicist W. Voigt (≈1900). He identified such tensorial entities

* A very special case of this approach, but with *rectilinear* coordinates, was used in analytic geometry, and led to *algebraic* geometry; but TC was designed to apply to general curvilinear, i.e., *nonlinear*, coordinates.

in his famous studies of crystal (anisotropic) elasticity.* To him is also due the physically motivated term *tensor*, from the Latin *tendere*, i.e., to stress. (Hamilton had already used the term tensor in the 1840s, in connection with his famous *quaternions*, but did not pursue it any further, especially to physics.) However, the full significance of tensors began to be fully appreciated, by nonmathematicians anyway, only after 1915, 1916: those were the heroic years when A. Einstein, single-handedly, finally succeeded in extending his *Special Theory of Relativity* (of 1905) to his general theory of space–time–gravitation, more commonly known as *General Theory of Relativity*. In that epoch-making achievement of his, Einstein was aided by his mathematician friend M. Grossman, who introduced him to the then new TC (1912 to 1914), and the latter turned out to be the natural mathematical tool for Einstein's relativistic ideas. As Einstein himself puts it: "In the light of the knowledge attained, the happy achievement seems almost a matter of course, and any intelligent student can grasp it without too much trouble. But the years of anxious searching in the dark, with their intense longing, their alternations of confidence and exhaustion, and the final emergence into the light — only those who have themselves experienced it can understand that," and "Imagine my joy at the feasibility of the general covariance and at the result that the equations yield the correct perihelion motion of Mercury. I was beside myself with ecstasy for days." The correction Einstein is referring to is just 43 seconds of arc per century, but the philosophical corrections of his theory to 20th century thinking were much more…finite!

In Relativity, tensors were an absolute necessity. We hope that this book will help persuade its readers that even in classical dynamics, tensors, although not an absolute necessity, are one of the best working tools for derivations, qualitative arguments, and overall *understanding of the formal structure of the equations of motion of constrained systems, especially their transformation properties under certain groups of coordinate transformations*. The latter, according to F. Klein, is the fundamental *ordering*, or *classifying*, principle of physical theories.

1.1.4 Tensors and Mechanics

TC is, roughly, divided into *holonomic* TC (HTC — Chapters 1 to 3), i.e., the calculus of Ricci and Levi-Civita, and *nonholonomic* TC (NHTC — Chapter 3 ff), to be described soon. Most expositions to date deal exclusively with HTC.

The earliest *systematic* applications of HTC to analytical dynamics are contained in the extensive and influential memoir "On the Geometry of Dynamics," by the famous Irish mathematical physicist J. L. Synge (1927).** He followed by a short and little-known monograph on tensorial dynamics (1936), including a brief section

---

* See, e.g., Voigt (1910).

** However, sporadic uses of HTC in mechanics can be found as far back as the 1890s, e.g., a paper by Levi-Civita on Lagrange's equations of motion, and another by A. Wassmuth on the Gauss "Principle of Least Constraint (or Compulsion)." Also, we must mention Somoff (1878, 1879; written during the first half of the 1870s), which, among other basic kinematico–inertial results, contains the earliest *vectorial treatment of the Christoffel symbols*! Unfortunately, this truly admirable and most original Russian work appeared before the era of the widespread use of vectors (even to elementary mechanics), and so its fundamental contributions went largely unnoticed. Its vectorial approach found its way to Western (German) expositions thanks to the works of the famous German mechanikers W. Schell and K. Heun.

on NHTC.* The connection between tensors and dynamics is based on the fact that, from the viewpoint of differential geometry, TC is really the study of differential invariants of the quadratic (metric) form of a Riemannian space, while in dynamics, the kinetic energy of a mechanical system becomes such a quadratic form in the "generalized velocities" of the system (to be detailed later).

The lesser-known classical NHTC was formulated between the two World Wars by such distinguished mathematicians, mathematical physicists, and mechanicians as (alphabetically): Hessenberg, Horák, Schouten, Struik, Synge, Vagner (or Wagner), Vranceanu, Wundheiler, et al. Its earliest *systematic* application to dynamics dates from the 1930s — the aforementioned Synge (1936), Vranceanu (1936), Dobronravov (1940s, fundamental comprehensive memoir of 1948) — and was further elaborated during the 1950s and 1960s by Bressan, Ferrarese, Udeschini-Brinis, Novoselov, et al. Also Kron (1936) utilized it to rotating electrical machines and, inspired by him, K. Kondo and his co-workers (1950s, 1960s), applied it to other areas of engineering science.

1.1.5 Exterior Forms (Cartan Calculus)

A third modern development that we must mention is the theory of *differential forms*, or *exterior calculus*, due to E. Cartan (early 20th century), and promoted enthusiastically (mostly by mathematicians and others peripherally related to engineering) as the only kind of applied geometry worth teaching to engineers and physicists today. This Cartanian calculus (CC) will *not* be presented here for the following reasons: after long and careful examination of the relevant literature, I have concluded that CC does not offer any particular advantage to nonmathematicians over classical TC. At this point, CC constitutes a rather *expensive* and *low-yield* investment, for engineers anyway. This writer believes that the tool should be commensurate to the task. So far, classical TC has been serving dynamics quite faithfully, efficiently and economically. Why, then, junk it to the voracious trash bin of history (actually a–history!) on such short, and of dubious usefulness, advertisement? Especially when the effect of CC on dynamics, and engineering in general, has so far been virtually nil (which brings to my mind comparisons with the disastrous "modern math" of the 1960s).**

The classical dynamics applications of CC that I have found in the literature are either remarkably trivial (of the chainsaw-to-cut-butter-variety, assuming that CC is the chainsaw) or have already been solved and understood by non-Cartanian means.*** In my involvement with engineering science and mechanics, over the past 30 or so years, ranging from civil engineering all the way to engineering relativistic physics, I have not found a single aspect/problem of those arts and sciences, in particular theoretical engineering dynamics, that classical TC could not handle clearly, competently, and effectively, and that CC could! The real danger here is that too much preoccupation with the latter may alienate, frustrate, and eventually

* A valuable summary of Synge's work of 1927 appeared in the last (4th) edition of the well-known (nontensorial) treatise of Whittaker (1937), while a more-extensive coverage of it was included in the monumental encyclopedic article of Prange (1935).

** (See following page for footnote.)

*** (See following page for footnote.)

repel from applied geometry many willing and able engineers/nonmathematicians, at a loss to everyone.

1.2 SOME ALGEBRA

> I have no doubt that to some such developments as the foregoing may appear to be mere symbol — juggling. The criticism is a healthy one and I feel some sympathy with it. But at the same time the search for the most compact and suggestive notation has always been of absorbing interest to some mathematicians; the history of mathematics shows that time spent in such refinements is by no means wasted.
>
> Synge (1936, p.34)

Echoing Synge and others, we request the utmost goodwill and patience of the reader throughout this rather dry and stiff section of the entire book. It is best to get over with this irreducibly painful, but indispensable, part of tensors as soon as possible.*

1.2.1 INDICES AND THEIR ORDER

From now on, and for reasons that will gradually become clear, we shall be making extensive use of *symbols* (or *kernels* — tensorial or not): upper- and/or lowercase, Latin and/or Greek; characterized by one or more *indices* (or suffixes): upper- and/or lowercase, Latin and/or Greek, up (*superscripts* — *not* to be taken for powers!) and/or down (*subscripts*), accented and/or unaccented; e.g.,

$$a_k,\ a^k{}_\beta,\ \alpha_{k\beta,\sigma},\ A_{k'k''},\ A^{k'}{}_k,\ A_{k,l\beta},\ A^{\beta\Gamma}|_\Delta,\ R^{K'}{}_{k\beta},\ \Theta_{k,\beta'\Delta}$$

$$(a,\ \alpha,\ A,\ R,\ \Theta\text{: kernels;}\quad k,\ k',\ k'',\ l,\ \beta,\ \beta',\ \sigma,\ l,\ \Gamma,\ \Delta,\ K'\text{: indices})\quad (1.2.1)$$

** The following is a candid, and rare, admission from a mathematician (who is doing his best to appeal to engineers, etc.), which confirms our misgivings about CC: "*When not to use [differential] forms.* It is time to correct the impression I may have given that differential forms are the solution to all mathematical problems....The formalism of differential forms and the exterior calculus is a highly structured language. This structure is both a strength and a limitation. In this language there are things we cannot say. This leads to an advantage: what can be said usually makes sense....This structure is also a disadvantage. Sometimes a calculation or proof involves steps in which terms are split into pieces that are separately not coordinate-independent. Such calculations are either impossible in differential-form language, or else involve arduous circumlocutions. I must admit that in several places in this book I first had to work things out in 'old tensor'" and "The moral: Use the right tool for the job, even if it is encrusted with indices" (Burke, 1985, pp. 268, 270).

And another, generally eager supporter of CC, makes this sobering evaluation: "The popularity of the theory of differential forms among theoretical physicists today attests to its value. At the same time, the difficulty in learning it and the lack of general applicability has discouraged its study by the wider community." (Vold, 1990, p. 703).

A rare example of a happy medium is Misner et al. (1973), where exterior calculus is used, mostly for *electromagnetism*, and always with translation to the "old tensor."

*** Only Galissot (1954) escapes this broad indictment; but even in his case, classical TC does the job just as well and without his expensive gear.

* For complementary reading, we recommend (alphabetically): Duschek and Hochrainer (1960, V.1), Hawkins (1963), Käestner (1960), Maxwell (1958), Mc Connell (1931), Sokolnikoff (1951), Veblen (1932).

When both sub- and superscripts occur, to avoid ambiguity with the raising of lower indices and lowering of upper indices (to be explained later), it is sometimes necessary to specify the *index order* — we should not place upper and lower indices in the same vertical line. This is also important in matrix multiplication; thus in $\mathbf{A} = (A^k{}_l)$ k is the *first* index (indicating *row* number) while l is the *second* index (indicating *column* number); whereas in $\mathbf{A}' = (A_l{}^k)$ l is the *first* index (indicating *row* number) while k is the *second* index (indicating *column* number). At other times, the index order will be unimportant; e.g., $A^{k'}{}_k = A_k{}^{k'}$.

1.2.2 Index Conventions

To simplify notation in the extensive subsequent summations, we introduce the following key conventions:

- The **summation convention** (introduced by Einstein), according to which we omit the (discrete) summation symbol(s) Σ, by agreeing to sum from 1 to n, or any other desired and clearly understood range of values, over the same lower and upper pair of "dummy" indices (see Remark iii below); for as many such pairs as they may appear in a multiple sum term. For example (with k, l, r, s: dummy indices, all ranging from 1 to n):

a. $$A_1 B^1 + A_2 B^2 + \ldots + A_n B^n = \sum A_k B^k \equiv A_k B^k = A_l B^l$$

$$= A_r B^r = A_s B^s; \qquad (1.2.1a)$$

b. $$\sum A_{kl} B^{ls} \equiv A_{kl} B^{ls} = A_{kr} B^{rs}; \qquad (1.2.1b)$$

c. $$\sum\sum A_{kl} B^{kl} = \sum\sum A_{lk} B^{lk} \equiv A_{kl} B^{kl} = A_{kr} B^{kr} = A_{rs} B^{rs}; \qquad (1.2.1c)$$

- The **range convention**: all free, i.e., nonpaired, indices are taken to run from 1 to n, or any other agreed upon value. Thus, an equation involving f free indices stands for n^f equations. The free indices on both sides of all tensorial equations should match.*

Remarks

i. The notations x^k, y^k, etc. will be used both for the entire set of variables

$$x \equiv (x^1, x^2, \ldots, x^n), \quad y \equiv (y^1, y^2, \ldots, y^n) \text{ etc.,}$$

respectively, *and* for a generic variable from these sets.

---

* Exceptions to this important rule may be made in some simple and well-understood cases involving equality of holonomic with nonholonomic components.

ii. In derivatives like $\partial A^l/\partial x^k$, l is considered as a *superscript*, and k as a *subscript*. Thus:

$$\partial A^k/\partial x^k \equiv \sum(\partial A^k/\partial x^k) = \partial A^1/\partial x^1 + \partial A^2/\partial x^2 + \ldots + \partial A^n/\partial x^n. \quad (1.2.2)$$

iii. A summed-over pair of indices is called *dummy*, or *umbral*; and the reader has, most likely, noticed that it does not matter which symbol (from an appropriate set) is used for such indices; recall the right side of Equation 1.2.1.a.* We also notice that from an equation like $A^kB_k = A^k C_k$ it does *not*, in general, follow that $B_k = C_k$; each side of the former equation represents the sum of n different terms, whereas the latter n equations follow only with *additional* specifications on the A^k. Failure to apply this simple fact correctly has led to significant errors and misunderstandings (e.g., difference between equations involving *actual* and *virtual* work of forces on a system).

iv. The sole drawback of this powerful and fertile index convention, *summation* (for dummy *indices*)/*range* (for free indices), is that one must carefully observe the *position* of each index, i.e., up or down, first or second, etc. On the other hand, this practice eventually becomes a helpful ally to its systematic user. As Eddington (1924, p. 50) puts it: "The convention is not merely an abbreviation but an immense aid to the analysis giving it an impetus which is nearly always in a profitable direction. Summations occur in our investigations without waiting for our tardy approval."** Also it may become difficult to express the *general term* of an expression like Equations 1.2.1a to c.

v. *Summation Suppression*: This may be achieved, for example, by putting parentheses around each index of the selected pair(s) of indices.*** Thus, $A_k = B_{k(l)} C^{(l)}$ (with $k, l = 1, \ldots, n$) stands for the n equations:

$$A_1 = B_{1(l)}C^{(l)}, \quad A_2 = B_{2(l)}C^{(l)}, \ \ldots, \quad A_n = B_{n(l)}C^{(l)}, \quad (1.2.2a)$$

and each one of them, in turn, stands for the n equations:

$$\begin{array}{lll} A_1 = B_{11}C^1, & A_1 = B_{12}C^2, \ \ldots, & A_1 = B_{1n}C^n, \\ \cdots\cdots & \cdots\cdots & \cdots\cdots \\ A_n = B_{n1}C^1, & A_n = B_{n2}C^2, \ \ldots, & A_n = B_{nn}C^n, \end{array} \quad (1.2.2b)$$

respectively; i.e., $A_k = B_{k(l)} C^{(l)}$ stands for a total of n^2 equations; whereas $A_k = B_{kl} C^l$ stands for the n equations:

* This is similar to the situation with *definite* integrals in calculus: it is the limits of integration that matter, not the particular label used for the integration variable(s).

** The same *cannot* be said of the currently popular direct/matrix notation!

*** This practice is not uniform; e.g., some authors insert small bars under those pairs of indices.

$$A_1 = B_{11}C^1 + B_{12}C^2 + \ldots + B_{1n}C^n,$$

$$\ldots\ldots\ldots\ldots\ldots\ldots\ldots\ldots\ldots\ldots\ldots\ldots\ldots$$

$$A_n = B_{n1}C^1 + B_{n2}C^2 + \ldots + B_{nn}C^n. \tag{1.2.2c}$$

vi. In those (rare) cases where the summation extends to more than two indices, we *keep* the summation symbol. Thus, we write: $\sum A_k B^{kl} C_{sk}$, instead of $A_k B^{kl} C_{sk}$.

1.2.3 Symmetry and Antisymmetry

i. A set of indexed quantities $A^{\cdots}{}_{\ldots}$ is said to be *symmetric* in any number of its indices (up and/or down) if their values remain unchanged by any permutation of that group of indices. Thus, if $A^r{}_{kl} = A^r{}_{lk}$, then the $A^r{}_{kl}$ are symmetric in their subscripts; and if $A^r{}_{kl} = A_k{}^r{}_l$, then the $A^r{}_{kl}$ are symmetric in their *first* and *second* indices.

ii. The $A^{\cdots}{}_{\ldots}$ are called *anti-* (*or skew-*) *symmetric*, or *alternating*, in a specified group of their indices (up and/or down), if they remain unchanged by an even permutation of these indices (something that can always be done by an even or odd, but *finite*, number of transpositions of *pairs* of these indices), and if they simply change their sign, i.e., are multiplied by –1, by an odd permutation of that index group. Thus, if $A^r{}_{kl} = -A^r{}_{lk}$, then the $A^r{}_{kl}$ are antisymmetric in their subscripts; and if $A^r{}_{kl} = -A_k{}^r{}_l$, then the $A^r{}_{kl}$ are antisymmetric in their *first* and *second* indices. Finally, if the $A^{\cdots}{}_{\ldots}$ are symmetric (antisymmetric) in *all* their indices, then they are called *completely* symmetric (antisymmetric). For example, to say that the A_{bcd} are completely antisymmetric means that

$$A_{bcd} = -A_{bdc} = -A_{cbd} = A_{cdb} = A_{dbc} = -A_{dcb}; \tag{1.2.3}$$

and similarly for A^{bcd} (see also Section 2.7).*

1.2.4 Special Symbols

1.2.4.1 The Kronecker Delta

The Kronecker Delta:

$$\delta_{kl} = \delta^{kl} = \delta_k{}^l = \delta^l{}_k = \delta_{lk} = \delta^{lk} = \delta_l{}^k = \delta^k{}_l, \tag{1.2.4a}$$

is defined by

$$\delta_{kl} = 1, \text{ if } k = l; \quad \text{e.g., } \delta_{11} = \delta^{33} = \delta_7{}^7 = \delta_{(n)(n)} = 1, \tag{1.2.4b}$$

$$= 0, \text{ if } k \neq l; \quad \text{e.g., } \delta_{12} = \delta^{23} = \delta_3{}^5 = \delta_{(n)}{}^{(n)} = 0. \tag{1.2.4c}$$

* We point out that, according to these definitions, symmetry (antisymmetry) of $A^k{}_l$ means that $A^k{}_l = A_l{}^k$ $(= -A_l{}^k)$; which, in general, is *not* the same thing with the symmetry (antisymmetry) of the (elements of the) *matrix* $(A^k{}_l)$, i.e., $A^k{}_l = A^l{}_k$ $(= -A^l{}_k)$.

Since $\delta^k{}_l A_k = A_l$ and $\delta^k{}_l A^l = A^k$, $\delta^k{}_l$ is also called the *substitution symbol*, or operator.*

Examples of the use of $\delta^k{}_l$:

- If the x^k are independent variables, then $\partial x^k/\partial x^l \equiv x^k{}_{,l} = \delta^k{}_l$.**
- Let $y = a_k x^k$ (a_k: constant coefficients). Then, $y_{,l} = a_k x^k{}_{,l} = a_k \delta^k{}_l = a_l$.
- If $Q = a_{kl} x^k x^l \equiv 0$ for all values of the variables x^k (a_{kl}: constant coefficients), then $a_{kl} = -a_{lk}$. Proof: We have $Q_{,s} = a_{kl} (\delta^k{}_s)\, x^l + a_{kl} x^k (\delta^l{}_s) = a_{sl} x^l + a_{ks} x^k$ (= 0), and so

$$(Q_{,s})_{,r} = (Q_{,r})_{,s} \equiv Q_{,sr} = Q_{,rs} = a_{sl}\, \delta^l{}_r + a_{ks}\, \delta^k{}_r = a_{sr} + a_{rs} = 0, \text{ q.e.d.}$$

- Similarly, let the reader show that, if $a_{kls} x^k x^l x^s \equiv 0$, then $a_{kls} + a_{lsk} + a_{skl} = 0$.

1.2.4.2 The Levi-Civita Permutation Symbols

These symbols, ε_{klr} and ε^{klr} (sometimes also denoted by e_{klr} and e^{klr}; but here the latter are reserved for a slightly different purpose, see Section 2.5) are defined, for k, l, r = 1, 2, 3, by

$$\varepsilon_{klr} = \varepsilon^{klr} = (k-l)(l-r)(r-k)/|k-l|\,|l-r|\,|r-k| = (k-l)(l-r)(r-k)/2$$

= 0; if k, l, r are *not* all distinct; e.g., $\varepsilon_{122} = \varepsilon^{313} = \varepsilon_{121} = \varepsilon^{211} = \ldots = 0$;

= + 1; if $k \neq l \neq r \neq k$ and k, l, r constitute an *even* permutation of 1, 2, 3; e.g., $\varepsilon_{123} = \varepsilon_{312} = \varepsilon^{231} = \varepsilon^{123} = \ldots = 1$;

= −1; if $k \neq l \neq r \neq k$ and k, l, r constitute an *odd* permutation of 1, 2, 3; e.g., $\varepsilon_{132} = \varepsilon_{321} = \varepsilon^{213} = \varepsilon^{132} = \ldots = -1$;

(Generally: $\varepsilon_{klr} = \varepsilon_{lrk} = \varepsilon_{rkl} = -\varepsilon_{lkr} = -\varepsilon_{krl} = -\varepsilon_{rlk}$, and similarly for ε^{klr}). (1.2.5)

It is not hard to show that, for any free index choices,

$$\varepsilon_{kls}\, \varepsilon^{kls} = 6 \qquad (\textit{zero} \text{ free pairs}), \qquad (1.2.6a)$$

---

* Generally: (anything)$_l$ $\delta^l{}_k$ = (anything)$_k$, provided that the range of l includes k. The Kronecker delta can be viewed as the *discrete* counterpart of the famous Dirac–delta "function": using standard calculus notations we have — with $f(\ldots)$ any sufficiently smooth function, x: dummy variable, x_o: free variable:

$$\int_a^b \delta(x - x_o) f(x)\,dx = f(x_o), \qquad \text{if } a < x_o < b,$$
$$= 0, \qquad \text{otherwise.} \qquad (1.2.4d)$$

** Here we have introduced the most convenient notation "*subcomma for (usually) partial derivative,*" relative to well-understood variables, which shall be used repeatedly throughout the rest of the book.

$$\varepsilon_{kls}\varepsilon^{klh} = 2\delta^h{}_s \qquad (\textit{one} \text{ free pair}), \qquad (1.2.6b)$$

$$\varepsilon_{kls}\,\varepsilon^{kbh} = \delta^b{}_l\,\delta^h{}_s - \delta^h{}_l\,\delta^b{}_s \quad (\textit{two} \text{ free pairs}). \qquad (1.2.6c)$$

1.2.4.3 The Generalized Kronecker Delta

This symbol of order, or rank, $2k$, $\delta_{a'a''\ldots}{}^{b'b''\ldots} \equiv \delta^{b'b''\ldots}{}_{a'a''\ldots}$, has k subscripts: a', a'', … and k superscripts: b', b'', …, each running from 1 to n (k and n being unrelated) and is *antisymmetric* in both these sets. Its n^{2k} components are defined as

> + 1 or −1 according as an even or odd permutation is needed to arrange its super- (sub-) scripts in the same order as its sub- (super-) scripts, provided that those indices are all *distinct* from each other *and* both kinds consist of the same set of (positive) indices; 0 in all other cases; also, for $k > n$, $\delta_{a'a''\ldots}{}^{b'b''\ldots} = 0$.

Thus,

- For $n = 3$ and $k = 2$:

$$\delta_{33}{}^{12} = \delta_{13}{}^{21} = \delta_{12}{}^{33} = \ldots = 0,$$

$$\delta_{12}{}^{12} = \delta_{13}{}^{13} = \delta_{31}{}^{31} = \ldots = +1,$$

$$\delta_{21}{}^{12} = \delta_{31}{}^{13} = \delta_{12}{}^{21} = \ldots = +1;$$

- For $n = 3$ and $k = 3$:

$$\delta_{163}{}^{173} = \delta_{123}{}^{122} = \delta_{221}{}^{123} = \delta_{323}{}^{312} = \ldots = 0,$$

$$\delta_{231}{}^{123} = \delta_{123}{}^{123} = \ldots = +1,$$

$$\delta_{213}{}^{123} = \ldots = -1;$$

- For $n = 8$ and $k = 4$:

$$\delta_{1438}{}^{1348} = -1.$$

- Any 2×2 determinant of the matrix

$$\begin{pmatrix} x^1\, x^2 \ldots x^n \\ y^1\, y^2 \ldots y^n \end{pmatrix}$$

  equals $x^k y^l - x^l y^k = \delta_{rs}{}^{kl} x^r y^s$.

- If $A_1, \ldots, A_n$ are functions of $x^1, \ldots, x^n$, then, with the identifications from the preceding example: $x^k \to (\ldots)_{,k}$ and $y^k \to A_k$,

$$A_{k,l} - A_{l,k} = \delta_{kl}{}^{rs} A_{r,s} = \delta_{kl}{}^{sr} A_{s,r}. \tag{1.2.7a}$$

- Any 3×3 determinant of the $3 \times n$ matrix (with $k, l, s; a, b, c$: 1, …, n)

$$\begin{pmatrix} x^1 & x^2 \ldots x^n \\ y^1 & y^2 \ldots y^n \\ z^1 & z^2 \ldots z^n \end{pmatrix}$$

equals the determinant

$$\begin{vmatrix} x^k & x^l & x^s \\ y^k & y^l & y^s \\ z^k & z^l & z^s \end{vmatrix} = \delta_{abc}{}^{kls} x^a y^b z^c; \tag{1.2.7b}$$

and, generally,

$$\delta_{abc}{}^{kls} T^{abc} = T^{kls} + T^{lsk} + T^{skl} - T^{ksl} - T^{slk} - T^{lks}. \tag{1.2.7c}$$

- If $A_1, \ldots, A_n$ and $B_1, \ldots, B_n$ are functions of $x^1, \ldots, x^n$, then (with the identifications from Equation 1.2.7b): $x^k \to A_k$, $y^k \to B_k$, and $z^k \to \partial/\partial x^k \equiv (\ldots)_{,k}$; or $y^k \to (\ldots)_{,k}$, $z^k \to B_k$):

$$\delta_{kls}{}^{abc} A_a B_{b,c} = A_k (B_{l,s} - B_{s,l}) + A_l (B_{s,k} - B_{k,s}) + A_s (B_{k,l} - B_{l,k}). \tag{1.2.7d}$$

Such expressions appear as *holonomicity* (i.e., *integrability*) conditions of linear differential, or Pfaffian, equations (see Sections 3.13 ff and 6.7).

- If $A_{\ldots}{}^{\cdots}$ is *symmetric* in any two or more of its *subscripts*, then

$$\delta_{ij\ldots k}{}^{ab\ldots c} A_{ab\ldots c}{}^{pq\ldots r} = 0; \tag{1.2.7e}$$

and, similarly, if $A_{\ldots}{}^{\cdots}$ is *symmetric* in any two or more of its *superscripts*, then

$$\delta_{ij\cdots k}{}^{ab\ldots c} A_{pq\ldots r}{}^{ij\ldots k} = 0. \tag{1.2.7f}$$

- If $A_{\ldots}$ is *completely antisymmetric*, then

$$\delta_{ij\ldots l}{}^{ab\ldots c} A_{ab\ldots c} = k!\ A_{ij\ldots l} \quad (k\text{: number of } subscripts \text{ of } A_{\ldots}); \tag{1.2.7g}$$

and, similarly, if $A_{\ldots}$ is *completely antisymmetric*, then

$$\delta_{ij\ldots l}{}^{ab\ldots c} A^{ij\ldots l} = k!\ A^{ab\ldots c} \text{ (k: number of \textit{superscripts} of } A^{\cdots}). \quad (1.2.7h)$$

Problem 1.2.1

Show, for example, by counting the number of terms in the various sums, the following useful numerical identities:

- $$\delta_{b\ldots h}{}^{k\ldots r} = \begin{vmatrix} \delta^k{}_b \ldots \delta^k{}_h \\ \ldots\ldots\ldots \\ \delta^r{}_b \ldots \delta^r{}_h \end{vmatrix} = \begin{vmatrix} \delta^k{}_b \ldots \delta^r{}_b \\ \ldots\ldots\ldots \\ \delta^k{}_h \ldots \delta^r{}_h \end{vmatrix}; \quad (a)$$

and, therefore,

$$\delta_{rs}{}^{kl} = \ldots = \delta_r{}^k \delta_s{}^l - \delta_s{}^k \delta_r{}^l,\ \delta_{ml}{}^{kl} = (n-1)\delta_m{}^k,\ \delta_{kl}{}^{kl} = (n-1)n; \quad (a1)$$

and, generally,

$$\delta_{ij\ldots k}{}^{ij\ldots k} = n!/(n-r)! \quad [\boldsymbol{r}\text{: number of sub- (or super-) scripts of } \delta_{\ldots}{}^{\cdots}] \quad (a2)$$

- $\delta_{l'l''\ldots r'r''}{}^{k'k''\ldots r'r''\ldots} = [(n-p)!\,/\,(n-q)!]\delta_{l'l''\ldots}{}^{k'k''\ldots}$
  [$\boldsymbol{p}$: number of l's (or k's); $\boldsymbol{q}$: number of k's *and* r's (or l's and r's)] (b)

- $\delta_{l'l''\ldots r'r''\ldots}{}^{k'k''\ldots b'b''\ldots}\, \delta_{c'c''\ldots}{}^{r'r''\ldots} = [(q-p)!]\delta_{l'l''\ldots c'c''\ldots}{}^{k'k''\ldots b'b''\ldots}$
  [$\boldsymbol{q}$: number of r's (or b's, or c's); $\boldsymbol{p}$: number of l's (or k's)] (c)

- $\delta_{l'l''\ldots r'r''\ldots}{}^{k'k''\ldots b'b''\ldots}\, \delta_{b'b''\ldots}{}^{r'r''} = [(n-p)!\ (q-p)!\,/(n-q)!]\delta_{l'l''\ldots}{}^{k'k''\ldots}$
  [$\boldsymbol{p}$: number of l's (or k's); $\boldsymbol{q}$: number of r's (or b's)]. (d)

1.2.4.4 The Generalized Permutation Symbols

$\varepsilon_{\ldots}/\varepsilon^{\cdots}$ (with number of indices generally equal to, say, n) are defined by

- *Contravariant*:

$$\varepsilon^{a'a''\ldots} \equiv \delta^{a'a''\ldots}{}_{12\ldots n}\ (= \delta_{12\ldots n}{}^{a'a''\ldots}), \quad (1.2.8a)$$

- *Covariant*:

$$\varepsilon_{a'a''\ldots} \equiv \delta_{a'a''\ldots}{}^{12\ldots n}\ (= \delta^{12\ldots n}{}_{a'a''\ldots}); \quad (1.2.8b)$$

i.e., $\varepsilon_{\ldots}/\varepsilon^{\cdots}$ equals +1 or −1 according as its sub- (super-) scripts can be obtained from the positive integers 1, …, n by an *even* or *odd* permutation; and 0 in all other cases (two or more of its indices are equal).

Problem 1.2.2

Verify the following useful $\varepsilon_{...}/\varepsilon^{...}$ identities:

- $\varepsilon^{a'a''...} = \varepsilon_{a'a''...}$. (a)

- $\varepsilon^{a'a''...}\,\varepsilon_{a'a''...} = n!$ ($\boldsymbol{n}$: number of a's). (b)

- $\varepsilon^{a'a''...b'b''...}\,\varepsilon_{c'c''...b'b''...} = [(n-p)!]\delta_{c'c''...}{}^{a'a''...}$
  [$\boldsymbol{n}$: number of a's (c's) *and* b's; $\boldsymbol{p}$: number of a's = number of c's] (c)

  $\Rightarrow \varepsilon^{a'a''...}\,\varepsilon_{c'c''...} = \delta_{c'c''...}{}^{a'a''...}$ (number of a's = number of c's);

  e.g., $\varepsilon^{ijk}\,\varepsilon_{lmn} = \delta_{lmn}{}^{ijk} = \delta^i{}_l\,\delta^j{}_m\,\delta^k{}_n + \delta^i{}_m\,\delta^j{}_n\,\delta^k{}_l + \delta^i{}_n\,\delta^j{}_l\,\delta^k{}_m - \delta^i{}_l\,\delta^j{}_n\,\delta^k{}_m$

  $- \delta^i{}_n\,\delta^j{}_m\,\delta^k{}_l - \delta^i{}_m\,\delta^j{}_l\,\delta^k{}_n$ (by Equation a of Problem 1.2.1). (d)

- $\varepsilon^{a'a''...b'b''...}\,\delta_{b'b''...}{}^{c'c''...} = [(n-p)!]\varepsilon^{a'a''...c'c''...}$
  [$\boldsymbol{n}$: number of a's *and* b's = number of a's *and* c's; $\boldsymbol{p}$: number of a's]. (e)

- $\varepsilon_{a'a''...b'b''...j} = (-1)^p\,\varepsilon_{a'a''...jb'b''...}$
  [$\boldsymbol{p}$: number of indices $b'b''...$; and similarly for $\varepsilon^{...}$]. (f)

Here are some of the uses of $\varepsilon_{...}/\varepsilon^{...}$:

- A general $n \times n$ determinant $A \equiv |A^k{}_l| \equiv \mathrm{Det}(A^k{}_l)$ can be rewritten as (with n = number of a's):

  $$\varepsilon^{a'a''...}\,A^1{}_{a'}\,A^2{}_{a''}..., \quad \text{or as} \quad \varepsilon_{a'a''...}\,A^{a'}{}_1\,A^{a''}{}_2... . \tag{1.2.9a}$$

  For a 3×3 determinant, this yields

  $$A = \varepsilon_{klr}\,A^k{}_1\,A^l{}_2\,A^r{}_3 \quad (\textit{column}\ \text{expansion}) \tag{1.2.9b}$$

  $$= \varepsilon^{klr}\,A^1{}_k\,A^2{}_l\,A^3{}_r \quad (\textit{row}\ \text{expansion}). \tag{1.2.9c}$$

- Next, combining Equation 1.2.9a with the earlier $\varepsilon_{...}/\varepsilon^{...}$ properties we can show (say, by induction) that, for a general $n \times n$ determinant $A = |A^{k'}{}_k|$:

  $$A\ \varepsilon_{klr...} = \varepsilon_{k'l'r'...}\,A^{k'}{}_k\,A^{l'}{}_l\,A^{r'}{}_r... \quad (\textit{column}\ \text{permutation}), \tag{1.2.9d}$$

  $$A\ \varepsilon^{klr...} = \varepsilon^{k'l'r'}\,A^k{}_{k'}\,A^l{}_{l'}\,A^r{}_{r'}... \quad (\textit{row}\ \text{permutation}). \tag{1.2.9e}$$

These representations show immediately that *interchanging any two rows, or columns, of a determinant A changes its sign* ($\Rightarrow$ if any two rows or columns of A are equal, then $A = 0$).

- With the help of the above, the product of two $n \times n$ determinants, A and B, becomes

$$AB = A(\varepsilon_{b'b''\ldots} B^{b'}{}_1 B^{b''}{}_{2\ldots}) = (A \ \varepsilon_{b'b''\ldots})(B^{b'}{}_1 B^{b''}{}_{2\ldots})$$

$$= (\varepsilon_{a'a''\ldots} A^{a'}{}_{b'} A^{a''}{}_{b''}\ldots)(B^{b'}{}_1 B^{b''}{}_2\ldots) \quad \text{(after invoking Equation 1.2.9d)}$$

$$= \varepsilon_{a'a''\ldots} (A^{a'}{}_{b'} B^{b'}{}_1)(A^{a''}{}_{b''} B^{b''}{}_2) \ \ldots$$

$$\equiv \varepsilon_{a'a''\ldots} (C^{a'}{}_1)(C^{a''}{}_2) \ \ldots = (C^k{}_l) \equiv C, \tag{1.2.9f}$$

where

$$C^k{}_l \equiv A^k{}_r B^r{}_l = A^k{}_1 B^1{}_l + A^k{}_2 B^2{}_l + \ \ldots \ + A^k{}_n B^n{}_l. \tag{1.2.9g}$$

Problem 1.2.3

By definition, the *cofactor*, or *algebraic complement*, of the element $A^r{}_c$ of A ($\neq 0$, assumed), $\text{Cof}(A^r{}_c) \equiv C^c{}_r$ is what remains from A after dropping its (r)th row and (c)th column (that is, after striking out the row and column of $A^r{}_c$) and then multiplying the so resulting $(n-1) \times (n-1)$ subdeterminant by $(-1)^{c+r}$. Show that

$$A^r{}_{c'} C^{c''}{}_r = A \ \delta^{c''}{}_{c'} \quad \text{and} \quad A^{r'}{}_c C^c{}_{r''} = A \ \delta^{r'}{}_{r''}. \tag{a}$$

Remarks: The *first* of Equation a: (i) for $c' = c'' \equiv c$ gives the so-called *Laplace expansion* of A in terms of the elements of its (c)*th column*, while (ii) if $c' \neq c''$ states that the sum of the products of the elements of one column, the (c')th, by the cofactors of another, the (c'')th, vanishes. The *second* of Equation a does the same, but in terms of the elements of the *rows* of A and their cofactors.
For example, in terms of its *first* column elements, $A^r{}_1$, and their cofactors $C^1{}_r$, we have

$$A = A^r{}_1 C^1{}_r = A^r{}_1(\varepsilon_{rr''\ldots} A^{r''}{}_2\cdots) = \varepsilon_{r'r''\ldots} A^{r'}{}_1 A^{r''}{}_2\cdots,$$

as before. Further, if $\mathbf{A}^{-1}$ is the *inverse* matrix of $\mathbf{A} = (A^r{}_c)$ (assumed square and nonsingular), then, since $\mathbf{A}\,\mathbf{A}^{-1} = \mathbf{1}$ ($n \times n$ diagonal and unit, or *identity*, matrix) $\Rightarrow A^r{}_c (\mathbf{A}^{-1})^c{}_s = \delta^r{}_s$, and $\mathbf{A}^{-1}\mathbf{A} = \mathbf{1} \Rightarrow (\mathbf{A}^{-1})^r{}_c A^c{}_s = \delta^r{}_s$, we readily conclude that $(\mathbf{A}^{-1})^r{}_c = \text{Cof}(A^c{}_r)/A = C^r{}_c/A$.

Problem 1.2.4

Continuing from the preceding problem, show that

If $A^r{}_c B^c{}_s = \delta^r{}_s$, then $\mathrm{Det}(A^r{}_c)\,\mathrm{Det}(B^r{}_c) = 1$; and if $A^r{}_c = A^c{}_r$, then $C^c{}_r = C^r{}_c$. (a)

Problem 1.2.5

Let $A \equiv \mathrm{Det}(A_{kl})$, $C^{kl} \equiv \mathrm{Cof}(A_{kl})$ in A, $C \equiv \mathrm{Det}(C^{kl})$, and $D_{kl} \equiv \mathrm{Cof}(C^{kl})$ in C. Show that

i. $(n!)A = \varepsilon^{ab\ldots c}\,\varepsilon^{uv\ldots w}\,A_{au}A_{bv\ldots}\,A_{cw}$, $[(n-1)!]C^{au} = \varepsilon^{ab\ldots c}\,\varepsilon^{uv\ldots w}\,A_{bu\ldots}\,A_{cw}$ (a)

$\Rightarrow C = A^{n-1}$; (b)

ii. $D_{kl} = (A^{n-2})A_{kl}$, etc. (c)

Hence, verify that:

a. If $A_{kl} = A_{lk}$ (*symmetric* determinant), then $C^{kl} = C^{lk}$; (d)
b. If $A_{kl} = -A_{lk}$ (*antisymmetric* determinant) *and* n = *odd*, then $A = 0$; (e)
c. For a 3×3 determinant A: $C = A^2$, $C^{1i} = \varepsilon^{ijk}A_{2j}A_{3k}$, $2C^{bi} = \varepsilon^{bsh}\,\varepsilon^{ijk}A_{sj}A_{hk}$. (f)

1.2.5 Linear Equations, Cramer's Rule

To solve the linear nonhomogeneous system

$$A^k{}_l\, x^l = b^k \quad (A^k{}_l,\ b^k\text{: known; } x^l\text{: unknown}) \tag{1.2.10a}$$

we multiply it with $C^r{}_k \equiv \mathrm{Cof}(A^k{}_r)$ and sum over k:

$$(C^r{}_k A^k{}_l)\, x^l = C^r{}_k\, b^k \Rightarrow (A\ \delta^r{}_l)\, x^l = A\ x^r = C^r{}_k\, b^k \quad (\text{assuming that } A \neq 0)$$

$$\Rightarrow x^r = (C^r{}_k/A)\ b^k \equiv (\mathbf{A}^{-1})^r{}_k\, b^k \quad (\text{Cramer's rule}). \tag{1.2.10b}$$*

1.2.6 Functional Determinants (Jacobians)

The Jacobian of the n functions y in the n variables x: $y^k = y^k(x^l)$ $(k, l = 1, \ldots, n)$, J, is defined by

---

* If the elements of A are denoted as A_{cr}, and $\mathrm{Cof}(A_{cr}) \equiv C_{cr}$, instead of C_{rc}, then the earlier Laplace expansions are written, respectively, as $A_{cr}\,C_{cb} = A\ \delta_{rb}$ and $A_{rc}\,C_{bc} = \delta_{rb}$, while Cramer's rule becomes $A_{kl}\,x_l = b_k \Rightarrow x_l = (C_{kl}/A)\ b_k$.
Similarly, if $A \equiv |A_{kl}| \equiv \mathrm{Det}(A_{kl})$ and $\mathrm{Cof}(A_{kl}) \equiv C^{kl}$, Laplace's expansions and Cramer's rule assume the following forms, respectively,

$$A_{cr}\,C^{cb} = A\ \delta^b{}_r\ (\textit{column}\text{ expansion}) \quad \text{and} \quad A_{cr}\,C^{br} = A\ \delta^b{}_c\ (\textit{row}\text{ expansion});$$

$$A_{kl}\,x^l = b_k \Rightarrow x^l = (C^{kl}/A)\ b_k.$$

Finally, in some references one reads: $\mathrm{Cof}(A_{kl}) \equiv C_{lk}$, or $\mathrm{Cof}(A_{kl}) \equiv C^{lk}$. Then the above modify accordingly. For generalizations of these concepts and results, see, e.g., Veblen (1927, pp.9–11).

$$J \equiv |\partial y/\partial x| \equiv \partial(y^1, \ldots, y^n) / \partial(x^1, \ldots, x^n) = \varepsilon^{k'k''\cdots}(\partial y^1/\partial x^{k'})(\partial y^2/\partial x^{k''})\ldots \quad (1.2.11a)$$

If, further, $z^b = z^b(y^k)$ $(b = 1, \ldots, n)$ then, by chain rule, $\partial z^b/\partial x^l = (\partial z^b/\partial y^k)\,(\partial y^k/\partial x^l)$, and therefore, by the determinant multiplication theorem, their Jacobians satisfy (with some easily understood notations): $|\partial z/\partial x| = |\partial z/\partial y|\ |\partial y/\partial x|$. In particular, if $z^b(y^k) = x^b$, this yields

$$|\delta^b{}_l| = |\partial x^b/\partial y^k|\,|\partial y^k/\partial x^l| \Rightarrow |\partial y/\partial x| = 1/|\partial x/\partial y|. \quad (1.2.11b)$$

For a fixed l, the above may be viewed as a linear system in the unknown $\{\partial y^k/\partial x^l\}$, with the n^2 $\{\partial x^s/\partial y^k\}$. Hence, applying the earlier Cramer's rule to it, we obtain

$$\partial y^k/\partial x^l = \left[\mathrm{Cof}(\partial x^l/\partial y^k)\ \text{ in } |\partial x/\partial y|\right] / |\partial x/\partial y|. \quad (1.2.11c)$$

Finally, the main usefulness of Jacobians derives from the following

Theorem of Jacobi: The necessary and sufficent condition for the n functions of n variables $y^k(x^l)$ to be dependent is that $J = 0$. (For proof, see books on advanced calculus, analysis, etc.)

Problem 1.2.6

Show that

$$\delta_{k'k''\ldots}{}^{l'l''\cdots}\,|\partial y/\partial x| = \partial(y^{l'}, y^{l''}, \ldots) / \partial(x^{k'}, x^{k''}, \ldots), \quad \text{(a)}$$

and

$$\delta_{k'k''\ldots}{}^{l'l''\cdots} = \partial(x^{l'}, x^{l''}, \ldots) / \partial(x^{k'}, x^{k''}, \ldots). \quad \text{(b)}$$

1.2.7 Derivatives of Determinants

If the n^2 elements $A^k{}_l$ of a determinant A are functions of the n variables x^s, then, differentiating partially the representation $A = \varepsilon_{k'k''\ldots}A^{k'}{}_1 A^{k''}{}_2\ldots$, we find successively:

$$\partial A/\partial x^s = \varepsilon_{k'k''\ldots}\,[(\partial A^{k'}{}_1/\partial x^s)A^{k''}{}_2 \ldots + A^{k'}{}_1\,(\partial A^{k''}{}_2/\partial x^s) \ldots + \ldots]$$

$$= (\partial A^{k'}{}_1/\partial x^s)C^1{}_{k'} + (\partial A^{k''}{}_2/\partial x^s)C^2{}_{k''} + \ldots,$$

i.e.,

$$\partial A/\partial x^s = (\partial A^k{}_l/\partial x^s)C^l{}_k;\ \text{or via the comma notation: } A_{,s} = A^k{}_{l,s}\,C^l{}_k. \quad (1.2.12a)$$

Specialization: If A is the Jacobian of the transformation $x \to y$, i.e., $A = J = |\partial y/\partial x|$, then $A^k{}_l \to \partial y^k/\partial x^l$ and $\mathrm{Cof}(A^k{}_l) \to \mathrm{Cof}(\partial y^k/\partial x^l) = J\,(\partial x^l/\partial y^k)$, and so Equation 1.2.12a gives $\partial J/\partial x^s = [\partial/\partial x^s(\partial y^k/\partial x^l)]J\,(\partial x^l/\partial y^k)$, or finally

$$(1/J)(\partial J/\partial x^s) = \partial(\ln J) / \partial x^s = (\partial^2 y^k/\partial x^s\, \partial x^l)(\partial x^l/\partial y^k). \qquad (1.2.12b)$$

Problem 1.2.7

Let $A \equiv \mathrm{Det}(A_{kl})$, $C^{kl} \equiv \mathrm{Cof}(A_{kl})$ in A. Verify that, for a 3×3 determinant A:

i. $\partial A/\partial A_{kl} = C^{kl}$ (A_{kl}: independent elements); (a)

ii. $dA/dx = C^{kl}\,(dA_{kl}/dx)$ (A_{kl}: functions of x); (b)

iii. $d(\ln A)/dx = \mathrm{Trace}[\mathbf{A}^{-1}\,(d\mathbf{A}/dx)]$ (c)

[matrix $\mathbf{A} = (A_{kl})$: nonsingular, i.e., $A \neq 0$; $\mathbf{A}^{-1}$: inverse matrix of $\mathbf{A}$, $\ln(\ldots) \equiv \log_e(\ldots)$]. Extend the above to $n \times n$ determinants.

1.3 SOME GEOMETRY

The following is a compact, qualitative, and intuitive listing of some fundamental differential — geometric and tensorial terms and concepts.*

1.3.1 Primitive Concepts

Primitive concepts, or *undefined elements* (to be accepted here without further rigorous logical analysis) are those needed to construct a mathematical formalism, or model, that describes adequately a continuously differentiable manifold (see below). The properties of these elements are established by means of unproved propositions, or *axioms*, while the logical consequences of the latter constitute the theorems and corollaries of that model. Such concepts are: *point, place (or location), order, relation, correspondence (or mapping, or function); time, particle, rigid body, mass, and force.*

1.3.1.1 Set

A set, or collection, or aggregate, or class, is the totality of variables needed to describe a geometrical or physical system; for example, the set of (generally curvilinear) *position parameters*, or *Lagrangean coordinates*, needed to describe its location, or *configuration*:

$$q = (q^1, q^2, \ldots, q^n) \equiv (q^k;\ k = 1, \ldots, n). \qquad (1.3.1)\ **$$

1.3.1.2 Group (G)

A group is a set of elements $x, y, z, \ldots$ with the following rules of combination, to be denoted by $*$,

---

* For extensive and rigorous discussions, see, e.g., (alphabetically), Arnold (1988), Dubrovin et al. (1992), Kreyszig (1959), Lovelock and Rund (1975), Mishchenko and Fomenko (1988), Willmore (1959).

** Why general curvilinear coordinates are denoted as q^k, instead of q_k, is clarified later (Section 2.4).

i. The combinations $x * x$, $x * y$, etc. are also members of G (group property);
ii. $(x * y) * z = x * (y * z)$ (associativity);
iii. The set contains the *unit* element I, which is such that when it combines with any other member of G, x, it leaves it unchanged: $x* I = I* x = x$;
iv. Every element of G, x, has an *inverse*, x^{-1}, belonging to G; also $x * x^{-1} = x^{-1} * x = I$.
   - Properties i and ii are, by far, the most important. If $x, y, z, \ldots$ satisfy only them, the set is called *semigroup*.
   - If all the elements of G belong to a second group G', then G is called a *subgroup* of G'.
   - If G repeats itself, it is called *cyclic*.
   - If G has a finite (infinite) number of elements, it is called *finite* (*infinite*).
   - The elements of a group can be, not just numbers, but very abstract entities like translations, rotations, or general transformations. For example, the set of all reversible linear transformations, with combination rule the matrix multiplication, is called the full linear group, and by imposing restrictions on these transformations, such as preservation of length, or staying within a certain space (parity), we obtain various subgroups of it.
   - The groups that matter in tensor theory are, primarily, groups of *admissible* transformations of coordinates — to be elaborated upon shortly; and the fundamental concepts of *invariance, covariance,* and *contravariance* (Section 2.5 ff) refer to an entire admissible group, and not just a certain specific transformation.

1.3.1.3 Space

Space is a set of undefined elements called geometrical points. In mechanics, each such point represents a configuration of the system under discussion; alternatively, *space is the set of configurations of a system.*

1.3.1.4 Algebraic Definition

An n-dimensional (region of) space, V_n $(n < \infty)$, is a set of geometrical points that is, or can be put, in a one-to-one reciprocal correspondence with the n-dimensional *arithmetic* space V^*_n; that is with the totality of ordered n-tuples of real numbers (this book), called *algebraic points* $q = (q^1, \ldots, q^n) \equiv (q^k; k = 1, \ldots, n) \equiv (q^k)$ which obey the rules:

$$(q^1, \ldots, q^n) + (p^1, \ldots, p^n) = (q^1 + p^1, \ldots, q^n + p^n) \quad \text{(addition)}$$

$$\lambda\ (q^1, \ldots, q^n) = (\lambda q^1, \ldots, \lambda q^n) \quad \text{(multiplication with a number } \lambda\text{)}$$

$$|q^k - q_o{}^k| < c^k \quad (k = 1, \ldots, n;\ q_o{}^1, \ldots, q_o{}^n\text{: constants; } c^1, \ldots, c^n\text{: positive constants})$$

The numbers q^k representing a V_n-point P are its *coordinates.*

1.3.1.5 Examples

If $n = 1$, the points could be the frequencies of the spectrum of an oscillating particle; if $n = 4$, the points could be the *events* of space–time (Chapter 4); if n = arbitrary positive integer, they could represent the configurations of a discrete mechanical system.

1.3.2 Coordinate System(s) (CS)

A CS covering the set of geometrical points P of a V_n, is any one-to-one reciprocal correspondence between these points and their algebraic counterparts of V^*_n, q. The choice and construction of the particular CS, to be employed in a physical problem, is indicated by the latter's particular geometrical or physical aspects.

The *origin* of this CS is the V_n-point O corresponding to the values $q = 0$.

1.3.3 Manifold

Manifold is also termed *manyfold*; from the German *Mannigfaltigkeit* (Riemann, early 1850s), i.e., multiplicity, diversity, variety. Below we give several complementary definitions of this fundamental concept.

- A space that (1) is equipped with a CS, and (2) can support a *differentiable* structure; that is why its full name is *differentiable manifold.*
- An infinite dense point set, every point of which is determined by $n < \infty$ independent coordinates, or parameters (continuously variable real numbers).
- Any set X that can be continuously and differentiably parametrized via a finite number of independent (real-valued) coordinates, or parameters. Their number, say, n, is the *dimension* of X; which is then denoted as X_n.
- A continuous space that locally looks like Euclidean space (or flat space — to be detailed later), but globally can wrap, bend, and do "anything" as long as it stays continuous; i.e., *X_n is a generalization of two-dimensional surfaces in ordinary physical three-dimensional Euclidean space,* E_3. However, contrary to those surfaces, X_n is not necessarily viewed as embedded into some higher dimensional Euclidean space.*

---

* (1) Qualitatively: "Riemann's idea of a differentiable manifold (or of a surface imbedded [sic] in a Euclidean space) with geometric diversity can be described qualitatively as a patchwork or quiltwork of 'local' [meaning infinitesimally small] Euclidean hyperplanes. One imagines that the Riemannian surface is enveloped by a lot of very small tangent planes, each of which is Euclidean." (McCauley, 1997, p. 255); and more precisely "...The essential feature of a differentiable manifold is that it is covered by a set of coordinate neighborhoods, each having the same number of coordinates, with the property that two different systems of coordinates in a common region are related by a differentiable transformation of class not less than 1" [i.e., continuous and with up to first continuous derivatives] (Willmore, 1959, pp. 193 ff).
(2) In (most of) the older tensor/differential geometry literature (say, up until the early 1960s), the terms *space, manifold, variety, hyperspace* (for $n > 3$) were used virtually synonymously.

In sum, a manifold is a space that is both continuous and differentiable, and can, therefore, be studied by infinitesimal calculus. This is sufficiently general to cover practically all the geometrical needs of classical and relativistic (nonquantum) physics. So, from now on, *our most general spaces will be finite-dimensional differentiable manifolds.*

1.3.3.1 Examples

i. The set of all ellipses $(x/a)^2 + (y/b)^2 = 1$; a, b ($\neq a$): half-axes, constitutes a two-dimensional manifold X_2 with coordinates $q^1 = a$, $q^2 = b$. Each point of it represents a particular ellipse.
ii. The set of displacements of a rigid body moving in space with one of its points fixed; i.e., the set of its rotations about that point. Let the parameters of that set, X_3, be the three Eulerian angles of the body (relative to fixed axes): ϕ (precession) $\rightarrow$ θ (nutation) $\rightarrow$ ψ (proper spin). Each point of X_3 has as coordinates these three parameters and defines a particular rotation, or angular orientation, of the body.*

1.3.4 Coordinate Transformation(s) (CT)

The set of n functional equations

$$q^{k'} = q^{k'}(q^1, \ldots, q^n) \equiv q^{k'}(q^k)\ [k, k' = 1, \ldots, n] \quad \text{or, simply,} \quad q' = q'(q), \tag{1.3.2a}$$

is said to determine a *correspondence* between the sets of coordinates q and the numbers q', if the $q'(\ldots)$ are single valued and invertible (or reversible) for all points P of our $V_n \rightarrow X_n$, or some region of it, so as to yield the n inverse equations:

$$q^k = q^k(q^{k'}) \quad \text{or, simply,} \quad q = q(q'), \tag{1.3.2b}$$

where the $q(\ldots)$ are also single valued for all q terms given by Equation 1.3.2a; i.e., $q \leftrightarrow q'$ is one-to-one. Now, if we view each set of q, q' as the coordinates of the *same* geometrical point P of V_n, we have another way of attaching numbers to points, i.e., the q' are another CS; and Equations 1.3.2a and b define *a coordinate transformation* (CT); i.e., they serve as a "dictionary" enabling us to pass from the q "language" of describing the position of a V_n-point P to the q' "language;" and vice versa. Such a reading of Equations 1.3.2a and b, i.e., keeping P fixed and changing its mode of description (relabeling it), is called the *passive* interpretation of the transformation equations (Figure 1.1a). A *second* viewpoint (and, conceivably, there are more**), called the *active* interpretation of Equations 1.3.2a and b, is to look at $q \leftrightarrow q'$ as defining a transformation (or mapping) among V_n-points,

---

* For details on the Eulerian angles, see, e.g., Rosenberg (1977, pp. 76–82). It is shown there that (1) $0 \leq \phi$, $\psi \leq 2\pi$, $0 \leq \theta \leq \pi$; (2) the points $\psi = 0$, 2π are to be viewed as identical; and (3) the points $\theta = 0$, π do *not* correspond to a unique solution for the rates $d\phi/dt$, $d\psi/dt$. See also Example 6.5.1.

** See, for example, Schouten (1954b, p. 9).

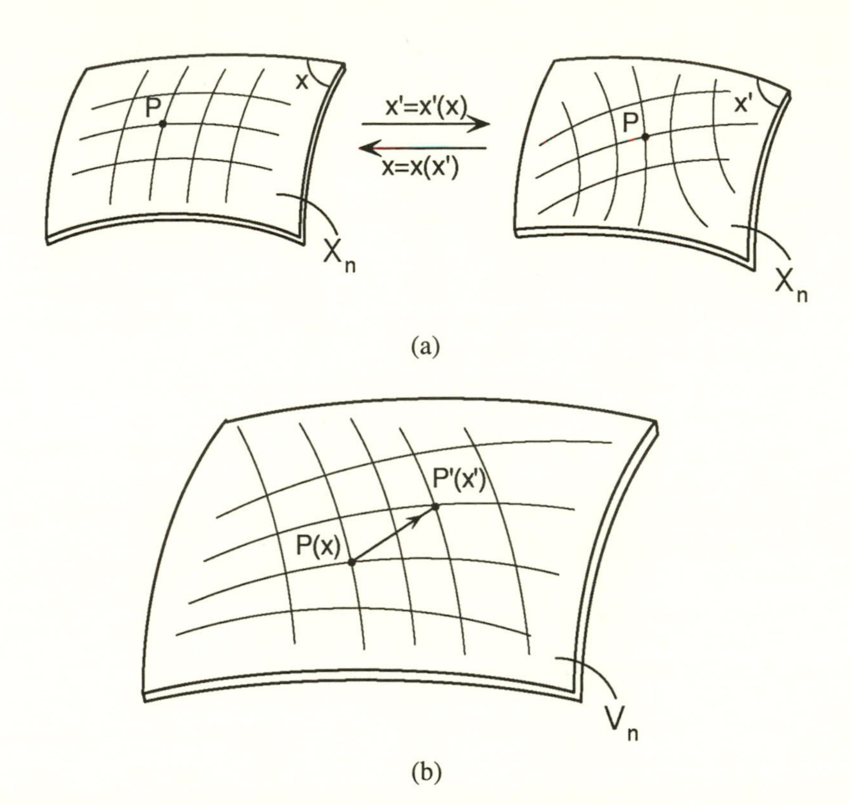

FIGURE 1.1 (a) *Passive* interpretation: *coordinate/axes transformation* (same point, different coordinates). (b) *Active* interpretation: *point transformation/mapping* (same coordinates, different points).

say, from $P(q)$ to $P(q') \equiv P'$, and vice versa, *both in the same CS* (Figure 1.1b). Then Equations 1.3.2a and b give the rules on how to correspond P to P' (to P'', etc.) The active interpretation is employed a lot in continuum mechanics (motion of solids, fluids etc. usually accompanied by deformation); although sometimes one uses both viewpoints. *In this book, and unless stated explicitly to the contrary, we shall mainly adopt the passive viewpoint.*

1.3.4.1 Systems of Coordinates vs. Frames of Reference

A CT is *not* to be confused with the more involved and *physical* concept of frame of reference transformation. For our purposes, a frame of reference is a conceptual framework of measuring rods (yardsticks) and clocks, extending into space and rigidly attachable to some reference-invariable body; i.e., a spatiotemporal coordinate system rigidly connected to that body.* Therefore, from the analytical viewpoint, *such frame of reference transformations can be studied as explicitly time dependent (i.e., kinematical) CT*; one CS rigidly attached, or embedded, to each frame, plus the *time* transformation:

* For more precise definitions, see for example Bergmann (1942) and (1962); also Section 4.1.

$$\{q' = q'(q, t), \quad t' = t'(q, t)\} \leftrightarrow \{q = q(q', t'), \quad t = t(q', t')\}. \quad (1.3.3)$$

In nonrelativistic (Newtonian) physics, we can take $t = t'$. In general, such frame transformations lead to *nontensorial* laws of transformation for the physical quantities involved. The relativistic way out of this difficulty is to modify ordinary space (and time) into a mathematical one in which physical quantities do transform tensorially.

1.3.5 Successive CT, Group Property

If the q variables are replaced by the q' variables via a Equation 1.3.2(a and b)–like CT, and then the q' variables are replaced by the q'' variables by a similar CT′: $q' \leftrightarrow q''$, then the direct transformation $q \leftrightarrow q''$ is called the *product* transformation: (CT) (CT′) (symbolically). In general (CT) (CT′) ≠ (CT′) (CT).

Now, we shall assume that *tensorial transformations form a group, G*; i.e.,

- Any two transformations of G carried out consecutively produce a transformation belonging to G (*closure* property);
- The *inverse* of every G-belonging transformation is also a G-transformation;
- G contains the *identity* transformation; and
- These CT are *associative*, i.e.,

$$(\mathrm{CT})\ [(\mathrm{CT}')(\mathrm{CT}'')] = [(\mathrm{CT})(\mathrm{CT}')]\ (\mathrm{CT}''), \quad \text{where: } \mathrm{CT}'': q'' \leftrightarrow q'''.$$

Thus, we have a perfectly definite *family* of CS: the (accidentally, or conveniently, chosen) initial one and all those obtainable from it by admissible transformations (defined below); and so *results in tensor theory are independent of the choice of the initial CS, as long as that system belongs to that family.*

1.3.6 Admissible CT

On physical grounds, such CT must be *real, single valued,* and *invertible.* This leads to the following mathematical restrictions:

- The functions $q'(\ldots)$, along with their first partial derivatives (at least), must be continuous in some X_n-region R, i.e., be at least of class C^1 there,* and (see *manifold orientation* below)
- $J \equiv |\partial q'/\partial q| \neq 0, \infty$ (if $J > 0$: *proper* CT; if $J < 0$: *improper* CT) (1.3.4) except possibly at certain special points of R.

Then the single-valued inverse $q = q(q')$ exists, and the $q^k(\ldots)$ are also of class C^1. So restricted transformations we shall call *admissible*, in R, and for physically

---

* **Definition**: A function is of class C^k, in some region R of a manifold X_n, if it is continuous and has continuous derivatives up to and including order k at each point of R. Thus a continuous function is of class C^0, while one with continuous derivatives of all orders is of class C^∞. This property is independent of the particular choice of CS in R.

obvious reasons they are the only ones to be discussed here; and, from now on, when we say CS we will mean one of a family related by transformations of that group.

Some Relevant Theorems

- Since $|\partial q''/\partial q| = |\partial q''/\partial q'|\,|\partial q'/\partial q|$ (transitivity of the Jacobians), the resultant of two admissible transformations is also admissible.*
- Since $|\partial q'/\partial q| = 1/|\partial q/\partial q'|$, the inverse of any restricted transformation is one of the same type; i.e., *the restricted transformations also form a group.*

1.3.7 INVARIANCE

Invariance is the constancy of some entity, or object, or form, under a CT.

1.3.7.1 Entity, or Object, Invariance

For example, a point P is such an invariant; i.e., under a CT: $q \leftrightarrow q'$, P is described by $P(q) \rightarrow P(q')$, but the CT does not affect the point itself. Similarly, a pair of points P_1 and P_2 (and the *distance* between them), or a *set of points* making up a curve or a surface, are invariant; and, accordingly, the same holds for any *point function.* If the latter is represented by $f(q)$ in the q CS, then under $q \rightarrow q'$ it will be represented by $f'(q') = f[q(q')]$. Such a quantity we shall call an *absolute scalar,* and $f(q)$ its (sole) component in the CS q.**

1.3.7.2 Form Invariance

A mathematical form that preserves itself in all admissible CS: If in addition to object invariance $f(q) = f'(q')$, we also have $f(q) = f(q')$, i.e., *invariance in the functional form.**** For example, the form giving the velocity or acceleration components of a particle; in some CS some of its terms may vanish.

With the help of the above we can now reiterate, and make more precise, the manifold definition: *A simple coordinate manifold X_n (of class C^k) is the entity consisting of (i) a space V_n of geometrical points P, plus (ii) the totality of allowable CS, i.e., CS mutually related by allowable CT (of class C^k) covering V_n.*

---

* It is not hard to see that such a transitivity property also holds for the kth *power* of the transformation Jacobians, where k is *any integer* (this last restiction does not come from mathematics, but it covers all physically useful cases).
For applications of this property, see *relative tensors*: Sections 2.5.4.

** As detailed in Section 2.5.4, every such invariant is a scalar function, but not every scalar function is invariant under admissible CT. [Incidentally, we shall use precise notations like $y = f(x)$ only when absolutely necessary (see Form Invariance, below). For most other purposes we will rely on such common "engineering" notations as $y = y(x)$; and let the precise meaning become clear from the context.

*** From these *two* kinds of invariance, i.e., in the value of a funtion (its dependent variable) and in the form of the funtion itself, it is the *first* one that is, by far, the more important to tensor theory and its applications; and so will be the one understood here, unless specified otherwise.

1.3.8 Manifold Orientation

Starting with an allowable CS q, for our X_n, we consider all other allowable CS that, in addition, are related to q via *proper CT*; i.e., for every other such CS, say, q', the $q \leftrightarrow q'$ Jacobian $J = |\partial q'/\partial q|$ (and therefore also $J^{-1} = |\partial q/\partial q'|$) is everywhere *positive* (or, at least, in some common neighborhood). Such a restricted X_n is called an *oriented* simple coordinate manifold of class C^k. Schematically,

$$(V_n\text{: space of geometrical points}) + (\text{CS: allowable and proper, of class } C^k)$$

$$= (X_n\text{: oriented and simple manifold, of class } C^k)$$

Thus, we do not only have one X_n, but *two*: one *positively oriented,* $X_n^{(+)}$ (*proper* X_n) and one *negatively oriented,* $X_n^{(-)}$ (*improper* X_n). Manifold orientation is needed in connection with such concepts as *relative* tensors; e.g., polar vs. axial vectors (Sections 2.5 and 9).*

1.3.9 Tensor Calculus (TC)

The study of abstract geometrical objects, tensors, as *functions of position, P*; i.e., of points described by numerical coordinates, in (a certain region of) a manifold. Although tensors are concretely represented in a particular CS q by an ordered set of numerical functions of q, i.e., their *components* in that CS (just like the components of a vector in vector calculus), *yet TC establishes relations and properties among tensors that are independent of the particular CS used for their concrete description.* Thus, such a tensorial point function $\mathbf{T}(\mathrm{P}) = \mathbf{T}(q^1, \ldots, q^n)$ will be described, or represented, by an ordered set of $n^{\cdots}$ ($< \infty$) components: $T(i_1, \ldots, i^{\cdots}; q^1, \ldots, q^n) \equiv T(i; q) \equiv T^{\cdots}{}_{\ldots}(q)$, where each *index* $i \ldots$ runs from 1 to n.

This association $\mathbf{q} \rightarrow \mathbf{T}^{\cdots}{}_{\ldots}(q)$ is called *q-scheme of measurements for* $\mathbf{T}(\mathrm{P})$ (and all other point functions of that class). In general, certain of the indices of each component of $\mathbf{T}$ will appear as **superscripts** ($i_{\mathbf{up}}$: $u, u', u'', \ldots$) and certain as **subscripts** ($i_{\mathbf{down}}$: $d, d', d'', \ldots$):

$$\mathbf{T}(\mathrm{P}) \rightarrow \{T(i; q)\} \rightarrow T^{uu'u''\cdots}{}_{dd'd''\ldots}(q^1, \ldots, q^n)$$

or

$$T_{dd'd''\ldots}{}^{uu'u''\cdots}(q_1, \ldots, q^n) \text{ etc.,} \tag{1.3.5}$$

a total of $n^{\textit{number of up and down indices}}$ (since all indices run from 1 to n), and this number is the same for all considered schemes of measurement. This **indicial-component** notation goes hand in hand with the summation convention.

---

* See also Lovelock and Rund (1975, p. 334).

The components of **T**(P) in any two admissible CS, q and q', $T^{\cdots}{}_{\cdots}(q)$ and $T'^{\cdots}{}_{\cdots}(q')$, respectively, are related by

$$T'^{\cdots}{}_{\cdots}(q') = T'^{\cdots}{}_{\cdots}[T^{1\ldots1}{}_{1\ldots1}(q), \ldots, T^{n\ldots n}{}_{n\ldots n}(q)], \qquad (1.3.6)$$

a functional equation to be detailed in Chapter 2.

Remarks

i. According to the modern viewpoint (advanced by Weyl, Veblen, et al.), the name tensor is reserved for the invariant point function **T**(P), i.e., for geometrical objects independent of a particular scheme of measurements. Nevertheless, here we shall frequently follow the more applied tradition or slang of TC (initiated by Einstein) of calling tensor the components of **T**(P) in a particular scheme of measurement; i.e., say, … the tensor $T^{kl}{}_{rs}(q)$ ….
ii. Whether a given set of indexed quantities represents a tensor, in a certain group of CS transformations, depends on the geometrical and/or physical rules of transformation of these quantities among the CS of that group. For example, that the components of stress, strain, moment of inertia, etc. transform tensorially follows from the equations of mechanics; *then* they become (properly) indexed quantities!
iii. Tensor *equations* are *form invariant* relative to a given group of CS transformations, and this makes them ideal for formulating physical laws (differential equations).* Thus the concept of invariance, under this or that group of CS transformations, is fundamental to TC.

1.3.10 Subspaces in a Manifold; Curves, Surfaces, etc.

If our coordinates $q^1, \ldots, q^n$ are *independent*, i.e., if no functional relation(s) of the form $f(q^1, \ldots, q^n) = 0$ has been presumed to exist among them, then they represent an n-dimensional manifold X_n. But if we couple them by such *constraining* equations (which is the common theme of engineering Lagrangean mechanics), then we create *submanifolds* inside X_n. Thus, the m ($< n$) independent *equations of constraint*, or condition,

$$f^D(q^1, \ldots, q^n) \equiv f^D(q) = 0 \qquad [D = 1, \ldots, m(< n)] \qquad (1.3.7)$$

define the *$(n - m)$-dimensional sub-manifold* $X_n^{n-m} \equiv X_n^I$ immersed in X_n. Indeed, solving Equation 1.3.7 for, say, the first m qs, or q^D (D for *dependent*), in terms of the remaining $n - m$ qs, or q^I (I for *independent*), yields the X_n^I-representation:

* As Sokolnikoff (1951, p. 51) puts it: "...whether a logical deduction based on a conglomerate of observational facts deserves the name of a natural law is often determined by the generality of such a deduction, and by its validity in sufficiently wide class of reference systems."

$$q^D = q^D(q^{m+1}, \ldots, q^n) \equiv q^D(q^I) \quad [I = m + 1, \ldots, n]; \tag{1.3.8a}$$

or, for extra generality, introducing the $n - m$ new independent parameters $u^I \equiv (u^{m+1}, \ldots, u^n)$, via the admissible CT: $q^I = q^I(u^I)$ (e.g., $q^I = u^I$), finally produces the equivalent $X_n{}^I$-representation:

$$q^D = q^D(u^I) \Rightarrow q^k = q^k(u^I) \quad (k = 1, \ldots, n). \tag{1.3.8b}$$

1.3.10.1 Special Cases

- If $\boldsymbol{m = n - 1}$, the above yield a *one-dimensional* subspace $X_n{}^1$, called a *curve* in X_n; the totality of points given by $q^k = q^k(u^1 \equiv u)$. If $u \to q^s$ ($1 \leq s \leq n$), the curve is the *(s)th coordinate line* of the system q.
- If $\boldsymbol{m = n - 2}$, the submanifold $X_n{}^I \to X_n{}^2$ is called a *surface*; while
- If $\boldsymbol{m = 1}$, the submanifold $X_n{}^I \to X_n{}^{n-1}$ is called a *hypersurface*; say, with equation $f(q) = 0$. Such equations divide X_n into *two* parts: one where f is *positive* (say, up or left) and one where f is *negative* (down or right). The equation $f(q) = \text{constant} \equiv c$ represents a *family of hypersurfaces*, one for each value of c — and only one through each X_n-point, if $f(\ldots)$ is *single valued.* Here, however, we shall be using the terms surface and hypersurface in X_n for any value of m: $1 \leq m < n$.

Finally, let both neighboring points $P(q)$ and $P'(q + dq)$ belong, or lie, on the surface $f^D = 0$. Then, expanding $f^D(q + dq) = 0$ à la Taylor, etc., we conclude that

$$df^D = \sum (\partial f^D/\partial q^k)_P \, dq^k \equiv f^D{}_{,k} \, dq^k = 0, \tag{1.3.9}$$

i.e., the line $PP'(dq)$ belongs to the *tangent space* of $X_n{}^I$ at P, $T_{n-m}(P) \equiv T_I(P)$. The latter is the plane through P with Equation 1.3.9.* (More in Sections 2.12 and 6.8.)

Conversely, the system of m *Pfaffian* equations $f^D{}_{,k} \, dq^k = 0$ defines a family of ∞^m spaces $X_n{}^I$ through the m equations $f^D = \text{constants of integration} \equiv c^D$, one such space through each X_n-point $P(q)$. These concepts are fundamental to both tensor theory/differential geometry and constrained system dynamics (i.e., Lagrangean analytical mechanics, Chapters 6 and 7).

Problem 1.3.1

Let q^k and $q^{k'}$ be general coordinates related by an admissible transformation. Show that the following identities hold among their gradients:

i. $(\partial q^{k'}/\partial q^l)(\partial q^l/\partial q^{r'}) = \delta^{k'}{}_{r'}, \quad (\partial q^{k'}/\partial q^l)(\partial q^k/\partial q^{k'}) = \delta^k{}_l;$ (a)

---

* We recall, from elementary analytic geometry, that the equation $a\,x + b\,y + c\,z = 0$, where a, b, c: constant coefficients, and x, y, z: rectangular Cartesian coordinates with origin O, represents a *plane through O.*

ii. $(\partial^2 q^{k'}/\partial q^k \partial q^l)(\partial q^l/\partial q^{l'}) + (\partial q^{k'}/\partial q^l)(\partial^2 q^l/\partial q^{l'} \partial q^{s'})(\partial q^{s'}/\partial q^k) = 0;$ (b)

iii. $(\partial^2 q^{k'}/\partial q^l \partial q^s)(\partial q^s/\partial q^{s'})(\partial q^l/\partial q^{l'}) + (\partial q^{k'}/\partial q^l)(\partial^2 q^l/\partial q^{l'} \partial q^{s'}) = 0;$ (c)

iv. $\partial^2 q^k/\partial q^{l'} \partial q^{s'} = -(\partial q^s/\partial q^{s'})(\partial q^l/\partial q^{l'})(\partial q^k/\partial q^{k'})(\partial^2 q^{k'}/\partial q^l \, \partial q^s)$, etc. (d)

Hint: Multiply Equation c with $\partial q^k/\partial q^{k'}$ and sum over k'.
For example, if $y = f(x) \equiv y(x)$, then $d^2x/dy^2 = -(dx/dy)^3 \, (d^2y/dx^2)$.

2 Tensor Algebra

This chapter presents the fundamental concepts, definitions, and rules of combination (algebra) of tensors at one point of a manifold (the particular structure of which is not important here), such as covariance and contravariance, as well as the metric tensor and its uses. The understanding of this fundamental chapter necessitates a moderate familiarity with Chapter 1; frequent reference to the latter would be quite beneficial.*

2.1 INTRODUCTION: AFFINE AND EUCLIDEAN, OR METRIC, VECTOR SPACES

This section is a handbook-like presentation of some fundamentals from the theory of *linear vector spaces.***

2.1.1 VECTORS

Vectors are mathematical objects with well-defined rules of *local* composition (algebra) and *spatiotemporal variation* (analysis). We assume that the reader is familiar with their *geometrical* aspects.

Let us consider the set S of vectors $\mathbf{u}$, $\mathbf{v}$, $\mathbf{w}$, ... (at this point understood as the *translation*, or *displacement*, vectors of elementary geometry, i.e., as differences of the position vectors of their end points) and the real numbers λ, μ, ν, ... and assume that they obey the following *axioms*:

2.1.1.1 0. Sum

To each pair of S-vectors $\mathbf{u}$ and $\mathbf{v}$ there corresponds a *unique* S-vector called their *sum*, denoted by $\mathbf{u} + \mathbf{v}$, and having the following properties:

1. $\mathbf{u} + \mathbf{v} = \mathbf{v} + \mathbf{u}$ (commutativity)
2. $\mathbf{u} + (\mathbf{v} + \mathbf{w}) = (\mathbf{u} + \mathbf{v}) + \mathbf{w} = \mathbf{u} + \mathbf{v} + \mathbf{w}$ (associativity)
3. $\mathbf{v} + \mathbf{0} = \mathbf{v}$ (existence of *zero*, or *null*, vector $\mathbf{0}$, for every $\mathbf{v}$)
4. $\mathbf{v} + (-\mathbf{v}) = \mathbf{0}$ (existence of *negative* of $\mathbf{v}$: $-\mathbf{v}$)

2.1.1.2 0′. Product with a Scalar

To each S-vector $\mathbf{v}$ and each *real* number λ there corresponds a unique element of S called the *product* of λ and $\mathbf{v}$, denoted by $\lambda\,\mathbf{v}$, and having the following properties:

---

* For complementary reading, we recommend (alphabetically): Brand (1947), Brillouin (1938, 1964), Kästner (1960), Lur'e (1968), (1990), Sokolnikoff (1951), Tietz (1955).

** For extensive discussions see any good book on linear algebra; see, e.g., (alphabetically): Finkbeiner (1978), Gelfand (1961), Shilov (1961), Tietz (1955); also Weyl (1922).

5. $1\,\mathbf{v} = \mathbf{v}$, and $0\,\mathbf{v} = \mathbf{0}$
6. $\lambda\,(\mu\,\mathbf{v}) = (\lambda\,\mu)\,\mathbf{v} = \lambda\,\mu\,\mathbf{v}$ (associativity)
7. $(\lambda + \mu)\,\mathbf{v} = \lambda\,\mathbf{v} + \mu\,\mathbf{v}$ (distributivity for *scalar* addition)
8. $\lambda\,(\mathbf{u} + \mathbf{v}) = \lambda\,\mathbf{u} + \lambda\,\mathbf{v}$ (distributivity for *vector* addition)

2.1.1.3 0″. Scalar, or Dot, Product

To each pair of *S*-vectors **u** and **v** there corresponds a unique real number called their *scalar*, or *dot*, product, denoted by $\mathbf{u} \cdot \mathbf{v}$, and having the following properties:

9. $\mathbf{u} \cdot \mathbf{v} = \mathbf{v} \cdot \mathbf{u}$ (commutativity)
10. $(\lambda\,\mathbf{u}) \cdot \mathbf{v} = \mathbf{u} \cdot (\lambda\,\mathbf{v}) = \lambda\,(\mathbf{u} \cdot \mathbf{v})$ (associativity for multiplication with a scalar)
11. $\mathbf{u} \cdot (\mathbf{v} + \mathbf{w}) = \mathbf{u} \cdot \mathbf{v} + \mathbf{u} \cdot \mathbf{w}$ (associativity for scalar/dot multiplication)
12. If $\mathbf{u} \cdot \mathbf{v} = 0$ for arbitrary **u**, then $\mathbf{v} = \mathbf{0}$

Vectors satisfying 1 to 12 are called *Euclidean vectors*, or simply vectors, and their set (or vector space, over the field of real numbers) *Euclidean vector space*, **E**. A *proper* vector space is one where, in addition to the above, $\mathbf{u} \cdot \mathbf{u} > 0$ for (every nonzero vector). Our Euclidean vector spaces will be assumed proper, unless specified otherwise.

If our vectors **u**, **v**, **w**, ... satisfy only 1 to 8, i.e., if no scalar product is specified, they are called *affine vectors* and their set *affine vector space*, **A**. For example, the vectors of elementary vector calculus are Euclidean vectors; but the rows and columns of a matrix are affine vectors. As discussed below (also Section 2.12), generally, *the term affine means absence of metric,* i.e., inability to measure lengths, angles, areas, etc.

2.1.2 Affine and Euclidean Point Spaces

The physical space of our experience and its mathematical models are spaces of *points* (→ manifolds). In there, as in a scaffold, we introduce *vectors, tensors,* etc. that make up vector spaces. To make further progress we must, next, examine *the coupling of these two fundamental types of objects*: ***points*** (*of a manifold*) *and* ***vectors*** (*of a vector space*).

Let us consider a manifold *X* and a vector space **X**; and, next, let us define the following correspondence: to each pair of *X*-points *P* and *Q*, *taken in that order*, there corresponds, in **X**, a vector **PQ** (*P:* origin, *Q:* terminus) with the following properties:

13. $\mathbf{PQ} = -\,\mathbf{QP}$
14. $\mathbf{PQ} = \mathbf{PR} + \mathbf{RQ}$ (*R:* arbitrary *X*-point)
15. For any *X*-point *O*, and any **X**-vector **v** there exists a *unique* *X*-point *P* such that $\mathbf{OP} = \mathbf{v}$

If $\mathbf{X}$ is Euclidean (affine) then the manifold in correspondence to it is called Euclidean (affine) point space; i.e., schematically,

Vector Space [Euclidean (Affine)] $\rightarrow$ Point Space/Manifold [Euclidean (Affine)]

2.1.3 Linear Independence, Dimension and Basis

The n vectors $\mathbf{v}_1, \ldots, \mathbf{v}_n$ are called *linearly dependent* (LD) if n (real) numbers $\lambda_1, \ldots, \lambda_n$, *not all zero*, can be found, such that

$$\sum \lambda_k \mathbf{v}_k = \mathbf{0} \qquad (k = 1, \ldots, n). \tag{2.1.1}$$

If no such numbers can be found, i.e., if the above implies that *all the λ_k vanish*, the $\mathbf{v}_k$ are called *linearly independent* (LI) *of order n*; or, *they form a free system of order n*.

Definition: A vector space is *n-dimensional*, $\mathbf{X}_n$, if it can yield, or admit, only n LI vectors, and *any* other set of $n + 1$ (or more) vectors, in that space, are LD.*

So, for the $n + 1$ $\mathbf{X}_n$-vectors $\mathbf{v}, \mathbf{e}_1, \ldots, \mathbf{e}_n$ we can write (with $\lambda \neq 0$):

$$\lambda\, \mathbf{v} + \sum \lambda_k \mathbf{e}_k = \mathbf{0} \Rightarrow \mathbf{v} = \sum v^k \mathbf{e}_k, \text{ where } v^k \equiv -(\lambda_k/\lambda). \tag{2.1.2}$$

Definition: The n LI vectors $\mathbf{e}_k$, or any other such LI vector system of *maximal order n*, we call a *basis* in that $\mathbf{X}_n$, and the coefficients v^k *components*, or coodinates, of $\mathbf{v}$ relative to that basis. (The reason we write v^k, and not simply v_k, will become clear soon).

Thus, an $\mathbf{X}$... possesses an infinity of bases; and one of the tasks of tensor algebra is *the determination of the relations between the components of particular vectors and tensors in any two distinct such bases.*

Theorem: Given a basis in a vector space, any vector of it can be represented *uniquely* as a linear combination of the vectors of this basis.

Theorem: For a given system of vectors to constitute a basis in a vector space it is necessary and sufficient that they be LI, and that every other vector of it be expressible in one and only one way as a linear combination of the given vectors.

2.1.3.1 Local Basis, or Frame

A basis, all vectors of which, $\mathbf{e}_k$, originate at a manifold point O; to be denoted as $\{O, \mathbf{e}_k; k = 1, \ldots, n\}$. If the origin is unimportant, and the range of k is well understood, we shall simply write $\{\mathbf{e}_k\}$; or even $\mathbf{e}_k$.

* Only *finite* dimensional vector spaces are considered here; and this serves the needs of *discrete* mechanics, i.e., of systems with a *finite number of degrees of freedom* (Chapters 4 through 7). In *continuum* mechanics $n \rightarrow \infty$.

2.1.3.2 Vector Subspace of an $\mathbf{X}_n$

A vector space $\mathbf{X}_m$, where $m \leq n$. Then,

- Any set of LI $\mathbf{X}_m$-vectors is also an LI set for $\mathbf{X}_n$.
- If $\mathbf{e}_{k'}$ ($k' = 1, \ldots, m$) is a $\mathbf{X}_m$-basis, then any $\mathbf{X}_m$-vector can be expressed uniquely as $\sum v^{k'} \mathbf{e}_{k'}$, and belongs to $\mathbf{X}_m$ for *any* values of the (real) numbers $v^{k'}$; and, conversely, if the $\mathbf{e}_{k'}$ are LI in $\mathbf{X}_n$, then the set of all vectors that can be expressed as a linear combination of them constitutes an m-dimensional subspace of $\mathbf{X}_n$. If $m = n$, the $\mathbf{X}_m$ and $\mathbf{X}_n$ coincide.

2.1.3.3 Complementary Vector Subspaces

Given an LI set of m vectors, it is always possible to *complete* it, or *augment* it, with $n - m$ vectors so as to obtain a basis for $\mathbf{X}_n$. If the vector space of the first set is $\mathbf{X}_m$ and that of the second is $\mathbf{X}_{n-m}$, then we can say that these two vector spaces are *mutually complementary*, relative to $\mathbf{X}_n$; or that they have only the zero vector in common. Any $\mathbf{X}_n$-vector, $\mathbf{v}$, is expressible as

$$\mathbf{v} = \sum v^{k'} \mathbf{e}_{k'} + \sum v^{k''} \mathbf{e}_{k''}; \tag{2.1.3}$$

where $\mathbf{e}_{k'}$ ($k' = 1, \ldots, m$) is an $\mathbf{X}_m$-basis, and $\mathbf{e}_{k''}$ ($k'' = m + 1, \ldots, n$) is an $\mathbf{X}_{n-m}$-basis; or, $\mathbf{v}$ can be decomposed *uniquely* as $\mathbf{v} = \mathbf{v}_{(m)} + \mathbf{v}_{(n-m)}$, where $\mathbf{v}_{(m)}$ and $\mathbf{v}_{(n-m)}$ belong, respectively, to $\mathbf{X}_m$ and $\mathbf{X}_{n-m}$.

Theorem: To every subspace $\mathbf{X}_m$ of an $\mathbf{X}_n$ there corresponds a *unique* complementary subspace $\mathbf{X}_{n-m}$.*

2.2 VECTOR ALGEBRA IN A EUCLIDEAN VECTOR SPACE

2.2.1 Dot Product

Let the (proper) Euclidean vector space $\mathbf{E}_n$ be referred to an arbitrary basis $\{\mathbf{e}_k\}$, and let for any two $\mathbf{E}_n$-vectors $\mathbf{u}$ and $\mathbf{v}$ [from now on with use of the summation convention of Section 1.2) and all Latin indices running from 1 to n]

$$\mathbf{u} = u^k \mathbf{e}_k, \quad \mathbf{v} = v^k \mathbf{e}_k. \tag{2.2.1}$$

Then

$$\mathbf{u} \cdot \mathbf{v} = g_{kl} u^k v^l = g_{lk} u^l v^k, \tag{2.2.2}$$

---

* These considerations are fundamental to the geometrical interpretation of the *constraint reactions* of mechanical systems, in generalized spaces (Chapters 6 and 7).

where, due to property 9 and the properness of $\mathbf{E_n}$,

$$g_{kl} \equiv \mathbf{e}_k \cdot \mathbf{e}_l = \mathbf{e}_l \cdot \mathbf{e}_k \equiv g_{lk} \quad [\mathrm{Det}(g_{kl}) \equiv |g_{kl}| \equiv g > 0]; \tag{2.2.3}$$

i.e., the symmetric and nondegenerate bilinear form (2.2.2) gives the scalar or dot product of two $\mathbf{E_n}$-vectors.

- If $\mathbf{u} = \mathbf{v}$, then Equation 2.2.2 gives

$$\mathbf{v} \cdot \mathbf{v} \equiv \mathbf{v}^2 \equiv v^2 = g_{kl}\, v^k v^l \equiv |\mathbf{v}|^2 \quad (>0, \text{ if } \mathbf{v} \neq \mathbf{0})$$

$$= (\textit{Norm, or Length, of } \mathbf{v})^2 \quad (= 0, \text{ if } \mathbf{v} = \mathbf{0}) \tag{2.2.4}$$

i.e., *norm of* $\mathbf{v}$ = *length of* $\mathbf{v} \equiv |\mathbf{v}| = (\mathbf{v} \cdot \mathbf{v})^{1/2}$.

- If $|\mathbf{v}| = 1$, $\mathbf{v}$ is called *unit*, or normalized.
- If the m $(\leq n)$ $\mathbf{E_n}$-vectors $\mathbf{e}_k$ satisfy

$$\mathbf{e}_k \cdot \mathbf{e}_l = \delta_{kl} \quad [= 0, \text{ if } k \neq l; \quad = 1, \text{ if } k = l \quad (k, l = 1, \ldots, m)] \tag{2.2.5}$$

they are called *orthonormal* (i.e., orthogonal and normalized). Clearly, every such system is LI; and if $m = n$ its vectors form an orthonormal, or *rectangular Cartesian*, basis in $\mathbf{E_n}$. For such a basis $g_{kl} \to \delta_{kl}$, and so Equation 2.2.4 yields the *Pythagorean theorem* in $\mathbf{E_n}$:

$$|\mathbf{v}|^2 = (v^1)^2 + (v^2)^2 + \ldots + (v^n)^2 \quad (\geq 0). \tag{2.2.5a}$$

- Given an arbitrary number m $(\leq n)$ of linearly independent vectors, we can construct from them, if desired, another orthonormal system of m vectors.*

2.2.2 Covariant and Contravariant Components of a Vector

Let us consider an arbitrary $\mathbf{E_n}$-basis $\{\mathbf{e}_k\}$ and a vector $\mathbf{v}$ there:

$$\mathbf{v} = v^k \mathbf{e}_k. \tag{2.2.6}$$

The "measuring numbers" v^k are the *contravariant* components of $\mathbf{v}$ in that basis; while

$$v_k \equiv \mathbf{v} \cdot \mathbf{e}_k, \tag{2.2.7}$$

are its *covariant* components, in the same basis.

Let us find the relations between these two kinds of components:

* This is known as the *E. Schmidt orthogonalization procedure*. See books on linear algebra.

- From the v^k to the v_k: Dotting Equation 2.2.6 with $\mathbf{e}_l$ and invoking Equation 2.2.3 yields

$$v_l = g_{kl}\, v^k = g_{lk}\, v^k. \tag{2.2.8a}$$

- From the v_k to the v^k: Solving the linear nonhomogeneous system (2.2.8a) for the v^k, say, via Cramer's rule (Section 1.2), we find

$$v^k = g^{kl}\, v_l, \tag{2.2.8b}$$

where the $g^{kl} = g^{lk}$ are defined (uniquely) by

$$g_{kl}\, g^{lh} = \delta^h{}_k, \quad \text{or} \quad (g^{kl}) = (g_{kl})^{-1}; \tag{2.2.9}$$

i.e.,

$$g^{kl} = [\mathrm{Cof}(g_{lk}) \ \text{in} \ \mathrm{Det}(g_{kl}) \equiv g] \,/\, g \Rightarrow G \equiv \mathrm{Det}(g^{kl}) = g^{-1}; \tag{2.2.9a, b}$$

also $g^{kl} = \partial(\ln g)/\partial g_{kl}$. (More on g^{kl} in Sections 2.7 and 8). With the help of the above, the dot product of **u** and **v**, Equation 2.2.2, and the norm of **v**, Equation 2.2.4, become, respectively,

$$\mathbf{u} \cdot \mathbf{v} = g_{kl}\, u^k\, v^l = g^{kl} u_l\, v_k = u^k\, v_k = u_k\, v^k, \tag{2.2.10a}$$

$$|\mathbf{v}|^2 = g_{kl}\, v^k\, v^l = g^{kl} v_l\, v_k = v^k\, v_k = v_k\, v^k. \tag{2.2.10b}$$

- Finally, in an *orthonormal* basis: $g_{kl} = g^{kl} = \delta_{kl} \Rightarrow v^k = v_k$; this explains why such component distinctions did not arise in elementary vector algebra (i.e., rectangular Cartesian axes).

2.2.3 Dual of (or to) a Basis

The $\mathbf{E}_n$-basis $\{\mathbf{e}^k\}$ defined (uniquely) by

$$\mathbf{e}^k = g^{kl}\, \mathbf{e}_l \Leftrightarrow \mathbf{e}_k = g_{kl}\, \mathbf{e}^l, \tag{2.2.11}$$

is called the *dual* or *reciprocal* to the basis $\{\mathbf{e}_k\}$; and vice versa. Thus **v** has the representations

$$\mathbf{v} = v^k\, \mathbf{e}_k = v_k\, \mathbf{e}^k. \tag{2.2.12}$$

2.2.4 Change of Basis

Let us consider the admissible basis change $\{\mathbf{e}_k\} \rightarrow \{\mathbf{e}_{k'}\}$ ($k, k' = 1, \ldots, n$) in $\mathbf{E}_n$, defined analytically by

$$\mathbf{e}_{k'} = a_{k'}{}^{k}\,\mathbf{e}_k \Leftrightarrow \mathbf{e}_k = a^{k'}{}_k\,\mathbf{e}_{k'}; \quad (a_{k'}{}^{k})\text{: given,} \quad (a^{k'}{}_k) = (a_{k'}{}^{k})^{-1}. \tag{2.2.13}*$$

Then, from the equivalent representations

$$\mathbf{v} = v^k\,\mathbf{e}_k = v_k\,\mathbf{e}^k = v^{k'}\,\mathbf{e}_{k'} = v_{k'}\,\mathbf{e}^{k'}, \quad \mathbf{v}\text{: general } \mathbf{E}_n\text{-vector,} \tag{2.2.14}$$

and Equations 2.2.13 we easily obtain the following *component transformation* equations:

$$v_{k'} = a_{k'}{}^{k}\,v_k \Leftrightarrow v_k = a^{k'}{}_k\,v_{k'} \quad \text{and} \quad v^{k'} = a^{k'}{}_k\,v^k \Leftrightarrow v^k = a_{k'}{}^{k}\,v^{k'}. \tag{2.2.15}$$

Such transformation relations are fundamental to tensor algebra: as shown later in this chapter, vectors are special tensors, and Equations 2.2.15 constitute special examples of tensor transformations.

- The transformation equations $v^{k'} = a^{k'}{}_k\,v^k$ and $\mathbf{e}_{k'} = a_{k'}{}^{k}\,\mathbf{e}_k$ show that the components $v^{k'}$ transform in a way that is "contrary" to that of the associated basis $\mathbf{e}_{k'}$, hence, the name *contra-variant* for them; for similar reasons the $v_{k'}$ are called *co-variant*. More on this later.

2.3 INTRODUCTION TO COORDINATE TRANSFORMATIONS — AFFINE/RECTILINEAR COORDINATES

We have just seen how *vector components* transform under basis changes. Let us now begin to see how *coordinates* transform. As preparation for the general case of curvilinear coordinates, let us examine the special case of *rectilinear*, or *affine*, coordinates. We consider a point P referred to the general local frame $\{O, \mathbf{e}_k\}$. Its *position* vector there is (Figure 2.1):

$$\mathbf{OP} = z^k\,\mathbf{e}_k, \quad z^k\text{: } \textit{rectilinear} \text{ components, or coordinates, of } P \text{ relative to } \{O, \mathbf{e}_k\}. \tag{2.3.1a}$$

In another frame with the *same origin*, or center, $\{O, \mathbf{e}_{k'}\}$, we shall have

$$\mathbf{OP} = z^{k'}\,\mathbf{e}_{k'}, \quad z^{k'}\text{: } \textit{rectilinear} \text{ components, or coordinates, of } P \text{ relative to } \{O, \mathbf{e}_{k'}\}. \tag{2.3.1b}$$

It is not hard to see that if

$$\mathbf{e}_{k'} = a_{k'}{}^{k}\,\mathbf{e}_k \Leftrightarrow \mathbf{e}_k = a^{k'}{}_k\,\mathbf{e}_{k'} \quad [\text{assuming } \mathrm{Det}(a^{k'}{}_k) \neq 0], \tag{2.3.2}$$

---

* Normally, the *order* of k, k' in $a_{k'}{}^{k}$, $a^{k'}{}_k$ is unimportant; and that is the advantage of *accented indices*. But it does become important if we view them as elements of *matrices*.

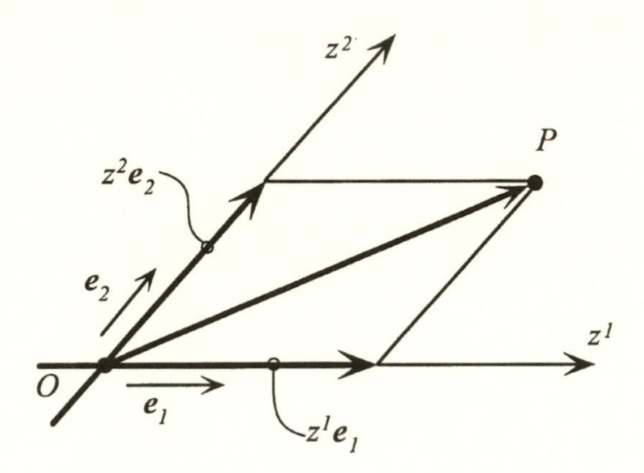

FIGURE 2.1 Position vector of a point P in an oblique (or skew) rectilinear frame (two-dimensional case).

then the z and z' are related by

$$z^{k'} = a^{k'}{}_k z^k = (\partial z^{k'}/\partial z^k) z^k \Leftrightarrow z^k = a_{k'}{}^k z^{k'} = (\partial z^k/\partial z^{k'}) z^{k'}, \tag{2.3.3}$$

where:

$$(\partial z^{k'}/\partial z^k)(\partial z^k/\partial z^{l'}) = a^{k'}{}_k a_{l'}{}^k = \delta^{k'}{}_{l'}, \ (\partial z^{k'}/\partial z^k)(\partial z^l/\partial z^{k'}) = a^{k'}{}_k a_{k'}{}^l = \delta^l{}_k. \tag{2.3.3a}$$

The transformation (2.3.3) (just like Equations 2.2.15) is fundamental; we shall soon see that *even in the case of the most general coordinate transformation $q \leftrightarrow q'$, their differentials, dq and dq', transform like Equations 2.3.3; i.e., locally, even nonlinear coordinate transformations behave as affine ones!*

Example 2.3.1

The *homogeneous* transformations (2.3.3) are called *central*, or *centered* —both frames have the same origin O (zero z and z'). If they do not, then we have additional *nonhomogeneous terms.* Let us take the two frames $\{O, \mathbf{e}_k\}$ and $\{O', \mathbf{e}_{k'}\}$ (Figure 2.2).

Then, with the earlier Equation 2.3.2, and

$$\mathbf{OO'} = a^k \mathbf{e}_k, \quad \mathbf{O'O} = a^{k'} \mathbf{e}_{k'}, \quad \text{and} \quad \mathbf{OP} = z^k \mathbf{e}_k, \quad \mathbf{O'P} = z^{k'} \mathbf{e}_{k'}, \tag{a}$$

and some simple geometry, we find, successively,

$$\mathbf{OP} = \mathbf{OO'} + \mathbf{O'P} = a^k \mathbf{e}_k + z^{k'} \mathbf{e}_{k'} = a^k \mathbf{e}_k + z^{k'}(a_{k'}{}^k \mathbf{e}_k) = z^k \mathbf{e}_k, \tag{b}$$

i.e.,

$$z^k = a_{k'}{}^k z^{k'} + a^k; \quad \text{similarly, } z^{k'} = a^{k'}{}_k z^k + a^{k'} \quad (a^k, a^{k'}\text{: nonhomogeneous terms}) \tag{c}$$

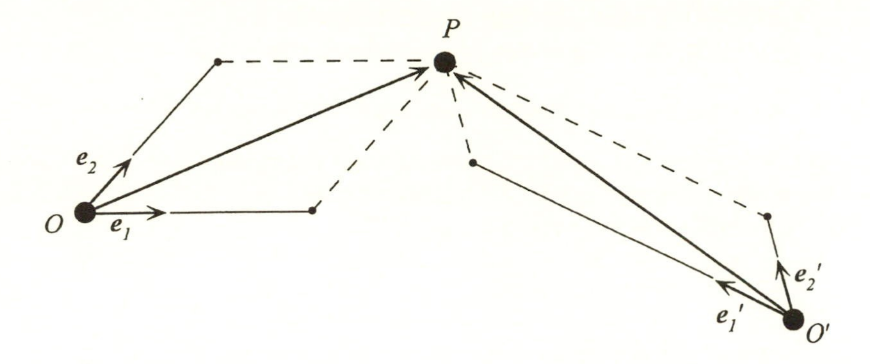

FIGURE 2.2 Position vectors in two *noncentral* frames: $\{O, \mathbf{e}_k\}$ and $\{O', \mathbf{e}_{k'}\}$ (two-dimensional case).

We readily see that the *differentials* of Equation c do transform *homogeneously*, as in the central case (Equation 2.3.3):

$$dz^k = a_{k'}{}^k\, dz^{k'} \Leftrightarrow dz^{k'} = a^{k'}{}_k\, dz^k. \tag{d}$$

- Coordinate systems derived from each other as in Equation c are called admissible *affine* systems; and
- The set of all such transformations constitutes the *affine group*.
- The *space* of points equipped with all admissible affine systems $z \to z' \to z'' \to \ldots$ is called an *affine space* A_n; and
- The study of all *invariant* properties of figures in an A_n, under the affine group, is called *affine geometry*.
- If we restrict ourselves to affine transformations such that $\mathrm{Det}(a^{k'}{}_k) = +1$, then the $z \to z' \to z'' \to \ldots$ create the *displacement group* (subgroup of the general affine group). Such transformations displace a figure into one congruent to it. Then, $a^{k'}{}_k a^{r'}{}_k = \delta^{k'r'}$, $a^{k'}{}_k a^{k'}{}_l = \delta_{kl}$.

From Equation c and recalling that all as are constant in the zs, we can easily show that in such a case, instead of the centered/homogeneous compatibility conditions (Equation 2.3.3a), we will have the following noncentered/nonhomogeneous ones:

$$a^{k'}{}_k\, a_{l'}{}^k = \delta^{k'}{}_{l'}, \quad a^{k'}{}_k\, a_{k'}{}^l = \delta^l{}_k; \quad \text{and} \quad a^{k'}\, a_{k'}{}^k = -a^k, \quad a^{k'}{}_k\, a^k = -a^{k'}; \tag{e}$$

from which we conclude that if $a^k = 0$ then $a^{k'} = 0$ (and vice versa.)*

Finally, if we adopt the *active* viewpoint for Equation c, i.e., view z^k and $z^{k'}$ as the initial and final coordinates of the same point P referred to the same coordinate system, then the general noncentered affine transformation maps points into points, straight lines into straight lines, and planes into planes — what is known in the theory of strain as *homogeneous* deformation.

* (See following page for footnote.)

2.4 GENERAL CURVILINEAR (NONLINEAR) COORDINATE TRANSFORMATIONS (CT)

2.4.1 Curvilinear Coordinates and Their Differentials

Let us now consider the case of general curvilinear coordinates $q, q', q'', \dots$ describing the points $\{P\}$ of a manifold X_n, relative to some origin O, and related to each other by generally nonlinear and admissible CT:

$$q^k = q^k(q^{k'}) \leftrightarrow q^{k'} = q^{k'}(q^k) \quad (k',\ k = 1,\ \dots,\ n). \tag{2.4.1}$$

These coordinates, however, do not represent covariant or contravariant components of the position vectors of these points $\{\mathbf{OP}\}$ relative to a frame, just geometrical objects; i.e., we may write $\mathbf{OP} = \mathbf{R} = R^k\,\mathbf{e}_k$, relative to a local frame $\{P,\ \mathbf{e}_k(q_P)\}$ or $\{O,\ \mathbf{e}_k(q_O)\}$, but $R^k \neq q^k$. Here is why: in a general manifold like X_n, e.g., a two-dimensional *curved* surface in ordinary physical space E_3, the notions of *straight line, plane*, etc. (that is, the products of the affine group) do not exist over *finite* X_n-regions, i.e., *globally*; that is why the q^k, $q^{k'}$, ... are called *curvilinear = curved.* However, as ordinary differential calculus shows, *their differentials* dq, dq', dq'', ..., *around a generic* X_n*-point P, do represent contravariant vectors*; that is, they transform like Equations 2.2.15; indeed, we find from Equation 2.4.1

$$dq^{k'} = (\partial q^{k'}/\partial q^k)dq^k \equiv A^{k'}{}_k\, dq^k \Leftrightarrow dq^k = (\partial q^k/\partial q^{k'})dq^{k'} \equiv A^k{}_{k'}\, dq^{k'}, \tag{2.4.2}$$ **

---

* Equations (e) can be brought to *homogeneous* forms with the following identifications:

$$a^{k'} \equiv a^{k'}{}_{n+1}, \quad a^k \equiv a^k{}_{(n+1)'}, \quad a^{(n+1)'}{}_k \equiv \delta^{(n+1)'}{}_k, \quad a^{(n+1)'}{}_{n+1} = 1;$$

$$a_{k'}{}^{n+1} \equiv \delta_{k'}{}^{n+1}, \quad z^{n+1} \equiv z^{(n+1)'} = 1. \tag{f}$$

Then, and with all Greek indices running from 1 to $n + 1$, Equations e assume the Equation 2.3.3a–like forms:

$$a^{\lambda'}{}_\lambda\, a_{\sigma'}{}^\lambda = \delta^{\lambda'}{}_{\sigma'}, \quad a^{\lambda'}{}_\lambda\, a_{\lambda'}{}^\sigma = \delta^\sigma{}_\lambda. \tag{g}$$

For example, the *first* of the above yields, for $\lambda' \to k'$ and $\sigma' \to l'$,

$$a^{k'}{}_\lambda\, a_{l'}{}^\lambda = a^{k'}{}_k\, a_{l'}{}^k + a^{k'}{}_{n+1}\, a_{l'}{}^{n+1} = a^{k'}{}_k\, a_{l'}{}^k + a^{k'}\, \delta_{l'}{}^{n+1}$$

$$= a^{k'}{}_k\, a_{l'}{}^k + a^{k'}\,(0) = \delta^{k'}{}_{l'}, \quad \text{i.e., Equation } e_1; \tag{h}$$

and for $\lambda' \to k'$ and $\sigma' \to (n + 1)'$:

$$a^{k'}{}_\lambda\, a_{(n+1)'}{}^\lambda = a^{k'}{}_k\, a_{(n+1)'}{}^k + a^{k'}{}_{(n+1)}\, a_{(n+1)'}{}^{(n+1)}$$

$$= a^{k'}{}_k\, a^k + a^{k'}\,(1) = \delta^{k'}{}_{(n+1)'} = 0, \text{ i.e., Equation } e_4; \tag{i}$$

and similarly for Equations $e_{2,3}$.

** And that is why we write q^k, dq^k and $q^{k'}, dq^{k'}$; instead of q_k, dq^k and $q_{k'}, dq^{k'}$, respectively.

where the partial derivatives are evaluated at P; i.e., the *transformation coefficients* (or *generalized direction cosines*) $A^{k'}{}_k \equiv A_k{}^{k'}$ and $A^k{}_{k'} \equiv A_{k'}{}^k$, contrary to their affine counterparts $a^{k'}{}_k \equiv a_k{}^{k'}$ and $a^k{}_{k'} \equiv a_{k'}{}^k$, which are constant, are functions of P:

$$A^{k'}{}_k = A^{k'}{}_k(q) \quad \text{or} \quad A^{k'}{}_k(q'), \text{ and } \quad A^k{}_{k'} = A^k{}_{k'}(q) \quad \text{or} \quad A^k{}_{k'}(q'); \quad (2.4.3)$$

and, clearly, satisfy the following Equation 2.3.3a-like compatibility conditions:

$$(\partial q^{k'}/\partial q^k)(\partial q^k/\partial q^{l'}) \equiv A^{k'}{}_k A^k{}_{l'} = \delta^{k'}{}_{l'}, \quad (\partial q^{k'}/\partial q^k)(\partial q^l/\partial q^{k'}) \equiv A^{k'}{}_k A^l{}_{k'} = \delta^l{}_k; \quad (2.4.4)$$

$$A^{k'}{}_k \equiv \text{Cofactor}(A^k{}_{k'}) \,/\, \text{Det}(A^l{}_{l'}), \quad A^k{}_{k'} \equiv \text{Cofactor}(A^{k'}{}_k) \,/\, \text{Det}(A^{l'}{}_l). \quad (2.4.4a)$$

As long as $\text{Det}(A^{k'}{}_k) \neq 0 \Leftrightarrow \text{Det}(A^k{}_{k'}) \neq 0$ (admissible transformation), these equations allow us to find the $A^{k'}{}_k$ from the $A^k{}_{k'}$, and vice versa.

Let us summarize our findings:

- At a generic point P of a general manifold X_n, every admissible CT: $q \leftrightarrow q'$ induces one and only one affine group transformation $dq \leftrightarrow dq'$. If the latter is integrable, or *holonomic* (see below) the converse is also true: to every such $dq \leftrightarrow dq'$ at least one $q \leftrightarrow q'$ can be found.
- Even in the nonlinear $q \leftrightarrow q'$ case, their differentials dq and dq' obey a linear and homogeneous (or centered affine) transformation law; and, hence, we can use *linear* mathematics (algebra, vector spaces). As discussed in Section 2.12, at each X_n-point P, transformations (2.4.2) define a *centered affine "tangent" point space*; and an associated "natural" tangent vector space.

2.4.2 Integrability, Holonomic Coordinates

By calculus, we have

$$\partial^2 q^{k'}/\partial q^r \partial q^s = \partial^2 q^{k'}/\partial q^s \partial q^r \quad \text{and} \quad \partial^2 q^k/\partial q^{r'} \partial q^{s'} = \partial^2 q^k/\partial q^{s'} \partial q^{r'}, \quad (2.4.5)$$

or in terms of their transformation coefficients (and with sub*commas* for partial derivatives):

$$\partial A^{k'}{}_s/\partial q^r = \partial A^{k'}{}_r/\partial q^s, \quad \text{or} \quad A^{k'}{}_{s,r} = A^{k'}{}_{r,s}; \quad (2.4.6a)$$

and

$$\partial A^k{}_{s'}/\partial q^{r'} = \partial A^k{}_{r'}/\partial q^{s'}, \quad \text{or} \quad A^k{}_{s',r'} = A^k{}_{r',s'}. \quad (2.4.6b)$$

Conversely, if the given coefficients of an Equation 2.4.2–like $dq \leftrightarrow dq'$ transformation satisfy the *integrability* conditions (2.4.6a and b) unconditionally (identically) and, for all values of their indices, the dq and dq' are guaranteed to be genuine differentials of true, or holonomic, coordinates; i.e., *the local measurements are part* *of a larger, or global, network of a measuring CS. Otherwise, the dq and dq′ are differentials of local, or pseudo-, or quasi-, or nonholonomic, coordinates.**

2.5 TENSOR DEFINITIONS

2.5.1 Tensors of Rank One (Vectors)

Motivated from the preceding considerations, and the results of Section 2.2, we now introduce the following basic definitions:

- If, in passing from the general CS q to the also general q' by the admissible CT: $[q' = q'(q) \leftrightarrow q = q(q')]$, the quantities $\{T^k(q) \equiv T^k\}$ transform to $\{T^{k'}(q') \equiv T^{k'}\}$ according to the prototype *contravariant* rule (2.4.2):

$$T^{k'} = (\partial q^{k'}/\partial q^k)T^k \equiv A^{k'}{}_k T^k \Leftrightarrow T^k = (\partial q^k/\partial q^{k'})T^{k'} \equiv A^k{}_{k'} T^{k'}, \qquad (2.5.1)$$

  then we say that the systems $\{T^k$, or $T^{k'}$, etc. $(k, k', \ldots = 1, \ldots, n)\}$ constitute the components of a *contravariant tensor of rank one*, or first rank in the CS $q, q', \ldots$; or, using the earlier-mentioned tensorial slang of "confusing" a tensor with its components (relative to well-understood coordinates) we simply state that T^k is a *contravariant vector.* In view of Equations 2.4.2, we may also say that

  vector components transforming as coordinate differentials are contravariant.

- Similarly, if the T_k transform to $T_{k'}$ as:

$$T_{k'} = (\partial q^k/\partial q^{k'})T_k \equiv A^k{}_{k'} T_k \Leftrightarrow T_k = (\partial q^{k'}/\partial q^k)T_{k'} \equiv A^{k'}{}_k T_{k'}, \qquad (2.5.2)$$

  they are called components of a *covariant tensor of rank one*, etc.; or we simply say that T_k is a *covariant vector.* The natural prototype of a covariant vector is the *gradient*, or natural derivative, of a scalar field $T = T(q) = T'(q') = T'$:

$$\partial T'/\partial q^{k'} = (\partial T/\partial q^k)(\partial q^k/\partial q^{k'}), \quad \text{or, simply,} \quad T'_{,k'} \equiv A^k{}_{k'} T_{,k}; \qquad (2.5.3)$$

  i.e., vector components transforming as the gradient of a scalar are covariant.

The scalar itself is actually a *tensor of rank zero*, and has the single component $T'(q') = T(q)$ in both CS; i.e., T and T' are numerically equal. (In this case T is also an *invariant*; but, as shown below (Section 2.5.4), a scalar is not always an invariant!)

---

* Here we remark that: (1) It is no accident that these concepts originated in problems of geodesy (Pfaff, early 19th century, who was Gauss' teacher; and Gauss was the founder of differential geometry, and a lot more); (2) Holonomic/nonholonomic *coordinates* are *not* to be confused with holonomic/nonholonomic *constraints*. These topics are detailed, respectively, in Section 3.12ff. and Chapter 6.

Remarks

i. The totality of contravariant vectors T^k (or dx^k) at an X_n-point P constitutes a local n-dimensional space; for example, for $n = 2$, X_2 is a (portion of a) surface, and the associated space of contravariant vectors dx^k at any X_2-point P is the *tangent plane,* there $T_2(P)$. And, since as we know from vector analysis, the covariant vector $\partial T/\partial q^k$ is *normal* to the surface $T(q)$ = constant, at P, we can say that these vectors form the normal space at that point.

ii. A *given* X_n-vector (field) $\mathbf{V} = \mathbf{V}(q) = \{V^k(q^l)\}$ can be visualized by a *congruence* of curves, that is, by the family of curves, or *vector lines*, it determines. These curves are such that only one of them passes through each X_n-point $P = P(q^k)$ and has $\mathbf{V}$ ($\neq \mathbf{0}$) tangent to it there; e.g., the longitude and latitude on the earth's surface, excluding the two poles. Hence, if $\mathbf{R}$ is the position vector of a generic vector line point (from some origin), the differential equations of these (invariant) lines are: $d\mathbf{R} \times \mathbf{V} = \mathbf{0}$, from which we obtain either the $n - 1$ scalar equations

$$dq^1/V^1 = dq^2/V^2 = \ldots = dq^n/V^n \tag{2.5.4a}$$

(= du; u: a curve parameter, e.g., arc length)

or

$$d\mathbf{R} = \mathbf{V}\, du \Rightarrow dq^k = (dq^k/du)du = V^k du = V^k[q^l(u)]du \equiv V^k(u)du. \tag{2.5.4b}$$

Integrating the above, we get $q^k = q^k(u;\, c_P)$, where c_P is the "name" of the particular curve through P (i.e., which curve). Thus at each such point, where not all V^k vanish, we obtain a *unique direction.* For example, for $n = 3$ Equations 2.5.4a,b give the *two* independent equations:

$$dq^1/V^1 = dq^2/V^2 = dq^3/V^3,$$

or

$$dq^2/dq^1 = V^2/V^1 \equiv f(q^1,\, q^2,\, q^3) \quad \text{and} \quad dq^3/dq^1 = V^3/V^1 \equiv g(q^1,\, q^2,\, q^3), \tag{2.5.4c}$$

from which we obtain, by integration, the two surfaces

$$F(q^1,\, q^2,\, q^3) = 0 \quad \text{and} \quad G(q^1,\, q^2,\, q^3) = 0,$$

or

$$q^2 = q^2(q^1) \quad \text{and} \quad q^3 = q^3(q^1); \tag{2.5.4d}$$

and since this introduces *two* constants of integration, the above represent a ∞^2 family of curves, intersections of the surfaces $F = 0$ and $G = 0$. In the general case, the $n-1$ independent equations (2.5.4a, b) integrate to the $n-1$ independent solutions:

$$F^k(q^1, \ldots, q^n) = \text{constant} \equiv C^k \quad (k = 1, \ldots, n-1), \tag{2.5.4e}$$

where the $n-1$ constants C^k are found by inserting in Equation 2.5.4e the coordinates of the X_n-point P. Since $q^1 = q^1(u), \ldots, q^n = q^n(u)$ the resulting equations define a unique curve (a *one-parameter* manifold) through P. By varying the C^ks we obtain different curves through different points; there are ∞^{n-1} curves in an X_n-congruence.* Next, n linearly independent vector fields $\mathbf{V}_1, \ldots, \mathbf{V}_n$ determine, through each X_n-point, n congruences; or, simply, an *n-tuple*, or *ennuple*, or *n-leg* of curves. A special such case is an *orthogonal* ennuple, i.e., n mutually orthogonal congruences of curves. The corresponding unit tangent vectors $\{\mathbf{u}_K\}$ have contravariant and covariant components $(u_K{}^k)$ and (u_{Kk}), respectively.**

2.5.2 Tensors of Rank Two

Extending the rules (2.5.1) and (2.5.2), we define such tensors by the following "product-of-two-vectors" — like, or *bilinear*, formulae:

- *Contravariant Tensor* $\{T^{kl}\} \leftrightarrow \{T^{k'l'}\}$:

$$T^{k'l'} = (\partial q^{k'}/\partial q^k)(\partial q^{l'}/\partial q^l)T^{kl} \equiv A^{k'}{}_k A^{l'}{}_l\, T^{kl}$$

$$\Leftrightarrow T^{kl} = (\partial q^k/\partial q^{k'})(\partial q^l/\partial q^{l'})T^{k'l'} \equiv A^k{}_{k'} A^l{}_{l'}\, T^{k'l'}; \tag{2.5.5a}$$

- *Covariant Tensor* $\{T_{kl}\} \leftrightarrow \{T_{k'l'}\}$:

$$T_{k'l'} = (\partial q^k/\partial q^{k'})(\partial q^l/\partial q^{l'})T_{kl} \equiv A^k{}_{k'} A^l{}_{l'}\, T_{kl}$$

$$\Leftrightarrow T_{kl} = (\partial q^{k'}/\partial q^k)(\partial q^{l'}/\partial q^l)T_{k'l'} \equiv A^{k'}{}_k A^{l'}{}_l\, T_{k'l'}; \tag{2.5.5b}$$

- *Mixed Tensors* $\{T^k{}_l \text{ or } T_l{}^k\} \leftrightarrow \{T^{k'}{}_{l'} \text{ or } T_{l'}{}^{k'}\}$:

$$T^{k'}{}_{l'} = (\partial q^{k'}/\partial q^k)(\partial q^l/\partial q^{l'})T^k{}_l \equiv A^{k'}{}_k A^l{}_{l'}\, T^k{}_l$$

$$\Leftrightarrow T^k{}_l = (\partial q^k/\partial q^{k'})(\partial q^{l'}/\partial q^l)T^{k'}{}_{l'} \equiv A^k{}_{k'} A^{l'}{}_l\, T^{k'}{}_{l'}, \tag{2.5.5c}$$

$$T_{l'}{}^{k'} = (\partial q^{k'}/\partial q^k)(\partial q^l/\partial q^{l'})\; T_l{}^k \equiv A^{k'}{}_k A^l{}_{l'}\, T_l{}^k$$

$$\Leftrightarrow T_l{}^k = (\partial q^k/\partial q^{k'})(\partial q^{l'}/\partial q^l)T_{l'}{}^{k'} \equiv A^k{}_{k'} A^{l'}{}_l\, T_{l'}{}^{k'}. \tag{2.5.5d}$$

---

* Symbolically: $\infty^{n-1} = \infty^n$ (number of points in X_n) $- \infty^1$ (number of points on curve).

** For further details, see, e.g., Eisenhart (1926, p. 40ff.). The above apply to both holonomic and nonholonomic vector fields alike (Section 3.12).

Problem 2.5.1

Using Equations 2.5.5a through d and the theorems on multiplication of determinants (Section 1.2), show that

$$\mathrm{Det}(T^{k'l'}) = |\partial q'/\partial q|^2\, \mathrm{Det}(T^{kl}),$$

$$\mathrm{Det}(T_{k'l'}) = |\partial q/\partial q'|^2\, \mathrm{Det}(T_{kl}),$$

$$\mathrm{Det}(T^{k'}{}_{l'}) = \mathrm{Det}(T^{k}{}_{l}). \qquad \text{(a)}$$

2.5.3 General Tensors

If, under the admissible CT $q' = q'(q) \leftrightarrow q = q(q')$,* the n^{u+d} quantities $T^{a\,\ldots\,e}{}_{g\,\ldots\,k}(q)$ (u: number of indices *up*, d: number of indices *down*) transform in the following "product-of-(u + d)-vectors," or *multilinear*, fashion:

$$T^{b'\ldots e'}{}_{g'\ldots k'} = (\partial q^{b'}/\partial q^{b}) \ldots (\partial q^{e'}/\partial q^{e})(\partial q^{g}/\partial q^{g'}) \ldots (\partial q^{k}/\partial q^{k'})T^{b\ldots e}{}_{g\ldots k}$$

$$\equiv (A^{b'}{}_{b} \ldots A^{e'}{}_{e})(A^{g}{}_{g'} \ldots A^{k}{}_{k'})T^{b\ldots e}{}_{g\ldots k},$$

$$\Leftrightarrow T^{b\ldots e}{}_{g\ldots k} = (\partial q^{b}/\partial q^{b'}) \ldots (\partial q^{e}/\partial q^{e'})(\partial q^{g'}/\partial q^{g}) \ldots (\partial q^{k'}/\partial q^{k})T^{b'\ldots e'}{}_{g'\ldots k'}$$

$$\equiv (A^{b}{}_{b'} \ldots A^{e}{}_{e'})(A^{g'}{}_{g} \ldots A^{k'}{}_{k})T^{b'\ldots e'}{}_{g'\ldots k'}, \qquad (2.5.6)$$

we call them components of a *tensor of rank* (or order, or valence, or type) **u** + **d**; *contravariant of rank u and covariant of rank d;* or, simply, a **r** (*-ank*) ≡ (**u, d**) type tensor. If $d = 0$, then $T^{b\,\ldots\,e}$ is called contravariant of rank u, or (u, 0) type; if $u = 0$, $T_{g\ldots k}$ is called covariant of rank d, or (0, d) type; and if d, $u \neq 0$, $T^{b\ldots e}{}_{g\ldots k}$ is called *mixed* of rank $u + d$, or (u, d) type. (We point out that the rank of a tensor is unrelated to the rank of a matrix.) This is a very compact notation: for example, in an X_3 (i.e., $n = 3$), the tensor equation

$$T^{k'}{}_{l'r's'} = (\partial q^{k'}/\partial q^{k})(\partial q^{l}/\partial q^{l'})(\partial q^{r}/\partial q^{r'})(\partial q^{s}/\partial q^{s'})T^{k}{}_{lrs}, \qquad (2.5.7)$$

stands for $3^{1+3} = 81$ *distinct* equations, each of them with a sum of 81 terms on its right side!

Remarks

i. The *order* of the indices in Equation 2.5.6 (and in Equations 2.5.5a through d) *is* important because it determines the position of each component $T^{\cdots}{}_{\cdots}$.**

---

* Here, and unless stated otherwise, such CT are understood in the *passive* sense (Section 1.3). Sometimes (e.g., continuum theory of deformation) we define tensors in the same CS but at *two* different points of space; e.g., if P' and P'' are two such points, we write

$$T_{k'k''}(P', P'') = A_{k'}{}^{l'}(P')\, A_{k''}{}^{l''}(P'')T_{l'l''}(P', P'').$$

** Some authors indicate that by putting overdots and subdots, but we shall avoid this practice.

ii. It is not hard to see that, for the above definitions, Equations 2.5.1 through 2.5.3, 2.5.5a through d, and 2.5.6, to be physically (dimensionally) meaningful: (1) all n products $T_k q^k$ (and, accordingly, all $T_{k'} q^{k'}$; no sum on k, k') must have the same dimensions; i.e., they must be of the same degree in the fundamental physical units (mass or force, length, time, etc.); otherwise we could not sum $(\partial q^1/\partial q^{1'})\, T_1$, $(\partial q^2/\partial q^{1'})\, T_2$, etc. For the same reason, all n ratios T^k/q^k (and, hence, all $T^{k'}/q^{k'}$; no sum on k, k') must have the same dimensions; and, (2) all n^2 products $T_{kl}\, q^k q^l$ ($\Rightarrow T_{k'l'}\, q^{k'} q^{l'}$; no sum on k, l, k', l') must have the same physical dimensions with each other; and so must all n^2 ratios $T^{kl}/q^k q^l$ ($\Rightarrow T^{k'l'}/q^{k'} q^{l'}$; no sum on k, l, k', l'); and all n^2 ratios $T^k{}_l\, q^l/q^k$ ($\Rightarrow T^{k'}{}_{l'}\, q^{l'}/q^{k'}$; no sum on k, l, k', l'), etc.; and similarly for higher rank tensors.

iii. The distinction between a *tensor* and a *tensor field* is the same as that between a vector and a vector field; i.e., tensor means tensor *algebra* (i.e., same spatial point — this chapter) while tensor field means tensor differential calculus → *analysis* (i.e., spatial variation — Chapter 3). However, we shall often simply say tensor and let the precise meaning become clear from the context.

iv. If only *rectilinear*, or *affine*, coordinates are used, then, clearly, all transformation coefficients $A^{k'}{}_k$, $A_{k'}{}^k$ are constant.

2.5.4 Relative Tensors (RT)

Quantities that transform like Equation 2.5.6 but with the *additional* factor

$$|\partial q/\partial q'|^w = (J^{-1})^w = J^{-w} = |\partial q'/\partial q|^{-w}, \tag{2.5.8}$$

where: $J \neq 0, \infty$; and $w = 0, \pm 1, \pm 2, \ldots$ (so that J^{-w} is single valued), on their *right* side are called components of a *relative* tensor or tensor *of weight* w (and rank $u + d$, as before).*

- Hence the tensors encountered so far, like Equation 2.5.6, are *relative tensors of weight zero.* Such tensors are called *absolute tensors*, or simply tensors;
- Relative tensors of weight +1 (–1) are called *densities (capacities)*;
- Relative tensors of rank *one* (*zero*) are called relative *vectors* (*scalars*).

Explicitly:

$$T^{k'} = |\partial q/\partial q'|^w A^{k'}{}_k T^k\text{: relative } contravariant \text{ vector,} \tag{2.5.8a}$$

$$T_{k'} = |\partial q/\partial q'|^w A_{k'}{}^k T_k\text{: relative } covariant \text{ vector,} \tag{2.5.8b}$$

$$T' = |\partial q/\partial q'|^w T\text{: } relative \text{ scalar, if } w = 1;\ absolute \text{ scalar or invariant, if } w = 0. \tag{2.5.8c}$$

* Relative tensors are, sometimes, called *pseudo* tensors, although, as Lodge (1974, p. 96) correctly points out, "the prefix 'pseudo' suggests a lowering of status that is hardly justified."

Definition: The *rank* (r), *kind* (number of indices *up* and *down*), *weight* (w), and *manifold point* (P) of a tensor $\mathbf{T}(P) = \{T^{\cdots}_{\cdots}(P)\}$ are its *four characteristics.*

2.5.4.1 Some Motivation for *RTs*

Such tensors are useful in expressing surface and volume elements, alternating/exterior products, and invariant differential operators like div(...) and curl(...) (⇒ Theorems of Gauss–Green, Stokes).* For example, *scalar densities* help us build *invariant integrals* (field theories); in general, the integral of an invariant scalar field $f = f(q)$ over a q-region R is *not* an invariant. Indeed: if $I \equiv \int_R f(q)\, dq^1 \ldots dq^n$, then under the admissible $q \to q' \Rightarrow R \to R'$: $f(q) = f'(q') = f'$, as advanced calculus shows:

$$I \to I' \equiv \int_{R'} f'(q')dq^{1'} \ldots dq^{n'}$$

$$\equiv \int_R f(q)\,|\partial q'/\partial q|\; dq^1 \ldots\; dq^n \equiv \int_R f(q)\,|\partial q/\partial q'|^{-1}\; dq^1 \ldots\; dq^n, \qquad (2.5.9a)$$

i.e., $I \neq I'$ (unless $|\partial q'/\partial q| = 1 \Rightarrow |\partial q/\partial q'| = 1$, e.g., proper orthogonal transformation, see below). But if $f' = |\partial q'/\partial q|^{-1} f = |\partial q/\partial q'|^{+1} f$ (i.e., a scalar *density*), then $I = I' =$ *invariant!* Applying this when $f \to$ *mass density* $\equiv \rho$ and $I \to$ *mass* $\equiv m$ we obtain (with y: rectangular Cartesian coordinates):

$$m = \int \rho(y)dy_1\, dy_2\, dy_3 = \int \rho(y)dy^1\, dy^2\, dy^3$$

$$= \int \rho[y(q)]\; |\partial y/\partial q|\; dq^1\, dq^2\, dq^3 = \int \rho(q)dq^1\, dq^2\, dq^3 \qquad (2.5.9b)$$

$$\Rightarrow \rho(q) = |\partial y/\partial q|^{+1}\, \rho(y)\text{: relative scalar of weight + 1 (i.e., a density!);} \quad (2.5.9c)$$

and between two general CSs $q \leftrightarrow q'$:

$$m = \int \rho(q)dq^1\, dq^2\, dq^3 = \int \rho'(q')dq^{1'}\, dq^{2'}\, dq^{3'} \Rightarrow \rho'(q') = |\partial q/\partial q'|^{+1}\, \rho(q). \quad (2.5.9d)$$

Example 2.5.1

Let $T \equiv \mathrm{Det}(T_{kl})$, where T_{kl} is a second-order covariant absolute tensor. We will show that: (1) T is a relative scalar invariant of weight +2, and (2) $\sqrt{T}$ is an invariant density.

Indeed, taking the determinant of both sides of $T_{k'l'} = A^k_{\ k'} A^l_{\ k'}\, T_{kl}$, and applying the theorems on multiplication of determinants (Section 1.2), we obtain:

---

* See Sections 2.7 and 3.6, and Examples/Problems below. RTs do not appear so much in discrete mechanics (this book) as they do in *field theories*, e.g., *continuum physics, relativity.*

$$T' = |\partial q/\partial q'|^2\, T \Rightarrow \sqrt{T'} = |\partial q/\partial q'|\, \sqrt{T} = J^{-1}\sqrt{T}, \quad \text{q.e.d.} \tag{a}$$

Example 2.5.2

Similarly, for an absolute contravariant vector, i.e., $T^{k'} = A^{k'}{}_k T^k$, and the new quantity $S^k \equiv \sqrt{U}\, T^k$, where $U \equiv \mathrm{Det}(U_{kl})$, U_{kl}: second-order covariant absolute tensor, using the results of Example 2.5.1, we find

$$\begin{aligned} S^{k'} &\equiv \sqrt{U'}\, T^{k'} = \left[|\partial q/\partial q'|\sqrt{U}\right]\left(A^{k'}{}_k T^k\right) = \left[|\partial q/\partial q'|\, A^{k'}{}_k\right]\left(\sqrt{U}\, T^k\right) \\ &= |\partial q/\partial q'|\, A^{k'}{}_k S^k, \end{aligned} \tag{a}$$

i.e., $S^k \equiv \sqrt{U}\, T^k$ is a contravariant vector density.

Specializations

If $U_{kl} \to g_{kl}$ (*metric* tensor — to be detailed in Section 2.8) and $q \to y$: rectangular Cartesian coordinates, then $g(y) \equiv \mathrm{Det}(g_{kl}) = +1$, and applying $(a)_2$ of the preceding Example 2.5.1 we find:

$$\sqrt{g(q)} = |\partial y/\partial q|\sqrt{g(y)} = |\partial y/\partial q| \;\Rightarrow\; g(q) = |\partial y/\partial q|^2; \tag{b1}$$

$$\sqrt{g(q')} = |\partial y/\partial q'|\sqrt{g(y)} = |\partial y/\partial q'|\left\{|\partial q/\partial y|\sqrt{g(q)}\right\} = |\partial q/\partial q'|\sqrt{g(q)}. \tag{b2}$$

Further, if $dV(y) \equiv dy^1 \cdots dy^n \equiv dV_n$: n-dimensional volume element in rectangular Cartesian coordinates, and $dV(q) \equiv dq^1 \cdots dq^n$, then using well known advanced calculus theorems and the above results, we get

$$\begin{aligned} dV(y) &= |\partial y/\partial q|\, dV(q) = [g(q)/g(y)]^{1/2}\, dV(q) = [\rho(q)/\rho(y)]\, dV(q) \\ &= |\partial y/\partial q'|\, dV(q') = [g(q')/g(y)]^{1/2}\, dV(q') = [\rho(q')/\rho(y)]\, dV(q'), \end{aligned} \tag{c}$$

and, therefore, and since $g(y) = +1$,

$$dV(q') = |\partial q'/\partial y|\, |\partial y/\partial q|\, dV(q) = |\partial q'/\partial q|\, dV(q) = |\partial q/\partial q'|^{-1}\, dV(q) \tag{d}$$

(i.e., $dV(q)$ is a relative scalar of weight -1), and

$$\sqrt{g(q)}\, dV(q) = \sqrt{g(q')}\, dV(q') = \ldots = \sqrt{g(y)}\, dV(y) = dV_n: \tag{e}$$

invariant *volume* element,

$$\rho(q)dV(q) = \rho(q')dV(q') = \ldots = \rho(y)dV(y) = dm: \text{ invariant } \textit{mass} \text{ element.} \tag{f}$$

Problem 2.5.2

Using the preceding results and the theorems on multiplication of determinants (Section 1.2), show that if $T^{k'l'}$, $T_{k'l'}$, $T^{k'}{}_{l'}$ are relative symmetric tensors of weight w, then:

$$\mathrm{Det}(T^{k'l'}) = |\partial q'/\partial q|^{w-2}\,\mathrm{Det}(T^{kl}), \qquad \mathrm{Det}(T_{k'l'}) = |\partial q/\partial q'|^{w+2}\,\mathrm{Det}(T_{kl}),$$

$$\mathrm{Det}(T^{k'}{}_{l'}) = |\partial q'/\partial q|^{w}\,\mathrm{Det}(T^{k}{}_{l}) \quad [\text{i.e., } \mathrm{Det}(T^{k}{}_{l}) \text{ is a relative scalar of weight } w] \quad (a)$$

Recall that symmetric means: $T^{kl} = T^{lk}$, $T_{kl} = T_{lk}$, $T^{k}{}_{l} = T_{l}{}^{k}$ (unlike matrix transposition!).

2.5.5 Geometrical Objects (GO)

Entities whose q-components $T^{\cdots}{}_{\cdots}(q)$ transform, under an admissible CS: $q \leftrightarrow q'$, so that the induced components $T'^{\cdots}{}_{\cdots}(q')$ are unique functions of the old ones, the transformation coefficients $\{A^{k'}{}_{k}\}$ and their gradients $\{A^{k'}{}_{k,l} \equiv A^{k'}{}_{kl}\}$, $\{(A^{k'}{}_{k,l})_{,s} \equiv A^{k'}{}_{kls}\}$, … ; i.e., $T'^{\cdots}{}_{\cdots}(q') = T'\,[T^{\cdots}{}_{\cdots};\, A^{k'}{}_{k'}, A^{k'}{}_{kl}, A^{k'}{}_{kls}, \ldots]$.*

Among the most important *GOs* are absolute tensors, tensor densities, Christoffel symbols, and Hamel coefficients (Section 3.13ff).

2.5.6 Rectangular Cartesian Coordinates (RCC)

Such coordinates, $y \leftrightarrow y'$, are related by transformations:

$$y^{k'} = a^{k'}{}_{k}\, y^{k} + a^{k'} \leftrightarrow y^{k} = a^{k}{}_{k'}\, y^{k'} + a^{k} \quad [\text{i.e., } A^{k'}{}_{k} \to a^{k'}{}_{k},\ A^{k}{}_{k'} \to a^{k}{}_{k'}] \quad (2.5.10a)$$

where

$$\partial y^{k'}/\partial y^{k} \equiv a^{k'}{}_{k} = \textit{cosine} \text{ of angle between } +y^{k'} \text{ and } +y^{k} \text{ (a constant in } y, y')$$

$$= \textit{cosine} \text{ of angle between } +y^{k} \text{ and } +y^{k'} = \partial y^{k}/\partial y^{k'} \equiv a^{k}{}_{k'}, \quad (2.5.10b)$$

and

$$a^{k'}{}_{k}\, a^{k}{}_{l'} = \delta^{k'}{}_{l'} \Rightarrow a^{k'}{}_{k}\, a^{l}{}_{k'} = \delta^{l}{}_{k} \text{ (\textit{orthogonal} i.e., length-preserving, transformation).} \quad (2.5.10c)$$

Hence, for such transformations:

- The difference between covariant and contravariant indices disappears; i.e., $y^{k'} = y_{k'}$, $y^{k} = y_{k}$, $a^{k'}{}_{k} = a_{k'k} = a^{k}{}_{k'} = a_{kk'}$, and so Equation 2.5.10a (assuming common origin for simplicity, i.e., $a^{k}, a^{k'} = 0$) reads:

---

* This concept and terminology seem to be due to Schouten (1954b, pp. 9 ff, 122).
Other authors restrict the term GO to nontensorial (i.e., nonphysical, noninvariant) quantities, describing properties of particular coordinare systems; while reserving the term tensor for *physical, invariant, objects*.

$$y_{k'} = a_{k'k}\, y_k \leftrightarrow y_k = a_{kk'}\, y_{k'}; \tag{2.5.10d}$$

and, accordingly, the summation convention and contraction (see below) for indices at the same level is a legitimate tensor operation.

2.5.7 Polar vs. Axial Vectors

Orthogonal transformations classify as follows:

General, or *full*, orthogonal:

$$J \equiv |\partial y'/\partial y| = |a^{k'}{}_{k}| = |\partial y/\partial y'| = |a^{k}{}_{k'}| \equiv J^{-1} = \pm 1; \tag{2.5.11a}$$

Proper orthogonal (rotations: from *right-handed* to *right-handed*):

$$J = +1, \qquad (2.5.11b)^*$$

Improper orthogonal (either *partial*, or *total,* i.e., *inversions* through the origin; or rotations and then reflections):

$$J = -1. \tag{2.5.11c}$$

Now,

- If, under a general orthogonal transformation among RCCs, the vector V_k transforms as

$$V_{k'} = (+1)a_{k'k}\, V_k \leftrightarrow V_k = (+1)a_{kk'}\, V_{k'}, \tag{2.5.11d}$$

  [i.e., $J^{-w} = (\pm 1)^{-w} = +1 \Rightarrow w = 0$, or *even*] it is called *polar* (or true, or genuine); whereas
- If it transforms as

$$V_{k'} = J\ a_{k'k}\, V_k \leftrightarrow V_k = J\ a_{kk'}\, V_{k'};\ J = +1 \text{ (proper) or } J = -1 \text{ (improper)} \tag{2.5.11e}$$

  [i.e., $J^{-w} = J$: $(\pm 1)^{-w} = \pm 1 \Rightarrow w = 1$, or *odd*] it is called *axial* (or pseudo-vector).

For example, if, under the inversion $y \to y' = -y$: $J = \text{Det}\,[\text{diagonal}\,(-1, -1, -1)] = -1$, and $a_{k'k} \to -\delta_{kl}$: $V_k \to V_{k'} = -V_k$, then V_k is polar; whereas if $V_k \to V_{k'} = V_k$, it is axial.

---

* Any right-handed RCCS y' may be obtained from another such CS y by a *continuous* motion. Hence, the corresponding orthogonal transformation must be able to change *continuously* into the *identity* transformation $y = y'$; and for the latter, clearly, $J = |\partial y'/\partial y| = +1$.

Hence, *if we stick with proper orthogonal transformations, the polar vs. axial vector difference (generally, the absolute vs. relative tensor difference) disappears; all such Cartesian tensors are absolute.**

2.5.8 Tensors vs. Matrices

Although these two are fundamentally different, they are frequently confused for each other, especially in the engineering literature. Let us clarify this:

- A *second-rank* tensor can be represented by an $n \times n$ (square) matrix; but a general $n \times m$ (nonsquare) matrix cannot be represented by a tensor! Under the *active* interpretation of a coordinate transformation (Section 1.3), the linear matrix equation:

$$U_k = A_{kl} V_l \quad (k = 1, \ldots, n \neq m; \quad l = 1, \ldots, m),$$

transforms a vector **V** from an m-dimensional space to another vector **U** in an n-dimensional space. Also, their rules of transformation under CT may be different.
- The "direction cosines" of a general tensor transformation, $A^{k'}{}_k$, $A_{k'}{}^{k}$, are *not* tensors, but can be represented by matrices.
- A tensor with rank higher than two cannot, in general, be represented by a single matrix.

2.5.9 Numerical Tensors

Numerical tensors are those tensors whose components are constant in all CS.

i. *Kronecker Delta* (Section 1.2) $\delta^k{}_l = \delta_l{}^k$ is a mixed second-order (absolute) tensor. Indeed, we have

$$\delta^{k'}{}_{l'} = (\partial q^{k'}/\partial q^k)(\partial q^l/\partial q^{l'})\delta^l{}_k$$

$$= (\partial q^{k'}/\partial q^b)(\partial q^b/\partial q^{l'}) = \partial q^{k'}/\partial q^{l'} = \delta^k{}_l = \delta^{k'}{}_{l'} = \delta^{k''}{}_{l''} = \ldots. \quad (2.5.12a)$$

ii. *Generalized Permutation Symbols* (Section 1.2). Since

$$\varepsilon^{b'c'd'\ldots} = |\partial q/\partial q'|\,(\partial q^{b'}/\partial q^b)(\partial q^{c'}/\partial q^c)(\partial q^{d'}/\partial q^d)\ \ldots\ \varepsilon^{bcd\,\ldots}, \quad (2.5.12b)$$

$$\varepsilon_{b'c'd'\ldots} = |\partial q/\partial q'|^{-1}\,(\partial q^b/\partial q^{b'})(\partial q^c/\partial q^{c'})(\partial q^d/\partial q^{d'})\ \ldots\ \varepsilon_{bcd\,\ldots}, \quad (2.5.12c)$$

---

* General tensor densities/capacities are to absolute tensors what axial vectors are to polar ones; i.e., such tensors transform as $T_{k'l'r'\ldots} = J\, a_{k'k}\, a_{l'l}\, a_{r'r}\ldots T_{klr\ldots}$. See also *axial vector of an antisymmetric tensor*, in tensor form of vector algebra, below.

the $\varepsilon^{\cdots}$ ($\varepsilon_{\cdots}$) are the components of a *contravariant* (*covariant*) tensor *density* (*capacity*).

iii. *Generalized Kronecker Delta* (Section 1.2), being the product of ε terms, one a relative tensor of weight –1 and the other of + 1 (with *contraction* — see below), are *absolute tensors:*

$$\delta_{cde\ldots}{}^{uvw\ldots} = (\varepsilon_{cde\ldots})(\varepsilon^{uvw\ldots}) \quad [\text{weight } (-1) \times \text{weight } (+1) = \text{weight } (0)]$$

Example 2.5.3

Let us prove Equations 2.5.12b and c for the three-dimensional case. We have successively:

$$\varepsilon_{bcd}\,(\partial q^b/\partial q^{b'})(\partial q^c/\partial q^{c'})(\partial q^d/\partial q^{d'}) = |\partial q/\partial q'|\,\varepsilon_{b'c'd'}$$

$$\Rightarrow \varepsilon_{b'c'd'} = |\partial q/\partial q'|^{-1}\,(\partial q^b/\partial q^{b'})(\partial q^c/\partial q^{c'})(\partial q^d/\partial q^{d'})\varepsilon_{bcd}, \tag{a}$$

i.e., ε_{bcd} is a *third-order covariant relative tensor capacity.*

Similarly, we show that ε^{bcd} is a *third-order contravariant tensor density:*

$$\varepsilon^{bcd}\,(\partial q^{b'}/\partial q^b)(\partial q^{c'}/\partial q^c)(\partial q^{d'}/\partial q^d) = |\partial q'/\partial q|\,\varepsilon^{b'c'd'}$$

$$\Rightarrow \varepsilon^{b'c'd'} = |\partial q/\partial q'|\,(\partial q^{b'}/\partial q^b)(\partial q^{c'}/\partial q^c)(\partial q^{d'}/\partial q^d)\varepsilon^{bcd}. \tag{b}$$

Example 2.5.4. Permutation Symbols and Tensors in Three Dimensions

We recall (Section 1.2) that these symbols have been defined as the following antisymmetric quantities:

$$\begin{aligned}
\varepsilon_{bcd} = \varepsilon^{bcd} \quad & [\text{with } b, c, d = 1, 2, 3] \\
= 0; \quad & \text{if any two of } b, c, d \text{ are equal,} \\
= +1; \quad & \text{if } b, c, d \text{ constitute an } \textit{even} \text{ permutation of } 1, 2, 3, \\
= -1; \quad & \text{if } b, c, d \text{ constitute an } \textit{odd} \text{ permutation of } 1, 2, 3.
\end{aligned} \tag{a}$$

Then we define the quantities:

$$e_{bcd} \equiv \sqrt{g}\;\varepsilon_{bcd} = \sqrt{g}\;\varepsilon^{bcd}, \quad e^{bcd} \equiv \left(1/\sqrt{g}\right)\varepsilon^{bcd} = \left(1/\sqrt{g}\right)\varepsilon_{bcd}$$

$$(\Rightarrow e_{bcd} = g\; e^{bcd}). \tag{b}$$

We shall show that *these (also antisymmetric) quantities are absolute tensors,* even though the εs are relative tensors.

Proof

As shown in Example 2.5.2 (or from the transformation law $g_{k'l'} = A^k{}_{k'} A^l{}_{l'} g_{kl}$, see Quotient Rule below), we have

$$\mathrm{Det}(g_{k'l'}) = |\partial q/\partial q'|^2 \,\mathrm{Det}(g_{kl}), \quad \text{or, simply, } g' = |\partial q/\partial q'|^2 g. \tag{c}$$

Therefore, and invoking the results of Example 2.5.3, we find, successively,

$$e_{b'c'd'} \equiv \sqrt{g'}\, \varepsilon_{b'c'd'} = \sqrt{g}\, |\partial q/\partial q'|\, \varepsilon_{b'c'd'} = \sqrt{g}\, |\partial q/\partial q'| \left(|\partial q/\partial q'|^{-1} A^b{}_{b'} A^c{}_{c'} A^d{}_{d'}\, \varepsilon_{bcd}\right)$$

$$= A^b{}_{b'} A^c{}_{c'} A^d{}_{d'} \left(\sqrt{g}\, \varepsilon_{bcd}\right) = A^b{}_{b'} A^c{}_{c'} A^d{}_{d'}\, e_{bcd}, \quad \text{q.e.d.} \tag{d}$$

Similarly,

$$e^{bcd} \equiv \left(1/\sqrt{g}\right) \varepsilon^{bcd} = \left[\left(1/\sqrt{g'}\right) |\partial q/\partial q'|\right] \left(|\partial q/\partial q'|^{-1} A^b{}_{b'} A^c{}_{c'} A^d{}_{d'}\, \varepsilon^{b'c'd'}\right)$$

$$= A^b{}_{b'} A^c{}_{c'} A^d{}_{d'} \left[\left(1/\sqrt{g'}\right) \varepsilon^{b'c'd'}\right] \equiv A^b{}_{b'} A^c{}_{c'} A^d{}_{d'}\, e^{b'c'd'}$$

$$\Rightarrow e^{b'c'd'} = A^{b'}{}_b A^{c'}{}_c A^{d'}{}_d\, e^{bcd}, \quad \text{q.e.d.} \tag{e}$$

In view of Equation c, the above can also be cast in the forms:

$$\varepsilon_{b'c'd'} = (g/g')^{1/2} A^b{}_{b'} A^c{}_{c'} A^d{}_{d'}\, \varepsilon_{bcd} = |\partial q/\partial q'|^{-1} A^b{}_{b'} A^c{}_{c'} A^d{}_{d'}\, \varepsilon_{bcd} \quad \text{(capacity)}, \tag{f1}$$

$$\varepsilon^{b'c'd'} = (g'/g)^{1/2} A^{b'}{}_b A^{c'}{}_c A^{d'}{}_d\, \varepsilon^{bcd} = |\partial q/\partial q'|\, A^{b'}{}_b A^{c'}{}_c A^{d'}{}_d\, \varepsilon^{bcd} \quad \text{(density)}. \tag{f2}$$

Finally, in rectangular Cartesian coordinates, $g = 1$, and so there: $\varepsilon_{bcd} = e_{bcd} = \varepsilon^{bcd} = e^{bcd}$.

2.6 PROPERTIES OF TENSOR TRANSFORMATION*

Reflexivity: The tensor components remain unchanged under the *identity* transformation:

$$q \leftrightarrow q' = q, \quad \Rightarrow T^{\cdots}{}_{\dots}(q) = T'^{\cdots}{}_{\dots}(q').$$

Symmetry: The tensor components return to their original values $T^{\cdots}{}_{\dots}$ under the *inverse* transformation $q' \to q$.

Transitivity (Group Property): The sequence of *CT:* $q \to q' \to q''$ gives the same final components $T''^{\cdots}{}_{\dots}(q'')$ as the direct transformation $q \to q''$. For

* The following hold for both absolute and relative tensors.

example, if $T^{k'} = A^{k'}{}_{k} T^{k}$, and $T^{k''} = A^{k''}{}_{k'} T^{k'}$ (where $A^{k'}{}_{k} \equiv \partial q^{k'}/\partial q^{k}$, $A^{k''}{}_{k'} \equiv \partial q^{k''}/\partial q^{k'}$), then $T^{k''} = A^{k''}{}_{k} T^{k}$ (where $A^{k''}{}_{k} \equiv \partial q^{k''}/\partial q^{k} = A^{k''}{}_{k'} A^{k'}{}_{k}$).

Linearity and Homogeneity → Quotient Rule: Each new $T'^{\cdots}{}_{\ldots}$, $T^{\cdots}{}_{\ldots}$, … is a linear and homogeneous combination of *all* the old $T^{\cdots}{}_{\ldots}$. Therefore,

- If all components of a tensor vanish in a given CS, of an admissible group G, then they will vanish in all other CS of G (in one-to-one correspondence with the given, or "initial" one); i.e.,

$$\text{If } T^{b\ldots e}{}_{g\ldots k}(q) = 0, \text{ then } T^{b'\ldots e'}{}_{g'\ldots k'}(q') = T^{b''\ldots e''}{}_{g''\ldots k''}(q'') = \ldots = 0; \quad (2.6.1)$$

  and vice versa; or, equivalently,
- If a geometrical or physical equation is tensorial (i.e., if it consists entirely of meaningful combinations of tensors), it remains form invariant under any admissible CT. For example,

$$\text{If } T_{kl}(q) = B_{kd}(q) C^{d}{}_{l}(q), \quad \text{then } T_{k'l'}(q') = B_{k'd'}(q') C^{d'}{}_{l'}(q'), \text{ etc.}$$

These fundamental results constitute what is known in tensor theory as the *quotient rule;* and it is this rule that makes tensors so useful in geometrical and physical applications.*

2.7 ALGEBRAIC OPERATIONS WITH TENSORS

These are based on the following theorem and rule:

- Tensors with the same four characteristics, i.e., space point (P), rank (r), kind (number of indices *u*p and *d*own), and weight (w), make up a linear space; while tensors differing from each other even in one of these characteristics belong to separate ("disjoint") linear spaces.
- Only such operations that produce a tensor from another are admissible.

Thus:

1. Multiplication of a tensor $T^{\cdots}{}_{\ldots}$ with a scalar λ produces a tensor $S^{\cdots}{}_{\ldots} = \lambda\, T^{\cdots}{}_{\ldots}$ with the same characteristics as $T^{\cdots}{}_{\ldots}$.
2. Addition or subtraction of two or more tensors with the same four characteristics produces a tensor with the same characteristics.
3. Multiplication of two or more tensors, at the same point, can be defined meaningfully (see inner and outer products below).

However, addition/subtraction of two scalars *at different space points* produces a scalar.

---

* For further details on the quotient rule, see the next section.

2.7.1 Symmetry Properties

(Also recall Section 1.2.) A general tensor $T^{\cdots}{}_{\cdots}$ is called *symmetric* (*antisymmetric,* or *skew-symmetric*) in its pair of indices k, l if

$$T^{\dots k \dots l \dots} = T^{\dots l \dots k \dots} \; (T^{\dots k \dots l \dots} = -T^{\dots l \dots k \dots}) \tag{2.7.1a}$$

$$T_{\dots k \dots l \dots} = T_{\dots l \dots k \dots} \; (T_{\dots k \dots l \dots} = -T_{\dots l \dots k \dots}) \tag{2.7.1b}$$

$$T^{\dots k \dots}{}_{\dots l \dots} = T^{\dots l \dots}{}_{\dots k \dots} \; (T^{\dots k \dots}{}_{\dots l \dots} = -T^{\dots l \dots}{}_{\dots k \dots}) \tag{2.7.1c}$$

i. For the common case of a *second*-rank tensor these definitions mean, respectively,

$$T^{kl} = T^{lk} \, (T^{kl} = -T^{lk}); \quad T_{kl} = T_{lk} \, (T_{kl} = -T_{lk}); \quad T^k{}_l = T_l{}^k \, (T^k{}_l = -T_l{}^k). \tag{2.7.1d}$$

In the mixed component case, symmetry (antisymmetry) should not be confused with matrix transposition. For example, if $\mathbf{A} = (A^k{}_l)$, then $\mathbf{A}^T \equiv (A^l{}_k)$, and from the *matrix* symmetry (antisymmetry): $A^k{}_l = A^l{}_k \, (= -A^l{}_k)$ it does not necessarily follow the *tensor* symmetry (antisymmetry): $A^k{}_l = A_l{}^k \, (= -A_l{}^k)$, if $A^k{}_l$ is a tensor, and vice versa.*

Theorem: Any *second*-rank tensor can be written uniquely as the sum of a symmetric and an antisymmetric (second-order) tensor:

$$T_{kl} = T_{(kl)} + T_{[kl]}, \tag{2.7.2}$$

$$2T_{(kl)} = 2T_{(kl)} \equiv T_{kl} + T_{lk} \quad \text{(symmetric part)}, \tag{2.7.2a}$$

$$2T_{[kl]} = -2T_{[lk]} \equiv T_{kl} - T_{lk} \quad \text{(antisymmetric part)}. \tag{2.7.2b}$$

The proof is obvious. Similarly, for contravariant and mixed components.

It is not hard to see that, *in an n-dimensional space, $T_{(kl)}$ has, at most, $n(n+1)/2$ independent (distinct) components, while for $T_{[kl]}$ that number is $n^2 - [n(n+1)/2] = n(n-1)/2$.* (As Equation 2.7.2b shows, the "diagonal elements" of T_{kl} — and, also, of T^{kl} and $T^k{}_l$ — vanish: $T_{kk} = 0$, no sum on k.)

For $T_{kl} \to \partial T_k / \partial q^l \equiv T_{k,l}$ the theorem yields the following useful decomposition:

$$T_{k,l} = T_{(k,l)} + T_{[k,l]}, \quad \text{where} \quad 2T_{(k,l)} = T_{k,l} + T_{l,k}, \quad 2T_{[k,l]} = T_{k,l} - T_{l,k}. \tag{2.7.2c)**$$

Problem 2.7.1

Let $T_{kl} = -T_{lk}$, $T^{kl} = -T^{lk}$, $R_{kl} = R_{lk}$, and S^{kl}: arbitrary. Show that:

---

* Almost all physically important second-rank tensors have symmetry/antisymmetry properties, e.g., tensors of stress and strain, of angular velocity and moment of inertia, of energy-momentum, etc.

** (See following page for footnote.)

$$T_{kl} S^{kl} = T_{kl} S^{[kl]}, \quad R_{kl} S^{kl} = R_{kl} S^{(kl)}, \quad R_{kl} T^{kl} = 0. \tag{a}$$

Equation a_3 shows that any quadratic form $Q_{kl} x^k x^l$ can always be rewritten as $Q_{(kl)} x^k x^l$.

Problem 2.7.2

Let T_{kl} and S^{kl} be general second-rank tensors (or indexed quantities). Show that

$$T_{[kl]} S^{kl} = T_{kl} S^{[kl]} = T_{[kl]} S^{[kl]}. \tag{a}$$

Problem 2.7.3

Let $R_{ab} \equiv P_a Q_b - P_b Q_a \equiv 2P_{[a} Q_{b]}$. Verify the "Plücker identity":

$$R_{ab} R_{cd} + R_{ac} R_{db} + R_{ad} R_{bc} = 0. \tag{a}$$

For further details on such "Plücker tensors," see Remark below and Duschek and Mayer (1930, v. 2, pp. 20–22).

ii. If the symmetry/antisymmetry properties of a *general* $(r)th$ rank tensor hold for the permutation of *any two* (and all) of its indices, adjacent or not, same level or mixed, or, equivalently, *for all its pairs of adjacent indices,* then that tensor is called *completely symmetric/antisymmetric*. To make sure that this definition does not lead to contradictions relative to the earlier definition (for an index pair) we supply a short proof:

Let us consider for concreteness the two arbitrary contravariant indices **k** and **l** separated by Σ other indices: $\ldots \mathbf{k}\, \alpha\, \beta\, \gamma \ldots \rho\, \sigma\, \mathbf{l} \ldots$. But this arrangement can be changed to $\ldots \mathbf{l}\, \alpha\, \beta\, \gamma \ldots \rho\, \sigma\, \mathbf{k} \ldots$ by a sequence of $\Sigma + \Sigma + 1 = 2\Sigma + 1$ *simple* permutations (i.e., among adjacent indices). Therefore, if $T^{\cdots}{}_{\cdots}$ is completely antisymmetric, then (and since $2\Sigma + 1$ is always an *odd* number):

$$T^{\ldots \mathbf{k}\alpha\beta\gamma\ldots\rho\sigma\mathbf{l}\ldots} = (-1)^{2\Sigma+1}\, T^{\ldots \mathbf{l}\alpha\beta\gamma\ldots\rho\sigma\mathbf{k}\ldots} = -T^{\ldots \mathbf{l}\alpha\beta\gamma\ldots\rho\sigma\mathbf{k}\ldots}.$$

The above consolidate into the following:

** Similarly, we define the completely symmetric and antisymmetric parts of the *third*-rank tensor T_{abc} by

$$3!\, T_{(abc)} \equiv (T_{abc} + T_{bca} + T_{cab}) + (T_{acb} + T_{cba} + T_{bac}), \tag{2.7.2d}$$

$$3!\, T_{[abc]} \equiv (T_{abc} + T_{bca} + T_{cab}) - (T_{acb} + T_{cba} + T_{bac}), \tag{2.7.2e}$$

respectively. However, here, $T_{abc} \neq T_{(abc)} + T_{[abc]}$. Instead, as the reader may verify,

$$T_{abc} = T_{(abc)} + T_{[abc]} + (2/3)\,(T_{[ab]c} + T_{[cb]a}) + (2/3)\,(T_{(ab)c} - T_{c(ab)}). \tag{2.7.2f}$$

For further details, see, e.g., Pozniak and Shikin (1990, pp. 150 ff), Stephani (1982, pp. 31 ff).

Theorem: In an n-dimensional space, a *completely antisymmetric* (r)th rank tensor can have at most $\nu = {}_nC_m \equiv n!/r!\,(n-r)! = n(n-1)(n-2)\ldots(n-r+1)/r!$ distinct components (= number of *repetitionless* combinations of n things, r at a time); and (1) if all particular indices are different, the (group of) corresponding components differ from each other only in sign; while (2) if even two such indices are equal, the components vanish.

Similarly, we can show that the number of distinct components of a *completely symmetric* (r)th rank n-tensor is $N = [(n-1)+r]!\,/\,(n-1)!\;r! = n(n+1)(n+2)\ldots(n+r-1)/r!$ (= number of combinations, *with repetition*, of n things, r at a time).

Hence, (1) if $n = 3$, $r = 3$, then $\nu = 1$, $N = 10$; (2) if $n = 3$, all components of such tensors of rank $r \geq 4$ vanish; and (3) if $n = r$, then $\nu = 1$ (since $0! \equiv 1$) (see also Problem 2.7.4, below). Further, using the laws of tensor transformation (Section 2.5), we can verify the following important theorem.

Theorem: The partial or complete symmetry/antisymmetry of a tensor is a *coordinate independent* property; that is why these concepts are useful to begin with. Thus, if $T^k{}_{lr} = \pm T^k{}_{rl}$, then $T^{k'}{}_{l'r'} = \pm T^{k'}{}_{r'l'}$; and if $T^k{}_{lr} = \pm T_{rl}{}^k$, then $T^{k'}{}_{l'r'} = \pm T_{r'l'}{}^{k'}$.

However, such a theorem does not, generally, hold for transposed (swapped) components; e.g., if $T^k{}_{lr} = \pm T^r{}_{lk}$, it does not necessarily follow that $T^{k'}{}_{l'r'} = \pm T^{r'}{}_{l'k'}$.

In view of the above, the transformation laws of the antisymmetric tensors T_{kl}, T^{kl} and T_{kls}, T^{kls} become

$$T_{k'l'} = A^k{}_{k'} A^l{}_{l'} T_{kl} = (A^k{}_{k'} A^l{}_{l'} - A^l{}_{k'} A^k{}_{l'}) T_{kl} \equiv 2A^{[k}{}_{k'} A^{l]}{}_{k'} T_{kl} \quad (\boldsymbol{k < l})$$

$$\equiv [(\partial q^k/\partial q^{k'})(\partial q^l/\partial q^{l'}) - (\partial q^l/\partial q^{k'})(\partial q^k/\partial q^{l'})] T_{kl},$$

or (recalling Equation 1.2.11a)

$$T_{k'l'} = [\partial(q^k,\, q^l)\,/\,\partial(q^{k'},\, q^{l'})] T_{kl} \Rightarrow T_{kl} = [\partial(q^{k'},\, q^{l'})\,/\,\partial(q^k,\, q^l)] T_{k'l'}, \quad (2.7.3a)$$

$$T^{k'l'} = [\partial(q^{k'},\, q^{l'})\,/\,\partial(q^k,\, q^l)] T^{kl} \Rightarrow T^{kl} = [\partial(q^k,\, q^l)\,/\,\partial(q^{k'},\, q^{l'})] T^{k'l'}; \quad (2.7.3b)$$

and

$$T_{k'l's'} = \ldots = [\partial(q^k,\, q^l,\, q^s)\,/\,\partial(q^{k'},\, q^{l'},\, q^{s'})] T_{kls} \Rightarrow T_{kls} = \ldots, \quad (2.7.3c)$$

$$T^{k'l's'} = \ldots = [\partial(q^{k'},\, q^{l'},\, q^{s'})\,/\,\partial(q^k,\, q^l,\, q^s)] T^{kls} \Rightarrow T^{kls} = \ldots, \quad (2.7.3d)$$

where in all summations, on the right sides of the above, *no Jacobian is repeated,* and similarly for higher rank completely antisymmetric tensors.

Example 2.7.1

We have already seen that, if $n = 3$, the completely antisymmetric tensor T^{kls} can have only one independent nonvanishing component: $T^{123} = T^{231} = T^{312} = -T^{132} = -T^{213} = -T^{321} \equiv T$, or, compactly, $T^{kls} = \varepsilon^{kls} T$.

Let us find how T^{123} transforms under a CT: $q \to q'$. We have

$$T' \equiv T^{1'2'3'} = A^{1'}{}_b A^{2'}{}_c A^{3'}{}_d T^{bcd} = (A^{1'}{}_1 A^{2'}{}_2 A^{3'}{}_3 + A^{1'}{}_2 A^{2'}{}_3 A^{3'}{}_1 + A^{1'}{}_3 A^{2'}{}_1 A^{3'}{}_2$$

$$- A^{1'}{}_1 A^{2'}{}_3 A^{3'}{}_2 - A^{1'}{}_2 A^{2'}{}_1 A^{3'}{}_3 - A^{1'}{}_3 A^{2'}{}_2 A^{3'}{}_1) T^{123} = \mathrm{Det}(A^{k'}{}_k) T^{123},$$

or

$$T' = \mathrm{Det}(A^{k'}{}_k) T \equiv |\partial q'/\partial q|\, T = |\partial q/\partial q'|^{-1}\, T; \tag{a}$$

i.e., $w = -1 \Rightarrow T^{kls}$ is a *scalar capacity* (or, a numerical pseudotensor).

Let the reader show that, again for $n = 3$, the completely antisymmetric tensor T_{kls} is a *scalar density;* i.e., $T_{kls} = \varepsilon_{kls} T$ transforms as $T' \equiv |\partial q/\partial q'|\, T = |\partial q'/\partial q|^{-1}\, T$.

(From the above we easily conclude that *any completely antisymmetric tensor is proportional to* ε.)

Example 2.7.2

Let us find how $1/\sqrt{g}$ transforms. We have

$$g_{kl} = A^{k'}{}_k A^{l'}{}_{l'} g_{k'l'} \Rightarrow |g_{kl}| = |A^{k'}{}_k|\, |A^{l'}{}_l|\, |g_{k'l'}|, \tag{a}$$

or

$$g = |A^{k'}{}_k|^2 g' \Rightarrow 1/\sqrt{g'} = |A^{k'}{}_k| \left(1/\sqrt{g}\right), \tag{b}$$

and comparing with Equation a of Example 2.7.1: $T^{123} \to 1/\sqrt{g}$, we see that the latter behaves like a completely antisymmetric third-order tensor e^{bcd} (recalling Example 2.5.4):

$$e^{bcd} = 1/\sqrt{g}, \quad \text{for } b, c, d \text{ cyclically as } 1, 2, 3;$$

$$= -1/\sqrt{g}, \quad \text{for } b, c, d \text{ cyclically as } 3, 2, 1;$$

$$= 0, \quad \text{otherwise.} \tag{c}$$

Similarly, its "conjugate" contravariant third-order tensor is found to equal $e_{bcd} = g_{bk} g_{cl} g_{dm} e^{klm}$:

$$e_{bcd} = \sqrt{g}, \quad \text{for } b, c, d \text{ cyclically as } 1, 2, 3;$$

$$= -\sqrt{g}, \quad \text{for } b, c, d \text{ cyclically as } 3, 2, 1;$$

$$= 0, \quad \text{otherwise.} \tag{d}$$

Remark

As pointed out by H. Grassmann, E. Cartan, et al. one can *build completely antisymmetric tensors from ordinary vectors*, in the form of determinants:

$$\text{Two vectors: } T^{kl} = \begin{vmatrix} u^k & u^l \\ v^k & v^l \end{vmatrix} = u^k v^l - u^l v^k \quad (= -T^{lk}). \tag{2.7.4a}$$

Generally,

$$T^{klr\ldots} = \begin{vmatrix} u^k & u^l & u^r \ldots \\ v^k & v^l & v^r \ldots \\ \ldots\ldots & \ldots\ldots & \ldots\ldots \\ \ldots\ldots & \ldots\ldots & \ldots\ldots \end{vmatrix}; \quad T^{klr\ldots} = -T^{lkr\ldots} = -T^{krl\ldots} \tag{2.7.4b}$$

Such tensors are called *exterior products* of vectors, or *multivectors,* and they equal areas, volumes, etc. of the region formed by their vectors (see Equations 2.9.22a ff and Example 3.6.1: Stokes' theorem).

Problem 2.7.4

Consider the *number of independent* (*distinct*) *components* of the n-dimensional tensor T_{kls} (or T^{kls}), #, for the following three cases:

$$\text{a. } k \neq l \neq s \neq k; \quad \text{b. } k = l \neq s; \quad \text{c. } k = l = s. \tag{a}$$

Show that

i. If T_{kls} (or T^{kls}) is *completely symmetric* (# → **N**), then

$$\mathbf{N}_a = n(n-1)(n-2)/6, \quad \mathbf{N}_b = n(n-1), \quad \mathbf{N}_c = n$$

$$\Rightarrow \mathbf{N}_{a+b+c} = n(n+1)(n+2)/6. \tag{b}$$

ii. If T_{kls} (or T^{kls}) is *completely antisymmetric* (# → ν), then

$$\nu_a = n(n-1)(n-2)/6, \quad \nu_b = 0, \quad \nu_c = 0 \Rightarrow \nu_{a+b+c} = n(n-1)(n-2)/6. \quad \text{(c)}$$

2.7.2 Outer (or Exterior, or Direct) Multiplication of Tensors

Juxtaposing two tensors of respective ranks $r_1 = (u_1, d_1)$ and $r_2 = (u_2, d_2)$, and weights w_1 and w_2 produces a new tensor, their *outer product*, of rank $r = r_1 + r_2 = (u_1 + u_2, d_1 + d_2)$ and weight $w = w_1 + w_2$. In particular, if $w_1 = w$ and $w_2 = -w$, the new product tensor has weight $w = 0$; i.e., it is an absolute tensor, e.g., the product of a density ($w_1 = +1$) with a capacity ($w_2 = -1$). Conversely, a tensor of rank r and weight w can be "decomposed" into an outer product of an absolute tensor of rank $r_1 = r$ and $w_1 = 0$ with a relative scalar of rank $r_2 = 0$ and $w_2 = w$. For example, if $T^k{}_l$ and S^{bc} are tensors of respective weights w_T and w_S, then the tensor: $T^k{}_l\, S^{bc} \equiv R^k{}_l{}^{bc}$, a relative tensor of rank 2 + 2 = 4 and weight $w_R = w_T + w_S$, is called their outer product. Other examples of outer products are

$$T_{kl} \equiv u_k v_l - u_l v_k = 2u_{[k} v_{l]} = \delta_{kl}{}^{bc} u_b v_c \quad \text{(a second order covariant tensor);} \quad (2.7.5a)$$

which, for $n = 3$ yields the contravariant vector identity:

$$T^k = \varepsilon^{klm} u_l v_m = (1/2)\varepsilon^{klm} T_{lm} \quad \text{(see also Equations 2.9.22a ff).} \quad (2.7.5b)$$

We notice that T^k and T^{kl} have the same (nonzero) components:

$$u_2 v_3 - u_3 v_2, \quad u_3 v_1 - u_1 v_3, \quad u_1 v_2 - u_2 v_1. \quad (2.7.5c)$$

Generally, the outer product of m covariant vectors $v^{(1)}{}_k, v^{(2)}{}_k, \ldots, v^{(m)}{}_k$ is defined as

$$\delta^{kk'k''\ldots}{}_{ll'l''\ldots}\, v^{(1)}{}_k\, v^{(2)}{}_{k'}\, v^{(3)}{}_{k''}\ldots \equiv T_{ll'l''\ldots} \quad \text{(covariant tensor).} \quad (2.7.5d)$$

2.7.3 Contraction of Tensors*

Setting a subscript and a superscript equal to each other, thus creating a pair of dummy indices (one contra- and one covariant), in a tensor of rank $r = (u, d)$ and weight w, and then summing according to the summation convention, produces a new tensor of rank $r - 2 = (u - 1, d - 1)$ (assuming $u, d \geq 1$) and weight w; i.e., *contraction does not affect the weight.* For example, contracting $T^k{}_{lb}$ over k and b yields the covariant vector $(3 - 2 = 1)$ $T^k{}_{lk} = T^b{}_{lb} \equiv T_l$; contracting $T^k{}_l$ yields the invariant $T^k{}_k = T^l{}_l =$ *Trace of* $(T^k{}_l)$.

2.7.4 Inner Multiplication or Transvection of Tensors

This is a combination of outer multiplication and contraction. For example, contracting the outer product $T_{kl} S^b{}_c$, of the tensors T_{kl} and $S^b{}_c$, yields the following inner products: $T_{kl} S^k{}_c \equiv R_{lc}$, or $T_{kl} S^l{}_c \equiv R_{kc}$; and $T_{kl}{}^b S^{kdl} = T_{ef}{}^b S^{edf} = R^{bd}$ is the inner product of

* In German (probably due to A. Einstein): *Verjüngung* = rejuvenation!

$T_{kl}{}^b$ and S^{edf} over k and l (e and f). Also, as with the outer product, if $T_{...}{}^{...}$ and $S_{...}{}^{...}$ have respective weights w_T and w_S, their inner products have weight $w_T + w_S$; i.e., in both outer and inner products, the weights of the factors add together.

In particular, the invariant $U_k V^k$, formed by the inner product of the vectors U_k and V^k is called their *dot* or *scalar* product; and, if we define $(U_k V^k)' \equiv U_{k'} V^{k'}$ and $(T^{kl} R_{kl})' \equiv T^{k'l'} R_{k'l'}$, then all functions of the forms $U_k V^k$ and $T^{kl} R_{kl}$ are *invariant*. However, we should point out that $\sum V_k T^b{}_k$ is *not* a (contravariant) vector; k is down on both factors!

2.7.5 Quotient Rule (QR)

With the help of the inner multiplication, we can now formulate a very useful theorem known as QR, which allows us to detect the tensorial character, or absence thereof, of indexed quantities. Let us start with a concrete case of *absolute* tensors: If in the given set of equations $T^k{}_{hb} R^{hb} = S^k$ we know that $T^k{}_{hb}$ and R^{hb} are tensors of the type indicated by their indices, then it is not hard to show that S^k is a contravariant vector. In applications, however, we need its converse; i.e., *if in the given (inner product-type) relations: $T(k, h, b)\, R^{hb} = S^k$, S^k is known to be a (first-order contravariant tensor) vector and R^{hb} an arbitrary second-order contravariant tensor, then the n^3 quantities $T(k, h, b)$, functions of the coordinates q, constitute the components, in q, of a mixed third-order tensor, to be represented by $T^k{}_{hb}$.*

Proof: Under an admissible CT: $q \to q'$, the above relation should become successively:

$$T'(k', h', b')R^{h'b'} = S^{k'} = A^{k'}{}_k S^k = A^{k'}{}_k [T(k, h, b)R^{hb}]. \tag{2.7.6a}$$

But $R^{hb} = A^h{}_{h'} A^b{}_{b'} R^{h'b'}$, and so Equation 2.7.6a transforms to

$$[T'(k', h', b') - A^{k'}{}_k A^h{}_{h'} A^b{}_{b'} T(k, h, b)]R^{h'b'} = 0, \tag{2.7.6b}$$

from which since R^{hb} and therefore also $R^{h'b'}$ are arbitrary, it follows that $T'(k', h', b') = A^{k'}{}_k A^h{}_{h'} A^b{}_{b'} T(k, h, b)$, i.e., $T(k, h, b)$ deserves to be represented as $T^k{}_{hb}$, q.e.d.

Thus, if in $V^k = g^{kl} V_l$, V^k is a contravariant vector and V_l an arbitrary covariant vector, then g^{kl} must be a second-order contravariant tensor; and if u_k, v^k, w^k are three arbitrary vectors and $T^k{}_{bc} u_k v^b w^c$ is an invariant, then we deduce that $T^k{}_{bc}$ is a third-order mixed tensor, i.e., the coefficients of an invariant *trilinear* form in the variables u_k, v^k, w^k constitute a third-order tensor. This can be generalized to the following alternative definition of tensors: *an invariant multilinear (inner product) form, in the set of distinct and arbitrary r vectors $u^k, v^{k'}, w^{k''}, \ldots$*

$$T = T_{kk'k''\ldots}\, u^k v^{k'} w^{k''\ldots}, \tag{2.7.7}$$

is called an (r)th rank tensor, with the n^r $\{T_{...}\}$ as its covariant components; and similarly for $\{T^{...}\}$ and $\{T_{...}{}^{...}\}$. For example, a covariant vector V_k can be defined by the *linear* invariant:

$$V = V_k u^k = (V_k u^k)' \equiv V_{k'} u^{k'} = \ldots; \tag{2.7.7a}$$

and a covariant tensor T_{kl} by the *bilinear* invariant

$$T = T_{kl} u^k v^l = (T_{kl} u^k v^l)' \equiv T_{k'l'} u^{k'} v^{l'} = \ldots. \tag{2.7.7b}$$

However, if $T = T_{kl} u^k u^l$ is invariant for arbitrary u^k, then, not T_{kl}, but its *symmetric* part $T_{(kl)}$ is a second-order covariant tensor; and since $T_{[kl]} u^k u^l = 0$, $T = T_{(kl)} u^k u^l$.

2.7.5.1 QR for Relative, and General Tensors

In this case, the earlier theorem becomes as follows: If the given relations $T(k, h, b) R^{hb} = S^k$, where S^k is known to be a relative tensor of weight w_S and R^{hb} an arbitrary relative tensor of weight w_R, then $T(k, h, b)$ is a relative tensor of weight w_{S-R}, to be represented by $T^k{}_{hb}$ (law of exponents); and similarly for the invariant tensor definition (Equation 2.7.7). The extension of QR to the most general tensor case does not offer any difficulty. Last, we state the (by now) "self-evident" rule: *All terms of a tensor equation must be of the same rank, kind, and weight;* this requirement may, sometimes (e.g., fluid mechanics), help one make correct guesses about the form of the solutions of these equations.

2.7.6 Conjugate (or Associated) Tensors

The conjugate tensor T^{kl}, to the covariant second order tensor T_{kl}, is defined uniquely by (recalling Cramer's rule from Section 1.2):

$$T_{kl} T^{lh} = \delta^h{}_k \Rightarrow T^{lh} \equiv \mathrm{Cof}(T_{hl}) / T: \ \text{reduced cofactor of } T_{hl} \text{ in } T \equiv \mathrm{Det}(T_{kl}) \neq 0, \tag{2.7.8}$$

$$[\text{or } T_{kl} T^{hl} = \delta^h{}_k, \quad T^{hl} \equiv \mathrm{Cof}(T_{hl})/T; \quad \text{or } T_{kl} T^{kh} = \delta^h{}_l; \quad \text{or } T_{kl} T^{hk} = \delta^h{}_l].$$

This is a reciprocal relation: T_{kl} is also conjugate to $T^{kl;}$ i.e., T_{kl} and T^{kl} are conjugate to each other. Last, it is not hard to show that *if T_{kl} is symmetric (or diagonal), so is T^{kl}, and vice versa.*

2.8 THE METRIC TENSOR

2.8.1 The Fundamental (Covariant) Metric Tensor

Let us consider a rectangular Cartesian CS: $y \equiv (y^k; k = 1, \ldots, n)$ covering an n-dimensional *Euclidean space E_n*. Roughly, Euclidean means that it is possible to build in there a coordinate system such that the length, or distance, between any two of its points, *no matter how far apart they may be*, is given by the n-dimensional version of the Pythagorean formula. Then, the square of the distance between the neighboring E_n-points $P(y)$ and $P'(y + dy)$ is

$$ds^2 = (dy^1)^2 + (dy^2)^2 + \ldots + (dy^n)^2 = dy^k\, dy^k = \delta_{kl}\, dy^k\, dy^l. \qquad (2.8.1)$$

In any other general CS: $y \leftrightarrow q$, covering E_n that length will be given by the invariant *fundamental metric form (positive definite):*

$$ds^2 = \delta_{kl}\,[(\partial y^k/\partial q^r)dq^r]\,[(\partial y^l/\partial q^s)dq^s] \equiv g_{rs}\, dq^r\, dq^s \;\; (= g_{r's'}\, dq^{r'}\, dq^{s'} = \ldots); \quad (2.8.2)$$

where, due to the ds^2-invariance and the quotient rule, the coefficients,

$$g_{rs} = g_{sr} = (\partial y^k/\partial q^r)(\partial y^l/\partial q^s)\delta_{kl} = (\partial y^k/\partial q^r)(\partial y^k/\partial q^s), \qquad (2.8.3)$$

are the corresponding *covariant components of the metric tensor.* In view of the symmetry, only $n(n+1)/2$ of these components are independent (and potentially nonzero). Hence, between any two CS: $q \leftrightarrow q'$, we shall have, with $g_{r's'}(q') \equiv g_{r's'}$ and $g_{rs}(q) \equiv g_{rs}$:

$$g_{r's'} = (\partial q^r/\partial q^{r'})(\partial q^s/\partial q^{s'})g_{rs} \Leftrightarrow g_{rs} = (\partial q^{r'}/\partial q^r)(\partial q^{s'}/\partial q^s)g_{r's'}. \qquad (2.8.4)$$

2.8.2 Definitions

i. If $g_{kl}(q) = \delta_{kl}$, for all q, the CS is called *rectangular Cartesian*; and
ii. If $g_{kl}(q) = (\text{constant})_{kl}$, for all q, it is called *nonrectangular Cartesian,* or *rectilinear.*

If (g_{kl}) = diagonal $(d_1, d_2, \ldots, d_n)$, d_k: constants, the CS is *rectangular but non–Cartesian.*

Example 2.8.1

Let us calculate the distance between two points in E_3, P_1 and P_2, s_{12}, in both rectangular Cartesian (y) and rectilinear affine (z) coordinates, which are related by the central transformation: $y^k = a^k{}_l z^l$ (Figure 2.3).

With P_1: $(y_1{}^k)$ or $(z_1{}^l)$ and P_2: $(y_2{}^k)$ or $(z_2{}^l)$ $(k, l = 1, 2)$, we have

$$(s_{12})^2 = (\mathbf{P}_1\, \mathbf{P}_2)^2 = (y_2{}^1 - y_1{}^1)^2 + (y_2{}^2 - y_1{}^2)^2 \quad [\text{i.e., } g_{kl}(y) = \delta_{kl}]$$

$$= g_{kl}(z)(z_2{}^k - z_1{}^k)(z_2{}^l - z_1{}^l), \quad \text{where } g_{kl}(z) \equiv g_{kl} = a^s{}_k\, a^s{}_l,\ z\text{-metric.} \quad (a)$$

But from the figure we readily obtain

$$y^1 = (\cos\theta)z^1 + (\cos\phi)z^2, \quad y^2 = (\sin\theta)z^1 + (\sin\phi)z^2, \qquad (b)$$

$$\Rightarrow a^1{}_1 = \cos\theta,\ a^1{}_2 = \cos\phi,\ a^2{}_1 = \sin\theta,\ a^2{}_2 = \sin\phi; \qquad (c)$$

and, therefore,

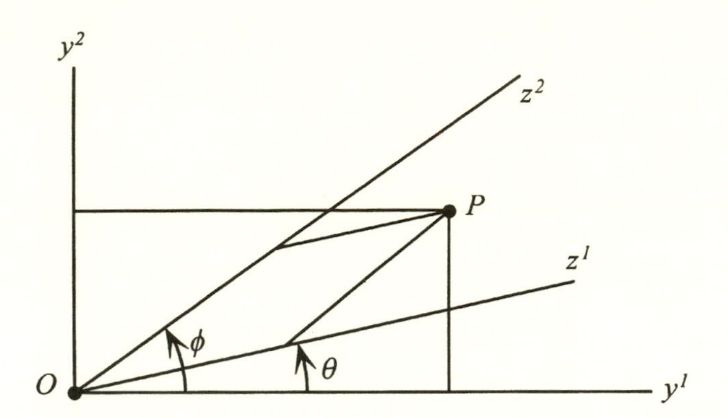

FIGURE 2.3 Length in rectangular Cartesian and rectilinear coordinates (two-dimensional case):

$$(s_{12})^2 \equiv (P_1 P_2)^2;\ y^k = a^k{}_l z^l, \quad g_{kl}(y) = \delta_{kl}, \quad g_{kl}(z) \equiv g_{kl} = a^s{}_k a^s{}_l.$$

$$g_{kl}(z) \equiv g_{kl} = a^1{}_k a^1{}_l + a^2{}_k a^2{}_l$$

$$\Rightarrow g_{11} = 1, \quad g_{12} = g_{21} = \ldots = \cos(\phi - \theta), \quad g_{33} = 1. \tag{d}$$

Hence, and with $P_1 \to O$ (origin): (0, 0), $P_2 \to P$: (y) or (z), $s_{12} \to s$,

$$s^2 = (\mathbf{OP})^2 = (y^1)^2 + (y^2)^2 = g_{11}(z^1)^2 + 2g_{12} z^1 z^2 + g_{22}(z^2)^2$$

$$= (z^1)^2 + 2\cos(\phi - \theta) z^1 z^2 + (z^2)^2\text{: “law” of cosines.} \tag{e}$$

For $\theta = 0$ and $\phi = \pi/2$ the above yields the Pythagorean formula.

Finally, let us assume that the units of length along the, say, three mutually orthogonal axes y^1, y^2, y^3, are not equal to each other (rectangular Cartesian case), but they are in the ratios 1: h: k. Then,

$$s^2 = (\mathbf{OP})^2 = (y^1)^2 + (hy^2)^2 + (ky^3)^2 = g_{11}(y^1)^2 + g_{22}(y^2)^2 + g_{33}(y^3)^2$$

where $g_{11} = 1$, $g_{22} = h^2$, $g_{33} = k^2$ (i.e., rectangular but non-Cartesian coordinates).

2.8.3 Conjugate (Contravariant) Metric Tensor

Having defined $g_{kl}(q) \equiv g_{kl} = g_{lk}$ we, next, define its *conjugate contravariant* metric tensor, $g^{kl}(q) \equiv g^{kl} = g^{lk}$, by the Equation 2.7.8-like equation, with the notation $g \equiv \mathrm{Det}(g_{kl}) \equiv |g_{kl}| \neq 0$:

$$g_{kl}\, g^{ls} = g_{kl}\, g^{sl} = \delta^s{}_k \quad (\text{or } g_{kl}\, g^{sk} = g_{kl}\, g^{ks} = \delta^s{}_l); \tag{2.8.5}$$

i.e.,

$$g^{ls} \equiv \left[\mathrm{Cof}(g_{sl}) / g\right] \quad \text{or} \quad (g^{kl}) = (g_{kl})^{-1}; \quad \text{also } g^{kl} = \partial(\ln g) / \partial g_{kl}. \tag{2.8.5a}$$

As stated earlier (Section 2.7), g_{kl} and g^{kl} are *conjugate* to each other, and are both symmetric. Further, since $\mathrm{Det}(\delta_{kl}) \equiv |\delta_{kl}| = 1$, we shall have:

$$g = |\partial y/\partial q|^2\, |\delta_{kl}| = |\partial y/\partial q|^2 > 0 \text{ (relative scalar of weight +2)}; \tag{2.8.5b}$$

and, therefore:

$$|g^{kl}| = |g_{kl}|^{-1} = 1/g = |\partial q/\partial y|^2 > 0 \text{ and } g^{kl} = (1/g)(\partial g/\partial g_{kl}). \tag{2.8.5c}$$

Problem 2.8.1

Verify that:

$$dg^{kl} = -g^{kr}\, g^{ls}\, dg_{rs}, \quad dg_{kl} = -g_{kr}\, g_{ls}\, dg^{rs}, \quad dg/g = g^{kl}\, dg_{kl} = -g_{kl}\, dg^{kl}, \tag{a}$$

i.e.,

$$g^{kl}{}_{,b} = -g^{kr}\, g^{ls}\, g_{rs,b}, \text{ etc.}$$

Problem 2.8.2

Verify that if $T^{\cdots}{}_{\cdots}$ is a *relative* tensor of weight w, then $\left(\sqrt{g}\right)^w T^{\cdots}{}_{\cdots}$ is an *absolute* tensor (usefulness of g in changing the weight of relative tensors).

2.8.4 Left-/Right-Handedness (Orientation) of a CS

A CS: q is called *right-handed* (*left-handed*) if the scalar density $\sqrt{g}$ is *positive* (*negative*). Further, since (from Equation 2.8.5a with $q \to q'$ and $y \to q$, and assuming that $|\partial q/\partial q'| > 0$: oriented manifold, Section 1.3)

$$\sqrt{g'(q')} = |\partial q / \partial q'|\sqrt{g(q)}, \tag{2.8.6}$$

an arbitrary sign choice for a *particular CS,* of a group of coordinates, defines the sign and therefore the right-/left-handedness of that group. Here we shall be using groups of right-handed CS.

2.8.5 Raising and Lowering of Indices

The main function of g_{kl} and g^{kl} is to *raise and lower indices* (usually but not always) of tensors.

- To *raise (lower)* an index of the tensor $T^{\cdots}{}_{\cdots}$, we form an inner product of it with the *contravariant (covariant)* metric tensor; for example (note order of indices),

Raising: $g^{kl}\, T_{ls}{}^{h} = T^{k}{}_{s}{}^{h}, \quad g^{kl}\, T_{kr} = T^{l}{}_{r}, \quad g^{kl}\, T_{rl} = T_{r}{}^{k};$ (2.8.7a)

Lowering: $g_{kl}\, T^{l}{}_{rs} = T_{krs}, \quad g_{kl}\, T^{rl} = T^{r}{}_{k}, \quad g_{kl}\, A^{kr} = T_{l}{}^{r};$ (2.8.7b)

Mixed: $g_{kl}\, T^{rs}{}_{ab}{}^{k} = g_{kl}\,(g^{kc}\, T^{rs}{}_{abc}) = \delta^{c}{}_{l}\, T^{rs}{}_{abc} = T^{rs}{}_{abl}, \quad g^{kl}\, T^{rs}{}_{abl} = T^{rs}{}_{ab}{}^{k}.$(2.8.7c)

In particular, for a second-rank tensor, we have the following equivalent forms:

$$T_{lm} = g_{sl}\, g_{km}\, T^{sk} = g_{sl}\, T^{s}{}_{m} = g_{km}\, T_{l}{}^{k}, \tag{2.8.7d}$$

$$T^{l}{}_{m} = g^{ls}\, T_{sm} = g_{km}\, T^{lk} = g^{kl}\, g_{ms}\, T_{k}{}^{s}, \tag{2.8.7e}$$

$$T_{m}{}^{l} = g^{sl}\, T_{ms} = g_{km}\, T^{kl} = g_{ms}\, g^{kl}\, T^{s}{}_{k}. \tag{2.8.7f}$$

Thus, these operations change the *type* of the tensor (if $T^{\cdots}{}_{\cdots}$ is a tensor, to begin with), but *preserve its rank and weight;* i.e., *they do not alter the tensor in any fundamental way — it is still the same tensor in a different representation.* Thus, the tensor equations $T_{klr} = R_{kl}\, S_r$ and $T^{k}{}_{lr} = R^{k}{}_{l}\, S_r$ are equivalent. Also, these are *inverse* operations: raising and then lowering the same index, and vice versa, leads to the original tensor.

2.8.5.1 Mixed Metric Tensor

By applying the above rules to g_{kl} and g^{kl}, we can easily verify that

$$g^{k}{}_{l} = g_{l}{}^{k} = \delta^{k}{}_{l} \Rightarrow g^{l}{}_{k}\, T^{k} = T^{l}, \quad g^{l}{}_{k}\, T_{l} = T_{k} \text{ (substitution operation)}; \tag{2.8.7g}$$

i.e., the mixed tensor $g^{k}{}_{l}$ is identical with $\delta^{k}{}_{l}$, and hence constant in all CS: $g^{k}{}_{l} = g^{k'}{}_{l'} = g^{k''}{}_{l''} = \ldots = \delta^{k}{}_{l}$; although δ_{kl} and δ^{kl} are the particular forms/values of the metric tensor in rectangular Cartesian coordinates.

Problem 2.8.3

i. For a general tensor T_{kl} $(k, l = 1, \ldots, n)$ verify that

$$T_{(kl)} = [T_{(kl)} - (1/n)(T^{b}{}_{b})g_{kl}] + (1/n)(T^{b}{}_{b})g_{kl}, \tag{a}$$

where $T^{b}{}_{b} \equiv$ *Trace of* $\mathbf{T} \equiv \mathrm{Tr}(\mathbf{T})$. Such representations are useful in solid mechanics.

ii. Let T_{kl}, T^{kl} be *antisymmetric*. Verify that

$$T_{kl} = (g_{ki}\, g_{lj} - g_{kj}\, g_{li})T^{ij} \quad (i < j). \tag{b}$$

Hint: Apply the procedure of Equations 2.7.3a and b to Equation 2.8.7d.

Problem 2.8.4

Show that for a general tensor T_{kl}:

$$T_{kl}\, dg^{kl} = -T^{kl}\, dg_{kl}. \tag{a}$$

2.9 TENSORIAL FORM OF ORDINARY VECTOR ALGEBRA

Here we summarize the basic formulae of elementary vector algebra in ordinary Euclidean space E_3. This will, it is hoped, help the reader understand and appreciate better the usefulness of the hitherto developed tensor formalism ("index juggling").

2.9.1 Bases in Curvilinear Coordinates

Let us consider a set of general *curvilinear* coordinates $q \equiv (q^1, q^2, q^3) \equiv (q^k)$ covering a region of E_3 and a fixed "origin" O in it. Then the *position* vector of an arbitrary E_3-point, relative to O, R, can be expressed as

$$\mathbf{R} = \mathbf{R}(q^k), \tag{2.9.1}$$

and the (generally) *nonunit*, *nonorthogonal*, and *noncoplanar* vectors:

$$\mathbf{e}_k \equiv \partial\mathbf{R}/\partial q^k, \quad \text{at some } E_3\text{-point } P = P(q), \tag{2.9.2}$$

define a fundamental "natural," or *gradient*, covariant local basis, or frame, there: $\{P; \mathbf{e}_k\}$ (Figure 2.4). Hence, under a differential coordinate variation: $q \to q + dq$, $\mathbf{R}$ changes by

$$d\mathbf{R} = (\partial\mathbf{R}/\partial q^k)dq^k = \mathbf{e}_k\, dq^k\text{: } \textit{elementary displacement vector.} \tag{2.9.3}$$

(We notice, however, that both $d\mathbf{R}$ and the $\mathbf{e}_k$ are independent of the origin O.)

Now,

- The (signed) *volume* of the elementary "tangential parallelopiped" formed by the three vectors $\mathbf{e}_{1,2,3}$, equals, as known from elementary vector algebra,

$$V = \mathbf{e}_1 \cdot (\mathbf{e}_2 \times \mathbf{e}_3) = \mathbf{e}_2 \cdot (\mathbf{e}_3 \times \mathbf{e}_1) = \mathbf{e}_3 \cdot (\mathbf{e}_1 \times \mathbf{e}_2)$$

$$= (\mathbf{e}_1, \mathbf{e}_2, \mathbf{e}_3) > 0 \quad \text{(for proper choice of naming of the basis vectors).} \tag{2.9.4}$$

- The local *contravariant*, *reciprocal*, or *dual*, basis to $\{P; \mathbf{e}_k\}$, $\{P; \mathbf{e}^k\}$, is defined by

$$\mathbf{e}^1 \equiv (\mathbf{e}_2 \times \mathbf{e}_3)/V, \quad \mathbf{e}^2 \equiv (\mathbf{e}_3 \times \mathbf{e}_1)/V, \quad \mathbf{e}^3 \equiv (\mathbf{e}_1 \times \mathbf{e}_2)/V; \tag{2.9.5a}$$

or, equivalently, by the $3 \times 3 = 9$ equations: $\mathbf{e}^k \cdot \mathbf{e}_l = \delta^k{}_l$

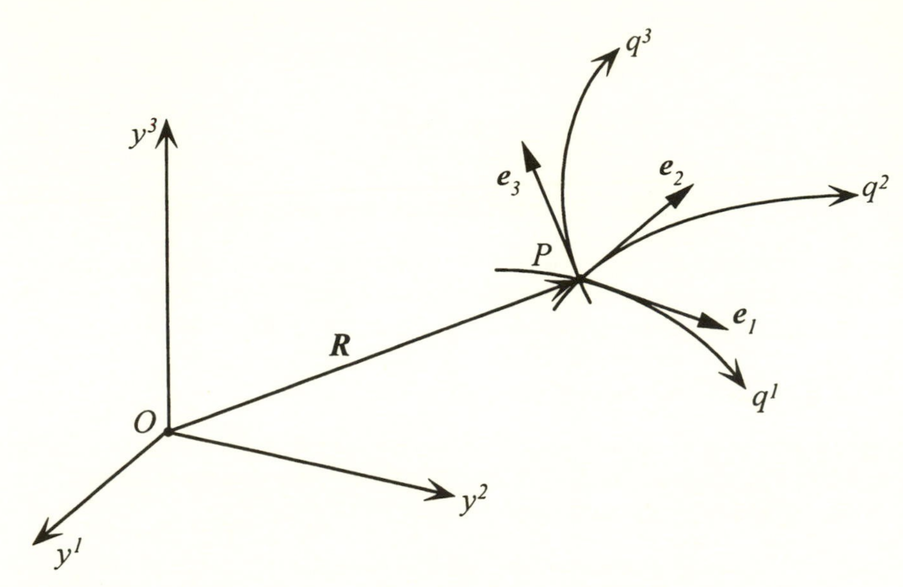

FIGURE 2.4 Local natural, or gradient, basis in general curvilinear coordinates in E_3.

$[\Rightarrow |\mathbf{e}^k|\ |\mathbf{e}_k| \cos(\mathbf{e}^k, \mathbf{e}_k) = 1 \Rightarrow |\mathbf{e}^k| = 1 / |\mathbf{e}_k| \cos(\mathbf{e}^k, \mathbf{e}_k)$; no sum over k]. (2.9.5b)*

The above show that $\mathbf{e}_1$ is tangent to the coordinate line q^1 (i.e., q^2, q^3 = constant), while $\mathbf{e}^1$ is *normal* to the surface q^1 = constant; and analogously for $\mathbf{e}_{2,3}$ and $\mathbf{e}^{2,3}$ (Figure 2.5):

$$\mathbf{e}^k = \partial q^k/\partial \mathbf{R} \equiv \text{grad } q^k = (\partial q^k/\partial q^l)\mathbf{e}^l = \delta^k{}_l\, \mathbf{e}^l = \mathbf{e}^k; \tag{2.9.5f}$$

i.e., Equations 2.9.5b,f simply express the chain rule:

---

* It is not hard to see that, generally, the bases $\mathbf{e}_{1,2,3}$ and $\mathbf{e}^{1,2,3}$ are *mutually reciprocal;* i.e., the reciprocal of the reciprocal basis $\mathbf{e}^{1,2,3}$ is the fundamental basis $\mathbf{e}_{1,2,3}$.

Direct Analytical Proof:

$$\mathbf{e}^2 \times \mathbf{e}^3 = (1/V^2)(\mathbf{e}_3 \times \mathbf{e}_1) \times (\mathbf{e}_1 \times \mathbf{e}_2) \quad \text{(expanding, using well known vector identities)}$$

$$= (1/V^2)\{\mathbf{e}_1 [\mathbf{e}_2 \cdot (\mathbf{e}_3 \times \mathbf{e}_1)] - \mathbf{e}_2 [\mathbf{e}_1 \cdot (\mathbf{e}_3 \times \mathbf{e}_1)]\} \quad \text{(the } \textit{second} \text{ term vanishes)}$$

$$= (1/V^2)(\mathbf{e}_1 V) = \mathbf{e}_1/V \Rightarrow \mathbf{e}_1 = V(\mathbf{e}^2 \times \mathbf{e}^3), \quad \text{etc. cyclically, q.e.d.;} \tag{2.9.5c}$$

and therefore:

$$V^* \equiv \mathbf{e}^1 \cdot (\mathbf{e}^2 \times \mathbf{e}^3) = (1/V)(\mathbf{e}^1 \cdot \mathbf{e}_1) = 1/V; \tag{2.9.5d}$$

so that finally:

$$\mathbf{e}_1 \equiv (\mathbf{e}^2 \times \mathbf{e}^3) / V^*, \quad \mathbf{e}_2 \equiv (\mathbf{e}^3 \times \mathbf{e}^1) / V^*, \quad \mathbf{e}_3 \equiv (\mathbf{e}^1 \times \mathbf{e}^2) / V^*. \tag{2.9.5e}$$

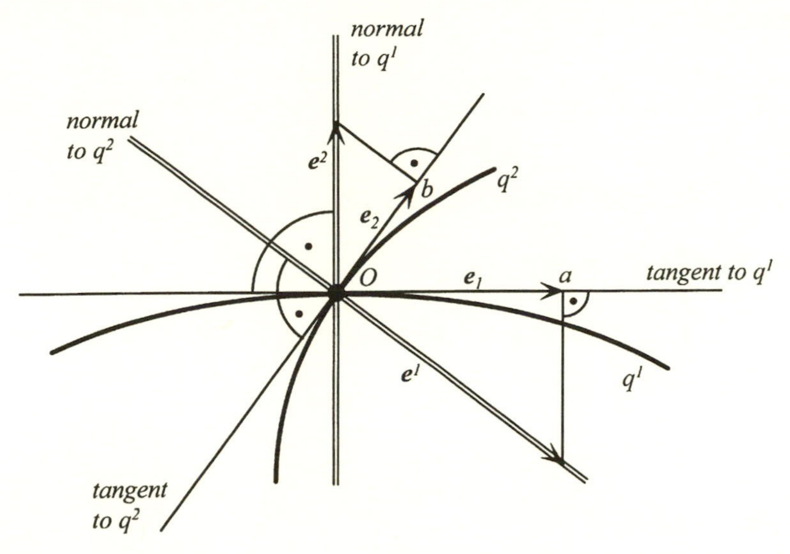

Figure 2.5 Mutually reciprocal, or dual, bases: $\{\mathbf{e}_k$: *natural*, or *gradient*$\}$, $\{\mathbf{e}^k$: *dual*$\}$ (two-dimensional case): $\mathbf{Oa} = \mathbf{e}_1$ and $\mathbf{Ob} = \mathbf{e}_2$ are chosen as (covariant) unit vectors: $\mathbf{e}_1 \cdot \mathbf{e}^1 = \delta^1{}_1 = 1 \Rightarrow |\mathbf{e}_1|\,|\mathbf{e}^1| \cos(\mathbf{e}_1, \mathbf{e}^1) = 1 \Rightarrow |\mathbf{e}^1| = 1/\cos(\mathbf{e}_1, \mathbf{e}^1)$, and similarly for $|\mathbf{e}^2|$.

$$(\partial q^k/\partial \mathbf{R}) \cdot (\partial \mathbf{R}/\partial q^l) = (\text{normal})^k \cdot (\text{tangent})_l = (\partial q^k/\partial q^l) = \delta^k{}_l. \quad (2.9.5g)$$

- The following dot products are of fundamental importance:

$$\mathbf{e}_k \cdot \mathbf{e}_l = g_{kl} = g_{lk}; \quad \mathbf{e}^k \cdot \mathbf{e}^l = g^{kl} = g^{lk}; \quad \mathbf{e}^k \cdot \mathbf{e}_l = g^k{}_l = \delta^k{}_l. \quad (2.9.6)$$

2.9.1.1 Special Cases

i. If $\mathbf{e}_k \cdot \mathbf{e}_l = g_{kl}(q)$: diagonal [i.e., $g_{kl} = 0$ $(k \neq l)$, $g_{kk} \neq 0$ (no sum over k)], the basis is *orthogonal*, and so are the associated (generally curvilinear) coordinates; and vice versa. Then (recalling Equations 2.2.11): $\mathbf{e}^k = (\mathbf{e}^k \cdot \mathbf{e}^l)\,\mathbf{e}_l = g^{kl}\,\mathbf{e}_l = g^{kk}\,\mathbf{e}_k \neq \mathbf{e}_k$ (no sum over k).

ii. If $g_{kl} = \delta_{kl}$, the bases are called *orthogonal Cartesian*, or *orthonormal*. Such bases are *self-reciprocal*, i.e., $\mathbf{e}_k = \mathbf{e}^k$. Given an orthogonal basis $\{\mathbf{e}_k\}$, the corresponding orthonormal basis is

$$\left\{\mathbf{e}_k/\sqrt{g_{kk}} = \mathbf{e}_k\sqrt{g^{kk}}\right\} \text{ (no sum over } k).$$

iii. In a Euclidean space, like E_3, it is possible to have g_{kl}: *constant throughout space*, i.e., introduce *global rectilinear*, or *affine*, coordinates there; further, by orthonormalizing the latter, we can always obtain *global rectangular Cartesian* coordinates, i.e., have $g_{kl} = \delta_{kl}$ *throughout space*. In non-Euclidean (e.g., Riemannian) spaces that can be done only *locally* (Sections 2.12 and 5.6).

Example 2.9.1

With the help of the above let us show, by direct proof, that g_{kl} and g^{kl} are indeed mutually conjugate (Section 2.8); i.e., that the matrices (g_{kl}) and (g^{kl}) are mutually inverse:

$$g^{kl} = [\mathrm{Cof}(g_{kl})/g] \equiv (G^{kl}/g) = g^{-1}(\partial g/\partial g_{kl}); \; g \equiv |g_{kl}|, \; g^* \equiv |g^{kl}| = g^{-1}. \quad \text{(a)}$$

Proof: We have successively:

$$g_{ks}\, g^{kl} = (\mathbf{e}_k \cdot \mathbf{e}_s)(\mathbf{e}^k \cdot \mathbf{e}^l) = \mathbf{e}_s \cdot [\mathbf{e}_k(\mathbf{e}^k \cdot \mathbf{e}^l)]$$

$$= \mathbf{e}_s \cdot [\mathbf{e}^l \times (\mathbf{e}_k \times \mathbf{e}^k) + \mathbf{e}^k(\mathbf{e}^l \cdot \mathbf{e}_k)]$$

$$= \mathbf{e}_s \cdot [\mathbf{e}^l \times (\mathbf{e}_k \times \mathbf{e}^k)] + g^l{}_s; \quad \text{(b)}$$

but by Equations 2.9.5a,

$$\mathbf{e}_k \times \mathbf{e}^k = (1/V)[\mathbf{e}_1 \times (\mathbf{e}_2 \times \mathbf{e}_3) + \mathbf{e}_2 \times (\mathbf{e}_3 \times \mathbf{e}_1) + \mathbf{e}_3 \times (\mathbf{e}_1 \times \mathbf{e}_2)] = \mathbf{0}, \quad \text{(c)}$$

and therefore $g_{ks}\, g^{kl} = g^l{}_s = \delta^l{}_s$, q.e.d.

[Or, using well known vector identities, we find:

$$\mathbf{e}^l \times (\mathbf{e}_k \times \mathbf{e}^k) = (\mathbf{e}^l \cdot \mathbf{e}^k)\, \mathbf{e}_k - (\mathbf{e}^l \cdot \mathbf{e}_k)\, \mathbf{e}^k = g^{lk}\, \mathbf{e}_k - \delta^l{}_k\, \mathbf{e}^k = \mathbf{e}^l - \mathbf{e}^l = \mathbf{0}]. \quad \text{(d)}$$

For example, we have

$$g^{12} \equiv \mathbf{e}^1 \cdot \mathbf{e}^2 = (1/V^2)(\mathbf{e}_2 \times \mathbf{e}_3) \cdot (\mathbf{e}_3 \times \mathbf{e}_1)$$

$$= (1/V^2)[(\mathbf{e}_2 \cdot \mathbf{e}_3)(\mathbf{e}_3 \cdot \mathbf{e}_1) - (\mathbf{e}_1 \cdot \mathbf{e}_2)(\mathbf{e}_3 \cdot \mathbf{e}_3)]$$

$$= (1/V^2)(g_{23}\, g_{31} - g_{12}\, g_{33}) = G^{12}/V^2; \quad \text{(e)}$$

and, generally,

$$V^2\, g^{kl} = g_{rp}\, g_{sq} - g_{rq}\, g_{sp}, \quad \text{where: } k, r, s;\ l, p, q\text{: cyclic permutations of } 1, 2, 3. \quad \text{(f)}$$

Finally, comparing the above with Equations 2.9.4, 2.9.5f, and Equation a we conclude that

$$V \equiv (\mathbf{e}_1, \mathbf{e}_2, \mathbf{e}_3) = \sqrt{g}\,, \quad V^* \equiv (\mathbf{e}^1, \mathbf{e}^2, \mathbf{e}^3) = \sqrt{g^*} = V^{-1} = \left(\sqrt{g}\right)^{-1}. \quad \text{(g)}$$

2.9.2 Transformation of Basis Vectors

Under a CT: $q \to q'$,

$$\mathbf{R} = \mathbf{R}(q) = \mathbf{R}[q(q')] = \mathbf{R}(q'), \tag{2.9.7a}$$

and therefore, since $d\mathbf{R}$ is an invariant,

$$d\mathbf{R} = (\partial\mathbf{R}/\partial q^k)dq^k \equiv \mathbf{e}_k\, dq^k = (\partial\mathbf{R}/\partial q^{k'})dq^{k'} \equiv \mathbf{e}_{k'}\, dq^{k'} = \mathbf{e}_{k'}\, [(\partial q^{k'}/\partial q^k)dq^k],$$

$$\Rightarrow \mathbf{e}_k = (\partial q^{k'}/\partial q^k)\mathbf{e}_{k'} \equiv A^{k'}{}_k\, \mathbf{e}_{k'} \leftrightarrow \mathbf{e}_{k'} = (\partial q^k/\partial q^{k'})\mathbf{e}_k \equiv A^k{}_{k'}\, \mathbf{e}^k. \tag{2.9.7b}$$

Similarly, we have

$$\mathbf{e}^{b'} = g^{b's'}\, \mathbf{e}_{s'} = (A^{b'}{}_b A^{s'}{}_s\, g^{bs})(A^k{}_{s'}\, \mathbf{e}_k) = \dots = A^{b'}{}_b(g^{bk}\, \mathbf{e}_k) = A^{b'}{}_b\, \mathbf{e}^b,$$

and

$$\mathbf{e}^b = A^b{}_{b'}\, \mathbf{e}^{b'}. \tag{2.9.7c}$$

The above allow us to express the following geometrical quantities.

2.9.2.1 Length of Elementary Displacement Vector

Using Equation 2.9.3 we find

$$ds^2 = d\mathbf{R} \cdot d\mathbf{R} = (\mathbf{e}_k \cdot \mathbf{e}_l)dq^k\, dq^l \equiv g_{kl}\, dq^k\, dq^l. \tag{2.9.8}$$

Accordingly, if at a point P, we take q^2, q^3 = constant, and vary only q^1, i.e., take dq^2, $dq^3 = 0$,

$$ds^2 \to (ds_{(1)})^2 = g_{11}(dq^1)^2$$

$$\Rightarrow ds_{(1)} = \sqrt{g_{11}}\, dq^1: \tag{2.9.8a}$$

elementary arc length along coordinate line q^1, at P (Figure 2.6a);

while the corresponding elementary displacement vector becomes

$$d\mathbf{R} \to d\mathbf{R}_{(1)} = \mathbf{e}_1\, dq^1 = \mathbf{e}_1 \left(ds_{(1)}/\sqrt{g_{11}}\right) \equiv \mathbf{e}_{<1>}\, ds_{(1)}, \tag{2.9.8b}$$

where

$$\mathbf{e}_{<1>} \equiv \mathbf{e}_1/\sqrt{g_{11}} : \text{unit vector tangent to coordinate line } q^1, \text{ at } P. \tag{2.9.8c}$$

Similarly for the coordinate lines q^2 and q^3.

2.9.2.2 Elementary Area

For the elementary area, say, on the surface q^1 = constant, dA_1, we have successively (Figure 2.6b),

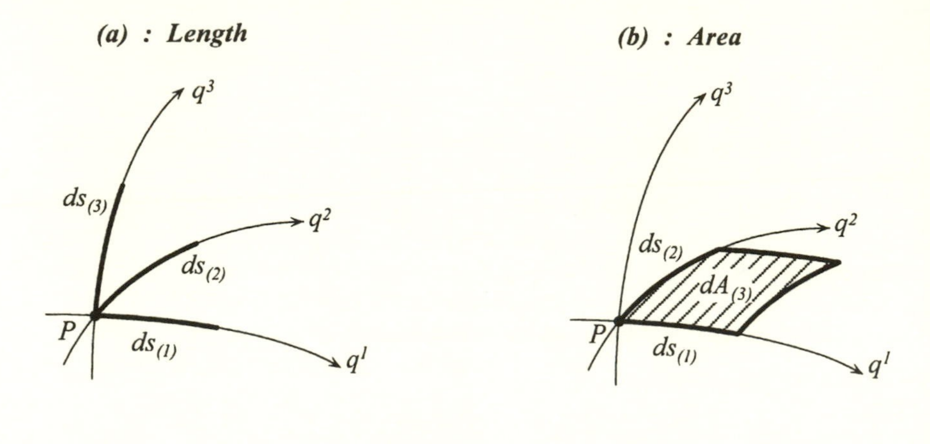

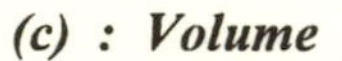

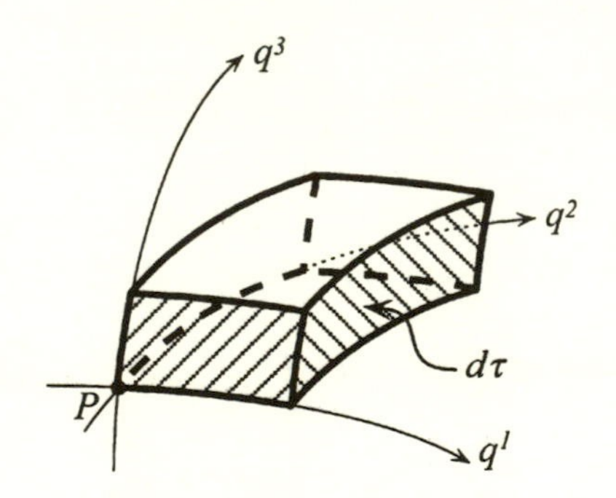

FIGURE 2.6 Elementary Length (a), Area (b), and Volume (c) in curvilinear coordinates (three-dimensional case).

$$dA_1 = |d\mathbf{R}_{(2)} \times d\mathbf{R}_{(3)}| = |\mathbf{e}_2 \times \mathbf{e}_3|\, dq^2\, dq^3 = [(\mathbf{e}_2 \times \mathbf{e}_3) \cdot (\mathbf{e}_2 \times \mathbf{e}_3)]^{1/2}\, dq^2\, dq^3$$

$$= [(\mathbf{e}_2 \cdot \mathbf{e}_2)(\mathbf{e}_3 \cdot \mathbf{e}_3) - (\mathbf{e}_2 \cdot \mathbf{e}_3)(\mathbf{e}_3 \cdot \mathbf{e}_2)]^{1/2}\, dq^2\, dq^3 \quad \text{(using vector identities)}$$

$$= [g_{22}\, g_{33} - (g_{23})^2]^{1/2}\, dq^2\, dq^3 \equiv \left(\sqrt{g g^{11}}\right) dq^2\, dq^3; \tag{2.9.9}$$

and similarly for dA_2 and dA_3.

Generally, *the element of area on the (k)th, or q^k-coordinate surface* equals (see Equations 2.9.11a to c below)

$$d\mathbf{A}_k = d\mathbf{R}_{(l)} \times d\mathbf{R}_{(m)} = (\mathbf{e}_l \times \mathbf{e}_m)dq^l\, dq^m = \left(\sqrt{g}\, \varepsilon_{klm}\mathbf{e}^k\right) dq^l\, dq^m \quad (k \neq l \neq m \neq k)$$

i.e.,

$$dA_k = (g\, g^{kk})^{1/2}\, dq^l\, dq^m \quad \text{(no sum on } k;\ k \neq l \neq m \neq k). \tag{2.9.9a}$$

(For extensions to n dimensions, in connection with Stokes' theorem, see Example 3.6.1.)

2.9.2.3 Elementary Volume

We have successively (Figure 2.6c):

$$d(\text{volume}) \equiv d\tau = d\mathbf{R}_{(1)} \cdot (d\mathbf{R}_{(2)} \times d\mathbf{R}_{(3)}) = \mathbf{e}_1 \cdot (\mathbf{e}_2 \times \mathbf{e}_3) dq^1\, dq^2\, dq^3$$

$$= \sqrt{g}\, dq^1\, dq^2\, dq^3 \equiv \sqrt{g}\, dV, \; g \equiv |g_{kl}|, \tag{2.9.10a}$$

and therefore integrating over the q^k-domain that describes the volume in question, we find

$$\tau = \int d\tau = \int \sqrt{g}\, dV. \tag{2.9.10b}$$

Then the elementary mass dm, occupying $d\tau$, equals (recalling relevant discussion in Section 2.5):

$$dm = \rho\, d\tau = \rho(q)\sqrt{g(q)}\, dV \quad [(\text{scalar density})\; dV], \tag{2.9.10c}$$

and integrating as in Equation 2.9.10b, we obtain

$$m = \int dm = \int \rho \sqrt{g}\, dV. \tag{2.9.10d}$$

2.9.3 Permutation Tensors

Combining the results of Example 2.5.4 with Equations 2.9.4ff and Example 2.9.1, we easily conclude that, if k, r, s are as 1, 2, 3 (or are obtainable from them circularly):

$$e_{krs} = \mathbf{e}_k \cdot (\mathbf{e}_r \times \mathbf{e}_s) = V = \sqrt{g} \quad \text{and}$$

$$e^{krs} = \mathbf{e}^k \cdot (\mathbf{e}^r \times \mathbf{e}^s) = V^{-1} = 1/\sqrt{g}; \tag{2.9.11a}$$

and so the basis relations (Equations 2.9.5a through g) assume the forms:

$$\mathbf{e}_k \times \mathbf{e}_l = e_{klb}\, \mathbf{e}^b \quad \text{and} \quad \mathbf{e}^k \times \mathbf{e}^l = e^{klb}\, \mathbf{e}_b. \tag{2.9.11b}$$

Finally, we recall that in rectangular Cartesian coordinates y^k: $g \to g(y) = 1 \Rightarrow V = 1$ and $e_{klb} \to \varepsilon_{klb}$, $e^{klb} \to \varepsilon^{klb}$ and, therefore,

$$e_{klb}(q) = \sqrt{g(q)}\,\varepsilon_{klb} = \sqrt{g(q)}\,e_{klb}(y),$$
$$e^{klb}(q) = \varepsilon^{klb}/\sqrt{g(q)} = e^{klb}(y)/\sqrt{g(q)}. \tag{2.9.11c}$$

2.9.4 Components of Vectors in Various Bases

Let us consider the four E_3-vectors **a**, **b**, **c**, and **V**, of which the first three are linearly independent (not necessarily unit) and the last is arbitrary. Then, and with the notation,

$$\mathbf{a}\cdot(\mathbf{b}\times\mathbf{c}) = \mathbf{b}\cdot(\mathbf{c}\times\mathbf{a}) = \mathbf{c}\cdot(\mathbf{a}\times\mathbf{b}) \equiv (\mathbf{a},\ \mathbf{b},\ \mathbf{c})$$

which equals the oriented volume of parallelepiped spanned by **a**, **b**, **c** (*positive*, if they build a *right*-hand system), the following identity holds (see texts on vectors):

$$\mathbf{a}(\mathbf{V},\ \mathbf{b},\ \mathbf{c}) + \mathbf{b}(\mathbf{a},\ \mathbf{V},\ \mathbf{c}) + \mathbf{c}(\mathbf{a},\ \mathbf{b},\ \mathbf{V}) = \mathbf{V}(\mathbf{a},\ \mathbf{b},\ \mathbf{c}); \tag{2.9.12a}$$

from which, since here $(\mathbf{a}, \mathbf{b}, \mathbf{c}) \neq 0$, we obtain

$$\mathbf{V} = [(\mathbf{V},\ \mathbf{b},\ \mathbf{c})\ /\ (\mathbf{a},\ \mathbf{b},\ \mathbf{c})]\mathbf{a} + [(\mathbf{a},\ \mathbf{V},\ \mathbf{c})\ /\ (\mathbf{a},\ \mathbf{b},\ \mathbf{c})]\mathbf{b}$$
$$+ [(\mathbf{a},\ \mathbf{b},\ \mathbf{V})\ /\ (\mathbf{a},\ \mathbf{b},\ \mathbf{c})]\mathbf{c}. \tag{2.9.12b}$$

Applying this to the mutually reciprocal bases $\mathbf{e}_{1,2,3}$ and $\mathbf{e}^{1,2,3}$ and the general vector **V**, we find

$$\mathbf{V} = [(\mathbf{V},\ \mathbf{e}_2,\ \mathbf{e}_3)\ /\ (\mathbf{e}_1,\ \mathbf{e}_2,\ \mathbf{e}_3)]\mathbf{e}_1 + [(\mathbf{e}_1,\ \mathbf{V},\ \mathbf{e}_3)\ /\ (\mathbf{e}_1,\ \mathbf{e}_2,\ \mathbf{e}_3)]\mathbf{e}_2$$
$$+ [(\mathbf{e}_1,\ \mathbf{e}_2,\ \mathbf{V})\ /\ (\mathbf{e}_1,\ \mathbf{e}_2,\ \mathbf{e}_3)]\mathbf{e}_3 = V^k\mathbf{e}_k, \tag{2.9.13a}$$
$$= [(\mathbf{V},\ \mathbf{e}^2,\ \mathbf{e}^3)\ /\ (\mathbf{e}^1,\ \mathbf{e}^2,\ \mathbf{e}^3)]\mathbf{e}^1 + [(\mathbf{e}^1,\ \mathbf{V},\ \mathbf{e}^3)\ /\ (\mathbf{e}^1,\ \mathbf{e}^2,\ \mathbf{e}^3)]\mathbf{e}^2$$
$$+ [(\mathbf{e}^1,\ \mathbf{e}^2,\ \mathbf{V})\ /\ (\mathbf{e}^1,\ \mathbf{e}^2,\ \mathbf{e}^3)]\mathbf{e}^3 = V_k\,\mathbf{e}^k. \tag{2.9.13b}$$

Also, from the earlier definition of reciprocal bases, we readily obtain

$$\mathbf{e}^1 = (\mathbf{e}_2\times\mathbf{e}_3)/(\mathbf{e}_1, \mathbf{e}_2, \mathbf{e}_3), \quad \mathbf{e}^2 = (\mathbf{e}_3\times\mathbf{e}_1)/(\mathbf{e}_1, \mathbf{e}_2, \mathbf{e}_3), \quad \mathbf{e}^3 = (\mathbf{e}_1\times\mathbf{e}_2)/(\mathbf{e}_1, \mathbf{e}_2, \mathbf{e}_3),$$

$$\mathbf{e}_1 = (\mathbf{e}^2\times\mathbf{e}^3)/(\mathbf{e}^1, \mathbf{e}^2, \mathbf{e}^3), \quad \mathbf{e}_2 = (\mathbf{e}^3\times\mathbf{e}^1)/(\mathbf{e}^1, \mathbf{e}^2, \mathbf{e}^3), \quad \mathbf{e}_3 = (\mathbf{e}^1\times\mathbf{e}^2)/(\mathbf{e}^1, \mathbf{e}^2, \mathbf{e}^3),$$

where

$$(\mathbf{e}_1,\ \mathbf{e}_2,\ \mathbf{e}_3)(\mathbf{e}^1,\ \mathbf{e}^2,\ \mathbf{e}^3) = 1; \tag{2.9.14a}$$

and with their help **V** can also be represented as:

$$\mathbf{V} = [\mathbf{V}\cdot\mathbf{e}_3/(\mathbf{e}_1, \mathbf{e}_2, \mathbf{e}_3)](\mathbf{e}_1\times\mathbf{e}_2) + [\mathbf{V}\cdot\mathbf{e}_1/(\mathbf{e}_1, \mathbf{e}_2, \mathbf{e}_3)](\mathbf{e}_2\times\mathbf{e}_3)$$

$$+ [\mathbf{V}\cdot\mathbf{e}_2/(\mathbf{e}_1, \mathbf{e}_2, \mathbf{e}_3)](\mathbf{e}_3\times\mathbf{e}_1) \tag{2.9.14b}$$

$$= [\mathbf{V}\cdot\mathbf{e}^3/(\mathbf{e}^1, \mathbf{e}^2, \mathbf{e}^3)](\mathbf{e}^1\times\mathbf{e}^2) + [\mathbf{V}\cdot\mathbf{e}^1/(\mathbf{e}^1, \mathbf{e}^2, \mathbf{e}^3)](\mathbf{e}^2\times\mathbf{e}^3)$$

$$+ [\mathbf{V}\cdot\mathbf{e}^2/(\mathbf{e}^1, \mathbf{e}^2, \mathbf{e}^3)](\mathbf{e}^3\times\mathbf{e}^1). \tag{2.9.14c}$$

The various components of $\mathbf{V} = V^k\mathbf{e}_k = V_k\mathbf{e}^k$ are related as follows (Figure 2.7):

$$V_k = \mathbf{V}\cdot\mathbf{e}_k = (V^l\mathbf{e}_l)\cdot\mathbf{e}_k = V^l g_{kl} = (V_l\mathbf{e}^l)\cdot\mathbf{e}_k = V_l\, g^l{}_k, \tag{2.9.15a}$$

$$V^k = \mathbf{V}\cdot\mathbf{e}^k = (V^l\mathbf{e}_l)\cdot\mathbf{e}^k = V^l g_l{}^k = (V_l\mathbf{e}^l)\cdot\mathbf{e}^k = V_l\, g^{lk}, \tag{2.9.15b}$$

i.e.,

$$V_k = g_{kl}\, V^l \Leftrightarrow V^k = g^{kl}\, V_l. \tag{2.9.15c}$$

In another CS: $q \to q'$, with $dq^{k'} = A^{k'}{}_k dq^k \Leftrightarrow dq^k = A^k{}_{k'} dq^{k'}$, where $A^{k'}{}_k A^k{}_{l'} = \delta^{k'}{}_{l'}$ and $A^{k'}{}_k A^l{}_{k'} = \delta^l{}_k$, the basis vectors transform as

$$\mathbf{e}^k \to \mathbf{e}^{k'} = A^{k'}{}_k\mathbf{e}^k \Leftrightarrow \mathbf{e}^k = A^k{}_{k'}\mathbf{e}^{k'}, \quad \mathbf{e}_k \to \mathbf{e}_{k'} = A^k{}_{k'}\mathbf{e}_k \Leftrightarrow \mathbf{e}_k = A_k{}^{k'}\mathbf{e}_{k'}, \tag{2.9.16a}$$

while the components of a general vector $\mathbf{V}$ transform as

$$V^k \to V^{k'} = A^{k'}{}_k V^k \Leftrightarrow V^k = A^k{}_{k'} V^{k'}, \quad V_k \to V_{k'} = A^k{}_{k'} V_k \Leftrightarrow V_k = A_k{}^{k'} V_{k'}. \tag{2.9.16b}$$

The above and the equivalent $\mathbf{V}$-representations:

$$\mathbf{V} = V^k\mathbf{e}_k = V_k\mathbf{e}^k = V^{k'}\mathbf{e}_{k'} = V_{k'}\mathbf{e}^{k'}, \tag{2.9.16c}$$

explain why the V_k are called *covariant*: they vary just like the $\mathbf{e}_k$ (co-vary); and the V^k are called *contravariant*: they vary contrary to the $\mathbf{e}_k$, and like the $\mathbf{e}^k$. The V_k and V^k are called mutually *contragredient.*

In rectangular Cartesian coordinates: g_{kl}, $g^{kl} \to \delta^k{}_l = \delta_{kl} = \delta^{kl} \Rightarrow \mathbf{e}_k = \mathbf{e}^k$, i.e., the bases $\{\mathbf{e}_k\}$, $\{\mathbf{e}^k\}$ coincide (or they are *self-reciprocal*), and so the difference between V_k and V^k disappears! Then Equation g of Example 2.9.1 yields:

$$V \equiv (\mathbf{e}_1, \mathbf{e}_2, \mathbf{e}_3) = \sqrt{g} = V^* \equiv (\mathbf{e}^1, \mathbf{e}^2, \mathbf{e}^3) = \sqrt{g^*} = V^{-1} = \left(\sqrt{g}\right)^{-1} \Rightarrow V^2 = 1$$

$$\Rightarrow V = +1 \text{ [\textit{right-hand} (dextral) system] or } -1 \text{ [\textit{left-hand} (sinistral) system]} \tag{2.9.16d}$$

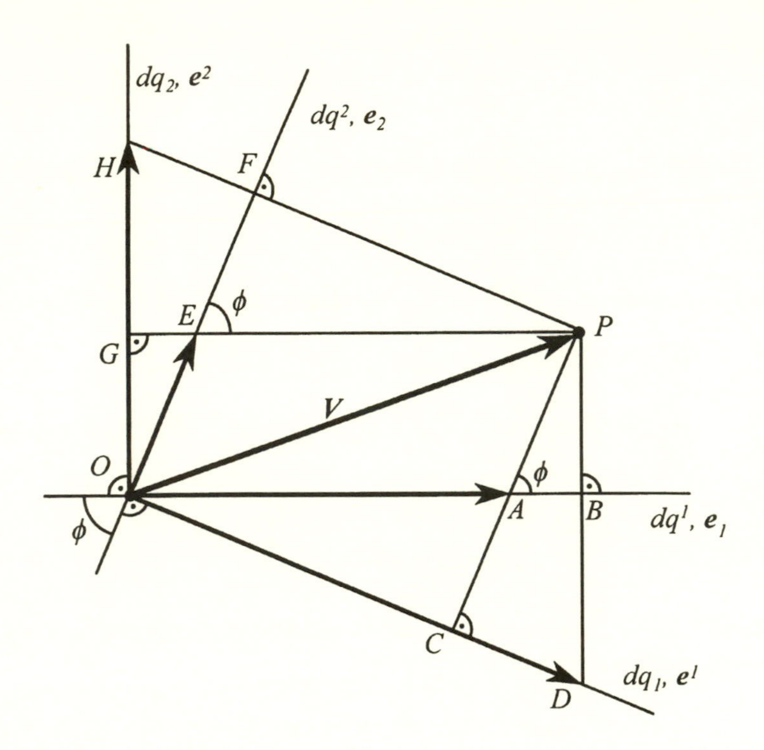

FIGURE 2.7 Contravariant (V^k) and covariant (V_k) components of a vector **V** (two-dimensional case):

$\mathbf{OA} = V^1 \mathbf{e}_1$, $\mathbf{OE} = V^2 \mathbf{e}_2$, $\mathbf{OD} = V_1 \mathbf{e}^1$, $\mathbf{OH} = V_2 \mathbf{e}^2$; $\mathbf{u}_k$, $\mathbf{u}^k$: *unit* vectors along $\mathbf{e}_k$, $\mathbf{e}^k$, respectively.

$$\mathbf{V} = V^1\mathbf{e}_1 + V^2\mathbf{e}_2 = V^1\left(\sqrt{g_{11}}\,\mathbf{u}_1\right) + V^2\left(\sqrt{g_{22}}\,\mathbf{u}_2\right) = \left(\sqrt{g_{11}}\,V^1\right)\mathbf{u}_1 + \left(\sqrt{g_{22}}\,V^2\right)\mathbf{u}_2$$

$$V_1\mathbf{e}^1 + V_2\mathbf{e}^2 = V^1\left(\mathbf{u}^1/\sqrt{g_{11}}\right) + V_2\left(\mathbf{u}^2/\sqrt{g_{22}}\right) = \left(V_1\sqrt{g_{11}}\right)\mathbf{u}^1 + \left(V_2\sqrt{g_{22}}\right)\mathbf{u}^2$$

Then, and since $g_{12} \equiv \mathbf{e}_1 \cdot \mathbf{e}_2 = |\mathbf{e}_1|\,|\mathbf{e}_2| \cos\phi = \left(\sqrt{g_{11}}\right)\left(\sqrt{g_{22}}\right)\cos\phi$ [$\phi \equiv$ angle(dq^1, dq^2)]:

$$V_1 = g_{1k} V^k = g_{11} V^1 + g_{12} V^2$$

$$\Rightarrow V_1 / \sqrt{g_{11}} = \sqrt{g_{11}}\, V^1 + \left(g_{12}/\sqrt{g_{11}}\right)V^2 = \sqrt{g_{11}}\,V^1 + \left(\sqrt{g_{22}}\,V^2\right)\cos\phi;$$

similarly for V_2.

- *Contravariant* components: $OA = V^1\,|\mathbf{e}_1| = V^1\sqrt{g_{11}}$, $OC = V^1 / |\mathbf{e}^1| = V^1/\sqrt{g^{11}}$ (recall Equation 2.9.5b$_2$),
- *Covariant* components: $OB = V_1 / |\mathbf{e}_1| = V_1/\sqrt{g_{11}}$, $OD = V_1\,|\mathbf{e}^1| = V^1\sqrt{g_{11}}$; similarly for V_2.

[If $|\mathbf{e}_1| = |\mathbf{e}_2| = 1 \Rightarrow \sqrt{g_{11}} = \sqrt{g_{22}} = 1$, $g_{12} = \cos\phi \Rightarrow V_1 = V^1 + V^2\cos\phi$, $V_2 = V^2 + V^1\cos\phi$.]

The *scalar, or dot, product* of two vectors, **U** and **V**, is defined by

$$\mathbf{U}\cdot\mathbf{V} = U^k V_k = U_k V^k = g_{kl}\, U^k V^l = g^{kl}\, U_k V_l = g^k{}_l\, U_k V^l = g^k{}_l\, U^l V_k; \quad (2.9.17a)$$

and for **U** = **V** it gives the (square of the) *magnitude*, or *norm* of **V** (recall Equations 2.2.4 ff):

$$|\mathbf{V}|^2 \equiv \mathbf{V}\cdot\mathbf{V} = \mathbf{V}^2 = V^2 = V_k V^k = g_{kl}\, V^k V^l = g^{kl}\, V_k V_l = g^k{}_l\, V_k V^l. \quad (2.9.17b)$$

The *vector, or cross, product* of two vectors, **U** and **V**, is defined by

$$\mathbf{U}\times\mathbf{V} = -(\mathbf{V}\times\mathbf{U}) = U^k V^l (\mathbf{e}_k \times \mathbf{e}_l) = U^k V^l (e_{klb}\, \mathbf{e}^b) \equiv W_b\, \mathbf{e}^b = \mathbf{W}, \quad (2.9.18a)$$

i.e.,

$$W_b = e_{klb}\, U^k V^l = \sqrt{g}\, \varepsilon_{klb}\, U^k V^l = \sqrt{g}\, \varepsilon_{klb}\, (U^k V^l - U^l V^k)\,/\,2, \quad (2.9.18b)$$

and similarly

$$W^b = e^{klb}\, U_k V_l = \left(\varepsilon^{klb}/\sqrt{g}\right) U_k V_l = \varepsilon^{klb}(U_k V_l - U_l V_k)\,/\,2\,\sqrt{g}\,; \quad (2.9.18c)$$

or in compact (symbolic) determinant form:

$$\mathbf{W} = \mathbf{U}\times\mathbf{V} = \left(1/\sqrt{g}\right)\begin{vmatrix} \mathbf{e}_1 & U_1 & V_1 \\ \mathbf{e}_2 & U_2 & V_2 \\ \mathbf{e}_3 & U_3 & V_3 \end{vmatrix} = \sqrt{g}\begin{vmatrix} \mathbf{e}^1 & U^1 & V^1 \\ \mathbf{e}^2 & U^2 & V^2 \\ \mathbf{e}^3 & U^3 & V^3 \end{vmatrix} \quad (2.9.18d)$$

to be expanded according to their first column:

$$W_1 = \sqrt{g}\,(U^2 V^3 - U^3 V^2), \quad W^1 = \left(1/\sqrt{g}\right)(U_2 V_3 - U_3 V_2), \quad (2.9.18e)$$

etc. cyclically.

2.9.4.1 Applications

i. With the identifications: $\mathbf{U} \rightarrow \nabla \times (\ldots) \equiv \mathrm{curl}(\ldots)$, we obtain from Equations 2.9.18a ff:

$$\mathbf{W} = \nabla \times \mathbf{V} \equiv \mathrm{curl}\mathbf{V} \quad [\text{or sometimes: } (\partial/\partial\mathbf{R}) \times \mathbf{V}], \quad (2.9.19a)$$

or, in components (with commas denoting $\partial/\partial q$-derivatives):

$$W^b = e^{bkl}(\partial/\partial q^k)V_l \equiv e^{bkl}V_{l,k} = (e^{bkl}/2)(V_{l,k} - V_{k,l}) \equiv e^{bkl}V_{[l,k]} \tag{2.9.19b}$$

$$= \left(\varepsilon^{bkl}/\sqrt{g}\right)V_{l,k} = \left(\varepsilon^{bkl}/2\sqrt{g}\right)(V_{l,k} - V_{k,l}) \equiv \left(\varepsilon^{bkl}/\sqrt{g}\right)V_{[l,k]} = (\text{curl}\mathbf{V})^b. \tag{2.9.19c}$$

ii. Similarly, the *gradient* of an absolute scalar (i.e., invariant) S is

$$\text{grad } S \equiv \partial S/\partial \mathbf{R} = (\partial S/\partial q^k)(\partial q^k/\partial \mathbf{R}) = (\partial S/\partial q^k)\mathbf{e}^k = g^{lk}(\partial S/\partial q^k)\mathbf{e}_l; \tag{2.9.20a}$$

and for $S \rightarrow q^k$ it gives:

$$\text{grad } q^k = \mathbf{e}^k \quad \text{(recall the earlier Equation 2.9.5c).} \tag{2.9.20b)*}$$

iii. For the *mixed*, or *triple, dot* product of **U**, **V**, **W**, where: $D \equiv \mathbf{U} \cdot (\mathbf{V} \times \mathbf{W}) \equiv (\mathbf{U}, \mathbf{V}, \mathbf{W})$, we find, successively,

$$D = e_{klb}U^k V^l W^b = \sqrt{g}\,\varepsilon_{klb}U^k V^l W^b = e^{klb}U_k V_l W_b = \varepsilon^{klb}U_k V_l W_b/\sqrt{g}, \tag{2.9.21}$$

(which can also be easily put in determinant form, like Equation 2.9.18d). For **U**, **V**, **W** $\rightarrow \mathbf{e}_{1,2,3}$, $D \rightarrow \sqrt{g}$, and for **U**, **V**, **W** $\rightarrow \mathbf{e}_{1,2,3}$, $D \rightarrow 1/\sqrt{g}$, and the above yield the earlier Equations 2.9.11a ff.

iv. *Axial vector of an (absolute) antisymmetric tensor.* In three dimensions, such a tensor has, at most, *three* independent components: in covariant form (and similarly for contravariant and/or mixed forms):

$$\mathbf{T} = (T_{kl} = -T_{lk}) = \begin{pmatrix} 0 & T_{12} & T_{13} \\ -T_{12} & 0 & T_{23} \\ -T_{13} & -T_{23} & 0 \end{pmatrix} \tag{2.9.22a}$$

Therefore, by the quotient rule, the three quantities

$$T^1 \equiv T_{23},\ T^2 \equiv -T_{13},\ T^3 \equiv T_{12};\ \text{or, compactly: } T^x = e^{xkl}T_{kl}/2 = -e^{kxl}T_{kl}/2, \tag{2.9.22b}$$

constitute a contravariant vector called the *axial vector of the antisymmetric tensor.* Conversely, *to any ordinary vector there corresponds, uniquely, an antisymmetric tensor,* given by the solution of Equation 2.9.22b for T_{kl}:

$$T_{kl} = e_{klx}T^x = -e_{kxl}T^x. \tag{2.9.22c)**}$$

* For the *divergence,* see Section 3.6.

** To go from Equation 2.9.22c to 2.9.22b, we dot it with e^{bkl} and then apply ε/e-identities (Section 1.2):

$$e^{bkl}T_{kl} = e^{bkl}(e_{klx}T^x) = 2\delta^b{}_x T^x = 2T^b. \tag{2.9.22d}$$

Similarly, in the *contravariant* case, i.e., $T_{kl} \rightarrow T^{kl}$, the equations corresponding to Equations 2.9.22b and c are

$$T_x = e_{xkl}\, T^{kl}/2 = -e_{kxl} T^{kl}/2 \quad \text{and} \quad T^{kl} = e^{klx}\, T_x = -e^{kxl}\, T_x. \qquad (2.9.22e)$$

Mixed forms, i.e., $T^k{}_l$, $T_l{}^k$, can be reduced to the above with the help of the metric tensor; e.g.,

$$T^k{}_l = g^{kb}\, T_{bl} = g^{kb}\, e_{blx}\, T^x, \quad T_l{}^k = g^{kb}\, T_{lb} = g^{kb}\, e_{lbx}\, T^x = -g^{kb}\, e_{lbx}\, T^x = -T^k{}_l. \quad (2.9.22f)*$$

In view of the above, we can say that the earlier Equation 2.9.18 show that the cross product $\mathbf{U} \times \mathbf{V}$ equals either (1) the *absolute antisymmetric tensor* (or *bivector*) $W_{kl} \equiv U_k V_l - U_l V_k$ (or $W^{kl} = \ldots$) or (2) the axial vector W^k (or W_k).

[In an n-dimensional space, W_{kl} is called the *exterior product* of the vectors **U** and **V**, and has $n(n-1)/2$ distinct, potentially nonzero, components (recalling Equations 2.7.4a to 2.7.5d). They may be viewed as the *areas resulting by parallel projections, on the coordinate planes, of the two-dimensional parallelepiped with sides the given vectors, both emanating from a common point* (recall Equations 2.9.9a and b; also, see Stokes' theorem, Example 3.6.1). This explains geometrically why *a necessary and sufficient condition for* ***U*** *and* ***V*** *to be independent, is that all* W_{kl} *vanish.]*

Problem 2.9.1

i. Let the contravariant vectors $\mathbf{V}^k$ have weights w_k ($k = 1, 2, 3$). Find the weight of $\mathbf{V}_1 \cdot (\mathbf{V}_2 \times \mathbf{V}_3)$.
ii. Let the covariant vectors $\mathbf{V}_k$ have weights w_k ($k = 1, 2, 3$). Find the weight of $\mathbf{V}^1 \cdot (\mathbf{V}^2 \times \mathbf{V}^3)$.

---

* These results can be put into a more memorable, *direct*, form: taking rectangular Cartesian coordinates ($e \rightarrow \varepsilon$), for convenience, and dotting Equation 2.9.22c or 2.9.22e$_2$ with the arbitrary vector **v** yields:

$$T_{kl}\, v_l = \varepsilon_{klx}\, T_x\, v_l = -\varepsilon_{kxl}\, T_x\, v_l; \qquad (2.9.22g)$$

or, in direct notation, with $\mathbf{T} = (T_{kl})$, $\mathbf{t} = (T_k)$, and $\mathbf{v} = (v_k)$ (see below):

$$\mathbf{T}\,\mathbf{v} = -(\mathbf{t} \times \mathbf{v}) = \mathbf{v} \times \mathbf{t}. \qquad (2.9.22h)$$

Remark: To avoid the "asymmetric" minus sign in the direct notation (Equation 2.9.22h), *some authors define the axial vector as the negative of ours.* Then the above are changed to

$$T_x = -e_{xkl}\, T^{kl}/2 = e_{kxl}\, T^{kl}/2 \Rightarrow T^{kl} = e^{kxl}\, T_x = -e^{klx}\, T_x \quad \text{(and similarly for } T^x\text{)}, \qquad (2.9.22i)$$

and

$$\mathbf{T}\,\mathbf{v} = \mathbf{t} \times \mathbf{v}\ [= -(\mathbf{v} \times \mathbf{t})]. \qquad (2.9.22j)$$

But now the indicial notation (Equation 2.9.22i) looks a bit asymmetric! The reader should be aware of such sign differences when comparing references. For a discussion of the direct notation and its applications to rigid body dynamics, see, e.g., Lagally (1964) and Simmonds (1994).

2.10 PHYSICAL COMPONENTS OF VECTORS AND TENSORS

For a general vector **V** we have the following equivalent representations (recalling Equation 2.9.8c):

$$\mathbf{V} = V^k \mathbf{e}_k = \sum V^k \left(\sqrt{g_{kk}}\, \mathbf{e}_{<k>}\right) \equiv V^{<k>} \mathbf{e}_{<k>} \quad \text{(sum over } k) \tag{2.10.1a}$$

$$= V_k \mathbf{e}^k = \sum V_k \left(\sqrt{g^{kk}}\, \mathbf{e}^{<k>}\right) \equiv V_{<k>} \mathbf{e}^{<k>} \quad \text{(sum over } k) \tag{2.10.1b}$$

where

$$\mathbf{e}_{<k>} \equiv \mathbf{e}_k / |\mathbf{e}_k| = \mathbf{e}_k / \sqrt{g_{kk}}, \quad \mathbf{e}^{<k>} \equiv \mathbf{e}^k / |\mathbf{e}^k| = \mathbf{e}^k / \sqrt{g^{kk}} \quad \text{(no sum over } k) \tag{2.10.1c}^*$$

unit vectors. The numbers (no sum over k):

$$V^{<k>} \equiv V^k \sqrt{g_{kk}} = \sqrt{g_{kk}}\,(g^{kl} V_l) \quad \left(\neq \mathbf{V} \cdot \mathbf{e}^{<k>} = V^k / \sqrt{g^{kk}}, \text{ in general}\right), \tag{2.10.2a}$$

$$V_{<k>} \equiv V_k \sqrt{g^{kk}} = \sqrt{g^{kk}}\,(g_{kl} V^l) \quad \left(\neq \mathbf{V} \cdot \mathbf{e}_{<k>} = V_k / \sqrt{g_{kk}}, \text{ in general}\right), \tag{2.10.2b}$$

are, respectively, the (nontensorial!) contravariant and covariant *physical components* of **V** along the unit basis vectors $\mathbf{e}_{<k>}$, $\mathbf{e}^{<k>}$, respectively, and, clearly, have the *same sign* as the corresponding vector components.

Proceeding in a similar fashion, we define the physical components of a second-rank tensor as follows (again, no sum over k, l):

$$T^{<kl>} \equiv T^{kl} \sqrt{g_{kk}} \sqrt{g_{ll}} = \ldots, \tag{2.10.3a}$$

$$T_{<kl>} \equiv T_{kl} \sqrt{g^{kk}} \sqrt{g^{ll}} = \ldots, \tag{2.10.3b}$$

$$T^{<k}{}_{l>} \equiv T^k{}_l \sqrt{g_{kk}} \sqrt{g^{ll}} = \ldots, \tag{2.10.3c}$$

$$T_{<l}{}^{k>} \equiv T_l{}^k \sqrt{g_{kk}} \sqrt{g^{ll}} = \ldots, \tag{2.10.3d}$$

and analogously for higher-rank tensors.**

* Recall Figure 2.7, with $\mathbf{u}_k \to \mathbf{e}_{<k>}$.

Remarks

i. In *orthogonal* curvilinear coordinates: $g_{kk} = 1/g^{kk}$ (no sum on k) $\Rightarrow \mathbf{e}_{<k>} = \mathbf{e}^{<k>}$, and so Equations 2.10.2a and b yield: $V^{<k>} = V_{<k>} = V^k \sqrt{g_{kk}} = V_k \sqrt{g^{kk}} = V_k / \sqrt{g_{kk}} \Rightarrow V_k = g_{kk} V^k = V^k/g^{kk}$ ($\neq V^k$, in general; no sum on k); i.e., there the physical components of a vector are its Cartesian components in a local orthonormal basis with axes tangent to the coordinate lines.
ii. In applied problems, the passage to physical components should always be done at the final stage of the calculations. We shall not use such components systematically here.

2.11 ON DIRECT, OR DYADIC (POLYADIC, ETC.), OR INVARIANT, REPRESENTATIONS OF TENSORS

So far, only vectors have been expressed directly, that is, in terms of basis vectors and components: $\mathbf{V} = V^k \mathbf{e}_k = V_k \mathbf{e}^k = V^{k'} \mathbf{e}_{k'} = V_{k'} \mathbf{e}^{k'}$; while general tensors have been represented only in terms of their components, the reason for this being the ease of visualization of the former as compared with the difficulty (near impossibility) of envisioning the latter. However, since in some theoretical and conceptual discussions the direct representation of tensors might prove useful, we sketch it briefly below.

2.11.1 General Tensors

A tensor $\mathbf{T}$ may be represented in a general CS q and a local frame $\{P, \mathbf{e}_k\}$ there as follows:

$$\mathbf{T} = T^{a\ldots e}{}_{g\ldots l}\, \mathbf{e}_a \ldots \mathbf{e}_e\, \mathbf{e}^g \ldots \mathbf{e}^l, \quad \text{etc.} \tag{2.11.1}$$*

The $2n$ basis vectors $\{\mathbf{e}_k, \mathbf{e}^k\}$ can be viewed as defining the CS q, locally, and thus bring out the invariant character of $\mathbf{T}(P)$ (more in next section). Conversely, if the quantities $T^{\cdots}{}_{\ldots}(q; P)$ and $T^{\cdots}{}_{\ldots}(q'; P)$ are related by the transformation equation (2.5.6), we can view them as components of a general tensor $\mathbf{T}$ along the bases $\{P, \mathbf{e}_k, \mathbf{e}^k\}$ and $\{P, \mathbf{e}_{k'}, \mathbf{e}^{k'}\}$, respectively.
For a second-rank tensor, Equation 2.11.1 gives

$$\mathbf{T} = T^{kl}\, \mathbf{e}_k\, \mathbf{e}_l \quad \text{(contravariant representation)} \tag{2.11.2a}$$

$$= T_{kl}\, \mathbf{e}^k\, \mathbf{e}^l \quad \text{(covariant representation)} \tag{2.11.2b}$$

$$= T^k{}_l\, \mathbf{e}_k\, \mathbf{e}^l \quad \text{(mixed representation: } k\text{: first, } l\text{: second)} \tag{2.11.2c}$$

$$= T_l{}^k\, \mathbf{e}^l\, \mathbf{e}_k \quad \text{(mixed representation: } k\text{: second, } l\text{: first);} \tag{2.11.2d}$$

** For additional, related, definitions of physical components, see (alphabetically) Malvern (1969, pp. 606–613), Simmonds (1994, Ch. 3), Synge and Schild (1949, p. 144), Truesdell (1953, 1954).

* Some authors, to stress the *order /noncommutativity* of the indices in Equation 2.11.1 use between the e terms symbols like ⊗: *tensor product*; i.e., they write Equation 2.11.1 as

$$\mathbf{T} = T^{a\ldots e}{}_{g\ldots l}\, \mathbf{e}_a \otimes \ldots \otimes \mathbf{e}_e \otimes \mathbf{e}^g \otimes \ldots \otimes \mathbf{e}^l. \tag{2.11.1a}$$

and if $\mathbf{T} = T^{kl}\,\mathbf{e}_k\,\mathbf{e}_l$ is antisymmetric, i.e., $T^{kl} = -T^{lk}$, $T^{kk} = 0$ (no sum on k), then, successively,

$$\mathbf{T} = T^{kl}\,\mathbf{e}_k\,\mathbf{e}_l + T^{ij}\,\mathbf{e}_i\,\mathbf{e}_j \quad (k < l,\ i \geq j)$$

$$= T^{kl}\,(\mathbf{e}_k\,\mathbf{e}_l - \mathbf{e}_l\,\mathbf{e}_k) \quad (k < l) \tag{2.11.3}$$

[recalling (2.7.2b, c)] while, if $\mathbf{T} \to \mathbf{e}_k$ or $\mathbf{e}^k$, then Equation 2.11.1 yields, respectively,

$\mathbf{e}_k = \delta_k{}^l\,\mathbf{e}_l = \delta_{kl}\,\mathbf{e}^l$, i.e., $\mathbf{e}_k$ has *contravariant* components $(\delta_k{}^1, \ldots, \delta_k{}^n)$

and *covariant* components $(\delta_{k1}, \ldots, \delta_{kn})$, (2.11.3b)

$\mathbf{e}^k = \delta^k{}_l\,\mathbf{e}^l = \delta^{kl}\,\mathbf{e}_l$, i.e., $\mathbf{e}^k$ has *contravariant* components $(\delta^{k1}, \ldots, \delta^{kn})$

and *covariant* components $(\delta^k{}_1, \ldots, \delta^k{}_n)$. (2.11.3c)

2.11.2 Dyadics

These are tensors of rank two,* i.e., as in Equations 2.11.2a through d. Every dyadic is expressible as the sum of the following n dyads:

$$\mathbf{D} = \mathbf{u}_k\,\mathbf{v}^k = (u_k{}^b\,\mathbf{e}_b)(v^k{}_h\,\mathbf{e}^h) = (u_k{}^b\,v^k{}_h)\mathbf{e}_b\,\mathbf{e}^h \equiv D^b{}_h\,\mathbf{e}_b\,\mathbf{e}^h$$

$$= \mathbf{e}_b(D^b{}_h\,\mathbf{e}^h) \equiv \mathbf{e}_b\,\mathbf{D}^b = \mathbf{e}^h(D^b{}_h\,\mathbf{e}_b) \equiv \mathbf{e}^h\,\mathbf{D}_h \tag{2.11.4}$$

(sum of outer, or tensor, product of two arbitrary but linearly independent vectors: an *antecedent* $\mathbf{u}_k$ and a *consequent* $\mathbf{v}^k$, etc.).

Dyadics combine with vectors and/or each other in various ways. Here are some of their most physically useful such operations ($\mathbf{D}$, $\mathbf{B}$, $\mathbf{C}$: dyads, $\mathbf{u}$, $\mathbf{v}$, $\mathbf{w}$: vectors, and $\cdot$ for *dot* product):

- The *scalar* of a dyadic $\mathbf{D} = D^{kl}\,\mathbf{e}_k\,\mathbf{e}_l$ is

$$\text{Trace of } \mathbf{D} \equiv \text{Tr}\,\mathbf{D} = \mathbf{v}_k \cdot \mathbf{v}^k = D^k{}_k = D_k{}^k = g_{kl}\,D^{kl} = g^{kl}\,D_{kl}. \tag{2.11.5a}$$

- The *determinant* of $\mathbf{D}$ is

$$\text{Determinant of } \mathbf{D} \equiv \text{Det}\,\mathbf{D} = \text{Det}(D^k{}_l) = \text{Det}(D_l{}^k). \tag{2.11.5b}$$

We notice that Det $\mathbf{D} \neq \text{Det}(D_{kl}) \neq \text{Det}(D^{kl}) \neq$ Det $\mathbf{D}$, unless $\text{Det}(g_{kl}) = 1$.

- The *inner product* of the two dyadics $\mathbf{B}$ and $\mathbf{C}$, $\mathbf{B} \cdot \mathbf{C}$, is

* ΔΥΟ (*dío*) is Greek for two. This terminology is due to J. W. Gibbs, one of the founders of vector analysis (late 19th century).

$$\mathbf{D} = \mathbf{B} \cdot \mathbf{C} = (B^k{}_l \, \mathbf{e}_k \, \mathbf{e}^l) \cdot (C^h{}_j \, \mathbf{e}_h \, \mathbf{e}^j) = (B^k{}_l \, C^h{}_j) \, [\mathbf{e}_k \, (\mathbf{e}^l \cdot \mathbf{e}_h) \, \mathbf{e}^j]$$

$$= (B^k{}_l \, C^h{}_j)(\delta^l{}_h \, \mathbf{e}_k \, \mathbf{e}^j) = (B^k{}_l \, C^l{}_j)(\mathbf{e}_k \, \mathbf{e}^j) \equiv D^k{}_j \, \mathbf{e}_k \, \mathbf{e}^j, \text{ i.e., } D^k{}_j = B^k{}_l \, C^l{}_j; \quad (2.11.5c)$$

also

$$D_{kj} = B_{kl} \, C^l{}_j = B_k{}^l \, C_{lj}, \quad \text{etc.} \quad (2.11.5d)$$

- The *outer*, or *tensor*, product of the two vectors **u** and **v**, **u v**, is (in contravariant components):

$$(\mathbf{u} \, \mathbf{v})^{kl} = u^k \, v^l \quad [\text{sometimes also denoted as } (\mathbf{u} \otimes \mathbf{v})^{kl}]. \quad (2.11.5e)$$

- The linear and homogeneous transformations $\mathbf{D} \cdot \mathbf{v}$ and $\mathbf{v} \cdot \mathbf{D}$ are, respectively,

$$\mathbf{u} = \mathbf{D} \cdot \mathbf{v} = (D^k{}_l \, \mathbf{e}_k \, \mathbf{e}^l) \cdot (v^b \, \mathbf{e}_b) = (D^k{}_l \, v^b)[\mathbf{e}_k \, (\mathbf{e}^l \cdot \mathbf{e}_b)]$$

$$= (D^k{}_l \, v^b)(\delta^l{}_b \, \mathbf{e}_k) = (D^k{}_l \, v^l) \, \mathbf{e}_k \equiv u^k \, \mathbf{e}_k, \quad \text{i.e., } u^k = D^k{}_l \, v^l; \quad (2.11.5f)$$

also

$$u^k = D^{kl} \, v_l \Rightarrow u_k = D_k{}^l \, v_l = D_{kl} \, v^l; \quad (2.11.5g)$$

$$\mathbf{u} = \mathbf{v} \cdot \mathbf{D} = (v_b \, \mathbf{e}^b) \cdot (D^k{}_l \, \mathbf{e}_k \, \mathbf{e}^l) = (v_b \, D^k{}_l)[(\mathbf{e}^b \cdot \mathbf{e}_k)\mathbf{e}^l]$$

$$= (v_b \, D^k{}_l)(\delta^b{}_k \, \mathbf{e}^l) = (D^k{}_l \, v_k)\mathbf{e}^l \equiv u_l \, \mathbf{e}^l, \quad \text{i.e., } u_l = D^k{}_l \, v_k, \quad \text{etc.} \quad (2.11.5h)$$

Similarly,

$$\mathbf{u} \cdot (\mathbf{D} \cdot \mathbf{v}) = u^k(D_{kl} \, v^l) = u^k(D_k{}^l \, v_l) = u_k(D^{kl} \, v_l) = u_k(D^k{}_l \, v^l); \quad (2.11.5i)$$

and $\mathbf{w} = \mathbf{u} \times (\mathbf{D} \cdot \mathbf{v})$ reads, in rectangular Cartesian coordinates (for convenience),

$$w_k = \varepsilon_{klb} \, u_l \, D_{bh} \, v_h = (\varepsilon_{klb} \, u_l \, D_{bh}) v_h, \quad \text{i.e., } \mathbf{u} \times (\mathbf{D} \cdot \mathbf{v}) = (\mathbf{u} \times \mathbf{D}) \cdot \mathbf{v}; \quad (2.11.5j)$$

while $\mathbf{w} = \mathbf{D} \cdot (\mathbf{u} \times \mathbf{v})$ becomes

$$w_k = D_{kl} \, (\varepsilon_{lbh} \, u_b \, v_h) = (\varepsilon_{lbh} \, D_{kl} \, u_b) v_h, \quad \text{i.e., } \mathbf{D} \cdot (\mathbf{u} \times \mathbf{v}) = (\mathbf{D} \times \mathbf{u}) \cdot \mathbf{v}. \quad (2.11.5k)^*$$

---

* The above show that, at each manifold point, a dyadic **D** associates to each vector **v** another vector **u**, generally *noncollinear* to **v**. The standard mechanics examples of such dyadics are those of *inertia, stress,* and *strain*.
For further details on dyadics and their applications to mechanics (rigid bodies, continua, etc.), see e.g., Gontier (1969), Lagally (1964), Lotze (1950), Simmonds (1994); also Pipes (1963).

Problem 2.11.1

Let $\mathbf{D} = \mathbf{u}_k \mathbf{v}^k$ (Equation 2.11.4) and $\mathbf{d} \equiv (\mathbf{u}_k \times \mathbf{v}^k)/2$ (*axial* vector of $\mathbf{D}$). Show that: (i) If $\mathbf{D}$ is *symmetric,* then $\mathbf{d} = 0$ and (ii) if $\mathbf{D}$ is *antisymmetric,* then for every vector $\mathbf{x}$:

$$\mathbf{x} \cdot \mathbf{D} = -\mathbf{D} \cdot \mathbf{x} = \mathbf{d} \times \mathbf{x}. \tag{a}$$

i.e., *vector multiplication is equivalent to an inner multiplication with an antisymmetric dyadic.*

2.12 INTRODUCTION TO RIEMANNIAN SPACES

As an introduction to the next chapter on tensor analysis, let us examine briefly the fundamental concepts of Riemannian spaces, tangent spaces, and affinely connected spaces.

2.12.1 Riemannian Space, R_n

This is an n-dimensional differentiable manifold X_n, at each point $P \equiv P(q)$ of which, in a region of interest described by the curvilinear coordinates $q \equiv (q^1, \ldots, q^n)$, there is associated a (real) *fundamental* (nonsingular, at least) *differential quadratic form*:

$$(ds)^2 \equiv ds^2 = ds^2(q) = g_{kl}\, dq^k\, dq^l, \tag{2.12.1}$$

where $g_{kl} = g_{kl}(q) = g_{lk}(q)$: *metric coefficients* (given real, single-valued, continuous, and twice differentiable (C^2 class) functions of the qs; and, since ds^2 is quadratic, there is no real loss in generality in assuming them *symmetric* and

$$g_{(n)} \equiv g \equiv \mathrm{Det}(g_{kl}) \neq 0 \quad (k, l = 1, 2, \ldots, n) \tag{2.12.1a}$$

[$\Rightarrow$ it cannot be reduced, by a (real, nonsingular) linear transformation $dq^k \leftrightarrow dq^{k'}$, to a form with fewer than n variables], such that *the line element ds represents the distance (arc length) between the adjacent R_n-points $P(q)$ and $P'(q + dq)$,* and is assumed invariant under (admissible) coordinate changes.

In this book, as in most nonrelativistic physics, we also stipulate that $g_{kl}\, dq^k\, dq^l$ be *positive-definite*; i.e., positive for all nonzero values of the dqs.* Then R_n is called *properly* (or purely) *Riemannian*, and ds is real. If ds^2 is *sign-indefinite*, then R_n is called *pseudo*-, or semi-, Riemannian.

At each R_n-point, the $g_{kl}(q)$ terms define a standard, or *unit*, of length; i.e., they characterize all metric properties there: with their help we can calculate lengths, angles, etc. (see below), and that is why we refer to R_n as a special differentiable and metric manifold.

* As is well known, the necessary and sufficient conditions for this are $g_{(k)} > 0$, $k = 1, \ldots, n$.

2.12.2 Tangent (Point and Vector) Spaces

These occur at a point P of an X_n (later to specialize to an R_n). These constitute the generalization of the tangent plane to a point of an ordinary two-dimensional surface. But here we must define them by *intrinsic*, or *internal*, means; that is, in terms of the available structure of X_n, without using the properties of the surrounding space. To this end, we, first, define the *tangent vector* to an X_n-lying curve through P, $C_n(P)$, by such means (i.e., definitions of the type "tangent as the limiting position of a secant PP', as P' tends to P" will not do — *PP' does not lie in R_n*]. If we represent that curve, in the neighborhood of P, by

$$\{q^k(u) = q^k(P) + (dq^k/du)_P\, u;\ k = 1, \ldots, n\} \tag{2.12.2}$$

(u: curve parameter, e.g., $u = s$: arc length)

(assuming that, at each P, not all dq^k/du vanish, i.e., assuming regularly parametrized smooth curves) then, its tangent vector there, $\mathbf{t}(P)$, is defined by the ordered n-ple: $\{(dq^k/du)_P\} \equiv (t^1, \ldots, t^n)$. Each such curve specifies a tangent vector; and conversely, each n-ple $\{(dq^k/du)_P\}$ constitutes the tangent vector to the X_n-curve (Equations 2.12.2). Now, (i) the set of all these n-ples makes up the n-dimensional *tangent point space,* or *plane,* to X_n at P, $T_P(X_n)$ or $T_n(P)$, while (ii) the corresponding set of vectors $\{\mathbf{t}(P)\}$, tangent to all $C_n(P)$, makes up the n-dimensional (and hitherto affine) *tangent vector space* to X_n at P, $\boldsymbol{T_P(X_n)}$ or $\boldsymbol{T_n(P)}$.*

Next, the special curve $C_n(P)$ obtained by *keeping all qs constant, except $q*$,* is called a $q*$-parametric (or -coordinate) line, through that point, $C_n(P, *)$; i.e., analytically (using standard calculus notations, and with k, $* = 1, \ldots, n$, as usual):

$$C_n(P, *):\ \{q^k(u) = q^k(P) + \delta^k{}_* u\};$$

or

$$q^*(u) = q^*(P) + u, \quad q^k(u) = q^k(P), \quad k \neq *; \tag{2.12.2a}$$

and, accordingly, its corresponding tangent vector $\mathbf{t}_* \to \mathbf{e}_*$ has the following component representation: $\mathbf{e}_* \equiv (e^k{}_*) = (\delta^k{}_*) \equiv (\delta^1{}_*, \delta^2{}_*, \ldots, \delta^n{}_*)$ (recall Equations 2.11.3b and c); or, *in extenso*,

$$\mathbf{e}_1 = (1, 0, \ldots, 0), \quad \mathbf{e}_2 = (0, 1, \ldots, 0), \quad \ldots, \quad \mathbf{e}_n = (0, 0, \ldots, 1). \tag{2.12.2b}$$

Since these vectors are linearly independent, they form a (covariant) "natural" frame, or basis, for $\mathbf{T_n(P)}$ (that is, one that is naturally associated to the CS q there):

* As becomes clear in what follows, the manifold itself does not carry, or hold, any vectors (tensors, etc.); it is the tangent vector space, at each of its points P, that does that.

$\{P, \mathbf{e}_k \equiv \partial\mathbf{P}/\partial q^k\}$; and so we can write: $\mathbf{t} = t^k \mathbf{e}_k$. Equivalently, we may define the $\mathbf{e}_k$s by the invariant expression:

$$d\mathbf{P} \equiv \mathbf{e}_k \, dq^k \equiv \text{Displacement vector } \mathbf{PP}', \tag{2.12.2c}$$

determined by $P(q)$ and $P'(q + dq)$, and resolved in the natural frame at P.

Under a CT: $q \to q'$, about P, the components dq^k/du transform as Equation 2.4.2 ff:

$$(dq^k/du)_P \to (dq^{k'}/du)_P = (\partial q^{k'}/\partial q^k)_P \, (dq^k/du)_P \equiv A^{k'}{}_k(P)(dq^k/du)_P$$

$$[\Rightarrow (dq^k/du)_P = (\partial q^k/\partial q^{k'})_P \, (dq^{k'}/du)_P \equiv A^k{}_{k'}(P)(dq^{k'}/du)_P]; \tag{2.12.2d}$$

while the associated new natural frame $\{P, \mathbf{e}_{k'}\}$, induced by this CT, is found from Equation 2.12.2c, with Equation 2.12.2d:

$$d\mathbf{P} = \mathbf{e}_k \, dq^k = \mathbf{e}_{k'} \, dq^{k'} \Rightarrow \mathbf{e}_{k'} = A^k{}_{k'}(P) \, \mathbf{e}_k \quad [\Leftrightarrow \mathbf{e}_k = A^{k'}{}_k(P) \, \mathbf{e}_{k'}]. \tag{2.12.2e}$$*

Finally, we define the *dual* or *reciprocal* (contravariant) frame $\{P, \mathbf{e}^k\}$ to the (covariant) frame $\{P, \mathbf{e}_k\}$, both in $\mathbf{T_n(P)}$, by $\mathbf{e}^k \cdot \mathbf{e}_l = \delta^k{}_l \, (k, l = 1, \ldots, n)$.

With the help of the above, we can write

$$\mathbf{t} = t^k \, \mathbf{e}_k = t^{k'} \, \mathbf{e}_{k'} \Rightarrow t^{k'} = A^{k'}{}_k \, t^k, \; t^k = A^k{}_{k'} \, t^{k'}; \tag{2.12.2f}$$

and similarly for any tensor expressed, à la Equations 2.11.1 ff: $\mathbf{T}(P) = T^{a \ldots e}{}_{g \ldots l}(P) \, \mathbf{e}_a \ldots \mathbf{e}_e \, \mathbf{e}^g \ldots \mathbf{e}^l$. Next, to be able to define the lengths of vectors, etc. in these tangent spaces, we must equip them with a metric, either the earlier Riemannian $g_{kl}(P)$ or the metric of the surrounding space. Let us choose the former: this transforms $T_n(P)$ $[\mathbf{T_n(P)}]$ into the *tangent Euclidean point [vector] space to R_n at P.* Let us examine briefly the resulting metric properties. If, for a particular direction, we choose $u = s \Rightarrow du = ds$, then $\{dq^k/du\} \to \{dq^k/ds\}$: *direction parameters* of that direction, say from P to P' (analogous to the Euclidean direction cosines); and if PP' is specified, so are the corresponding direction parameters, and vice versa. Now, (i) a given set $\{(dq^k/ds)_P\}$ and a ds, as calculated from the fundamental form (2.12.1), i.e., $ds = (g_{kl} \, dq^k \, dq^l)^{1/2}$, define the $\mathbf{T_n(P)}$-vector of elementary displacement $d\mathbf{P}$, of magnitude $ds \equiv |d\mathbf{P}|$, while, those direction parameters along with the positive scalar V define the $\mathbf{T_n(P)}$-vector $\mathbf{V}$, of magnitude V, and (ii) if $d_1\mathbf{P}$, $d_2\mathbf{P}$ are two elementary displacements along d_1s, d_2s, respectively, emanating from P, then the angle θ between them is defined by

---

* Occasionally, admissible coordinate differential transformations, at P, lead to "nonnatural," or *nonholonomic*, i.e., nonglobal, "coordinates" (recall Equations 2.4.5 and 2.4.6a and b). More on this in Section 3.12 ff.

$$\cos(d_1\mathbf{P},\ d_2\mathbf{P}) \equiv \cos\theta = g_{kl}\,(d_1q^k/d_1s)(d_2q^l/d_2s) \qquad (2.12.3)^*$$

$$= g_{kl}\left[d_1q^k/(g_{ij}\,d_1q^i\,d_1q^j)^{1/2}\right]\left[d_2q^l/(g_{ab}\,d_2q^a\,d_2q^b)^{1/2}\right]$$

(= ±1, according as $d_1\mathbf{P}$, $d_2\mathbf{P}$ are in the same or in opposite directions);

which also establishes the *angular* metric of R_n.

Then, the *dot,* or *scalar,* product of $d_1\mathbf{P}$ and $d_2\mathbf{P}$ is defined by

$$d_1\mathbf{P}\cdot d_2\mathbf{P} = |d_1\mathbf{P}|\ |d_2\mathbf{P}|\ \cos\theta \equiv d_1s\ d_2s\ \cos\theta$$

$$= g_{kl}\,d_1q^k\,d_2q^l \quad \text{(invoking Equation 2.12.3)} \qquad (2.12.3a)$$

and that of the two $\mathbf{T_n(P)}$-vectors $\mathbf{V}$ (along $d_1\mathbf{P}$) and $\mathbf{U}$ (along $d_2\mathbf{P}$) by

$$\mathbf{V}\cdot\mathbf{U} = U\ V\cos(\mathbf{V},\ \mathbf{U}) = U\ V\ g_{kl}\,(d_1q^k/d_1s)(d_2q^l/d_2s)$$

$$= g_{kl}\,[U\,(d_1q^k/d_1s)]\ [V(d_2q^l/d_2s)] \quad (= \mathbf{U}\cdot\mathbf{V}); \qquad (2.12.3b)$$

and if $\mathbf{V}\cdot\mathbf{U} = 0$, while $U, V \neq 0$, then these vectors are called *mutually perpendicular.* For $\mathbf{V} = \mathbf{U}$, the above yields the (square of the) *length,* or *norm,* of $\mathbf{V}$: $\mathbf{V}\cdot\mathbf{V} \equiv \mathbf{V}^2 \equiv V^2$. Applying the above to $d\mathbf{P} = dq^k\,\mathbf{e}_k$, while assuming distributivity, comparing with Equation 2.12.1, and since the dqs are arbitrary, we get

$$ds^2 = d\mathbf{P}\cdot d\mathbf{P} = (dq^k\,\mathbf{e}_k)\cdot(dq^l\,\mathbf{e}_l) = (\mathbf{e}_k\cdot\mathbf{e}_l)\ dq^k\,dq^l \qquad (2.12.3c)$$

$$\Rightarrow g_{kl} = \mathbf{e}_k\cdot\mathbf{e}_l = e_k\,e_l\cos(\mathbf{e}_k,\ \mathbf{e}_l) \Rightarrow |\mathbf{e}_k| \equiv e_k = \sqrt{g_{kk}} \quad \text{(no sum on } k\text{)},$$

while applying it to $\mathbf{V} = V^k\,\mathbf{e}_k$ and $\mathbf{U} = U^k\mathbf{e}_k$, where $V^k \equiv \mathbf{V}\cdot\mathbf{e}^k$, etc., we find

$$\mathbf{V}\cdot\mathbf{U} = g_{kl}\,V^k\,U^l = g^{kl}\,V_k\,U_l = V_k\,U^k = V^k\,U_k, \qquad (2.12.3d)$$

$$\mathbf{V}\cdot\mathbf{V} \equiv V^2 = g_{kl}\,V^k\,V^l = g^{kl}\,V_k\,V_l = V_k\,V^k, \quad \text{where } V_k \equiv g_{kl}\,V^l, \quad g_{kl}\,g^{lb} = \delta^b{}_k.$$

The above make also clear that, in an R_n, the $\{\mathbf{e}_k\}$ are not necessarily defined by some gradient-type equations, like the Euclidean (2.9.2 ff): $\mathbf{e}_k \equiv \partial\mathbf{r}/\partial q^k$; in general, no such relations exist; but, as explained earlier, *their lengths and angles are determined from the given* $\{g_{kl}\}$. Accordingly, *neither* $d\mathbf{P} = \mathbf{e}_k\,dq^k$ *nor* $d\mathbf{e}_k = (\ldots)_{kl}\,dq^l$ *need be exact differentials;* i.e., there, in general, $\partial\mathbf{e}_k/\partial q^l \neq \partial\mathbf{e}_l/\partial q^k$, $\partial^2\mathbf{e}_k/\partial q^b\partial q^h \neq \partial^2\mathbf{e}_k/\partial q^h\partial q^b$; and although we may, as an aid to visualization, think of $\mathbf{r}$ as the position vector of P relative to some origin O, in or out of R_n, *that vector will not lie wholly in* R_n — we cannot write $\mathbf{r} = r^k\,\mathbf{e}_k$; i.e., *the* R_n*-points have no position vector lying in the space they define;* (see Affinely Connected Manifolds, below; and Section

* For a proof that, $|\cos\theta| \leq 1 \Rightarrow \theta$: real; see, e.g., Sokolnikoff (1964, pp. 203–204).

5.6). Also, we recall (Section 2.9) that even in a Euclidean space, *both* $d\mathbf{P}$ *and the* $\mathbf{e}_k$ *are independent of the origin.*

2.12.3 Flatness and Curvature

A differentiable metric space, say an R_n, is *flat* if and only if it admits a *global* system of rectangular Cartesian coordinates $y = (y^1, \ldots, y^n)$, such that the (square of the) arc length, $(\Delta s)^2 \equiv \Delta s^2$, between *any* two of its points, $P(y)$ and $Q(y + \Delta y)$, i.e., *no matter how far apart they may be,* is given by the generalized (global) Pythagorean theorem:

$$\Delta s^2 = \varepsilon_1(\Delta y^1)^2 + \varepsilon_2(\Delta y^2)^2 + \ldots + \varepsilon_n(\Delta y^n)^2, \tag{2.12.4a}$$

where the ε_k $(k = 1, \ldots, n)$ are all constant and equal to +1 or –1.

Definition: A flat Riemannian space is called Euclidean, E_n, if all its ε_k equal +1 (i.e., there is an infinity of possible Riemannian spaces, but only one Euclidean space.)*

On the other hand, every generally nonflat, or curved, R_n admits, at any of its points P, local rectangular Cartesian coordinates y such that the (square) of arc length ds^2, between its neighboring points $P(y)$ and $P'(y + dy)$, is given by the differential version of Equation 2.12.4a:

$$ds^2 = \varepsilon_1(dy^1)^2 + \varepsilon_2(dy^2)^2 + \ldots + \varepsilon_n(dy^n)^2. \tag{2.12.4b}$$

In sum, *every Riemannian space is locally flat; and if* ds^2 *is positive definite, it is locally (properly, or purely) Euclidean.*

---

* i. We are reminded from the theory of quadratic forms that

- If $\text{rank}(ds^2) = n$ $[\Rightarrow \text{Det}(g_{kl}) \neq 0]$, no zeros occur among the ε_ks; and, further, a real transformation $dy \leftrightarrow dY$ exists which brings ds^2 to the *normal* form:

$$ds^2 = (dY^1)^2 + (dY^2)^2 + \ldots + (dY^h)^2 - (dY^{h+1})^2 - \ldots - (dY^n)^2 \quad [0 \leq h \leq n]$$

- At a given point q, the *signature* of ds^2, S, is defined by

$$S \equiv (\text{\# positive coefficients of } ds^2) - (\text{\# negative coefficients of } ds^2) = h - (n - h) = 2h - n.$$

Sylvester's "Law of Inertia": h, and hence also S, remain the same for all such transformations, i.e., the dimension (n) and signature (S) are invariants of ds^2; and if $S = n$, for every q, ds^2 is positive definite. For details, see texts on linear algebra; or Tietz (1955, pp. 132–133), Weatherburn (1938, pp. 11 ff.)

ii. The possibility that some ε_ks equal –1 is needed in Relativity. *In classical (nonrelativistic) mechanics (this book), if all the* y^k *are spatial coordinates, all* ε_k *equal +1* and, accordingly, Δs^2 is positive definite; i.e., in classical mechanics, the terms flat and Euclidean are synonymous.

Now, for a manifold to be useful in physical applications, in addition to being *differentiable* and *possess a metric,* it must also be *connected.* Let us say a few words about this third property; this, also, will help us understand better the position of Riemannian spaces relative to more general manifolds.

2.12.4 Linearly, or Affinely, Connected Manifolds, Integrability

We begin with a differentiable manifold, like X_n (Section 1.3); that is, one where tensors are defined by their local transformation properties, i.e., tensor *algebra.* But to be able to do tensor *analysis,* we need to be able to compare vectors and tensors attached to neighboring (i.e., different!) manifold points, say, the X_n-points P and $P' \equiv P + dP$, i.e., lying on the neighboring tangent spaces $\mathbf{T_n(P)}$ and $\mathbf{T_n(P')}$, we need to add to X_n a certain structure called *connection,* or connectedness. This is *a rule, or mapping, relating the corresponding neighboring bases* $\{\mathbf{e}_k(P)\}$ *and* $\{\mathbf{e}_k(P') = \mathbf{e}_k(P + dP) = \mathbf{e}_k(P) + d\mathbf{e}_k\}$, and then, by successive application of it, be able to compare vectors and tensors located at arbitrary manifold points.

Here (and this covers most needs of classical physics) we assume that the $d\mathbf{e}_k$:

i. *have no component outside* $\mathbf{T_n(P)}$, i.e., $d\mathbf{e}_k = (\ldots)_k{}^l\, \mathbf{e}_l(P)$ (not true in theory of surfaces, see Section 5.5 ff); and
ii. *they depend linearly on the dqs.* Hence, we write: $d\mathbf{e}_k = \Lambda^b{}_{kl}\, dq^l\, \mathbf{e}_b$, where the *components*, or *coefficients*, of the affine connection, or simply *affinities* (of the second kind) $\Lambda^b{}_{kl} = \Lambda^b{}_{kl}(q)$ are, generally, *asymmetric* in k, l, i.e., $\Lambda^b{}_{kl} \neq \Lambda^b{}_{lk}$. Such a *linearly,* or *affinely, connected* manifold we denote as L_n.

If, in addition, L_n is equipped with a *metric,* $g_{kl}(q) = g_{lk}(q)$, then it becomes an affinely *connected metric manifold.* Then, $d(\ldots)$-differentiating $\mathbf{e}^k \cdot \mathbf{e}_l = \delta^k{}_l$ and, invoking the above $d\mathbf{e}_k = \ldots$, etc., we easily deduce that $d\mathbf{e}^k = -\Lambda^k{}_{bs}\, \mathbf{e}^b\, dq^s$. (However, tensorial differentiation *can* be defined in terms of the affinities alone, that is, with no reference to metric.) As shown in Section 3.8 ff, *the affinities are not tensors* but, their *antisymmetric* part, $2\Lambda^b{}_{[kl]} \equiv \Lambda^b{}_{kl} - \Lambda^b{}_{lk}$, is a new tensor, called the manifold *torsion* (E. Cartan, 1922).

Further, as also shown in Section 3.8 ff, *the integrability conditions for* $d\mathbf{P} = \mathbf{e}_k\, dq^k$ (i.e., $\mathbf{e}_{k,l} = \mathbf{e}_{l,k}$) *require that the affinities be symmetric,* i.e., that the manifold be torsionless; and when this is combined with the above (i.e., metric g_{kl} + affine connection $d\mathbf{e}_k = \Lambda^b{}_{kl}\, dq^l\, \mathbf{e}_b$), these affinities are related to the metric and its gradients exactly like the (symmetric) *Christoffel symbols* (Section 3.3). *Our Riemannian* R_n *will be such a metric and torsionless manifolds.*

Finally, as shown in Section 5.6, the integrability conditions for $d\mathbf{e}_k = (\ldots)_{kl}\, dq^l = (\ldots)_k{}^l\, \mathbf{e}_l$ (i.e., $\mathbf{e}_{k,a,b} = \mathbf{e}_{k,b,a}$) require that curvilinear coordinates q (covering the Euclidean space E_n) exist such that the given $g_{kl}(q)\, dq^k\, dq^l$ be the (line element)^2, ds^2, of that space; i.e., *the* $g_{kl}(q)$ *terms have to satisfy certain compatibility conditions (which, as shown there, express the requirement that the manifold be flat, or Euclidean);* while, if $\mathbf{e}_{k,a,b} \neq \mathbf{e}_{k,b,a}$, the form $g_{kl}(q)\ dq^k\, dq^l$ defines an R_n.

The most important manifolds used in this book are summarized below, in order of decreasing generality (or increasing structure):

- V_n: general space (possibly discrete, i.e., *nondifferentiable*).
- X_n: general *differentiable manifold* (with curvilinear coordinates q^k and invertible + continuous + differentiable transformations $q \to q'$); *no connection, no metric.*
- M_n: *connected* manifold (i.e., a connected X_n), but no metric.
- L_n: *linearly,* or *affinely, connected* manifold (in general with *torsion*); i.e., a linearly connected X_n; no metric.
- R_n: *Riemannian* manifold: an L_n with symmetric affinities (i.e., no torsion) + (real C^2-differentiable + symmetric +) positive definite *metric* $g_{kl}(q)$.
- E_n: *Euclidean* manifold: a *metric, linear,* and *flat* manifold, i.e., can introduce global rectilinear coordinates: g_{kl} = constant; or rectangular Cartesian coordinates: $g_{kl} \to \delta_{kl}$; e.g., tangent space to an R_n-point plus (local) metric. In sum, a flat R_n.

2.12.5 Differences Between Affine (Nonmetric) and Metric Spaces

In an affine space we have n axes of rectilinear coordinates on each of which we have defined a particular unit of length (whereas Cartesian means that all axes have the same unit), and no possibility of comparing these various units with each other. Thus, in an affine space we cannot define the length of a vector (in the corresponding vector space, $\mathbf{T_n}$) because there is no common measure among its components; and, similarly, in there *we cannot define distances:* the vector $\mathbf{PP'}$, where $P(q^k)$ and $P'(q^k + dq^k)$, will have (contravariant) components dq^k but *we cannot form a combination of them that represents the length of* $\boldsymbol{PP'}$. Despite these drawbacks, affine spaces do have physical applications. For example, in the thermodynamics of gases we build diagrams with three orthogonal but non-Cartesian axes: one for the pressure p, one for the volume v, and one for the temperature θ. The variables p, v, θ have no common measure, and, of course, different dimensions; and if we change their fundamental units, these variables will change numerically in *arbitrary proportions.* As a result, it is impossible to define distances in that p–v–θ space; that is, if a state of the gas is represented by a point P and a "neighboring" one by P' then there is no distance between these two points. For example, we can choose, along the three axes, 1 cm to represent 1 atmosphere (p), 1 cm to represent 1 cm^3 (v), and 1 cm to represent 1 Celsius degree (θ), but this choice is arbitrary because another diagram built with different p, v, θ units would do (theoretically) just as well. Further, since the orthogonality of the axes is only a convention, *the affine properties must be invariant under arbitrary inclinations among these axes.* Hence, we cannot measure angles either — an angle $d\phi$ is the ratio of two lengths; say, one along the circumference of a circle, ds, and its radius R: $ds = R\, d\phi$; similarly for trigonometric functions, such as sin(...), cos(...), because they are also ratios of lengths.

In sum, *in affine geometry we are only interested in properties that remain invariant under arbitrary scale and angle changes of the axes.*

Finally, in some applied areas, like meteorology, *mixed,* or nonpure, spaces are employed; i.e., spaces with some of their coordinates affine and some metric — see e.g., Brillouin (1964, pp. 19–22); and, for a more elaborate and quantitative space classification, see Laugwitz (1968, pp. 154–156).

3 Tensor Analysis

So far we have seen tensor algebra, that is, the rules of combination of tensor components *at one and the same point of a manifold.* In this chapter we shall examine tensor analysis, that is, *how tensors vary from point to point* of that manifold, or the application of differential calculus to tensors — better, tensor fields. More specifically, and since, as shown below, ordinary and partial derivatives of tensor components do *not* transform as tensors under general admissible coordinate transformations (CT), we will formulate special invariant (or absolute, or *covariant*) derivatives of these components that do transform as tensors. As Weyl (1922, p. 58) puts it, "...Tensor analysis tells us how, by differentiating with respect to the space co–ordinates, a new tensor can be derived from the old one in a manner entirely independent of the co–ordinate system." Thus, tensor analysis is the ideal tool for the mathematical description of the states of a spatially and temporally extended system via form–invariant differential equations, consisting entirely of such covariant derivatives.*

3.1 INTRODUCTION

In order for this new "absolute differential calculus" to be a self-consistent and invariant generalization of ordinary vector analysis/field theory, we require that

- The differential of an (absolute or relative) tensor **T**, d**T**, be a tensor of the same rank (or type) and weight as **T**. In particular, if **T** is an absolute scalar T, then d**T** must reduce to the ordinary differential dT; and
- For **T**, **S**; **u**, **v**; f, respectively, (addable) tensors, vectors, and scalar:

$$\mathrm{d}(\mathbf{T} + \mathbf{S}) = \mathrm{d}\mathbf{T} + \mathrm{d}\mathbf{S}, \quad \mathrm{d}(\mathbf{T}\ \mathbf{S}) = (\mathrm{d}\mathbf{T})\mathbf{S} + \mathbf{T}(\mathrm{d}\mathbf{S}), \tag{3.1.1a}$$

$$\mathrm{d}(f\,\mathbf{T}) = df\,\mathbf{T} + f\,\mathrm{d}\mathbf{T}, \quad d(\mathbf{u}\cdot\mathbf{v}) = d\mathbf{u}\cdot\mathbf{v} + \mathbf{u}\cdot d\mathbf{v}. \tag{3.1.1b}$$

To get a first, qualitative, glimpse into the general structure of d**T** we differentiate $\mathbf{T} = T^{\cdots}{}_{\ldots}\,\mathbf{e}\ldots$ (recall Equation 2.11.1) while using the chain rules (Equation 3.1.1b):

$$\mathrm{d}\mathbf{T} = dT^{\cdots}{}_{\ldots}\,\mathbf{e}\ldots + T^{\cdots}{}_{\ldots}\,d\mathbf{e}\ldots. \tag{3.1.2}$$

Now, the *first* term in Equation 3.1.2, $dT^{\cdots}{}_{\ldots}\,\mathbf{e}\ldots$, is the ordinary differential part: the components $T^{\cdots}{}_{\ldots}$ change, the basis vectors **e**... do not. The *second* term, $T^{\cdots}{}_{\ldots}\,d\mathbf{e}\ldots$, is the key to tensor analysis: the $T^{\cdots}{}_{\ldots}$ do not change but the **e**... do, since we move

---

* For complementary reading, we recommend (alphabetically): Brand (1947), Eringen (1971), Schouten (1954 b), Sokolnikoff (1951, 1964); also Lodge (1974, Ch. 8).

from a manifold point $P(q)$ to a neighboring manifold point $P + dP(q + dq)$; i.e., to the first order:

$$\mathbf{e}\ldots(P + dP) = \mathbf{e}\ldots(P) + d\mathbf{e}\ldots(P); \tag{3.1.2a}$$

unless we can use global *rectilinear* coordinates in our manifold, in which case $d\mathbf{e}\ldots(P) = 0$. Now, continuing the train of thought initiated in Section 2.12 (or, borrowing from ordinary three-dimensional differential geometry, e.g., Frenet–Serret formulae, Section 5.3; or rigid body kinematics, e.g., Example 3.8.3) we write down the following fundamental representation, or *transport law*:

$$d\mathbf{e}\ldots(P) = (\textit{coefficients of rotation} \text{ of basis vectors, from } P \textit{ to } P + dP)_P\, \mathbf{e}\ldots(P). \tag{3.1.2b}$$

These rotation coefficients, to be detailed later, are, in turn, proportional to certain fundamental nontensorial geometrical objects called Christoffel symbols (generally, *coefficients of connection,* or *Affinities*), $\Gamma^{\cdots}{}_{\ldots}$, i.e., $d\mathbf{e}\ldots(P) = \Gamma^{\cdots}{}_{\ldots}\, dq^{\cdots}\, \mathbf{e}\ldots(P)$, so that Equation 3.1.2 becomes

$$d\mathbf{T} = DT^{\cdots}{}_{\ldots}\, \mathbf{e}\ldots, \tag{3.1.2c}$$

where

$$DT^{\cdots}{}_{\ldots\,\text{(absolute differential, } \textit{tensorial})} \equiv dT^{\cdots}{}_{\ldots\,\text{(ordinary differential, } \textit{nontensorial})}$$

$$+\, \Gamma^{\cdots}{}_{\ldots}\, dq \ldots_{\text{(~ Christoffels/affinities, } \textit{nontensorial})}. \tag{3.1.2d}$$

Finally, at this point, the precise nature of the manifold to be used is immaterial, as long as it is general enough to accommodate all the proposed operations. That issue is addressed in Section 3.8 ff.

3.2 DIFFERENTIATION OF TENSOR COMPONENTS

Let $T^{\cdots}{}_{\ldots}(P) = T^{\cdots}{}_{\ldots}(q)$ be the components of a general relative or absolute tensor field (as in vector field theory) at a generic point P, in a region of interest in our manifold, in a general coordinate system (CS) $q \equiv (q^k;\ k = 1, \ldots, n)$. Also, we will assume that these components are as many times differentiable as needed and will be using the popular comma-followed-by-a-subscript notation for $(\partial/\partial q^k)$-derivatives: $\partial T^{\cdots}{}_{\ldots}/\partial q^k \equiv T^{\cdots}{}_{\ldots,k}$; $\partial T'^{\cdots}{}_{\ldots}/\partial q^{k'} \equiv T'^{\cdots}{}_{\ldots,k'}$; or, sometimes, the operator notation: $T^{\cdots}{}_{\ldots,k} \equiv \partial_k T^{\cdots}{}_{\ldots}$; and similarly for higher such derivatives: $\partial^2 T^{\cdots}{}_{\ldots}/\partial q^k \partial q^l = \partial^2 T^{\cdots}{}_{\ldots}/\partial q^l \partial q^k \equiv T^{\cdots}{}_{\ldots,k,l} = T^{\cdots}{}_{\ldots,l,k} = \partial_l \partial_k T^{\cdots}{}_{\ldots} = \partial_k \partial_l T^{\cdots}{}_{\ldots}$; also the earlier-introduced notations: $\partial q^{k'}/\partial q^k \equiv A^{k'}{}_k = A_k{}^{k'}$ and $\partial q^k/\partial q^{k'} \equiv A^k{}_{k'} = A_{k'}{}^k$. We begin by noticing that if T is an absolute scalar, then $T_{,k}$ is an absolute covariant vector, i.e., a tensor. Indeed, differentiating $T = T(q) = T'(q') = T'$ we obtain

$$\partial T'/\partial q^{k'} = (\partial T/\partial q^k)(\partial q^k/\partial q^{k'}), \text{ or } T'_{,k'} = A^k{}_{k'}\, T_k. \tag{3.2.1}$$

Unfortunately, this property does *not* extend to other tensors $T^{\cdots}{}_{\cdots}$: in general, *partial differentiation is not a tensor operation*, i.e., it does not create new tensors. Let us show this, for simplicity, for the absolute covariant vector T_k. Differentiating its transformation law: $T_{k'} = A^k{}_{k'}\, T_k$, we find

$$\begin{aligned}
(\partial T_k/\partial q^r)' &\equiv \partial T_{k'}/\partial q^{r'} \equiv T_{k',r'} \\
&= \{\partial/\partial q^r\, [(\partial q^k/\partial q^{k'})T_k]\}\, (\partial q^r/\partial q^{r'}) \\
&= (\partial q^k/\partial q^{k'})(\partial q^r/\partial q^{r'})(\partial T_k/\partial q^r) + (\partial^2 q^k/\partial q^{k'}\partial q^{r'})\, T_k \\
&\equiv A^k{}_{k'}\, A^r{}_{r'}\, T_{k,r} + A^k{}_{k',r'}\, T_k;
\end{aligned} \tag{3.2.2a}$$

which shows that *if the second term on the right side of Equation 3.2.2a were zero, $T_{k,r}$ would transform as an absolute second-rank covariant tensor.* It is this nontensorial term, specifically its part involving *second* coordinate transformation derivatives, $\partial^2 q^k/\partial q^{k'}\partial q^{r'} \equiv A^k{}_{k',r'}$ (unlike tensor transformation terms which involve only *first* such derivatives) that destroys the tensor character of Equation 3.2.2a. Similarly, for an absolute contravariant vector T^k we find

$$\begin{aligned}
(\partial T^k/\partial q^r)' &\equiv \partial T^{k'}/\partial q^{r'} \equiv T^{k'}{}_{,r'} \\
&= \{\partial/\partial q^r\, [(\partial q^{k'}/\partial q^k)T^k]\}\, (\partial q^r/\partial q^{r'}) \\
&= (\partial q^{k'}/\partial q^k)(\partial q^r/\partial q^{r'})(\partial T^k/\partial q^r) + (\partial q^r/\partial q^{r'})(\partial^2 q^{k'}/\partial q^k\partial q^r)T^k \\
&\equiv A^{k'}{}_k\, A^r{}_{r'}\, T^k{}_{,r} + A^r{}_{r'}\, A^{k'}{}_{k,r}\, T^k;
\end{aligned} \tag{3.2.2b}$$

i.e., again, the second term, specifically $A^{k'}{}_{k,r}$, destroys the tensor character of $T^k{}_{,r}$. In sum, unless $A^k{}_{k',r'}$ ($A^{k'}{}_{k,r}$) vanish (e.g., in rectilinear/affine coordinates), the gradients $T_{k,r}$ ($T^k{}_{,r}$) do *not* transform as absolute second-rank tensors, and, consequently, the respective differentials: $dT_k = T_{k,r}\, dq^r$ and $dT^k = T^k{}_{,r}\, dq^r$, do *not* transform as absolute vectors; and similarly for the gradients of higher rank tensors.

In what follows, by judiciously combining first coordinate and vector/tensor component gradients, and other geometrical objects, we shall build special derivatives, called covariant derivatives, which do not contain the undesired second coordinate gradients $A^k{}_{k',r'}$ and yet transform as a tensors!

Example 3.2.1

As shown later (Chapters 5 and 6), the contravariant components of the *velocity* vector of a particle P are $dq^k/dt \equiv v^k$. Differentiating its law of transformation $v^{k'} =$

$A^{k'}{}_k v^k$ with respect to time t [obtained by $d(\ldots)/dt$-differentiating $q^{k'} = q^{k'}(q^k)$] we find, successively,

$$\begin{aligned} dv^{k'}/dt &= (dA^{k'}/dt)v^k + A^{k'}{}_k(dv^k/dt) \\ &= (\partial A^{k'}/\partial q^l)(dq^l/dt)v^k + A^{k'}{}_k(dv^k/dt) \\ &= A^{k'}{}_k(dv^k/dt) + A^{k'}{}_{k,l}\, v^k\, v^l; \end{aligned} \tag{a}$$

i.e., due to the second term, $A^{k'}{}_{k,l}\, v^k\, v^l$, $d^2q^k/dt^2 = dv^k/dt$ *cannot*, in general, be used to represent the contravariant components of the *acceleration* vector of P; and similarly if $q^{k'} = q^{k'}(q^k, t)$.

Example 3.2.2. Rectangular Cartesian Coordinates

In this case (recalling Equations 2.5.10a ff):

$$q^{k'} \to y^{k'} = y_{k'} = a_{k'k}\, y_k + a_{k'}, \tag{a}$$

$$\begin{aligned} A^{k'}{}_k \to a_{k'k} &\equiv \partial y_{k'}/\partial y_k = \cos(y_{k'}, y_k) = \cos(y_k, y_{k'}) = \partial y_k/\partial y_{k'} \\ &\equiv a_{kk'}\ (\leftarrow A^k{}_{k'}), \quad a_{k'}\text{: constant} \Rightarrow A^k{}_{k',r'} \to a_{kk',r'} = 0, \end{aligned} \tag{b}$$

and, therefore, Equation 3.2.2a specializes to the tensor transformation:

$$T_{k',l'} = a_{k'k}\, a_{l'l}\, T_{k,l} = a_{kk'}\, a_{ll'}\, T_{k,l}. \tag{c}$$

Similarly for rectilinear/affine coordinates (Section 2.3).

Problem 3.2.1

i. Recalling the integrability conditions (Equations 2.4.6a and b):

$$A^{k'}{}_{l,r} = A^{k'}{}_{r,l} \quad \text{and} \quad A^k{}_{l',r'} = A^k{}_{r',l'}, \tag{a}$$

show that the *antisymmetric* combination (components of curl of vector **T**):

$$T_{k,l} - T_{l,k} \equiv 2T_{[k,l]} \quad \text{(or, sometimes, } \partial_l T_k - \partial_k T_l \equiv 2\partial_{[l}T_{k]}) \tag{b}$$

is an absolute *second*-rank covariant tensor.

ii. Consider an absolute second-rank antisymmetric covariant tensor, i.e., $T_{kl} = -T_{lk}$. Verify that the combination:

$$T_{klr} \equiv T_{kl,r} + T_{lr,k} + T_{rk,l}, \tag{c}$$

is an absolute *third*-rank covariant tensor (see also Equation 3.5.7b).

iii. Consider a contravariant vector density T^k. Verify that $\partial T^k/\partial q^k \equiv T^k{}_{,k}$ is an invariant density.

iv. Consider the second-rank antisymmetric contravariant tensor density $T^{kl} = -T^{lk}$. Verify that $T^{kl}{}_{,l}$ is a contravariant vector density.

3.3 THE CHRISTOFFEL SYMBOLS

3.3.1 DEFINITIONS

The (holonomic) Christoffel symbols of the *first* ($\Gamma_{k,lr}$) and *second* ($\Gamma^k{}_{lr}$) kind, associated with the CS q, are defined, by

$$\Gamma_{k,lr} \equiv (1/2)(\partial g_{kl}/\partial q^r + \partial g_{kr}/\partial q^l - \partial g_{lr}/\partial q^k)$$

$$\equiv (1/2)(g_{kl,r} + g_{kr,l} - g_{lr,k}), \tag{3.3.1a}$$

$$\Gamma^k{}_{lr} \equiv g^{ks}\,\Gamma_{s,lr} = (g^{ks}/2)(g_{sl,r} + g_{sr,l} - g_{lr,s}), \tag{3.3.1b}$$

$$\Rightarrow \Gamma_{k,lr} = g_{ks}\,\Gamma^s{}_{lr}. \tag{3.3.1c)*$$

It is shown later (Section 3.7) that *the Christoffels are not tensors*, despite appearances, and even though we use g^{ks} (g_{ks}) to raise (lower) their indices.**

3.3.2 PROPERTIES

- Both kinds of Christoffels are *symmetric in their last two subscripts*, i.e.

$$\Gamma_{k,lr} = \Gamma_{k,rl} \quad \text{and} \quad \Gamma^k{}_{lr} = \Gamma^k{}_{rl}. \tag{3.3.2a}$$

- From the above, we readily find

$$g_{kl,r} = \Gamma_{k,lr} + \Gamma_{l,kr}\ (= g_{ks}\,\Gamma^s{}_{lr} + g_{ls}\,\Gamma^s{}_{kr}), \tag{3.3.2b}$$

$$[\Rightarrow g_{kl,r} - g_{kr,l} = \ldots = \Gamma_{l,kr} - \Gamma_{r,kl} = \Gamma_{l,rk} - \Gamma_{r,lk}]$$

and

---

* *Not* the derivatives of some Γ_k with respect to q^l, q^r.
Other common notations for them are

$$\Gamma_{k,lr} \equiv \Gamma_{lr,k} \equiv \Gamma_{klr} \equiv \Gamma_{lrk} \equiv \Gamma_{k;lr} \equiv \Gamma_{lr;k} \equiv [k,\ lr] \equiv [lr,\ k] \equiv [k;\ lr] \equiv [lr;\ k],$$

$$\Gamma^k{}_{lr} \equiv [{}_l{}^k{}_r] \equiv [{}^l{}_k{}^r] \equiv \{{}_l{}^k{}_r\} \equiv \{{}^l{}_k{}^r\}.$$

Our notation, Equations 3.3.1a and b, seems to be due to O. Veblen ("Princeton School") and looks simpler than those shown above.

** This, also, becomes clear from their geometrical interpretation, which is given in Section 3.8.

$$g^{kl}{}_{,r} = -g^{ks}\,\Gamma^{l}{}_{sr} = -g^{ls}\,\Gamma^{k}{}_{sr}; \tag{3.3.2c}$$

which express the gradients of the metric tensor in terms of the Christoffels.

- Contracting Equation 3.3.1b (while assuming that $\mathrm{Det}(g_{kl}) \equiv g > 0$, and recalling that $\partial g/\partial g_{kl} = g\, g^{kl}$ (Equations 2.8.5a ff) yields the useful formulae:

$$\Gamma^{k}{}_{lk} = \Gamma^{k}{}_{kl} = (g^{ks}/2)(g_{ks,l} + g_{ls,k} - g_{kl,s}) = (g^{ks}/2)g_{ks,l}$$

$$= (1/2g)(\partial g/\partial g_{ks})(\partial g_{ks}/\partial q^{l}) = (1/2g)(\partial g/\partial q^{l})\,.$$

$$= \left(1/\sqrt{g}\right)\left(\partial\sqrt{g}/\partial q^{l}\right) = \left(\ln\sqrt{g}\right)_{,l}. \tag{3.3.2d}$$

- As mentioned earlier, the Christoffels are *not* (third-rank) tensors; they are just geometrical objects associated with the CS q. It is shown in Section 3.7 ff that given any two CS, q and q', the corresponding second-kind Christoffels, $\Gamma(q) \equiv \Gamma$ *and* $\Gamma'(q') \equiv \Gamma'$, are related by the following *transformation equations:*

$$\Gamma^{k'}{}_{l'r'} = (\partial q^{k'}/\partial q^{k})(\partial q^{l}/\partial q^{l'})(\partial q^{r}/\partial q^{r'})\;\Gamma^{k}{}_{lr} + (\partial q^{k'}/\partial q^{k})(\partial^2 q^{k}/\partial q^{l'}\partial q^{r'})$$

$$\equiv A^{k'}{}_{k}\,A^{l}{}_{l'}\,A^{r}{}_{r'}\,\Gamma^{k}{}_{lr} + A^{k'}{}_{k}\,A^{k}{}_{l',r'} \quad \text{(primary result)} \tag{3.3.2e}$$

$$= A^{k'}{}_{k}\,A^{l}{}_{l'}\,A^{r}{}_{r'}\,\Gamma^{k}{}_{lr} + A^{k'}{}_{k}\,A^{k}{}_{r',l'}$$

$$[\text{if } A^{k}{}_{l',r'} = A^{k}{}_{r',l'} \quad \text{(holonomic coordinates)}] \tag{3.3.2f}$$

$$= A^{k'}{}_{k}\,A^{l}{}_{l'}\,A^{r}{}_{r'}\,\Gamma^{k}{}_{lr} - A^{l}{}_{l'}\,A^{r}{}_{r'}\,A^{k'}{}_{l,r} \quad \text{(from the primary result)} \tag{3.3.2g}$$ *

$$= A^{k'}{}_{k}\,A^{l}{}_{l'}\,A^{r}{}_{r'}\,\Gamma^{k}{}_{lr} - A^{l}{}_{l'}\,A^{r}{}_{r'}\,A^{k'}{}_{r,l}$$

$$[\text{if } A^{k'}{}_{l,r} = A^{k'}{}_{r,l} \quad \text{(holonomic coordinates)}] \tag{3.3.2h}$$

Clearly, the second group of terms in the above destroys the tensor character of the Christoffels: the new Christoffels, $\Gamma^{k'}{}_{l'r'}$, are linear but nonhomogeneous functions of the old Christoffels, $\Gamma^{k}{}_{lr}$; and vice versa.

---

* Differentiating the compatibility conditions (2.4.4): $A^{k'}{}_{l}A^{l}{}_{l'} = 0$, we obtain, successively,

$$A^{k'}{}_{l,r'}A^{l}{}_{l'} + A^{k'}{}_{l}A^{l}{}_{l',r'} = 0 \Rightarrow A^{k'}{}_{l}A^{l}{}_{l',r'} = -A^{l}{}_{l'}A^{k'}{}_{l,r'} = -A^{l}{}_{l'}(A^{k'}{}_{l,r}A^{r}{}_{r'}),\ \text{q.e.d.}$$

Problem 3.3.1

Show that the above *second*-kind Christoffel transformations can be put in the equivalent form:

$$(\partial q^k/\partial q^{k'})\Gamma^{k'}{}_{l'r'} = (\partial q^l/\partial q^{l'})(\partial q^r/\partial q^{r'})\Gamma^k{}_{lr} + (\partial^2 q^k/\partial q^{l'}\partial q^{r'}), \tag{a}$$

or, compactly,

$$A^k{}_{k'}\ \Gamma^{k'}{}_{l'r'} = A^l{}_{l'}\ A^r{}_{r'}\ \Gamma^k{}_{lr} + A^k{}_{l',r'};\ \text{etc.} \tag{b}$$

Problem 3.3.2

Using Equations 3.3.2e through h, show that the *first*-kind Christoffels obey the following (also nontensorial) transformation:

$$\Gamma_{k',l'r'} = (\partial q^k/\partial q^{k'})(\partial q^l/\partial q^{l'})(\partial q^r/\partial q^{r'})\Gamma_{k,lr} + g_{kl}\ (\partial q^k/\partial q^{k'})(\partial^2 q^l/\partial q^{l'}\partial q^{r'}) \tag{a}$$

$$\equiv A^k{}_{k'}\ A^{l'}{}_l\ A^{r'}{}_r\ \Gamma_{k,lr} + g_{kl}\ A^k{}_{k'}\ A^l{}_{l',r'} \tag{b}$$

$$= A^k{}_{k'}\ A^{l'}{}_l\ A^{r'}{}_r\ \Gamma_{k,lr} + g_{kl}\ A^k{}_{k'}\ A^l{}_{r',l'}, \quad \text{if } A^l{}_{l',r'} = A^l{}_{r',l'}. \tag{c}$$

Problem 3.3.3. Christoffels in Orthogonal Curvilinear Coordinates

Show that in this case, where $g_{kl} = 0$ for $k \neq l$, i.e., (g_{kl}) and, therefore, (g^{kl}) are *diagonal*, and $g^{kk} = 1/g_{kk}$ (no sum over k), the Christoffels reduce to the following *three* distinct cases:

i. $\Gamma_{k,lr} = 0 \Leftrightarrow \Gamma^k{}_{lr} = 0$

[$\mathbf{k \neq l \neq r \neq k}$. There are $n(n-1)/2$ such Christoffels (of each kind)].(a)

ii. $\Gamma_{k,ll} = -(1/2)(\partial g_{ll}/\partial q^k) \Leftrightarrow \Gamma^k{}_{ll} = -(1/2g_{kk})(\partial g_{ll}/\partial q^k)$

$= -(g^{kk}/2)(\partial g_{ll}/\partial q^k)$

[No sum, $\mathbf{k \neq l}$. There are $n(n-1)$ such Christoffels]. (b)

iii. $\Gamma_{k,kl} = \Gamma_{k,lk} = (1/2)(\partial g_{kk}/\partial q^l) \quad \Leftrightarrow \Gamma^k{}_{kl} = \Gamma^k{}_{lk}$

$= (1/2g_{kk})(\partial g_{kk}/\partial q^l) = (g^{kk}/2)(\partial g_{kk}/\partial q^l) = (1/2)[\partial \ln|g_{kk}| \,/\, \partial q^l]$

[No sum. If $\mathbf{k \neq l}$, there are $n(n-1)$ such Christoffels; if $\mathbf{k = l}$, there are n]. (c)

Problem 3.3.4

Verify that, in an n-space, there can be, at most, $N_C \equiv n^2(n + 1)/2$ *distinct* Christoffels $\Gamma^k{}_{lr} = \Gamma^k{}_{rl}$. For example, for $n = 2 \Rightarrow N_C = 6$; and for $n = 3 \Rightarrow N_C = 18$.

3.3.3 Successive Coordinate Transformations

Let us consider, first, the *successive* coordinate transformations $q \to q' \to q''$, and then the *direct* transformation $q \to q''$. It is not hard to show, using (Equations 3.3.2e through h) and the results of the preceding Problems, that the (first- and second-kind) Christoffels obtained by both the direct and the successive transformations coincide (transitivity). Further, and since the Christoffels pertain to a particular CS, it does not matter from which (initial) coordinates they came, via admissible coordinate transformations and by equations like those shown above.

In particular, if we start with rectilinear/affine coordinates z^k, then since: $g_{kl}(z)$, $g^{kl}(z)$ = constant $\Rightarrow \Gamma^k{}_{lr}(z), \Gamma_{k,lr}(z) = 0$, applying Equations 3.3.2e ff, with $q^k \to z^k$ and $q^{k'} \to q^k$ (general curvilinear system), we obtain, after some simple renaming: $\Gamma^{k'}{}_{l'r'}(q') \to \Gamma^k{}_{lr}(q) \equiv \Gamma^k{}_{lr}$, etc.,

$$\Gamma^k{}_{lr} = (\partial q^k/\partial z^{k'})(\partial^2 z^{k'}/\partial q^l \partial q^r) = -(\partial z^{l'}/\partial q^l)(\partial z^{r'}/\partial q^r)(\partial^2 q^k/\partial z^{r'}\partial z^{l'}), \quad (3.3.3a)$$

$$\Gamma_{k,lr} = g_{k'l'}(z)(\partial z^{k'}/\partial q^k)(\partial^2 z^{l'}/\partial q^l \partial q^r). \quad (3.3.3b)$$

The transformation formulae, Equations 3.3.2e through h, etc., can be proved directly from the Γ-definitions (Equations 3.3.1a and b) and the transformation properties of the metric tensor: $g_{k'l'} = A^k{}_{k'} A^l{}_{l'} g_{kl}$, $g^{k'l'} = A^{k'}{}_k A^{l'}{}_l g^{kl}$; but at this point such algebraic derivations (standard fare in older, nonvectorial, treatments) would be long and unmotivated exercises in partial differentiations. Simpler, geometrical, proofs will be presented in Section 3.7 ff.

3.3.4 Antisymmetric Part of the Christoffels

From Equations 3.3.2e we readily obtain:

$$\Gamma^{k'}{}_{l'r'} - \Gamma^{k'}{}_{r'l'} = A^{k'}{}_k A^l{}_{l'} A^r{}_{r'} (\Gamma^k{}_{lr} - \Gamma^k{}_{rl}) + A^{k'}{}_k(A^k{}_{l',r'} - A^k{}_{r',l'}), \quad (3.3.4a)$$

or, compactly,

$$\Gamma^{k'}{}_{[l'r']} = A^{k'}{}_k A^l{}_{l'} A^r{}_{r'} \Gamma^k{}_{[lr]} + A^{k'}{}_k A^k{}_{[l',r']}, \quad (3.3.4b)$$

$$= A^{k'}{}_k A^l{}_{l'} A^r{}_{r'} \Gamma^k{}_{[lr]} - A^l{}_{l'} A^r{}_{r'} A^{k'}{}_{[l,r]}; \quad (3.3.4c)$$

and similarly for $\Gamma_{k,[lr]}$. As mentioned earlier, in holonomic coordinates (the only kind examined so far), $A^k{}_{[l',r']} = 0$, $A^{k'}{}_{[l,r]} = 0$, and so for such coordinates, as Equations 3.3.4a through c show, $\Gamma^k{}_{[lr]}$ is a tensor, i.e., if $\Gamma^k{}_{[lr]} = 0$, then $\Gamma^{k'}{}_{[l'r']} = 0$.

3.4 THE COVARIANT DERIVATIVE (CD)

Although the "gradient, or partial derivative, operation" $(\ldots)_{,k}$ and the Christoffels, individually, are not tensors, yet when taken together, in a special combination, produce a very important tensorial quantity known as *covariant derivative* (CD), which constitutes the tensorial/invariant counterpart of the above nontensorial partial derivative of ordinary differential calculus (and reduces to it in rectilinear/affine coordinates, where, as explained earlier, the Christoffels vanish). Here, as with the Christoffels, we present the basic definitions and properties; while the geometrical interpretations etc. are presented later (Sections 3.7 ff).

3.4.1 DEFINITIONS, THEOREMS

The CD of the absolute vectors T^k and T_k are defined, respectively, by

$$T^k|_r \equiv T^k{}_{,r} + \Gamma^k{}_{lr}\, T^l \quad \text{and} \quad T_k|_r \equiv T_{k,r} - \Gamma^l{}_{kr}\, T_l. \tag{3.4.1}$$

Theorem: $T^k|_r$ and $T_k|_r$ are second-rank absolute tensors of the type indicated by their indices, i.e., with our usual notations,

$$T^{k'}|_{r'} = A^{k'}{}_k\, A^r{}_{r'}\, T^k|_r \quad \text{and} \quad T_{k'}|_{r'} = A^k{}_{k'}\, A^r{}_{r'}\, T_k|_r. \tag{3.4.2}$$

The proofs are straightforward. For example, to prove Equation $3.4.2_2$ we dot multiply $\Gamma^{k'}{}_{l'r'}$, Equations 3.3.2e ff, with $T_{k'}$ and then subtract it from $T_{l',r'}$, Equation 3.2.2a. The result is

$$T_{l',r'} - \Gamma^{k'}{}_{l'r'}\, T_{k'} = A^l{}_{l'}\, A^r{}_{r'}\, (T_{l,r} - \Gamma^k{}_{lr}\, T_k), \text{ q.e.d.}; \tag{3.4.2a}$$

and similarly for Equation $3.4.2_1$.

The CDs of the absolute second-rank tensors T_{kl}, etc. are defined, analogously to Equations 3.4.1, by

$$T_{kl}|_r \equiv T_{kl,r} - \Gamma^s{}_{kr}\, T_{sl} - \Gamma^s{}_{lr}\, T_{ks}, \tag{3.4.3a}$$

$$T^{kl}|_r \equiv T^{kl}{}_{,r} + \Gamma^k{}_{sr}\, T^{sl} + \Gamma^l{}_{sr}\, T^{ks}, \tag{3.4.3b}$$

$$T^k{}_l|_r \equiv T^k{}_{l,r} + \Gamma^k{}_{sr}\, T^s{}_l - \Gamma^s{}_{lr}\, T^k{}_s, \tag{3.4.3c}$$

$$T_l{}^k|_r \equiv T_l{}^k{}_{,r} + \Gamma^k{}_{sr}\, T_l{}^s - \Gamma^s{}_{lr}\, T_s{}^k. \tag{3.4.3d}$$

Their proofs follow the same steps as for (3.4.2); only, naturally, the algebra is slightly longer.* The extension to the CDs of general absolute tensors should be

* Or, we may take the CD of the absolute scalar (invariant): $I \equiv T_{kl}\, U^k\, V^l = T_{k'l'}\, U^{k'}\, V^{l'} \equiv I'$, while recalling Equation 3.2.1, i.e., $I|_r = I_{,r}$ = absolute covariant vector, that is, $I'_{,r'} = I_{,r}\, (\partial q^r/\partial q^{r'}) \equiv A^r{}_{r'}\, I_{,r}$, and that CDs must satisfy the ordinary chain rule.

clear from a careful study of the structure of Equations 3.4.3a through d (see also Equation 3.4.4 below, with $w = 0$).

Definition: The CD of a general relative tensor $T^{\cdots u \cdots}{}_{\ldots d \ldots}$ of weight w is defined by

$$T^{\ldots u \ldots}{}_{\ldots d \ldots}|_r \equiv T^{\ldots u \ldots}{}_{\ldots d \ldots,r} + \ldots + \Gamma^u{}_{\blacklozenge r}\, T^{\ldots \blacklozenge \ldots}{}_{\ldots d \ldots} - \ldots - \Gamma^{\blacklozenge}{}_{dr}\, T^{\ldots u \ldots}{}_{\ldots \blacklozenge \ldots}$$

$$- w\, \Gamma^{\blacklozenge}{}_{\blacklozenge r}\, T^{\ldots u \ldots}{}_{\ldots d \ldots} \quad (\blacklozenge = 1, \ldots, n). \tag{3.4.4}$$

It can be shown, as with Equations 3.4.2, that $T^{\ldots u \ldots}{}_{\ldots d \ldots}|_r$ *is a relative tensor of rank one more than* $T^{\ldots u \ldots}{}_{\ldots d \ldots}$, *and of the same weight* w.

For example, the CD of the relative scalar T of weight w (i.e., $T' = |\partial q/\partial q'|^w\, T$) is the following relative covariant vector of weight w: $T|_i = T_{,i} - w\, \Gamma^k{}_{ki}\, T$.

Remarks

i. Since CDs are tensors, all their indices can be raised/lowered with the help of the metric tensor. Thus, we define the "contravariant derivatives" as follows:

$$T_{kl}|^r \equiv g^{rs}\, T_{kl}|_s \Leftrightarrow T_{kl}|_s = g_{sr}\, T_{kl}|^r, \tag{3.4.5a}$$

$$T^{kl}|^r \equiv g^{rs}\, T^{kl}|_s \Leftrightarrow T^{kl}|_s = g_{sr}\, T^{kl}|^r. \tag{3.4.5b}$$

ii. Second, or consecutive, and higher CDs are defined similarly. For example, $(T^k{}_l|_r)|_s \equiv T^k{}_l|_{rs}$ is the second CD of $T^k{}_l$, *first* relative to q^r, and *then* relative to q^s. It should be pointed out that, in a general non-Euclidean manifold (e.g., a Riemannian space), *such CDs do not commute*, i.e., $T^k{}_l|_{rs} \neq T^k{}_l|_{sr}$, although, of course, $(T^k{}_{l,r})_{,s} = (T^k{}_{l,s})_{,r}$.*

3.5 THE ABSOLUTE, OR INTRINSIC, DERIVATIVE (AD)

This is the tensorial counterpart of the ordinary derivative of a tensor field $T^{\cdots}{}_{\ldots}(q)$ defined along a manifold curve C: $q = q(u)$ (u: curve parameter, e.g., arc–length), i.e., $dT^{\cdots}{}_{\ldots}[q(u)]/du = (\partial T^{\cdots}{}_{\ldots}/\partial q^r)(dq^r/du)$, which is *not* a tensor.

3.5.1 Definitions

i. The *absolute*, or *intrinsic*, *derivative* (AD) of a general (absolute or relative) tensor $T^{\cdots}{}_{\ldots}(q) \to T^{\cdots}{}_{\ldots}[q(u)] \to T^{\cdots}{}_{\ldots}(u)$, along a curve $q^k = q^k(u)$, is defined by

$$DT^{\cdots}{}_{\ldots}/Du \equiv DT^{\cdots}{}_{\ldots}/du \equiv (T^{\cdots}{}_{\ldots}|_k)(dq^k/du). \tag{3.5.1)**}$$

* More on this remarkable noncommutativity of CDs in Section 3.10, on Riemann's *curvature tensor.*

** The term seems to be due to Synge (1926–27) and McConnell (1931). ADs are also denoted as $\delta T^{\cdots}{}_{\ldots}/\delta u$, but here we have reserved $\delta(\ldots)$ for *virtual* variations (Sections 6.3 ff).

Thus, if T is a relative scalar of weight w, then $DT/Du = dT/du - w\, T\, \Gamma^k{}_{kl}\, (dq^l/du)$; and, therefore, if $w = 0$, $DT/du \to dT/du$: an invariant (i.e., if T is an absolute constant, then $DT/Du = 0$). Similarly, the ADs of the absolute vectors T^k and T_k are, respectively,

$$DT^k/Du = dT^k/du + \Gamma^k{}_{lr}\, T^l (dq^r/du),$$

$$DT_k/Du = dT_k/du - \Gamma^l{}_{kr}\, T_l (dq^r/du). \qquad (3.5.1a)$$

ii. The *absolute differential* (Ad) of a general tensor $T^{\cdots}{}_{\cdots}(q) \to$ etc. is defined by

$$DT^{\cdots}{}_{\cdots} \equiv (T^{\cdots}{}_{\cdots}|_k) dq^k; \qquad (3.5.2)$$

from which we readily conclude, that the Ad of the absolute vectors T^k and T_k, are

$$DT^k = dT^k + \Gamma^k{}_{lr}\, T^l\, dq^r, \quad DT_k = dT_k - \Gamma^l{}_{kr}\, T_l\, dq^r. \qquad (3.5.2a)$$

Here, as in the CD case, we have an instance of *a sum of two nontensorial quantities, e.g., dT^k and $\Gamma^k{}_{lr}\, T^l\, dq^r$, producing a tensorial one;* i.e., even though $dT^{k'} \neq A^{k'}{}_k\, dT^k$ etc., yet

$$(T^{k'} = A^{k'}{}_k\, T_k \Rightarrow) \quad dT^{k'} + \Gamma^{k'}{}_{l'r'}\, T^{l'}\, dq^{r'} = A^{k'}{}_k (dT^k + \Gamma^k{}_{lr}\, T^l\, dq^r). \qquad (3.5.2b)$$

3.5.2 Some Properties/Theorems on CDs and ADs

- If the CD vanishes, so does the AD; and vice versa.
- **Ricci's Lemma:** The metric tensor behaves as a constant under covariant and/or absolute differentiation; i.e.,

$$g_{kl}|_r = 0, \quad g^{kl}|_r = 0, \quad g^k{}_l|_r = 0; \qquad (3.5.3)$$

and, therefore,

$$[\mathrm{Det}(g^{kl})]|_r = 0, \ [\mathrm{Det}(g_{kl})]|_r = 0. \qquad (3.5.3a)$$

Proof: In rectangular Cartesian coordinates $y^k = y_k$ the metric components are constant ($= \delta^k{}_l$) and therefore the corresponding Christoffels vanish. Hence, in such coordinates the CD of the metric components is zero. But *since this is a tensor property, it holds in any other (admissible) coordinates,* q.e.d.; and similarly for the AD.

The above argument makes clear that the *Kronecker Deltas* and the *Permutation Tensors* ($\to$ *Permutation Symbols*, in rectangular Cartesian coordinates) also obey Ricci's lemma, i.e.,

$$\delta_{kl}|_r = \delta^{kl}|_l = \delta^k{}_l|_r = e_{kls}|_r = e^{kls}|_r = \varepsilon_{kls}|_r = \varepsilon^{kls}|_r = 0. \tag{3.5.3b}$$

- All customary rules of differential calculus hold for CDs and ADs; for example,

$$DT^{\cdots}{}_{\ldots}/Du = (DT^{\cdots}{}_{\ldots}/Dv)(dv/du) \quad [\text{chain rule: } v = v(u)] \tag{3.5.4a}$$

$$(T^{\cdots}{}_{\ldots} \pm R^{\cdots}{}_{\ldots} \pm S^{\cdots}{}_{\ldots} \pm \ldots)|_r = T^{\cdots}{}_{\ldots}|_r \pm R^{\cdots}{}_{\ldots}|_r \pm S^{\cdots}{}_{\ldots}|_r \pm \ldots, \tag{3.5.4b}$$

$$(T^{\cdots}{}_{\ldots}\, R^{\cdots}{}_{\ldots})|_r = (T^{\cdots}{}_{\ldots}|_r)\, R^{\cdots}{}_{\ldots} + T^{\cdots}{}_{\ldots}\, (R^{\cdots}{}_{\ldots}|_r). \tag{3.5.4c}$$

- In rectangular Cartesian coordinates (where all tensors are absolute — recalling Equations 2.5.10a ff) the CD and AD reduce, respectively, to *partial* and *ordinary* (directional) derivatives:

$$T|_r = T_{,r}, \quad (f\, T^k)|_r = f_{,r}\, T^k + f\, T^k{}_{,r}\ (f\text{: scalar}),$$

$$(T^k\, T_k)|_r = (T^k\, T_k)_{,r} = T^k{}_{,r}\, T_k + T^k\, T_{k,r} = 2T_k\, T_{k,r}. \tag{3.5.5}$$

- If $T^{\cdots}{}_{\ldots}$ depends implicitly *and* explicitly on u, i.e., if $T^{\cdots}{}_{\ldots} = T^{\cdots}{}_{\ldots}[q^k(u), u]$ (e.g., u: time, as in the so called *Eulerian* description of continuum physics/fluid mechanics), then

$$DT^{\cdots}{}_{\ldots}/Du \equiv DT^{\cdots}{}_{\ldots}/du \equiv (T^{\cdots}{}_{\ldots}|_k)(dq^k/du) + \partial T^{\cdots}{}_{\ldots}/\partial u. \tag{3.5.6}$$

- For a *covariant* vector, *due to the symmetry of the Christoffels*, we have:

$$S_{kl} \equiv T_k|_l - T_l|_k = T_{k,l} - T_{l,k} \quad (= -S_{lk}, \text{ i.e., an antisymmetric tensor}); \tag{3.5.7a}$$

and, similarly, for the antisymmetric covariant tensor $T_{ab} = -T_{ba}$:

$$S_{abc} \equiv T_{ab}|_c + T_{bc}|_a + T_{ca}|_b = T_{ab,c} + T_{bc,a} + T_{ca,b}; \tag{3.5.7b}$$

$$(= S_{abc} = S_{bca} = S_{cab} = -S_{acb} = -S_{cba} = -S_{bac}, \text{ i.e., an antisymmetric tensor})$$

while for a contravariant vector *density* T^k:

$$T^k|_k = T^k{}_{,k} \quad (\text{scalar density}); \tag{3.5.7c}$$

and for an *antisymmetric* contravariant tensor *density* $T^{kl} = -T^{lk}$:

$$T^{kl}|_l = T^{kl}{}_{,l} \quad (\text{contravariant vector density}). \tag{3.5.7d}$$

Problem 3.5.1

Let V_k, V^k be an absolute vector. Show that:

$$d/du(g^{kl}\ V_k\ V_l) = 2g^{kl}\ V_k(DV_l/Du) = d/du(g_{kl}\ V^k\ V^l) = 2g_{kl}\ V^k(DV^l/Du). \quad \text{(a)}$$

3.6 SOME VECTOR ANALYSIS IN TENSOR NOTATION

(Recall Equations 2.9.19a ff.)

- The *gradient* of an invariant T is

$$(\text{grad}T)^k = T|_k = T_{,k}; \quad (3.6.1)$$

- The *divergence* of a contravariant vector T^k, and of a second-rank tensor $T^k{}_s$, T^{ks}, are, respectively,

$$\text{div}\mathbf{T} = T^k|_k = \left(1/\sqrt{g}\right)\left(\sqrt{g}T^k\right)_{,k} \quad \left[\text{by Ricci's lemma: } \left(\sqrt{g}\right)\Big|_k = 0\right] \quad (3.6.2)$$

$$T^k{}_s|_k = \left(1/\sqrt{g}\right)\left(\sqrt{g}T^k{}_s\right)_{,k} - \Gamma^h{}_{sk}\ T^k{}_h; \quad T^{ks}|_s = \left(1/\sqrt{g}\right)\left(\sqrt{g}T^{ks}\right)_{,s} + \Gamma^k{}_{sb}\ T^{sb}. \quad (3.6.2a)$$

Clearly, if T^{ks} is antisymmetric (and $\Gamma^k{}_{sb} = \Gamma^k{}_{bs}$), the second sum in Equation 3.6.2a$_2$ vanishes.

- The *Laplacian* operator is

$$\nabla^2(\ldots) \equiv \Delta(\ldots) \equiv \text{div grad}(\ldots) = g^{kl}(\ldots)|_{kl}; \quad (3.6.3)$$

and for an invariant T it gives

$$\nabla^2 T \equiv \Delta T = g^{kl}(T|_{kl}) = \left(1/\sqrt{g}\right)\left(\sqrt{g}g^{kl}T_{,l}\right)_{,k}. \quad (3.6.3a)$$

- The curl(-ing), or rot(-ation), of a *three-dimensional covariant* vector T_k is (assuming symmetric Christoffels):

$$(\text{curl }\mathbf{T})^k = e^{klr}\ T_r|_l = (1/2)e^{klr}(T_r|_l - T_l|_r) = \left(\varepsilon^{klr}/2\sqrt{g}\right)(T_r|_l - T_l|_r)$$

$$= \left(\varepsilon^{klr}/2\sqrt{g}\right)(T_{r,l} - T_{l,r}) = \left(\varepsilon^{klr}/\sqrt{g}\right)T_{r,l}; \quad (3.6.4)$$

while the curl of a *contravariant* vector is obtained by raising indices in the above. [For an extension of the curl of a vector to an n-dimensional space, and in general nonholonomic coordinates ($\Rightarrow$ asymmetric Christoffels), see Problem 3.16.3.]

- Theorems of *Gauss–Green* (G-G, of divergence) and *Stokes* (S, of circulation) in *three* dimensions. Let τ be the volume enclosed by the *closed* surface σ (or $\partial\tau$), and let A be an *open surface* ending on the *closed line* (or contour) C. If $\mathbf{n} = n^k\, \mathbf{e}_k = n_k\, \mathbf{e}^k$ is an *outward* unit normal vector to σ and A, then (using standard integral calculus notations, and with $d\mathbf{C}$: elementary contour vector):

$$\textbf{G–G:} \quad \iiint_\tau \operatorname{div} \mathbf{T}\, d\tau = \iint_\sigma \mathbf{T} \cdot \mathbf{n}\, d\sigma$$

$$\Rightarrow \iiint_\tau T^k|_k\, d\tau = \iint_\sigma T^k\, n_k\, d\sigma, \tag{3.6.5a}$$

$$\textbf{S:} \quad \iint_A \operatorname{curl} \mathbf{T} \cdot \mathbf{n}\, dA = \int_C \mathbf{T} \cdot d\mathbf{C}$$

$$\Rightarrow \iint_A e^{rlk}\, T_k|_l\, n_r\, d\mathrm{A} = \int_C T_k\, dC^k. \tag{3.6.5b}$$

Example 3.6.1

Theorem of Stokes in n-Dimensions:* The flux of the rotation (or curling) of a vector field T_k through an (orientable) two-dimensional surface A, in a torsionless n-dimensional manifold (i.e., symmetric Christoffels – see Section 3.8), say, an R_n, equals the work, or circulation, of T_k along the simply connected boundary, or perimeter, of A, C:

$$J \equiv \int_C T_k\, dq^k = \iint_A (T_{k,l} - T_{l,k})\, d_1q^l\, d_2q^k \quad (k, l = 1, 2, \ldots, n;\ k < l) \tag{a}$$

where $d_1q^l \equiv (\partial q^l/\partial u)\, du$, $d_2q^l \equiv (\partial q^l/\partial v)\, dv$ (u, v: coordinates, or parameters, of surface A); or, with the help of

$$dA^{kl} \equiv \delta^{kl}{}_{rs}\, d_1q^r\, d_2q^s = (\delta^k{}_r\, \delta^l{}_s - \delta^k{}_s\, \delta^l{}_r)\, d_1q^r\, d_2q^s$$

$$= \ldots = d_1q^k\, d_2q^l - d_1q^l\, d_2q^k\ (= -dA^{lk}) \quad \text{(recalling Problem 1.2.1)} \tag{b}$$

(antisymmetric infinitesimal *two-dimensional surface element,* spanned by d_1q^k and d_2q^k)

* May be omitted in a first reading.

$$J \equiv \int_C T_k \, dq^k = (1/2) \iint_A (T_{k,l} - T_{l,k}) dA^{lk}$$

$$\left[= (1/2) \iint_A (T_k|_l - T_l|_k) dA^{lk}, \quad \text{since we assumed: } T_{k,l} - T_{l,k} = T_k|_l - T_l|_k\right]$$

$$= \iint_A T_{k,l} \, dA^{lk} = \iint_A T_{k,l} \, (d_1 q^l \, d_2 q^k - d_1 q^k \, d_2 q^l)$$

$$= \iint_A (d_1 T_k \, d_2 q^k - d_2 T_k \, d_1 q^k) = \iint_A (d_1 \mathbf{T} \cdot d_2 \mathbf{R} - d_2 \mathbf{T} \cdot d_1 \mathbf{R}), \tag{c}$$

where: $d_1\mathbf{R} \equiv (d_1 q^k)$ and $d_2\mathbf{R} \equiv (d_2 q^k)$ (see below).

Thus, Stokes' theorem states that: *the substitution* $dq^k \rightarrow (\partial .../\partial q^l) \, dA^{lk}$ *transforms a closed curve integral to one over the surface delimited by that curve.*

Let us elaborate on the earlier three-dimensional form. With $\mathbf{n} \, dA \equiv d\mathbf{A}$, the above yields:

$$J = \iint_A T_{k,l} \, dA^{lk} = \iint_A e^{rlk} T_{k,l} \, dA_r = \iint_A \text{curl } \mathbf{T} \cdot d\mathbf{A}, \tag{d}$$

where (recalling Equation 2.9.9):

$$(d\mathbf{A})_r = (d_1\mathbf{R} \times d_2\mathbf{R})_r \equiv dA_r = e_{rkl} \, d_1 q^k \, d_2 q^l = \sqrt{g} \, \varepsilon_{rkl} \, d_1 q^k \, d_2 q^l$$

is the vector normal to (oriented) surface element formed by the two linearly independent vectors $d_1\mathbf{R} = (d_1 q^k)$ and $d_2\mathbf{R} = (d_2 q^k)$, and of magnitude equal to the *area* of that element:

$$dA_r = (1/2) e_{rlk} \, dA^{lk} = \left(\sqrt{g}/2\right) \varepsilon_{rlk} \, dA^{lk}$$

$$\Rightarrow dA^{lk} = e^{lkr} \, dA_r = \left(\varepsilon^{lkr}/\sqrt{g}\right) dA_r, \tag{e}$$

i.e., dA_r is the *axial vector* of dA^{lk} (recall Equation 2.9.22a ff).

Remarks

Similarly, recalling Equation 2.9.10a, the (oriented) volume element $d\tau$, built from the three linearly independent vectors, or elements, $d_1\mathbf{R} = (d_1 q^k)$, $d_2\mathbf{R} = (d_2 q^k)$, $d_3\mathbf{R} = (d_3 q^k)$, equals

$$d\tau = d_1\mathbf{R} \cdot (d_2\mathbf{R} \times d_3\mathbf{R}) = e_{rkl}\, d_1q^r\, d_2q^k\, d_3q^l = \sqrt{g}\,(\varepsilon_{rkl}\, d_1q^r\, d_2q^k\, d_3q^l)$$

$$= (1/3!)e_{rkl}\, dA^{rkl} = \left(\sqrt{g}/3!\right)\varepsilon_{rkl}\, dA^{rkl}, \tag{f}$$

where (recalling Section 1.2): $dA^{rkl} = \delta^{rkl}{}_{ijs}\, d_1q^i\, d_2q^j\, d_3q^s = 3 \times 3$ Determinant with rows (in that order): (d_1q^r, d_1q^k, d_1q^l), (d_2q^r, d_2q^k, d_2q^l), (d_3q^r, d_3q^k, d_3q^l); just like

$$dA^{kl} = 2 \times 2 \text{ Determinant with rows (in that order): } (d_1q^k, d_1q^l),\ (d_2q^k, d_2q^l). \tag{g}$$

If, further, these line elements coincide with the coordinate lines, i.e., $(d_1q^1 = dq^1$, $d_1q^2 = 0$, $d_1q^3 = 0)$, $(d_2q^1 = 0$, $d_2q^2 = dq^2$, $d_2q^3 = 0)$, $(d_3q^1 = 0$, $d_3q^2 = 0$, $d_3q^3 = dq^3)$, then $d\tau = \sqrt{g}\, dq^1\, dq^2\, dq^3 \equiv \sqrt{g}\, dV(q)$ (recalling Example 2.5.2).

Hence, in such a unifying approach, *ordinary elementary areas and volumes are viewed, respectively, as two- and three-dimensional infinitesimal elements, on corresponding hypersurfaces, in an R_n;* i.e., on a more abstract level, their qualitative differences appear as quantitative ones; and this leads, later, to a unified formulation of the theorems of Gauss–Green and Stokes.

For further details and generalizations of these fundamental concepts and theorems, see, e.g., (alphabetically): Landau and Lifshitz (1964–1970, pp. 33–35), Lovelock and Rund (1975–1989, Ch. 5, esp. pp. 161–163), Prange (1933–1935, pp.657–689), Stephani (1980, pp. 59–64), Synge and Schild (1949, Ch. 7, esp. pp. 267–275); also Carmeli (1982, Ch. 2).

3.7 PARALLELISM, STRAIGHT LINES

The conditions of *parallelism*, or *parallel transport*, or *constancy*, of a tensor field $T^{\cdots}{}_{\cdots}(q)$* are

- Along a manifold *curve* $\{q^k(u), k = 1, \ldots, n; u$: curve parameter$\}$:

$$DT^{\cdots}{}_{\cdots}/Du = 0, \quad \text{or} \quad DT^{\cdots}{}_{\cdots} = 0; \tag{3.7.1a}$$

  for example, constancy of a *vector* along a curve means [with $d^*(\ldots)$ denoting a special differential]:

$$DT^k = dT^k + \Gamma^k{}_{lr}\, T^l\, dq^r = 0 \Rightarrow d^*T^k = -\Gamma^k{}_{lr}\, T^l\, dq^r, \tag{3.7.1b}$$

$$DT_k = dT_k - \Gamma^l{}_{kr}\, T_l\, dq^r = 0 \Rightarrow d^*T_k = \Gamma^l{}_{kr}\, T_l\, dq^r. \tag{3.7.1c}$$

- Throughout the *manifold*:

---

* In mechanics terms: *inertialness*; i.e., no forces ⇒ no accelerations (Chapters 4 ff).

$$T^{\cdots}{}_{\cdots}|_k = 0 \quad [\Rightarrow T^{\cdots}{}_{\cdots} = \text{constant, in rectilinear coordinates}]. \tag{3.7.1d}$$

3.7.1 On Geodesics

A curve $q^k = q^k(s)$ ($u \to s$: arc length) is a *straight(-est)* line if its unit tangent vector $\tau^k \equiv dq^k/ds$ remains parallel to itself, or, it remains constant; i.e.,

$$D\tau^k/Ds \equiv D/Ds(dq^k/ds) = d^2q^k/ds^2 + \Gamma^k{}_{lr}(dq^l/ds)(dq^r/ds) = 0. \tag{3.7.2}^*$$

Such curves are called *geodesics* (of that manifold), and their equations (Equation 3.7.2), turn out to be the *Euler–Lagrange* (or stationarity, or first-order) conditions of the following variational problem:

$$\int ds \equiv \int (g_{kl}\, dq^k\, dq^l)^{1/2} \equiv \int [g_{kl}\, (dq^k/ds)(dq^l/ds)]^{1/2}\, ds$$

$$\equiv \int \sqrt{F}\, ds = \text{stationary}; \tag{3.7.3a}$$

or, equivalently (to avoid square roots),

$$\int F\, ds \equiv \int [g_{kl}(dq^k/ds)(dq^l/ds)]ds = \text{stationary}; \tag{3.7.3b}$$

i.e.,

$$\delta \int F\, ds = 0 \Rightarrow E_k(F) \equiv d/ds[\partial F/\partial(dq^k/ds)]$$

$$- \partial F/\partial q^k = 0\text{: Equation 3.7.2.} \tag{3.7.4}^{**}$$

3.7.2 (First) Proof of the Christoffel Transformation Equations (Equations 3.3.2e ff)

This proof of Equations 3.3.2e ff uses the invariance of the geodesic equation (3.7.2) (see also Example 3.8.1).

---

* Since $g_{kl}(dq^k/ds)(dq^l/ds) = ds^2/ds^2 = 1$. If $u \neq s$, then $\tau^k = (dq^k/du)(du/ds) = (dq^k/du) / (ds/du)$, and so $D\tau^k/Ds = 0$, Equation 3.7.2, is replaced by

$$d^2q^k/du^2 + \Gamma^k{}_{rs}(dq^r/du)(dq^s/du) = [(d^2s/du^2) / (ds/du)](dq^k/du). \tag{3.7.2a}$$

More on geodesics in Section 5.5.

** For the equivalence of Equation 3.7.3a with 3.7.3b, and the derivation of Equation $3.7.4_2$ from Equation $3.7.4_1$, see, e.g., Eddington (1924, §28); or works on variational calculus/differential geometry.

Since the variational Equation 3.7.4$_1$ involves invariant quantities (like ds), its Euler–Lagrange Equation 3.7.4$_2$ are *form invariant*; i.e., in another CS $q' \Leftrightarrow q$ the geodesic Equation 3.7.2 will be

$$d^2q^{k'}/ds^2 + \Gamma^{k'}{}_{l'r'}\,(dq^{l'}/ds)\,(dq^{r'}/ds) = 0. \tag{3.7.5a}$$

Substituting into this the tensor equation: $dq^{k'}/ds = A^{k'}{}_k\,(dq^k/ds)$ and its $[d(\ldots)/ds]$-derivative:

$$d^2q^{k'}/ds^2 = (dA^{k'}{}_k/ds)(dq^k/ds) + A^{k'}{}_k(d^2q^k/ds^2) \quad \text{(then invoking Equation 3.7.2)}$$

$$= [(\partial A^{k'}{}_k/\partial q^r)\,(dq^r/ds)](dq^k/ds) + A^{k'}{}_k\,[-\Gamma^k{}_{lr}\,(dq^l/ds)(dq^r/ds)],$$

simplifying and changing some dummy indices, we finally obtain the transformation equations:

$$\Gamma^{k'}{}_{l'r'} = A^{k'}{}_k\,A^l{}_{l'}\,A^r{}_{r'}\,\Gamma^k{}_{lr} - A^l{}_{l'}\,A^r{}_{r'}\,A^{k'}{}_{l,r}, \quad \text{i.e., Equation 3.3.2g.} \tag{3.7.5b}$$

3.8 GEOMETRICAL INTERPRETATION OF CHRISTOFFELS; AFFINE MANIFOLDS, TORSION

3.8.1 Euclidean Manifolds

Let us begin by considering a vector field **V** at the point P of, say (with no loss in generality), a three-dimensional Euclidean manifold E_3, referred to the local covariant natural frame there $\{P, \mathbf{e}_k \equiv \partial\mathbf{R}/\partial q^k;\ k = 1, 2, 3\}$ where, recalling Section 2.9, $\mathbf{R} = \mathbf{R}(q^k)$: position vector of P relative to some E_3 – origin O; i.e., $\mathbf{V}(P) = V^k(P)\,\mathbf{e}_k(P)$. Then, at a neighboring E_3-point $P + dP(q^k + dq^k)$, and to the first order in the dqs, we shall have

$$\mathbf{V}(P + dP) = \mathbf{V}(P) + d\mathbf{V}(P) = [V^k(P) + dV^k(P)]\,[\mathbf{e}_k(P) + d\mathbf{e}_k(P)]$$

and therefore (from now on dropping the P-dependence):

$$d\mathbf{V} = (V^k + dV^k)(\mathbf{e}_k + d\mathbf{e}_k) - V^k\,\mathbf{e}_k \approx dV^k\,\mathbf{e}_k + V^k\,d\mathbf{e}_k \quad [= (\partial\mathbf{V}/\partial q^k)dq^k]$$

$$= [(\partial V^k/\partial q^l)dq^l]\,\mathbf{e}_k + V^k[(\partial\mathbf{e}_k/\partial q^l)dq^l] = [(\partial V^l/\partial q^k)\mathbf{e}_l + V^l(\partial\mathbf{e}_l/\partial q^k)]dq^k$$

$$\Rightarrow \partial\mathbf{V}/\partial q^k = (\partial V^l/\partial q^k)\mathbf{e}_l + V^l(\partial\mathbf{e}_l/\partial q^k)$$

$$= \text{apparent gradient} + \text{“spurious” change.} \tag{3.8.1}$$

In words, $\partial \mathbf{V}/\partial q^k \equiv \mathbf{V}_{,k}$ consists of a part arising from the variation of the components V^k, as the q^k change (apparent gradient), and another due to the basis vector change as $P \to P + dP$ (spurious change — absent in rectilinear coordinates). Let us examine this last term closely: from the definition $\mathbf{e}_k \equiv \partial \mathbf{R}/\partial q^k$ we readily obtain the *integrability*, or gradient(-ness), condition for $\{\mathbf{e}_k\}$:

$$\mathbf{e}_{k,l} = \partial/\partial q^l(\partial \mathbf{R}/\partial q^k) = \partial/\partial q^k(\partial \mathbf{R}/\partial q^l) = \mathbf{e}_{l,k}$$

$$\text{(although, generally, } \mathbf{e}^k{}_{,l} \neq \mathbf{e}^l{}_{,k}).^* \tag{3.8.2}$$

Now, in rectangular Cartesian coordinates $y_k = y^k = y^k(q^l)$, we have

$$d\mathbf{R} = d\mathbf{R}(y) = (\partial \mathbf{R}/\partial y^k)dy^k \equiv \mathbf{e}_{<k>}\, dy^k = d\mathbf{R}(q) = (\partial \mathbf{R}/\partial q^k)dq^k \equiv \mathbf{e}_k\, dq^k,$$

$$\Rightarrow \mathbf{e}_{<k>}\, dy^k = \mathbf{e}_l\, dq^l = \mathbf{e}_l(\partial q^l/\partial y^k)dy^k = \mathbf{e}_{<k>}(\partial y^k/\partial q^l)dq^l$$

$$\Rightarrow \mathbf{e}_{<k>} = (\partial q^l/\partial y^k)\mathbf{e}_l \Leftrightarrow \mathbf{e}_l = (\partial y^k/\partial q^l)\mathbf{e}_{<k>} \quad (\mathbf{e}_{<k>}\text{: } \textit{unit} \text{ vectors}) \tag{3.8.3a}$$

and, similarly,

$$\mathbf{e}_{<k>} = \mathbf{e}^{<k>} = (\partial y^k/\partial q^l)\mathbf{e}^l \Leftrightarrow \mathbf{e}^l = (\partial q^l/\partial y^k)\mathbf{e}^{<k>}. \tag{3.8.3b}$$

Next, differentiating Equation 3.8.3a$_2$, while noting that: $\mathbf{e}_{<k>,l} = \mathbf{0}$ (rectilinear basis), recalling the Christoffel transformation rule $z \to y$ (Equations 3.3.3a and b) and definitions (Equations 3.3.1a and b), we obtain

$$\mathbf{e}_{k,l} = (\partial^2 y^r/\partial q^l \partial q^k)\mathbf{e}_{<r>} = (\partial^2 y^r/\partial q^l \partial q^k)\,[(\partial q^s/\partial y^r)\mathbf{e}_s]$$

$$= [(\partial q^s/\partial y^r)(\partial^2 y^r/\partial q^l \partial q^k)]\,(g_{sh}\,\mathbf{e}^h)$$

i.e.,

$$\mathbf{e}_{k,l} = \Gamma^s{}_{kl}\,\mathbf{e}_s = \Gamma_{s,kl}\,\mathbf{e}^s \quad [\Rightarrow d\mathbf{e}_k = (\Gamma^s{}_{kl}\,dq^l)\mathbf{e}_s = (\Gamma_{s,kl}\,dq^l)\,\mathbf{e}^s]. \tag{3.8.4a}$$

The corresponding formula for the contravariant basis $\{\mathbf{e}^k\}$ is found either by repeating the above for Equation 3.8.3b; or by differentiating the defining relation $\mathbf{e}^k \cdot \mathbf{e}_s = \delta^k{}_s$. Indeed, the latter yields

$$\mathbf{e}^k{}_{,l} \cdot \mathbf{e}_s + \mathbf{e}^k \cdot \mathbf{e}_{s,l} = 0 \Rightarrow \mathbf{e}^k{}_{,l} \cdot \mathbf{e}_s = -\,\mathbf{e}_{s,l} \cdot \mathbf{e}^k$$

$$= -\,(\Gamma^r{}_{sl}\,\mathbf{e}_r) \cdot \mathbf{e}^k = -\Gamma^r{}_{sl}\,\delta^k{}_r = -\Gamma^k{}_{sl},$$

i.e.,

* Easily verified by differentiating their definition: $\mathbf{e}_k = g_{kr}\,\mathbf{e}^r \Rightarrow \mathbf{e}_{k,l} = \ldots.$

$$\mathbf{e}^k{}_{,l} = -\Gamma^k{}_{sl}\,\mathbf{e}^s = -g^{kr}\,\Gamma_{r,sl}\,\mathbf{e}^s = -g^{rs}\,\Gamma^k{}_{rl}\,\mathbf{e}_s\ [\Rightarrow d\mathbf{e}^k = (-\Gamma^k{}_{sl}\,dq^l)\mathbf{e}^s]. \quad (3.8.4b)$$

These basic results can be summarized in the following equivalent forms (compare with Equations 3.3.3a):

$$\Gamma^k{}_{lr} = \mathbf{e}^k \cdot \mathbf{e}_{l,r} = \mathbf{e}^k \cdot \mathbf{e}_{r,l} = \Gamma^k{}_{rl}\ (= -\mathbf{e}_l \cdot \mathbf{e}^k{}_{,r} = -\mathbf{e}_r \cdot \mathbf{e}^k{}_{,l}) \quad (3.8.5a)$$

$$\Gamma_{k,lr} = \mathbf{e}_k \cdot \mathbf{e}_{l,r} = \mathbf{e}_k \cdot \mathbf{e}_{r,l} = \Gamma_{k,rl}; \quad (3.8.5b)$$

$[\Rightarrow \Gamma^k{}_{kr} = \mathbf{e}^k \cdot \mathbf{e}_{k,r} = \mathbf{e}^k \cdot \mathbf{e}_{r,k} = -\mathbf{e}_k \cdot \mathbf{e}^k{}_{,r} = -\mathbf{e}_r \cdot \mathbf{e}^k{}_{,k} = \operatorname{div} \mathbf{e}_r$ (recall Equation 3.8.2)];

which clearly show that *the* Γ*s are the coefficients, or parameters, of the elementary rotation of the local basis* $\{\mathbf{e}_k\}$ in *going from P to the adjacent P* + *dP*.*

3.8.1.1 Relation of Equations 3.8.5a and b with the Earlier Christoffel Definitions (Equations 3.3.1a and b)

In view of Equation 3.8.5b, we find, successively,

$$\begin{aligned} 2\,\Gamma_{k,lr} &= \mathbf{e}_k \cdot \mathbf{e}_{l,r} + \mathbf{e}_k \cdot \mathbf{e}_{r,l} \\ &= [(\mathbf{e}_k \cdot \mathbf{e}_l)_{,r} - \mathbf{e}_l \cdot \mathbf{e}_{k,r}] + [(\mathbf{e}_k \cdot \mathbf{e}_r)_{,l} - \mathbf{e}_r \cdot \mathbf{e}_{k,l}] \\ &= [(\mathbf{e}_k \cdot \mathbf{e}_l)_{,r} + (\mathbf{e}_k \cdot \mathbf{e}_r)_{,l}] - (\mathbf{e}_l \cdot \mathbf{e}_{k,r} + \mathbf{e}_r \cdot \mathbf{e}_{k,l}) \\ &= [(\mathbf{e}_k \cdot \mathbf{e}_l)_{,r} + (\mathbf{e}_k \cdot \mathbf{e}_r)_{,l}] - (\mathbf{e}_l \cdot \mathbf{e}_{r,k} + \mathbf{e}_r \cdot \mathbf{e}_{l,k}) \\ &= (\mathbf{e}_k \cdot \mathbf{e}_l)_{,r} + (\mathbf{e}_k \cdot \mathbf{e}_r)_{,l} - (\mathbf{e}_l \cdot \mathbf{e}_r)_{,k} \\ &= g_{kl,r} + g_{kr,l} - g_{lr,k}; \quad \text{and similarly for } \Gamma^k{}_{lr}. \end{aligned} \quad (3.8.6)$$

3.8.1.2 Special Case: Moving Orthonormal Basis

In there:

$$\mathbf{e}_k = \mathbf{e}^k = \mathbf{e}_k(q) \quad \text{and} \quad \mathbf{e}_k \cdot \mathbf{e}_l = g_{kl} = \delta_{kl}\text{: constant;} \quad (3.8.7a)$$

and therefore $(\ldots)_{,r}$-differentiating Equation 3.8.7a$_2$ we get

$$\mathbf{e}_{k,r} \cdot \mathbf{e}_l + \mathbf{e}_k \cdot \mathbf{e}_{l,r} = \delta_{kl,r} = 0;$$

or, recalling Equation 3.8.5b,

* And this kinematic interpretation carries over to the cases of nongradient, or nonholonomic, bases; to be discussed in Section 3.15ff.

$$\Gamma_{l,kr} + \Gamma_{k,lr} = 0; \tag{3.8.7b}$$

i.e., the first kind Christoffels of a moving orthonormal basis are antisymmetric in their first two indices. (See also Example 3.8.3.)

Problem 3.8.1. Ricci's Lemma for the Basis Vectors

Verify that

$$\mathbf{e}_k|_l \equiv \mathbf{e}_{k,l} - \Gamma^r{}_{kl}\,\mathbf{e}_r = \mathbf{0} \quad \text{and} \quad \mathbf{e}^k|_l \equiv \mathbf{e}^k{}_{,l} + \Gamma^k{}_{rl}\,\mathbf{e}^r = \mathbf{0}; \tag{a}$$

i.e., *the basis vectors (like the metric tensor) behave as constant under covariant differentiation.* (*Hint:* Use Equations 3.8.4a and b.)

3.8.1.3 Geometrical Interpretation of CDs and ADs

Let us now return to Equation 3.8.1. In view of the preceding results, we can rewrite it as

$$\mathbf{V}_{,k} = V^l{}_{,k}\,\mathbf{e}_l + V^l\,\mathbf{e}_{l,k} = V^l{}_{,k}\,\mathbf{e}_l + V^l\,(\Gamma^r{}_{lk}\,\mathbf{e}_r) = (V^l{}_{,k} + \Gamma^l{}_{rk}\,V^r)\,\mathbf{e}_l, \tag{3.8.8a}$$

i.e.,

$$\partial\mathbf{V}/\partial q^k \equiv \mathbf{V}_{,k} = V^l|_k\,\mathbf{e}_l = V_l|_k\,\mathbf{e}^l$$

$$(\Rightarrow V^k|_k = \mathbf{V}_{,k}\cdot\mathbf{e}^k = \text{div}\,\mathbf{V}, \text{ recalling Equation 3.6.2}); \tag{3.8.8b}$$

the last equality either directly or by the Quotient Rule (Sections 2.6 and 2.7). Alternatively, using the results of Problem 3.8.1, above, we find

$$\mathbf{V}|_k \equiv (V^l\,\mathbf{e}_l)|_k = V^l|_k\,\mathbf{e}_l + V^l\,\mathbf{e}_l|_k = V^l|_k\,\mathbf{e}_l; \quad \text{i.e., } \mathbf{V}|_k \equiv \partial\mathbf{V}/\partial q^k \equiv \mathbf{V}_{,k}. \tag{3.8.8c}$$

Therefore, we can write, successively,

$$d\mathbf{V} = \mathbf{V}_{,k}\,dq^k = (V^l|_k\,dq^k)\mathbf{e}_l = (V_l|_k\,dq^k)\mathbf{e}^l \quad \text{(by Quotient Rule)} \tag{3.8.9a}$$

$$= DV^k\,\mathbf{e}_k = DV_k\,\mathbf{e}^k \quad \text{(recalling Equation 3.5.2)} \tag{3.8.9b}$$

i.e.,

$$(d\mathbf{V})^k \equiv d\mathbf{V}\cdot\mathbf{e}^k = DV^k \equiv dV^k + \Gamma^k{}_{lr}\,V^l\,dq^r,$$

$$\text{i.e., } (d\mathbf{V})^k \neq dV^k, \text{ in general} \tag{3.8.10a}$$

$$(d\mathbf{V})_k \equiv d\mathbf{V} \cdot \mathbf{e}_k = DV_k \equiv dV_k - \Gamma^l{}_{kr}\, V_l\, dq^r,$$

$$\text{(i.e., } (d\mathbf{V})_k \neq dV_k, \text{ in general).} \tag{3.8.10b}$$

These equations are of fundamental importance to rigid-body kinematics (moving frames etc.): there, $(d\mathbf{V})^k$ and $(d\mathbf{V})_k$ represent the *absolute* (i.e., inertial) change of $\mathbf{V}$, dV^k and dV_k represent the *relative* (or local) change of $\mathbf{V}$, while $+\,\Gamma^k{}_{lr}\, V^l\, dq^r$ and $-\,\Gamma^l{}_{kr}\, V_l\, dq^r$ represent the contribution of the motion of $\{\mathbf{e}_k\}$, or "frame transport," to $d\mathbf{V}$ — and are, thus, proportional to the inertial *angular velocity vector* $\rightarrow$ *tensor* of $\{\mathbf{e}_k\}$. Or, the absolute/intrinsic differential $DV^k \equiv (d\mathbf{V})^k$, etc. is the ordinary differential from the viewpoint of a local frame displaced, or transported, *parallel to itself* (i.e., no rotation) from $P(q)$ to $P + dP(q + dq)$.

Similarly, if $\mathbf{V}$ is the inertial velocity of a particle, then *parallel displacement*, or *transport*, of $\mathbf{V}$ (i.e., inertial constancy of it) means that $d\mathbf{V} = \mathbf{0}$; or, recalling Equations 3.7.1b and c,

$$d^*V^k \equiv (\partial V^k/\partial q^r)^*\, dq^r = -\Gamma^k{}_{lr}\, V^l\, dq^r,$$

$$d^*V_k \equiv (\partial V_k/\partial q^r)^*\, dq^r = \Gamma^l{}_{kr}\, V_l\, dq^r, \tag{3.8.10c}$$

from which

$$(\partial V^k/\partial q^r)^* = -\Gamma^k{}_{lr}\, V^l \quad \text{and} \quad (\partial V_k/\partial q^r)^* = \Gamma^l{}_{kr}\, V_l. \tag{3.8.10d}$$

(See Examples 3.8.2 and 3.8.3 below.) In Figure 3.1 we show the parallel transport of a vector $\mathbf{V}$ from point P_1 ($r = R$, $\phi = 0$) to point P_2 ($r = R$, $\phi = \pi/2$) along the (first) quadrant of a circle of radius R, on the plane E_2.*

3.8.1.4 Geometrical Meaning of dV^k, d^*V^k, and DV^k

To the first order, we have

$$dV^k \equiv V^k(q + dq) - V^k(q)$$

: V^k-difference *before* the transport of $\mathbf{V}$ from $P(q)$ to $P + dP \equiv Q(q + dq)$,

$$d^*V^k \equiv V^{*k}(q + dq) - V^k(q) \quad \text{(recalling Equations 3.8.10c and d)}$$

* As Landau and Lifshitz (1964–1970, §85) point out, the special differential DV^k, or DV_k, is needed because in *curvilinear coordinates, for the differential of a vector, etc. to be a vector, the two subtracted vectors* (which, to the first order produce this differential) *must both be at the same manifold point.* That is, one of the vectors must be, somehow, *transported* (or transplanted) to the nearby point of the other vector. And this transportation must be defined so that in rectilinear/affine coordinates (constant g_{kl}) it reduces to dV^k, dV_k; in such coordinates, due to the constancy of the basis vectors, parallel transport does not change the vector components; i.e., $DV^k = 0 \Rightarrow dV^k = 0$. More on this in the section on parallel transport in linearly, or affinely, connected and metric manifolds, below.

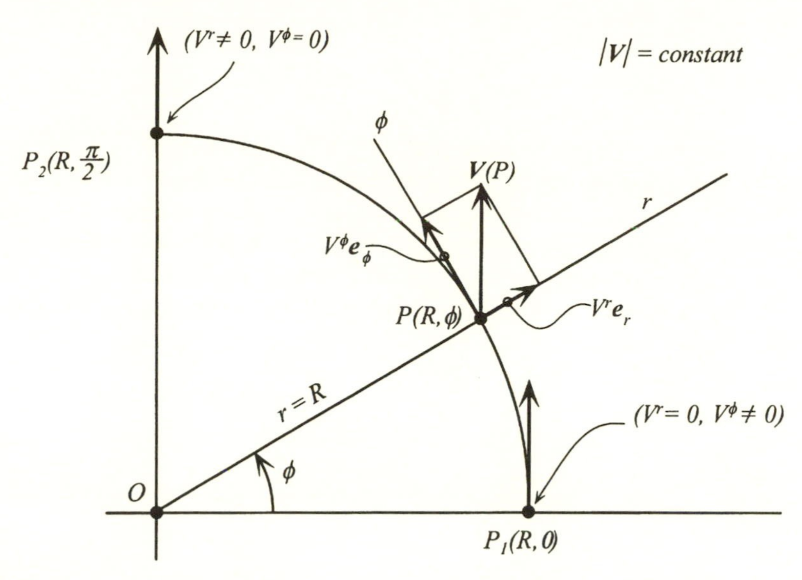

FIGURE 3.1 Parallel transport of a vector **V**, of constant length, along a circular arc.

: V^k-difference *after* the parallel transport of **V** from $P(q)$ to $P + dP$

$\equiv Q(q + dq)$; i.e., $V^{*k}(q + dq) = V^k$-component of $\mathbf{V}^*(Q)$ at Q (Figure 3.2)

Therefore,

$$dV^k - d^*V^k = [V^k(Q) - V^k(P)] - [V^{*k}(Q) - V^k(P)]$$

$$= V^k(Q) - V^{*k}(Q) = (V^k{}_{,r}\, dq^r) - (-\Gamma^k{}_{lr}\, V^l\, dq^r), \tag{3.8.11a}$$

$$\Rightarrow DV^k = dV^k - d^*V^k, \quad \text{i.e., (non-tensor) – (non-tensor) = tensor!} \tag{3.8.11b}$$

The extension of the above results to n-dimensional Euclidean spaces, E_n, is obvious.

Problem 3.8.2

Under parallel transport, the dot product of two absolute vectors, V^k and U_k, being an invariant, does not change; i.e., $d^*(V^kU_k) = 0$. From this and Equation 3.8.10c$_1$ deduce that

$$d^*U_k = \Gamma^l{}_{kr}\, U_l\, dq^r, \quad \text{i.e., Equations 3.8.10c}_2\text{, and 3.7.1c.} \tag{a}$$

3.8.2 General Linearly, or Affinely, Connected and Metric-Equipped Manifolds

Next, and continuing from Section 2.12, let us reexamine the above but in a general *linearly*, or *affinely, connected and metric manifold*, i.e., *an L_n with metric,*

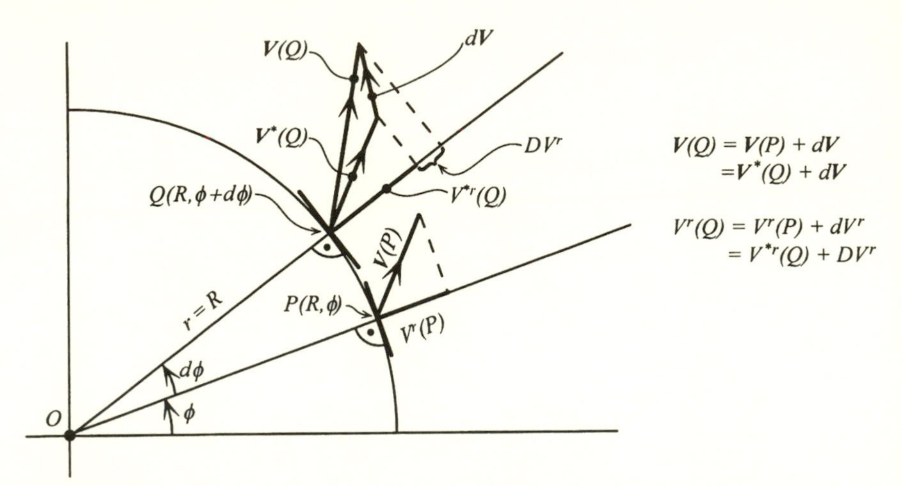

FIGURE 3.2 On the geometrical meaning of dV^k, d^*V^k, and DV^k (two dimensions, polar coordinates).

say g_{kl}. Here, the entire development is based on the following two basic assumptions:

1. *Adjacent* L_n*-bases,* $\{P, \mathbf{e}_k\}$ *and* $\{P + dP, \mathbf{e}_k + d\mathbf{e}_k\}$, *are connected by the linear mapping:*

$$d\mathbf{e}_k = \Lambda^s{}_{kl}\, dq^l\, \mathbf{e}_s \quad \text{(Transport law in a linearly, or affinely, connected manifold } L_n\text{)}, \tag{3.8.12a}$$

   or, equivalently,

$$\mathbf{e}_{k,l} = \Lambda^s{}_{kl}\, \mathbf{e}_s \Rightarrow \Lambda^s{}_{kl} = \mathbf{e}^s \cdot \mathbf{e}_{k,l}, \tag{3.8.12b}$$

   where *the* $\Lambda^k{}_{lr} = \Lambda^k{}_{lr}(q)$ *are the components (or coefficients, or parameters) of connection in* L_n *(of the second kind),* in the coordinates q; or, simply, its *affinities*; and
2. *The linear differential form* $d\mathbf{P} \equiv \mathbf{e}_k\, dq^k$ (Section 2.12) *is integrable.* As is well known from vector calculus, the necessary and sufficient conditions for this are

$$\mathbf{e}_{k,l} = \mathbf{e}_{l,k} \quad \text{(integrability conditions for } \mathbf{e}_k\, dq^k\text{, also in Problem 3.9.1)}, \tag{3.8.13}$$

Let us uncover the consequences of 1 and 2.

- To obtain $d\mathbf{e}^k$, we differentiate $\mathbf{e}^k \cdot \mathbf{e}_l = \delta^k{}_l = \text{constant}$, and then use Equation 3.8.12a for $d\mathbf{e}_l$. Thus, we find

$$0 = d(\mathbf{e}^k \cdot \mathbf{e}_l) = d\mathbf{e}^k \cdot \mathbf{e}_l + \mathbf{e}^k \cdot d\mathbf{e}_l = d\mathbf{e}^k \cdot \mathbf{e}_l + \mathbf{e}^k \cdot (\Lambda^s{}_{lb}\, dq^b\, \mathbf{e}_s)$$

$$= d\mathbf{e}^k \cdot \mathbf{e}_l + \delta^k{}_s(\Lambda^s{}_{lb}\, dq^b) = d\mathbf{e}^k \cdot \mathbf{e}_l + \Lambda^k{}_{lb}\, dq^b,$$

$$\Rightarrow d\mathbf{e}^k = -\Lambda^k{}_{hl}\, dq^l\, \mathbf{e}^h; \quad \text{or, equivalently, } \mathbf{e}^k{}_{,l} = -\Lambda^k{}_{hl}\, \mathbf{e}^h. \tag{3.8.14}$$

Similarly, differentiating $g_{kl} \equiv \mathbf{e}_k \cdot \mathbf{e}_l$, and then applying Equation 3.8.12a, we get

$$dg_{kl} = d\mathbf{e}_k \cdot \mathbf{e}_l + \mathbf{e}_k \cdot d\mathbf{e}_l = (\Lambda^s{}_{kb}\, dq^b\, \mathbf{e}_s) \cdot \mathbf{e}_l + \mathbf{e}_k \cdot (\Lambda^s{}_{lb}\, dq^b\, \mathbf{e}_s)$$

$$= (\Lambda^s{}_{kb}\, g_{sl} + \Lambda^s{}_{lb}\, g_{ks})\, dq^b,$$

or, equivalently,

$$\partial g_{kl}/\partial q^b \equiv g_{kl,b} = g_{sl}\, \Lambda^s{}_{kb} + g_{ks}\, \Lambda^s{}_{lb}, \tag{3.8.15}$$

or, with the definitions:

$\Lambda_{l,kb} \equiv g_{sl}\, \Lambda^s{}_{kb} = \mathbf{e}_l \cdot \mathbf{e}_{k,b}$: L_n-affinities of the *first* kind

$$[\Rightarrow \Lambda^s{}_{kb} = g^{sl}\, \Lambda_{l,kb}] \tag{3.8.16a}$$

$$[\Rightarrow d\mathbf{e}_k = \Lambda_{b,kl}\, dq^l\, \mathbf{e}^b \quad \text{or} \quad \mathbf{e}_{k,l} = \Lambda_{b,kl}\, \mathbf{e}^b] \tag{3.8.12c}$$

finally,

$$g_{kl,b} = \Lambda_{l,kb} + \Lambda_{k,lb}. \tag{3.8.16b}$$

- From Equation 3.8.12a and its consequence (Equation $3.8.14_1$), and invoking the reasoning of Section 3.1, we easily conclude that *the covariant and other invariant derivatives of L_n-vectors and general tensors are formally identical to their E_n-counterparts (Sections 3.4 and 3.5), but with their Christoffels replaced by the affinities.* (See Equations 3.8.18 through 3.8.19f, below; also Example 3.8.2.) Thus, we immediately obtain Ricci's lemma for the L_n-bases $\{\mathbf{e}_k\}$, $\{\mathbf{e}^k\}$:

$$D\mathbf{e}_k \equiv d\mathbf{e}_k - \Lambda^l{}_{ks}\, \mathbf{e}_l\, dq^s = \mathbf{0}, \quad D\mathbf{e}^k \equiv d\mathbf{e}^k + \Lambda^k{}_{ls}\, \mathbf{e}^l\, dq^s = \mathbf{0}. \tag{3.8.17}$$

- Utilizing Equation 3.8.12b in Equation 3.8.13 we easily conclude that *the affinities are symmetric in their subscripts:*

$$\mathbf{0} = \mathbf{e}_{k,l} - \mathbf{e}_{l,k} = (\Lambda^s{}_{kl} - \Lambda^s{}_{lk})\mathbf{e}_s \Rightarrow \Lambda^s{}_{kl} = \Lambda^s{}_{lk} \quad (\Rightarrow \Lambda_{s,kl} = \Lambda_{s,lk}). \tag{3.8.18}$$

(As explained later, these symmetry relations express the *torsionlessness* of our manifold.)

Now, and in analogy with the Euclidean case, we define the (infinitesimal) *parallel transport*, or *parallel displacement*, of an L_n-vector $\mathbf{V}$ by the constancy, or inertialness, requirement $d\mathbf{V} = \mathbf{0}$. This, with the help of Equation 3.8.12a, yields, using $d(\ldots) \to d^*(..)$ for such changes:

$$\mathbf{0} = d(V^k\,\mathbf{e}_k) = dV^k\,\mathbf{e}_k + V^k\,d\mathbf{e}_k = dV^k\,\mathbf{e}_k + V^k(\Lambda^s{}_{kl}\,dq^l\,\mathbf{e}_s)$$

$$= (dV^k + \Lambda^k{}_{ls}\,V^l\,dq^s)\mathbf{e}_k \equiv DV^k\,\mathbf{e}_k \quad [= d\mathbf{V} = D\mathbf{V}, \text{ by } (3.8.17)]; \tag{3.8.19a}$$

in sum,

$$d\mathbf{V} = \mathbf{0} \Rightarrow DV^k = 0 \tag{3.8.19b}$$

$$\Rightarrow d^*V^k = -\Lambda^k{}_{ls}\,V^l\,dq^s \quad [= -\Lambda^k{}_{sl}\,V^l\,dq^s, \text{ if } \Lambda^k{}_{ls} = \Lambda^k{}_{sl}] \tag{3.8.19c}$$

or, equivalently,

$$(\partial V^k/\partial q^s)^* = -\Lambda^k{}_{ls}\,V^l \quad [= -\Lambda^k{}_{sl}\,V^l, \text{ if } \Lambda^k{}_{ls} = \Lambda^k{}_{sl}]; \tag{3.8.19d}$$

and similarly, differentiating $V_k\,\mathbf{e}^k$, we get

$$d\mathbf{V} = DV_k\,\mathbf{e}^k = \mathbf{0} \Rightarrow DV_k \equiv dV_k - \Lambda^l{}_{ks}\,V_l\,dq^s = 0,$$

i.e.,

$$d^*V_k = \Lambda^l{}_{ks}\,V_l\,dq^s \quad [= \Lambda^l{}_{sk}\,V_l\,dq^s, \text{ if } \Lambda^l{}_{ks} = \Lambda^l{}_{sk}] \tag{3.8.19e}$$

or, equivalently,

$$(\partial V_k/\partial q^s)^* = \Lambda^l{}_{ks}\,V_l \quad [= \Lambda^l{}_{sk}\,V_l, \text{ if } \Lambda^l{}_{ks} = \Lambda^l{}_{sk}]. \tag{3.8.19f}$$

It is not hard to verify that *under such transport, the length of* $\mathbf{V}$ *remains invariant,* as in an E_n. Indeed, $d^*(\ldots)$-differentiating the square of the length of $\mathbf{V}$, $V^2 = g_{kl}\,V^k\,V^l$, we get

$$d^*(g_{kl}\,V^k\,V^l) \equiv d^*(\mathbf{V}\cdot\mathbf{V}) = d^*\mathbf{V}\cdot\mathbf{V} + \mathbf{V}\cdot d^*\mathbf{V}$$

$$= 0 \text{ (length invariance)}; \tag{3.8.20a}$$

or, expanding its first term and using in it Equation 3.8.19c,

$$0 = d(V^2) = (g_{kl,b}\,dq^b)V^k\,V^l + g_{kl}(d^*V^k)V^l + g_{kl}\,V^k(d^*V^l)$$

$$= (g_{kl,b} - g_{sl}\,\Lambda^s{}_{kb} - g_{ks}\,\Lambda^s{}_{lb})V^k\,V^l\,dq^b \equiv (Dg_{kl})V^k\,V^l \quad \text{(recall Equation 3.5.2)}$$

i.e.,

$$d(V^2) = (Dg_{kl})V^k\ V^l = 0; \tag{3.8.20b}$$

or, since the V^k, dq^k are arbitrary,

$$Dg_{kl} = 0 \Rightarrow g_{kl}|_b = 0 \Rightarrow g_{kl,b} = g_{sl}\ \Lambda^s{}_{kb} + g_{ks}\ \Lambda^s{}_{lb} \equiv \Lambda_{l,kb} + \Lambda_{k,lb}, \tag{3.8.20c}$$

i.e., Equation 3.8.16b. We notice that Equation 3.8.16b (or Equation 3.8.16c below) express the L_n-version of Ricci's lemma (Equation 3.5.3). Similar steps when applied to the scalar product of two parallel transported L_n-vectors, **V**, **U** also lead to (3.8.16b) and (3.8.20c) [Expand $0 = d^*(\mathbf{V} \cdot \mathbf{U}) \equiv d^*(g_{kl}\ V^k\ U^l) = \ldots$, etc.]

3.8.2.1 Fundamental Theorem of Riemannian Geometry

Now we are ready for the final and most important act of this subject: *bring together the affinities of the manifold (roughly, its deviation from rectilinearity) with its metric properties; i.e., relate its $\Lambda^k{}_{rs}$s with its g_{kl}s.* Indeed, as shown below, this is possible, and constitutes what is known as

The Fundamental Theorem of (local) Riemannian Geometry: Under the twin requirements of linear connectedness (Equations 3.8.12a and b) and integrability (or torsionlessness) (Equation 3.8.18), the affinities reduce (uniquely) to the earlier Christoffels (Section 3.3); i.e., then,

$$2\Lambda_{l,sk} = 2\Lambda_{l,ks} = 2\Gamma_{l,sk} = 2\Gamma_{l,ks} = g_{lk,s} + g_{ls,k} - g_{ks,l}, \tag{3.8.21a}$$

$$\Lambda^b{}_{sk} = \Lambda^b{}_{ks} = \Gamma^b{}_{sk} = \Gamma^b{}_{ks} \equiv g^{bl}\ \Lambda_{l,sk} = g^{bl}\ \Gamma_{l,sk}$$

$$= (g^{bl}/2)(g_{ls,k} + g_{lk,s} - g_{sk,l}). \tag{3.8.21b}$$

In view of the preceding results, the theorem can also be formulated as follows:

If, in addition to Equation 3.8.18, we require that the *length* of a manifold vector (or, equivalently, the scalar product of two manifold vectors), based on the given $n(n + 1)/2$ metric coefficients $g_{kl}(q) = g_{lk}(q)$ (say, positive definite, so that, eventually, $L_n \to R_n$), is *invariant under parallel transport* (for all manifold points in some region of interest, and along every direction, (i.e., C^1-manifold curve, through these points), then the Λs reduce to the earlier Christoffels;

or,

Torsionlessness plus length preservation under parallel transport determine the affinities uniquely (see proof below);

or,

*The only symmetric affinities that preserve vector lengths under parallel displacements are the Christoffels.**

The proof is simple: The $n' \equiv n^2(n+1)/2$ equations (3.8.16b) [$(n+1)/2$ equations, for each b] along with the $n'' \equiv n^2(n-1)/2$ equations (3.8.18) constitute a determinate system of $n' + n'' = n^3$ *linear* equations for the n^3 $\Lambda_{l,kb}$s. Indeed, by cyclic permutation of k, l, b in Equation 3.8.16b, we get

$$g_{lb,k} = \Lambda_{b,lk} + \Lambda_{l,bk} \tag{3.8.16c}$$

and

$$g_{bk,l} = \Lambda_{k,bl} + \Lambda_{b,kl}; \tag{3.8.16d}$$

then adding Equations 3.8.16b and c, and subtracting Equation 3.8.16d, we obtain

$$g_{lk,b} + g_{lb,k} - g_{bk,l} = (\Lambda_{l,kb} + \Lambda_{l,bk}) + (\Lambda_{b,lk} - \Lambda_{b,kl}) + (\Lambda_{k,lb} - \Lambda_{k,bl}),$$

or, since we assumed $\Lambda_{l,kb} = \Lambda_{l,bk}$, finally, $2\Lambda_{l,bk} = g_{lk,b} + g_{lb,k} - g_{bk,l} = 2\Gamma_{l,bk}$; and from this, $2\Lambda^h{}_{bk} \equiv g^{hl}\,(2\Lambda_{l,bk}) = g^{hl}(g_{lk,b} + g_{lb,k} - g_{bk,l}) = g^{hl}\,(2\Gamma_{l,bk}) \equiv 2\Gamma^h{}_{bk}$, which are none other than the earlier Equation 3.3.1b; q.e.d.**

In the light of the above, we can now restate our earlier definition of a Riemannian space (Section 2.12):

Definition: A linearly connected manifold is called *Riemannian*, R_n, (1) if it is equipped with a (real C^2-differentiable, symmetric, and) positive definite metric (most of this book) and (2) if its affinities are symmetric (i.e., no torsion ⇒ affinities = Christoffels); i.e., an R_n is an L_n with symmetric affinities and positive definite metric.

3.8.3 Asymmetric Affinities, Torsion

Since, as Equations 3.8.12b and 3.8.16a show, the general asymmetric affinities $\Lambda^k{}_{lb}$, $\Lambda_{k,lb}$ are related to the gradients of the L_n-bases just like their Euclidean and Riemannian counterparts, i.e., Equations 3.8.4a through 3.8.5b, it follows that these quantities transform among L_n-coordinate systems just like the Γs, i.e., Equation 3.3.2e ff, with the Γs replaced by Λs (and no use of symmetry in l, b). Further, as Equations 3.3.4a through c show, their antisymmetric part: $2\Lambda^k{}_{[lb]} \equiv \Lambda^k{}_{lb} - \Lambda^k{}_{bl} \equiv 2S^k{}_{lb}$: *torsion* of the manifold (E. Cartan, 1922), is a tensor; i.e., $S^{k'}{}_{l'b'} = A^{k'}{}_k\, A^l{}_{l'}\, A^b{}_{b'}\, S^k{}_{lb}$, even though the $\Lambda^k{}_{lb}$ are not! Hence, torsion (like curvature, (Section 3.10 ff) is a manifold property. These concepts are examined in the following sections.

---

* These alternative versions of the fundamental theorem (quite instructive, in our view) is what one finds in most prevectorial (i.e., components only) treatments of tensor calculus.

** See also the concise and classy derivations of the above in Lodge (1974, pp. 195–201).

Example 3.8.1. Transformation Equations of the Affinities

(Also of the Christoffels: Equations 3.3.2e ff.) Differentiating the defining equations between the bases $\{P, \mathbf{e}_l\}$ and $\{P, \mathbf{e}_{l'}\}$: $\mathbf{e}_l = A^{l'}{}_l\, \mathbf{e}_{l'} \Leftrightarrow \mathbf{e}_{l'} = A^l{}_{l'}\, \mathbf{e}_l$, yields

$$\mathbf{e}_{l',b'} = A^l{}_{l',b'}\, \mathbf{e}_l + A^l{}_{l'}\, \mathbf{e}_{l,b'} = A^l{}_{l',b'}\, \mathbf{e}_l + A^l{}_{l'}\, A^b{}_{b'}\, \mathbf{e}_{l,b}, \tag{a}$$

and, therefore, dotting this equation with $\mathbf{e}^{k'} = A^{k'}{}_k\, \mathbf{e}^k$, and then invoking Equation 3.8.12b, we find

$$\mathbf{e}^{k'} \cdot \mathbf{e}_{l',b'} = A^l{}_{l',b'}\, A^{k'}{}_k(\mathbf{e}^k \cdot \mathbf{e}_l) + A^l{}_{l'}\, A^b{}_{b'}\, A^{k'}{}_k(\mathbf{e}^k \cdot \mathbf{e}_{l,b})$$

$$= A^l{}_{l',b'}\, A^{k'}{}_k(\delta^k{}_l) + A^l{}_{l'}\, A^b{}_{b'}\, A^{k'}{}_k\, \Lambda^k{}_{lb}, \tag{b}$$

i.e.,

$$\Lambda^{k'}{}_{l'b'} = A^{k'}{}_l\, A^l{}_{l'}\, A^b{}_{b'}\, \Lambda^k{}_{lb} + A^{k'}{}_k\, A^k{}_{l',b'}. \tag{c}$$

Example 3.8.2. Total and Covariant Derivatives of Explicitly Time-Dependent L_n-Vectors and Tensors

Let $\mathbf{V} = \mathbf{V}(q, t) = V^k(q, t)\, \mathbf{e}_k(q) \equiv V^k\, \mathbf{e}_k$, where t is the time (or any other path variable). Then, we have, successively,

$$d\mathbf{V}/dt = (dV^k/dt)\mathbf{e}_k + V^k(d\mathbf{e}_k/dt)$$

$$= [\partial V^k/\partial t + (\partial V^k/\partial q^l)(dq^l/dt)]\, \mathbf{e}_k + V^k[(\partial \mathbf{e}_k/\partial q^l)(dq^l/dt)]$$

$$= (\partial V^k/\partial t)\mathbf{e}_k + [V^k{}_{,l}(dq^l/dt)]\mathbf{e}_k + V^k[(\Lambda^s{}_{kl}\, \mathbf{e}_s)(dq^l/dt)]$$

$$= \{\, \partial V^k/\partial t + [(V^k{}_{,l} + V^s\, \Lambda^k{}_{sl})(dq^l/dt)]\,\}\, \mathbf{e}_k \quad \text{(with some dummy index changes)},$$

or, finally,

$$d\mathbf{V}/dt = [\partial V^k/\partial t + V^k|_l\, (dq^l/dt)]\mathbf{e}_k \equiv (DV^k/Dt)\mathbf{e}_k, \tag{a1}$$

$$= [\partial V_k/\partial t + V_k|_l\, (dq^l/dt)]\mathbf{e}^k \equiv (DV_k/Dt)\mathbf{e}^k; \tag{a2}$$

$$= [dV^k/dt + \Lambda^k{}_{sl}\, V^s\, (dq^l/dt)]\mathbf{e}_k \equiv (d\mathbf{V}/dt)^k\, \mathbf{e}_k, \tag{b1}$$

$$= [dV_k/dt - \Lambda_{s,kl}\, V^s(dq^l/dt)]\mathbf{e}^k$$

$$= [dV_k/dt - \Lambda^s{}_{kl}\, V_s\, (dq^l/dt)]\mathbf{e}^k \equiv (d\mathbf{V}/dt)_k\, \mathbf{e}^k. \tag{b2}$$

The forms (a1, a2) find applications in *continuum* kinematics: if $\mathbf{V} = \mathbf{V}(q, t)$ is the velocity of a particle in the so-called Eulerian description, then, since $dq^l/dt = V^l$,

$$(d\mathbf{V}/dt)^k \to DV^k/Dt \equiv \partial V^k/\partial t + V^k|_l\, V^l: \text{ particle acceleration;} \qquad \text{(c)}$$

while (b1, b2) are more useful in *rigid-body* kinematics (see Examples 3.8.3 and 3.16.1).

Generalization: Similarly, for the second-rank tensor $\mathbf{T}(q, t) = T^{kl}(q, t)\, \mathbf{e}_k\, \mathbf{e}_l$ we can show that

$$\begin{aligned}(d\mathbf{T}/dt)^{kl} &= \partial T^{kl}/\partial t + T^{kl}|_s\,(dq^s/dt)\\ &= [\partial T^{kl}/\partial t + T^{kl}{}_{,s}(dq^s/dt)] + \Lambda^k{}_{rs}\, T^{rl}(dq^s/dt) + \Lambda^l{}_{rs}\, T^{kr}(dq^s/dt)\\ &= dT^{kl}/dt + (\Lambda^k{}_{rs}\, T^{rl} + \Lambda^l{}_{rs}\, T^{kr})(dq^s/dt). \qquad \text{(d)*}\end{aligned}$$

From this we can easily establish the corresponding formula for higher rank tensors.

Specialization: For the *absolute scalar* $T'(q', t) = T(q, t)$, where $q' = q'(q)$ and $t' = t$, the above reduce to

$$\partial T'/\partial t' = (\partial T/\partial t)(\partial t/\partial t') + (\partial T/\partial q^k)(\partial q^k/\partial t') = \partial T/\partial t, \qquad \text{(e)}$$

and

$$dT/dt = \partial T/\partial t + (\partial T/\partial q^k)(dq^k/dt) \equiv \partial T/\partial t + \text{grad}\, T\cdot \mathbf{V},\quad \mathbf{V} = (dq^k/dt)\mathbf{e}_k. \qquad \text{(f)}$$

Example 3.8.3. Kinematics of an Orthonormal Triad

$\{\mathbf{e}_k$; all Latin indices run from 1 to 3$\}$, moving in space under the *transport law*:

$$d\mathbf{e}_k = \mathbf{e}_{k,l}\, dq^l \equiv d\phi_{kl}\, \mathbf{e}^l = d\phi_{kl}\, \mathbf{e}_l \quad (\text{since here } \mathbf{e}^l = \mathbf{e}_l). \qquad \text{(a)}$$

Let us find the properties of $d\phi_{kl}$. We have, successively,

$$\begin{aligned}0 &= d(\delta_{kl}) = d(\mathbf{e}_k\cdot \mathbf{e}_l) = d\mathbf{e}_k\cdot \mathbf{e}_l + \mathbf{e}_k\cdot d\mathbf{e}_l\\ &= (d\phi_{kr}\,\mathbf{e}_r)\cdot \mathbf{e}_l + \mathbf{e}_k\cdot (d\phi_{lr}\,\mathbf{e}_r) = d\phi_{kr}(\delta_{rl}) + d\phi_{lr}(\delta_{kr}) = d\phi_{kl} + d\phi_{lk},\end{aligned}$$

i.e., $d\phi_{kl} \equiv d\mathbf{e}_k\cdot \mathbf{e}_l$ is antisymmetric. Comparing Equation a with Equation 3.8.4a, we conclude that

* This expression is useful in calculating the *absolute* rate of the moment of inertia tensor in terms of its components along *moving* axes.

$$d\phi_{kl} = \Gamma_{l,ks}\, dq^s \Rightarrow \Gamma_{l,ks} + \Gamma_{k,ls}$$

$$= 0 \;(\Rightarrow \Gamma^l{}_{ks} + \Gamma^k{}_{ls} = 0), \quad \text{i.e., Equation 3.8.7b.} \tag{b}$$

Then, Equation b2 of the preceding Example becomes, successively,

$$(d\mathbf{V}/dt)_k = dV_k/dt - [\Gamma_{s,kl}(dq^l/dt)]V^s = dV_k/dt + [\Gamma_{k,sl}(dq^l/dt)]V^s \tag{c}$$

$$\equiv dV_k/dt + \omega_{sk}\, V^s \equiv dV_k/dt + (e_{skh}\, \omega^h)V^s = dV_k/dt + (e_{khs}\, \omega^h)V^s,$$

where

$$d\phi_{ks}/dt \equiv \omega_{ks} = \Gamma_{s,kl}(dq^l/dt) = -\Gamma_{k,sl}(dq^l/dt) = -\omega_{sk} \tag{d}$$

$$= (d\mathbf{e}_k/dt) \cdot \mathbf{e}_s = -(d\mathbf{e}_s/dt) \cdot \mathbf{e}_k\text{: } \textit{tensor of angular velocity} \text{ of } \{\mathbf{e}_k\},$$

relative to "background/fixed" rectilinear (possibly rectangular Cartesian) axes, i.e., $d\mathbf{e}_k/dt = \omega_{kl}\, \mathbf{e}_l$; and (recalling Equations 2.9.22a ff) $\omega_{sk} \equiv e_{skh}\, \omega^h \Leftrightarrow \omega^h = (1/2)\, e^{hsk}\, \omega_{sk}$: *axial vector* of ω_{sk}, *in extenso*:

$$\omega_1 = \omega_{23} = -\omega_{32} = (d\mathbf{e}_2/dt) \cdot \mathbf{e}_3 = -(d\mathbf{e}_3/dt) \cdot \mathbf{e}_2, \tag{e1}$$

$$\omega_2 = \omega_{31} = -\omega_{13} = (d\mathbf{e}_3/dt) \cdot \mathbf{e}_1 = -(d\mathbf{e}_1/dt) \cdot \mathbf{e}_3, \tag{e2}$$

$$\omega_3 = \omega_{12} = -\omega_{21} = (d\mathbf{e}_1/dt) \cdot \mathbf{e}_2 = -(d\mathbf{e}_2/dt) \cdot \mathbf{e}_1, \tag{e3}$$

is the *vector of angular velocity* of $\{\mathbf{e}_k\}$: $\omega = (\omega_1, \omega_2, \omega_3)$. So, finally,

$$(d\mathbf{V}/dt)_k = dV_k/dt + (\omega \times \mathbf{V})_k \quad (\neq dV_k/dt, \text{ in general}), \tag{f}$$

which expresses the well-known *moving axes theorem* of dynamics (accelerated frames, rigid body kinematics/kinetics, etc.). See also Example 3.16.1. For a more general treatment of tensor calculus in time-dependent coordinate systems, see Kästner (1960, pp. 201–230); also Betten (1988, pp. 135–143).

3.9 GEOMETRICAL SIGNIFICANCE OF TORSION OF A MANIFOLD

Let us consider the two infinitesimal and independent displacement vectors $\mathbf{OA} = (d_1q^k)$ and $\mathbf{OB} = (d_2q^k)$ emanating from the generic point $O(q)$ of a linearly connected manifold L_n (Figure 3.3).

Now, let us *parallel transport* **OA** from $O(q)$ to $B(q + d_2q)$, and **OB** from $O(q)$ to $A(q + d_1q)$. In this way we obtain $\mathbf{BC}_1$ and $\mathbf{AC}_2$, where, recalling Equations 3.8.10c and d, to the second order and with the affinities evaluated at O:

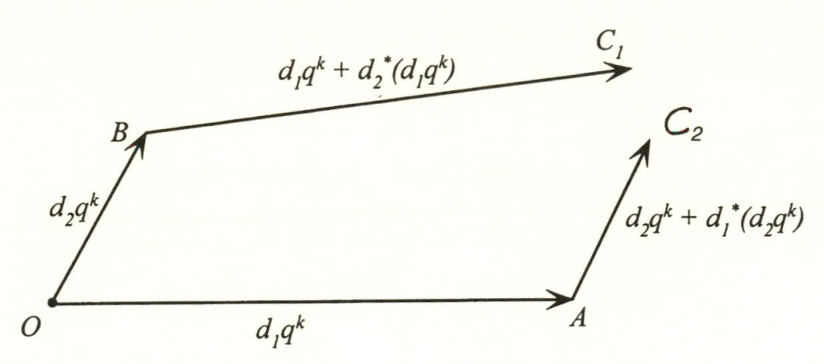

FIGURE 3.3 Parallel propagation of infinitesimal displacement vectors; in general, $C_1 \neq C_2$.

$$\mathbf{BC_1}:\ d_1q^k + d_2{}^*(d_1q^k) = d_1q^k + (-\Lambda^k{}_{lr}\, d_1q^l)d_2q^r, \tag{3.9.1a}$$

$$\mathbf{AC_2}:\ d_2q^k + d_1{}^*(d_2q^k) = d_2q^k + (-\Lambda^k{}_{lr}\, d_2q^l)d_1q^r. \tag{3.9.1b}$$

Accordingly, the (contravarint) components of the infinitesimal *closing displacement vector*

$$\mathbf{C_1C_2} = (\mathbf{OA} + \mathbf{AC_2}) - (\mathbf{OB} + \mathbf{BC_1}) = \mathbf{OC_2} - \mathbf{OC_1}, \tag{3.9.1c}$$

of the infinitesimal "quasi–parallelogram" OAC_2C_1BO, are

$$d_1{}^*(d_2q^k) - d_2{}^*(d_1q^k) = (\Lambda^k{}_{lr} - \Lambda^k{}_{rl})d_1q^l\, d_2q^r$$

$$\equiv 2\Lambda^k{}_{[lr]}\, d_1q^l\, d_2q^r \equiv 2S^k{}_{lr}\, d_1q^l\, d_2q^r; \tag{3.9.1d}$$

and, therefore, if $\Lambda^k{}_{lr} = \Lambda^k{}_{rl}$, or $S^k{}_{lr} = 0$, then $C_1 = C_2$ (to the second order); or, *infinitesimal closed parallelograms do exist in all directions around a manifold point O;* or, the parallel displacement of each of two infinitesimal line elements, of common origin, along each other produce closed (genuine) parallelograms, while, if $S^k{}_{lr} \neq 0$, the manifold is non–Riemannian. More generally, let f be a, say, absolute, scalar field, defined over some region of L_n. Then, the differential of f along the earlier direction d_1q equals: $d_1f = f_{,k}\, d_1q^k$; and the differential of the latter in the previous direction d_2q, where d_1q is to be parallel transported along d_2q, is

$$d_2{}^*(d_1f) = d_2{}^*(f_{,k}\, d_1q^k) = d_2{}^*(f_{,k})\, d_1q^k + f_{,k}\, d_2{}^*(d_1q^k)$$

$$= f_{,kl}\, d_2q^l\, d_1q^k + f_{,k}(-\Lambda^k{}_{lr}\, d_1q^l\, d_2q^r); \tag{3.9.2a}$$

and, similarly,

$$d_1{}^*(d_2f) = \ldots = f_{,lk}\, d_2q^l\, d_1q^k + f_{,k}(-\Lambda^k{}_{rl}\, d_1q^l\, d_2q^r), \tag{3.9.2b}$$

and subtracting the above side by side (and since $f_{,kl} = f_{,lk}$) we obtain the generalization of Equation 3.9.1d:

$$d_1{}^*(d_2 f) - d_2{}^*(d_1 f) = f_{,k}(\Lambda^k{}_{lr} - \Lambda^k{}_{rl})\, d_1 q^l\, d_2 q^r \equiv 2S^k{}_{lr}\, f_{,k}\, d_1 q^l\, d_2 q^r; \quad (3.9.2c)$$

i.e., for such differentiation of absolute scalars to be commutative, the manifold must be torsionless [$\Rightarrow (f|_k)|_l = (f|_l)|_k$ if and only if the manifold is torsionless, e.g., an R_n (see curvature, Sections 3.10ff)]. For further details on non-Riemannian spaces see (alphabetically): Eisenhart (1927), Eringen (1971, pp. 126–140), Synge and Schild (1949/1969, pp. 282–312), Tietz (1955, pp. 172–176).

Problem 3.9.1

Consider the independent infinitesimal displacement vectors $d_1\mathbf{P} \equiv d_1 q^k\, \mathbf{e}_k$ and $d_2\mathbf{P} \equiv d_2 q^k\, \mathbf{e}_k$, emanating from a manifold point P. Operating formally, and using $d_1\mathbf{e}_k = (\Lambda^l{}_{kb}\, d_1 q^b)\, \mathbf{e}_l$ and $d_2\mathbf{e}_k = (\Lambda^l{}_{kb}\, d_2 q^b)\, \mathbf{e}_l$, as in Equation 3.8.12a, show that

$$d_1(d_2\mathbf{P}) - d_2(d_1\mathbf{P}) = d_1(\mathbf{e}_k\, d_2 q^k) - d_2(\mathbf{e}_k\, d_1 q^k)$$

$$= \{[d_1(d_2 q^k) - d_2(d_1 q^k)] - (\Lambda^k{}_{lb} - \Lambda^k{}_{bl}) d_1 q^l\, d_2 q^b\}\mathbf{e}_k, \quad (a)$$

and, therefore, if $d_1(d_2 q^k) = d_2(d_1 q^k)$, then $d_1(d_2\mathbf{P}) - d_2(d_1\mathbf{P}) = (2S^k{}_{bl}\, d_1 q^l\, d_2 q^b)\, \mathbf{e}_k$; i.e., $\Lambda^k{}_{lb} = \Lambda^k{}_{bl}$ are the integrability conditions of $d\mathbf{P} \equiv \mathbf{e}_k\, dq^k$.

3.10 CURVATURE OF A MANIFOLD: GEOMETRICAL ASPECTS

Continuing in the spirit of the last section, let us examine the parallel transport of a, say, covariant vector V_k, in a torsionless manifold R_n, along the sides of the infinitesimal parallelogram $PP'QP''$ (Figure 3.4); first, along the route (1): $P \to P' \to Q' = Q$; and then along route (2): $P \to P'' \to Q'' = Q$. Since under such displacements, by Equations 3.8.10c,d, $d^*V_k = \Gamma^l{}_{kr}\, V_l\, dq^r$, we obtain successively (with $\Gamma_{\dots} \equiv \Gamma_{\dots}(P)$ and $V_{\dots} \equiv V_{\dots}(P)$, and other easily understood notations):

i. From P to P': $d_1{}^*V_k = \Gamma^l{}_{kr}\, V_l\, d_1 q^r$;
ii. From P to P'': $d_2{}^*V_k = \Gamma^l{}_{ks}\, V_l\, d_2 q^s$;
iii. From P' to $Q' = Q$:

$$d'V_k - d_1{}^*V_k = \Gamma^l{}_{ks}(P')\, V_l(P') d_1 q^s$$

$$= (\Gamma^l{}_{ks} + \Gamma^l{}_{ks,r}\, d_1 q^r)(V_l + d_1{}^*V_l) d_2 q^s \quad \text{(to the second order)}$$

$$= (\Gamma^l{}_{ks}\, V_l) d_2 q^s + (\Gamma^l{}_{ks,r}\, V_l) d_1 q^r\, d_2 q^s + (\Gamma^b{}_{ks}\, \Gamma^l{}_{br}\, V_l) d_1 q^r\, d_2 q^s;$$

iv. From P'' to $Q'' = Q$:

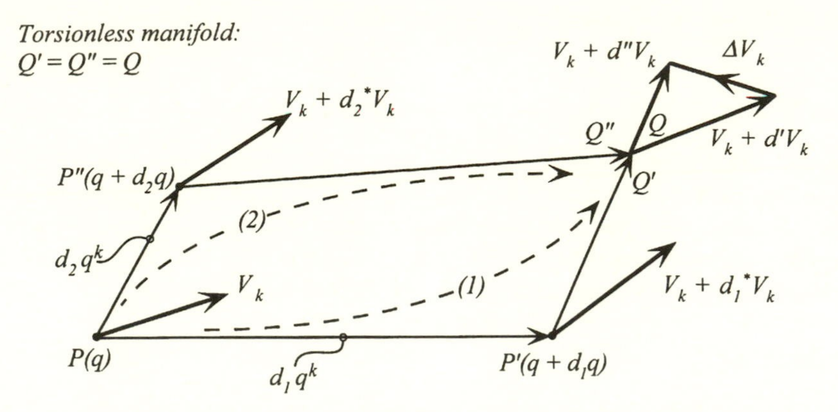

FIGURE 3.4 Parallel transport of a vector V_k, from P to Q, along routes (1) and (2).

$$d''V_k - d_2{}^*V_k = \Gamma^l{}_{kr}(P'')\,V_l(P'')d_1q^r$$

$$= (\Gamma^l{}_{kr} + \Gamma^l{}_{kr,s}\,d_2q^s)(V_l + d_2{}^*V_l)d_1q^r \quad \text{(to the second order)}$$

$$= (\Gamma^l{}_{kr}\,V_l)d_1q^r + (\Gamma^l{}_{kr,s}\,V_l)d_1q^r\,d_2q^s + (\Gamma^b{}_{kr}\,\Gamma^l{}_{bs}\,V_l)d_1q^r\,d_2q^s.$$

Therefore, the so-transported V_k, from P to Q via routes (1) and (2), will differ by

$$\Delta V_k \equiv d''V_k - d'V_k = \ldots = R^l{}_{ksr}\,V_l\,d_1q^r\,d_2q^s, \tag{3.10.1a}$$

where

$$R^l{}_{ksr} \equiv \Gamma^l{}_{kr,s} - \Gamma^l{}_{ks,r} + \Gamma^b{}_{kr}\,\Gamma^l{}_{bs} - \Gamma^b{}_{ks}\,\Gamma^l{}_{br}$$

: *Riemann–Christoffel Curvature tensor* (*R–C*). (3.10.1b)

Similarly, for a contravariant vector V^k, we find

$$\Delta V^k = -R^k{}_{lsr}\,V^l\,d_1q^r\,d_2q^s. \tag{3.10.1c}$$

From the above we readily conclude that if we parallel transport V_k (or V^k) from P to Q along route (1), and then from Q back to P *but along route* (2), i.e., if we parallel transport V_k around a small *closed* manifold-lying path, around the manifold point P, and denote the so-arising change of V_k by $\Delta_c V_k$ (c for closed), then,

$$\Delta_c V_k \equiv V_k(P \to P' \to Q \to P'' \to P) - V_k(P)$$

$$= (V_k + d'V_k) + [-(V_k + d''V_k)] = d'V_k - d''V_k = -\Delta V_k; \tag{3.10.2a}$$

and, similarly,

$$\Delta_c V^k = -\Delta V^k. \tag{3.10.2b}$$

Therefore, recalling the earlier surface element tensor $dA^{rs} \equiv d_1q^r\, d_2q^s - d_1q^s\, d_2q^r$ $(= -dA^{sr})$ (Example 3.6.1) and since, as Equation 3.10.1b shows, $R^l{}_{ksr} = -R^l{}_{krs,}$ we can replace Equations 3.10.1a and c with

$$\Delta_c V_k = (1/2)R^l{}_{ksr}\, V_l\, dA^{sr} \quad \text{and} \quad \Delta_c V^k = -(1/2)R^k{}_{lsr}\, V^l\, dA^{sr}. \tag{3.10.3)*}$$

Let us summarize our findings. *In a general torsionless manifold, the parallel transport of vectors is path dependent,* even locally. Such displacements become path independent, for all vectors and all possible infinitesimal parallelograms around a manifold point P (i.e., for arbitrary d_1q^r and d_2q^r), if and only if $R_{...}(P)$, Equation 3.10.1b, vanishes. If this holds identically, then $\Delta_c V_{...} = 0$ for arbitrary points P, Q, …; and we speak of *global* parallelism, or parallelism in the large (i.e., finite parallel transport) — see also Section 3.10.1, below. Figure 3.5 shows such a $\Delta_c V_{...}$ for the finite parallel transport of the tangent vector on an ordinary spherical surface, along its geodesic curves.** In a flat, or Euclidean, space, since $g_{kl} = \delta_{kl}$ (globally), we can cover the entire manifold with a single rectilinear CS with (constant orientation and) vanishing Christoffels; e.g., a global rectangular Cartesian system built by the unique parallel transport of n initially orthogonal vectors. Therefore, in such a manifold $R^k{}_{lsr} = 0$; and the converse is also true: if $R^k{}_{lsr} = 0$, then the manifold is flat (see also Section 5.6). In closing, we can say that the following statements are equivalent: R–C vanishes if and only if: (i) $g_{kl} = \delta_{kl} \Rightarrow \Gamma^k{}_{rs} = 0$ (globally); (ii) the parallel transport of vectors is path independent; and (iii) covariant derivatives of vectors (and tensors) commute: $(V_k|_r)|_s = (V_k|_s)|_r$, $(V^k|_r)|_s = (V^k|_s)|_r$ (see Section 3.11).

3.10.1 Additional Derivations of the Riemann–Christoffel (*R–C*) and Path Dependence

3.10.1.1 Exactness (or Perfect Differential) Conditions

Let P_o and P be two arbitrary points in a torsionless manifold, say a Riemannian R_n, and a given vector field V^k such that $V^k{}_o \equiv V^k(P_o)$. Let us join P_o and P with a manifold curve $C_{(1)}$, of parametric representation $q^k = q^k(u)$ (u: curve parameter), so that P_o: $q^k(u_o)$ and P: $q^k(u)$. Then, the vector $V^k(P) \equiv V^k$, resulting from $V^k{}_o$ via

---

* Equation 3.10.3$_2$ can also be derived from Equation 3.10.3$_1$ (and vice versa) as follows: since invariants do *not* vary under parallel transport, we shall have, successively,

$$0 = \Delta_c(V^k U_k) = \Delta_c V^k\, U_k + V^k\, \Delta_c U_k = \Delta_c V^k\, U_k + V^k[(1/2)R^l{}_{ksr}\, U_l\, dA^{sr})]$$

$$= \Delta_c V^l\, U_l + (1/2)R^l{}_{ksr}\, V^k\, U_l\, dA^{sr} = [\Delta_c V^l + (1/2)R^l{}_{ksr}\, V^k\, dA^{sr}]\, U_{l,}$$

from which, since U_l is arbitrary, Equation 3.10.3$_2$ follows.

** That is, i.e., its maximum circles. Generally, such a displacement means that *the angle of the transported vector with the geodesic tangent remains constant.* See also Section 5.5.

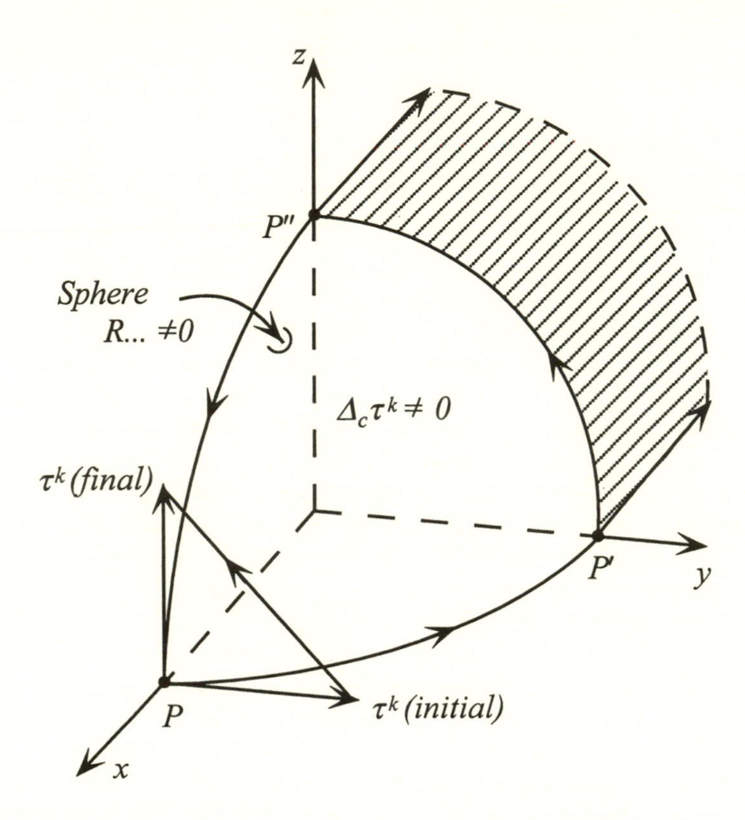

FIGURE 3.5 Parallel transport of tangent vector τ^k along the geodesics of a sphere. Since the latter is a curved manifold, the vector does not return to its original value.

parallel transport along $C_{(1)}$, is obtained by integrating the parallel transport Equation 3.8.10d:

$$(\partial V^k/\partial q^r)^* = -\Gamma^k{}_{lr}\, V^l \Rightarrow V^k{}_{(1)} = V^k{}_o - \int (\Gamma^k{}_{lr}\, V^l)(dq^r/du)du, \tag{3.10.4a}$$

where the integration extends from u_o to u. Clearly, V^k depends on P and on the particular curve connecting P_o with P, i.e., had we chosen a different curve $C_{(2)}$, then $V^k{}_{(1)} \neq V^k{}_{(2)}$; *unless the integrand of Equation 3.10.4a is a perfect, or exact, differential, i.e., unless it satisfies the following exactness conditions:*

$$\partial/\partial q^s(\Gamma^k{}_{lr}\, V^l) = \partial/\partial q^r(\Gamma^k{}_{ls}\, V^l). \tag{3.10.4b}$$

Expanding the above, using in it (3.8.10d), and then cancelling the arbitrary V^ls, we are readily led to the following path independent conditions:

$$\Gamma^k{}_{lr,s} + \Gamma^k{}_{bs}\, \Gamma^b{}_{lr} = \Gamma^k{}_{ls,r} + \Gamma^k{}_{br}\, \Gamma^b{}_{ls}; \tag{3.10.4c}$$

which, as a comparison with Equation 3.10.1b reminds us, are none other than: $R^k{}_{lsr} = 0$.

3.10.1.2 Via the Generalized Stokes' Theorem

The fundamental Equation a of Example 3.6.1 with the identification: $T_k\, dq^k \to d^*V^k = -\Gamma^k{}_{lr}\, V^l\, dq^r$ (k is now a *free* index, but this does not affect Stokes' theorem), yields

$$\Delta_c V^k \equiv \int_C d^*V^k = -\int_C (\Gamma^k{}_{lr}\, V^l) dq^r = -(1/2) \iint_A [(\Gamma^k{}_{lr}\, V^l)_{,s} - (\Gamma^k{}_{ls}\, V^l)_{,r}] dA^{sr}.$$

Expanding the last integrand, and recalling that here $V^l{}_{,s} = -\Gamma^l{}_{bs}\, V^b$, we finally obtain

$$\Delta_c V^k = \int_C d^*V^k = -(1/2) \iint_A R^k{}_{lsr}\, V^l\, dA^{sr} = (1/2) \iint_A R^k{}_{lrs}\, V^l\, dA^{sr}. \qquad (3.10.5a)$$

Similarly, we can show that

$$\Delta_c V_k = \int_C d^*V_k = (1/2) \iint_A R^l{}_{ksr}\, V_l\, dA^{sr} = -(1/2) \iint_A R^l{}_{krs}\, V_l\, dA^{sr}. \qquad (3.10.5b)$$

For *very small C and A*, these equations are none other than (3.10.3). Again, the path independence requirements $\Delta_c V^k, \Delta_c V^k = 0$, since $dA^{sr} \neq 0$ and V^k, V_k are arbitrary, lead to $R^k{}_{lsr} = 0$.

3.11 CURVATURE OF A MANIFOLD: ALGEBRAIC ASPECTS

The main advantage of covariant derivatives, over partial ones, is that they are tensors. On the negative side, however, such tensor derivatives are, in general, *noncommutative* ($\to$ path dependence!); i.e., for a covariant absolute vector V_k:

$$(V_{k,r})_{,s} \equiv V_{k,rs} = (V_{k,s})_{,r} \equiv V_{k,sr}; \quad \text{but} \quad (V_k|_r)|_s \equiv V_k|_{rs} \neq (V_k|_s)|_r \equiv V_k|_{sr}, \qquad (3.11.1)$$

even in a torsionless manifold, say a Riemannian R_n. Indeed, in such a space, we find, successively,

$$\begin{aligned} V_k|_{rs} &= (V_k|_r)_{,s} - \Gamma^b{}_{ks}(V_b|_r) - \Gamma^b{}_{rs}(V_b|_k) \\ &= (V_{k,rs} - \Gamma^b{}_{kr,s}\, V_b - \Gamma^b{}_{kr}\, V_{b,s}) \\ &\quad - (\Gamma^b{}_{ks}\, V_{b,r} - \Gamma^b{}_{ks}\, \Gamma^h{}_{br}\, V_h) - (\Gamma^b{}_{rs}\, V_{b,k} - \Gamma^b{}_{rs}\, \Gamma^h{}_{bk}\, V_h), \end{aligned} \qquad (3.11.2a)$$

and similarly (with $r \to s$ and $s \to r$ in the above),

$$V_k|_{sr} = \ldots = V_{k,sr} - \Gamma^b{}_{ks,r}\, V_b - \Gamma^b{}_{ks}\, V_{b,r}$$

$$- \Gamma^b{}_{kr}\, V_{b,s} + \Gamma^b{}_{kr}\, \Gamma^h{}_{bs}\, V_h - \Gamma^b{}_{sr}\, V_{b,k} + \Gamma^b{}_{sr}\, \Gamma^h{}_{bk}\, V_h, \tag{3.11.2b}$$

and therefore, subtracting Equation 3.11.2b from 3.11.2a, side by side while invoking the symmetry of the Christoffels, we readily find (note that *all terms containing V_k-derivatives cancel*!):

$$V_k|_{rs} - V_k|_{sr} = R^l{}_{krs}\, V_l \;(= R_{lkrs}\, V^l), \tag{3.11.3a}$$

where

$$R^l{}_{krs} \equiv \Gamma^l{}_{ks,r} - \Gamma^l{}_{kr,s} + \Gamma^b{}_{ks}\, \Gamma^l{}_{br} - \Gamma^b{}_{kr}\, \Gamma^l{}_{bs}, \tag{3.11.3b}$$

i.e., as in Equation 3.10.1b and

$$R_{lkrs} \equiv g_{lb}\, R^b{}_{krs} \;(\Leftrightarrow R^b{}_{krs} = g^{bl}\, R_{lkrs}). \tag{3.11.3c}$$

By the Quotient Rule, since the left side of Equation 3.11.3a is a third-rank tensor, the n^4 quantities $R^l{}_{krs}$ (R_{lkrs}), *although built from the nontensorial Christoffels and their derivatives*, they are the components of a fourth-rank and second-kind tensor, or mixed (first-kind, or covariant), the earlier called **R–C** curvature tensor. Further, *R–C* is built from the metric g_{kl} of the manifold under discussion and up to its second derivatives, and therefore it is an *internal* (or *intrinsic*) quantity of that manifold.

For a contravariant absolute vector V_k we, similarly, find

$$V^k|_{rs} - V^k|_{sr} = -R^k{}_{lrs}\, V^l. \tag{3.11.3d}$$

Problem 3.11.1

i. Show that in a general manifold with affinities $\Lambda^k{}_{rs}$ and torsion $S^k{}_{rs} \equiv \Lambda^k{}_{[rs]}$: $I|_{bc} - I|_{cb} = -2S^l{}_{bc}\, I|_l$ (I: an invariant),

$$V_k|_{bc} - V_k|_{cb} = R^l{}_{kbc}\, V_l - 2S^l{}_{bc}\, V_k|_l,$$

$$V^k|_{bc} - V^k|_{cb} = -R^k{}_{lbc}\, V^l - 2S^l{}_{bc}\, V^k|_l, \tag{a}$$

where $R^l{}_{kbc}$ (curvature tensor of that manifold) is given by a (3.11.3b)-like formula, but with the Γ terms replaced by the Λ terms. (For the corresponding result for general tensors $T^{\cdots}{}_{\ldots}$, which is due to G.C. Ricci, see Willmore, 1959, p. 215.)

ii. *Covariant R–C in terms of first kind Christoffels.* Verify that

$$R_{lkrs} = g_{lp}\, R^b{}_{krs} = [(g_{lb}\, \Gamma^b{}_{ks})_{,r} - g_{lb,k}\, \Gamma^b{}_{ks}] - [(g_{lb}\, \Gamma^b{}_{kr})_{,s} - g_{lb,s}\, \Gamma^b{}_{kr}]$$

$$+ g_{lb}\, \Gamma^b{}_{rm}\, \Gamma^m{}_{ks} - g_{lb}\, \Gamma^b{}_{sm}\, \Gamma^m{}_{kr}$$

$$= \ldots = (\Gamma_{l,ks})_{,r} - (\Gamma_{l,kr})_{,s} + \Gamma^b{}_{kr}\, \Gamma_{b,ls} - \Gamma^b{}_{ks}\, \Gamma_{b,lr}$$

$$= \ldots = (1/2)(g_{ls,kr} + g_{kr,ls} - g_{lr,ks} - g_{ks,lr}) + g^{bm}(\Gamma_{m,kr}\, \Gamma_{b,ls} - \Gamma_{m,ks}\, \Gamma_{b,lr})$$

$$= (1/2)(g_{ls,kr} + g_{kr,ls} - g_{lr,ks} - g_{ks,lr}) + g_{bm}(\Gamma^b{}_{kr}\, \Gamma^m{}_{ls} - \Gamma^m{}_{ks}\, \Gamma^b{}_{lr}). \quad \text{(b)}$$

3.11.1 Symmetries–Antisymmetries of *R–C*

Clearly, the *order* of the indices of *R–C* does matter — how, is answered by the following theorems, which are readily derived from the definitions (Equations 3.11.3b through d):

- *Antisymmetry* in *first* pair:
  $R_{lkrs} = -R_{klrs}$ [$\Rightarrow R_{kkrs} = 0$ (no sum on k)] (3.11.4a)
- *Antisymmetry* in *second* pair:
  $R_{lkrs} = -R_{lksr}$ [$\Rightarrow R_{lkss} = 0$ (no sum on s)] (3.11.4b)
  $R^l{}_{krs} = -R^l{}_{ksr}$ [$\Rightarrow R^l{}_{kss} = 0$ (no sum on s)] (3.11.4c)
- *Block* symmetry: $R_{lkrs} = R_{rslk}$ (3.11.4d)
- *Bianchi* identities:
  $R_{lkrs} + R_{lrsk} + R_{lskr} = 0$ [$\Rightarrow R^l{}_{krs} + R^l{}_{rsk} + R^l{}_{skr} = 0$](3.11.4e)
  Also *cyclically* in k, r, s: $k \to r \to s$, $r \to s \to k$, $s \to k \to r$; e.g., $R_{klrs} + R_{krsl} + R_{kslr} = 0$ etc.; i.e., given a group of four indices, there exists only one independent Bianchi identity.
- *R–C components with more than two equal indices vanish:* e.g.,
  $R_{1211} = 0$. (3.11.4f)

3.11.1.1 Number of Independent Components of *R–C*

Next, using the above, we prove that in an n-dimensional space, *R–C* has $n_R \equiv n^2(n^2 - 1)/12$ *independent (i.e., distinct) and potentially nonidentically vanishing components.* Indeed, first, we notice that the R_{lkrs} fall into the following *three* groups (no sum on repeated indices!):

i. R_{lklk} (with $l < k$, e.g., R_{1212}); in number equal to *the combinations of n numbers, 2 at a time* (for l and k): $n' = {}_nC_2 = n\,(n - 1)/2$;*
ii. R_{lklr}: *First subgroup:* $l < k < r$ (e.g., R_{1213}); in number equal to *the combinations of n numbers, 3 at a time* (for l, k, and r): $n''_1 = {}_nC_3 = n(n - 1)(n - 2) / (2)(3)$; *Second subgroup:* $k < l < r$ (e.g., R_{2123}); in number equal to *the combinations of n numbers, 3 at a time* (for k, l, and r): $n''_2 =$

* ${}_nC_m \equiv n!/m!\,(n - m)!$: combinations of n things, m at a time (see books on probability, etc.).

${}_nC_3 = \ldots$; *Third subgroup:* $k < r < l$ (e.g., R_{3132}); in number equal to *the combinations of n numbers, 3 at a time* (for k, r, and l): $n''_3 = {}_nC_3 = \ldots$; i.e., a total of: $n'' = n''_1 + n''_2 + n''_3 = 3 \cdot {}_nC_3 = n(n-1)(n-2)/2$;

iii. R_{lkrs} or R_{lrks} (with $l < k < r < s$); in number equal, for each subgroup, to *the combinations of n numbers, 4 at a time*; i.e., a total of $n''' = 2 \cdot {}_nC_4 = 2[n(n-1)(n-2)(n-3)/24]$. Hence, in an R_n, the total number of independent components of R_{lkrs} is

$$n_R \equiv n' + n'' + n''' = \ldots = n^2(n^2 - 1)/12, \quad \text{q.e.d.} \qquad (3.11.5)^*$$

This formula shows that n_R increases very rapidly with n.**

Problem 3.11.2

i. Show that in an ordinary two-dimensional surface with metric:

$$ds^2 = (dq^1)^2 + G^2(q^1, q^2)\,(dq^2)^2,$$

$$\text{i.e.,}\ \ g_{11} = 1,\ \ g_{22} = G^2(q^1, q^2),\ \ g_{12} = g_{21} = 0,$$

$$R_{1212} = R^1{}_{212} = -G[\partial^2 G / \partial(q^1)^2]. \qquad \text{(a)}$$

ii. Show that if $ds^2 = g_{11}\,(dq^1)^2 + 2g_{12}\,dq^1\,dq^2 + g_{22}\,(dq^2)^2$ (e.g., ordinary surface), then:

$$R_{1212} = \left(g\mathbf{y}2\sqrt{g}\right)\left\{\left[\left(g_{12}\mathbf{y}g_{11}\sqrt{g}\right)g_{11,2} - \left(1\mathbf{y}\sqrt{g}\right)g_{22,1}\right]_{,1} + \left[\left(2\mathbf{y}\sqrt{g}\right)g_{12,1} - \left(1\mathbf{y}\sqrt{g}\right)g_{11,2} - \left(g_{12}\mathbf{y}g_{11}\sqrt{g}\right)g_{11,1}\right]_{,2}\right\}, \qquad \text{(b)}$$

where $g \equiv \mathrm{Det}(g_{kl})$. (If $g_{12} = 0$, the above yields R_{1212} for *orthogonal* surface coordinates.)

Problem 3.11.3

Show that in a manifold *with torsion* the number of distinct (and potentially nonzero) components of R–C is $n^3(n-1)/2$.
Hint: Notice that here too: $R^l{}_{krs} = -R^l{}_{ksr}$.

---

* For other related proofs, see e.g., Bergmann (1942, pp. 172–174), Spain (1960, pp. 52–53).

** Here are the first few (n, n_R) pairs:
(1, 0): (trivial case); (2, 1): $R_{1212} = R_{2121} = -R_{1221} = -R_{2112}$ (ordinary surface); (3, 6): R_{3131}, R_{3232}, R_{1212}, R_{3132}, R_{3212}, R_{3112}; (4, 20): (General theory of relativity — one of the reasons why the latter is, analytically, so complicated); (5, 50); (6, 105); (7, 196); (8, 336); (9, 540); (10, 825).

3.11.2 Contraction(s) of R–C

Since $R^l{}_{lrs} = g^{lk} R_{klrs} = 0$ (by Equation 3.11.4a: $R_{lkrs} = -R_{klrs}$) and $R^l{}_{krl} = -R^l{}_{klr}$ (by Equation 3.11.4c), we only need the following contraction, called (symmetric) *Ricci* tensor of the *first* kind:

$$R_{kr} \equiv R^l{}_{krl} = g^{ls} R_{skrl} (= R_{rk}) = \Gamma^l{}_{kl,r} - \Gamma^l{}_{kr,l} + \Gamma^b{}_{kl} \Gamma^l{}_{br} - \Gamma^b{}_{kr} \Gamma^m{}_{bm}$$

$$= \left(\ln\sqrt{g}\right)_{,kr} - \Gamma^l{}_{kr,l} + \Gamma^b{}_{kl} \Gamma^l{}_{br} - \Gamma^b{}_{kr} \left(\ln\sqrt{g}\right)_{,b}. \tag{3.11.6a}$$

It is not hard to show that R_{kr} has $n(n + 1)/2$ independent components. Similarly, we define the, also symmetric, Ricci tensor of the *second* kind:

$$R^k{}_r \equiv g^{kl} R_{lr} (= R_r{}^k); \tag{3.11.6b}$$

and contracting this once more we obtain the *curvature invariant:*

$$R \equiv g^{kl} R_{lk} = R^k{}_k = R_k{}^k. \qquad (3.11.6c)^*$$

3.11.3 Riemannian, or Sectional, Curvature

This scalar is determined uniquely, at each point of an R_n and for each *pair* of vectors, i.e., directions, A^k and B^k, by

$$K(\mathbf{A}, \mathbf{B}) \equiv K \equiv R_{abcd} A^a B^b A^c B^d \,/\, G_{abcd} A^a B^b A^c B^d$$

$$\equiv R(\mathbf{A}, \mathbf{B}) \,/\, G(\mathbf{A}, \mathbf{B}), \tag{3.11.7a}$$

where

$$G_{abcd} \equiv g_{ac} g_{bd} - g_{ad} g_{bc}$$

$$\text{i.e., } G = (\mathbf{A} \cdot \mathbf{A})(\mathbf{B} \cdot \mathbf{B}) - (\mathbf{A} \cdot \mathbf{B})(\mathbf{A} \cdot \mathbf{B}) \qquad (3.11.7b)^{**}$$

In an ordinary two-dimensional surface, the choice $\mathbf{A} = (1, 0)$ and $\mathbf{B} = (0, 1)$ yields:

$$K = R_{1212} (g_{11} g_{22} - g_{12}{}^2)^{-1} = R_{1212}/g, \tag{3.11.7c}$$

which, clearly, is independent of directions.

From Equation 3.11.7a we deduce that

* Such quantities find applications in General Relativity ($n = 4 \Rightarrow R_{kr} = 0$, ten partial differential equations!).

** We notice that, in comparison, the (first) curvature of a curve, $k_{(1)}$, depends only on the position on the curve (Section 5.5ff).

- R_{klrs} and G_{klrs} have the same symmetry/antisymmetry properties;
- If $g_{kl} \rightarrow$ *diagonal,* then $G_{klkl} = g_{kk}\, g_{ll}$ ($k < l$, no sum on k, l); and

$$G_{abcd}\, A^a\, B^b\, C^c\, D^d = \ldots = (\mathbf{A} \cdot \mathbf{C})(\mathbf{B} \cdot \mathbf{D}) - (\mathbf{A} \cdot \mathbf{D})(\mathbf{B} \cdot \mathbf{C}). \quad (3.11.7d)$$

3.11.4 Curvature vs. Flatness (Riemannian vs. Euclidean Spaces)

If $K = 0$ (identically), the space is called *flat*. Clearly, if $R_{abcd} = 0$ (identically), the space is flat. Below we prove the converse: if $K = 0$, then $R_{abcd} = 0$; i.e., *the necessary and sufficient conditions for flatness is that the R–C vanish.*

Proof: From the K-definition (Equation 3.11.7a) we conclude that if $K = 0$, then (since for *independent* **A** and **B**, $G \neq 0$) $R_{abcd}\, A^a\, B^b\, A^c\, B^d = 0$, for *all* A^a and B^b; or, rearranging, $(R_{abcd}\, A^a\, A^c)\, B^b\, B^d = 0$, or $(R_{abcd}\, B^b\, B^d)\, A^a\, A^c = 0$. From the last two we conclude, respectively, that R_{abcd} is *antisymmetric* in the A indices a and c, and the B indices b and d, i.e.,

$$R_{abcd} + R_{cbad} = 0 \quad \text{and} \quad R_{abcd} + R_{adcb} = 0$$

$$\Rightarrow R_{cdab} + R_{adcb} = 0 \ \text{(by Equation 3.11.4d)}. \quad (3.11.8a)$$

Next, adding the *first* and *last* (third) of the above, we get

$$(R_{abcd} + R_{cdab}) + (R_{cbad} + R_{adcb}) = 0, \quad (3.11.8b)$$

or, since $R_{abcd} = R_{cdab}$ and $R_{cbad} = R_{adcb} = -R_{adbc}$, $R_{abcd} = R_{adbc}$. From this, by cyclic permutation of b, c, d we obtain, in addition, $R_{acdb} = R_{abcd}$ and $R_{adbc} = R_{acdb}$; in sum, $R_{abcd} = R_{adbc} = R_{acdb}$. Finally, combining these last two sets of equations with the Bianchi identities (Equation 3.11.4e): $R_{abcd} + R_{acdb} + R_{adbc} = 0$, we immediately conclude that $R_{abcd} = 0$, q.e.d.*

Problem 3.11.4 Mixed covariant derivatives of higher rank tensors $T^{\cdots}{}_{\cdots}$.

i. By direct calculation, show that (in a torsionless manifold):

$$T^{kl}|_{rs} - T^{kl}|_{sr} = -R^k{}_{brs}\, T^{bl} - R^l{}_{brs}\, T^{kb}, \quad (a)$$

$$T^k{}_l|_{rs} - T^k{}_l|_{sr} = -R^k{}_{brs}\, T^b{}_l + R^b{}_{lrs}\, T^k{}_b. \quad (b)$$

$$T_{kl}|_{rs} - T_{kl}|_{sr} = R^b{}_{krs}\, T_{bl} + R^b{}_{lrs}\, T_{kb}. \quad (c)$$

ii. Extend the above to the general case (Ricci's formulae); i.e.,

$$T^{a\ldots}{}_{b\ldots}|_{rs} - T^{a\ldots}{}_{b\ldots}|_{sr} = \ldots, \quad T_{ab\ldots}|_{rs} - T_{ab\ldots}|_{sr} = \ldots, \text{etc.} \quad (d)$$

* According to Spain (1960, preface and pp. 56–57), this clever proof is due to L. Lovitch.

Problem 3.11.5

By contraction of the results of the preceding problem and use of the symmetry/antisymmetry properties of R–C, show that

$$T^{kl}|_{kl} - T^{kl}|_{lk} = 0 \quad (T^{kl} \text{ not necessarily symmetric}); \tag{a}$$

$$R^k{}_{abc}|_d + R^k{}_{acd}|_b + R^k{}_{adb}|_c = 0, \quad R_{kabc}|_d + R_{kacd}|_b + R_{kadb}|_c = 0. \tag{b}$$

(*Note* cyclic permutation in *last three* indices: b, c, d; as in Bianchi identities, Equation 3.11.4e.)

Problem 3.11.6

Show that on a *two*-parameter manifold, immersed in a Riemannian R_n, i.e., $q^k = q^k(u, v)$, for any vector V^k, V_k:

$$D/Du(DV^k/Dv) - D/Dv(DV^k/Du) = R^k{}_{lrs}\, V^l(\partial q^r/\partial u)(\partial q^s/\partial v), \tag{a}$$

$$D/Du(DV_k/Dv) - D/Dv(DV_k/Du) = R_{klrs}\, V^l(\partial q^r/\partial u)(\partial q^s/\partial v). \tag{b}$$

Problem 3.11.7

Verify that in an R_n (with the usual notations, and recalling Figure 3.4):

$$d_2(d_1\mathbf{e}_l) - d_1(d_2\mathbf{e}_l) = (R^k{}_{lrs}\, d_2q^r\, d_1q^s)\mathbf{e}_k \quad \text{or} \quad (\mathbf{e}_{l,s})_{,r} - (\mathbf{e}_{l,r})_{,s} = R^k{}_{lrs}\, \mathbf{e}_k; \tag{a}$$

i.e., the integrability conditions of $d\mathbf{e}_l = (\Gamma^k{}_{ls}\, dq^s)\, \mathbf{e}_k$, $d_2(d_1\mathbf{e}_l) = d_1(d_2\mathbf{e}_l)$, yield $R^k{}_{lrs} = 0$.

Hint: Recall that $d\mathbf{e}_l = (\Gamma^k{}_{ls}\, dq^s)\, \mathbf{e}_k$, or $\mathbf{e}_{l,s} = \Gamma^k{}_{ls}\, \mathbf{e}_k$ (Equations 3.8.4a and 3.8.12a ff).

3.11.5 Closing Remarks

We may summarize the differences among the various manifolds encountered so far as follows. Let $d\mathbf{P} \equiv \mathbf{e}_k\, dq^k$ and $d\mathbf{e}_k = \Lambda^l{}_{ks}\, dq^s\, \mathbf{e}_l$. Then,

- *Euclidean space*, E_n: $d\mathbf{P}$: integrable (no torsion), $d\mathbf{e}_k$: integrable (no curvature);
- *Riemannian space*, R_n: $d\mathbf{P}$: integrable (no torsion), $d\mathbf{e}_k$: nonintegrable (curvature);
- *Linearly, or Affinely*, Connected and *Metric* space, L_n: $d\mathbf{P}$: nonintegrable (torsion), $d\mathbf{e}_k$: nonintegrable (curvature).

3.12 NONHOLONOMIC (NH) TENSOR ALGEBRA*

3.12.1 INTRODUCTION

Let us begin with a linearly, or affinely, connected and metric manifold, say an L_n (later to become an R_n), covered by curvilinear coordinates $q = (q^1, \ldots, q^n)$. As seen in Section 2.12 (and Sections 2.9 and 3.8 for an E_3), at each L_n-point, $P(q)$—with position vector $\mathbf{OP} = \mathbf{R}(P) \to \mathbf{R}(q) = \mathbf{R}$, relative to some origin O (in or out of L_n), although that is unimportant here—there corresponds a *natural* local covariant basis $\{P, \mathbf{e}_k;\ k = 1, \ldots, n\} \equiv \{P, \mathbf{e}_k\}$, or n-tuple (or ennuple, or n-leg), such that $d\mathbf{R} \equiv d\mathbf{P} \equiv \mathbf{e}_k\, dq^k$; and its *reciprocal*, or *dual*, contravariant basis $\{P, \mathbf{e}^k = g^{kl}\,\mathbf{e}_l\}$, where $g^{kl}\, g_{lr} = \delta^k{}_l$ and $g_{kl} \equiv \mathbf{e}_k \cdot \mathbf{e}_l$. Further, $\mathbf{e}_k$ is *tangent* to the coordinate line q^k, while $\mathbf{e}^k$ is *normal* to the hypersurface q^k = constant; for example, in an E_3:

$$\mathbf{e}_k \cdot \mathbf{e}^l = (\partial\mathbf{R}/\partial q^k) \cdot (\text{grad } q^l) \equiv (\partial\mathbf{R}/\partial q^k) \cdot (\partial q^l/\partial\mathbf{R}) = \partial q^k/\partial q^l = \delta^k{}_l. \quad (3.12.1)$$

We also recall (Equations 3.8.12a ff) that *if* $L_n \to R_n$, then the $\mathbf{e}_k$ satisfy the integrability, or gradientness, or *holonomicity*, conditions:

$$\mathbf{e}_{k,l} - \mathbf{e}_{l,k} = (\Lambda^b{}_{kl} - \Lambda^b{}_{lk})\mathbf{e}_b = (\Gamma^b{}_{kl} - \Gamma^b{}_{lk})\mathbf{e}_b = \mathbf{0} \Rightarrow \mathbf{e}_{k,l} = \mathbf{e}_{l,k}. \quad (3.12.2)$$

Next, the natural basis corresponding to the new holonomic (H), i.e., global, L_n-coordinates, at P, $q^{k'} = q^{k'}(q^k) \leftrightarrow q^k = q^k(q^{k'})$, will be given by $\mathbf{e}_{k'} = (\partial q^k/\partial q^{k'})\ \mathbf{e}_k \equiv A^k{}_{k'}\ \mathbf{e}_k$; and from this it easily follows that:

$$\mathbf{e}_{k',l'} - \mathbf{e}_{l',k'} = \ldots = (A^k{}_{k',l'} - A^k{}_{l',k'})\mathbf{e}_k + A^k{}_{k'}\, A^l{}_{l'}(\mathbf{e}_{k,l} - \mathbf{e}_{l,k}) \quad \text{(generally)} \quad (3.12.2a)$$

$$= (A^k{}_{k',l'} - A^k{}_{l',k'})\mathbf{e}_k \equiv 2A^k{}_{[k',l']}\ \mathbf{e}_k \quad \text{(if Equation 3.12.2 holds)}$$

$$= (\partial^2 q^k/\partial q^{l'}\partial q^{k'} - \partial^2 q^k/\partial q^{k'}\partial q^{l'})\mathbf{e}_k$$

$$= \mathbf{0} \quad \text{(recall Equations 2.4.6a and b).} \quad (3.12.2b)$$

These transformation formulae show that *if* $\mathbf{e}_{k,l}(q) = \mathbf{e}_{l,k}(q) \Rightarrow \{P, \mathbf{e}_k(q)\}$ *is holonomic, so will be any other basis* $\{P, \mathbf{e}_{k'}(q')\}$*, obtained from it by admissible* $q \leftrightarrow q'$ *transformations.*

3.12.2 NONHOLONOMIC COORDINATES AND BASES

Now, generalizing the above, let us examine the case where *not only the new basis is nonholonomic, but also the corresponding coordinates are nonholonomic.*** Let us subject the *coordinate differentials* dq^k to the local, admissible (invertible and

* For complementary reading, we recommend (alphabetically): Dobronravov (1948, 1970), Ferrarese (1963), Golab (1974), Hessenberg (1918), Novoselov (1979), Schouten (1954a, 1954b), Vranceanu (1936, 1961).

proper), homogeneous and linear, or affine centered in the differentials (recall Section 2.3) or *Pfaffian* transformations:

$$dq^k \to dq^K \equiv A^K{}_k \, dq^k \Leftrightarrow dq^k = A^k{}_K \, dq^K \; (k, l, \ldots; K, L, \ldots = 1, \ldots, n) \quad (3.12.3)$$

where

$$A^K{}_k = A_k{}^K = A^K{}_k(q)\text{: given functions of the } q\text{s (as well behaved as needed),} \quad (3.12.3a)$$

and the inverse local coefficients $A^k{}_K = A_K{}^k = A^k{}_K(q)$ are defined *uniquely,* by

$$A^k{}_K \, A^K{}_l = \delta^k{}_l \Leftrightarrow A^K{}_k \, A^k{}_L = \delta^K{}_L. \quad (3.12.3b)$$

The corresponding *covariant* bases $\{P, \mathbf{e}_k\}$ and $\{P, \mathbf{e}_K\}$ are related by the invariant equation:

$$d\mathbf{P} = \mathbf{e}_k \, dq^k = \mathbf{e}_K \, dq^K; \quad (3.12.4)$$

and this, upon utilization of Equations 3.12.3, yields the following transformation equations:

$$\mathbf{e}_K = A^k{}_K \, \mathbf{e}_k \Leftrightarrow \mathbf{e}_k = A^K{}_k \, \mathbf{e}_K. \quad (3.12.4a)$$

Then, as with the holonomic case, the associated *contravariant* bases $\{P, \mathbf{e}_k\}$ and $\{P, \mathbf{e}_K\}$ are defined by

$$\mathbf{e}^K = A^K{}_k \, \mathbf{e}^k \Leftrightarrow \mathbf{e}^k = A^k{}_K \, \mathbf{e}^K, \quad \mathbf{e}^k \cdot \mathbf{e}_l = \delta^k{}_l \Leftrightarrow \mathbf{e}^K \cdot \mathbf{e}_L = \delta^K{}_L; \quad (3.12.4b)$$

and, therefore,

$$A^K{}_k = \mathbf{e}^K \cdot \mathbf{e}_k = (\mathbf{e}^K)_k\text{: } \textit{covariant} \text{ components of } \mathbf{e}^K \text{ along } \mathbf{e}_k \quad (3.12.4c)$$

$$= (\mathbf{e}_k)^K\text{: } \textit{contravariant} \text{ components of } \mathbf{e}_k \text{ along } \mathbf{e}^K, \quad (3.12.4d)$$

$$A^k{}_K = \mathbf{e}^k \cdot \mathbf{e}_K = (\mathbf{e}_K)^k\text{: } \textit{contravariant} \text{ components of } \mathbf{e}_K \text{ along } \mathbf{e}^k \quad (3.12.4e)$$

$$= (\mathbf{e}^k)_K\text{: } \textit{covariant} \text{ components of } \mathbf{e}^k \text{ along } \mathbf{e}_K, \quad (3.12.4f)$$

$$A^K{}_L = \mathbf{e}^K \cdot \mathbf{e}_L = \delta^K{}_L. \quad (3.12.4g)$$

** (a) Such arbitrary bases are not connected with any global, or genuine, coordinates obtainable from the q^k by allowable transformations, but have other desirable properties. (b) The concept of nonholonomic *coordinates* is not to be confused with the more-involved concept of nonholonomic *constraints*; although the former can be profitably utilized in the study of the latter. For more details on this delicate matter, see Sections 6.4ff, especially Frobenius' theorem.

Next, let us check the local transformation (3.12.3) for holonomicity.* As the theory of Pfaffian *forms* (not equations!) teaches, for $dq^K \equiv A^K{}_k(q)\, dq^k$ to be an exact, or perfect, differential, i.e., *for a particular q^K to be a genuine, or holonomic, coordinate, like the $\{q^k\}$ and $\{q^{k'}\}$, it is necessary and sufficient that the following integrability conditions hold:*

$$\partial A^K{}_k/\partial q^s - \partial A^K{}_s/\partial q^k \equiv A^K{}_{k,s} - A^K{}_{s,k} \equiv 2A^K{}_{[k,s]} = 0, \qquad (3.12.5)$$

in some L_n-region of interest, for all values of k, s; and if that holds for all K, then the $\{q^K\}$ are holonomic coordinates, with the $\{\mathbf{e}_K\}$ along the corresponding tangents, at P, like the q^k's. Therefore, if Equations 3.12.5 do not hold, i.e., if $A^K{}_{k,s} \neq A^K{}_{s,k}$, the transformation $q^K = q^K(q^k)$ *cannot* be defined over a finite L_n-region; no finite q^K exist, only dq^K; or, the q^K are *nonglobal,* or *nongenuine,* or *nonholonomic,* or *anholonomic,* or *pseudo-,* or *quasi-coordinates,*** related to the holonomic coordinates q^k only through the nonexact Pfaffian transformations (Equations 3.12.3).

3.12.3 On Notation

Henceforth, uppercase Latin (and sometimes Greek) indices will be reserved for nonholonomic (NH) quantities/components. Accordingly, the notation dq^K would be logically sufficient to indicate that the q^K are no ordinary coordinates (like the q^k), i.e., that, in general, they do not exist. However, as an extra precautionary and eyestrain-lessening measure, from now on we shall denote such "coordinates" as θ^K; i.e., $q^K \equiv \theta^K$ and $dq^K \equiv d\theta^K$. Under this new, specialized, notation, Equations 3.12.3 and 3.12.4 become, respectively,

$$d\theta^K = A^K{}_k\, dq^k \Leftrightarrow dq^k = A^k{}_K\, d\theta^K, \qquad (3.12.6a)$$

$$d\mathbf{P} = \mathbf{e}_k\, dq^k = \mathbf{e}_K\, d\theta^K; \qquad (3.12.6b)$$

i.e., the $d\theta^K$ are the NH contravariant components of $d\mathbf{P}$ in the NH basis $\{P, \mathbf{e}_K\}$; and like their holonomic (H) counterparts dq^k, they generate the local *tangent point space (plane)* $T_n(P)$ to L_n, while the NH $\{\mathbf{e}_K\}$, just like the H $\{\mathbf{e}_k\}$, span the associated local *tangent vector space (plane)* $\mathbf{T}_n(\mathbf{P})$ (recall Section 2.12). As Cartan puts it, the Pfaffian forms (3.12.3) are the coordinates of the L_n-point $P'(q + dq)$ relative to the (generally) nonorthogonal axes at $P(q)$ formed by the tangents to the local NH directions, or congruences, $\{\mathbf{e}_K\}$.***

Finally, *successive* NH transformations possess the transitivity (group-like) property, just like H ones. In a new NH basis $\{P, \mathbf{e}_{K'}\}$, in which $d\mathbf{P} = \mathbf{e}_{K'}\, d\theta^{K'}$ and $d\theta^{K'} = A^{K'}{}_k\, dq^k$, we readily find

$$d\theta^{K'} = A^{K'}{}_k(A^k{}_K\, d\theta^K) \equiv A^{K'}{}_K\, d\theta^K, \qquad (3.12.7a)$$

* Subsequently (Sections 3.13 and 3.14), we develop the analytical machinery of NH tensor calculus, and then (Section 3.15) we test the local basis $\{P, \mathbf{e}_K\}$ for holonomicity.

** The last, a highly suggestive and fertile term; most likely, due to Whittaker (1904).

where:

$$A^{K'}{}_K \equiv [\partial(d\theta^{K'}) / \partial(dq^k)]\,[\partial(dq^k) / \partial(d\theta^K)] \equiv \partial(d\theta^{K'}) / \partial(d\theta^K); \qquad (3.12.7b)$$

and conversely:

$$d\theta^K = A^K{}_{K'}\, d\theta^{K'} \equiv [\partial(d\theta^K) / \partial(d\theta^{K'})]d\theta^{K'}. \qquad (3.12.7c)$$

3.12.4 NH Metric Tensor

From Equation 3.12.6b we obtain the invariant line element (squared) of L_n:

$$ds^2 = d\mathbf{P} \cdot d\mathbf{P} = \ldots = g_{kl}\, dq^k\, dq^l = g_{KL}\, d\theta^K\, d\theta^L, \qquad (3.12.8a)$$

($= g^{kl}\, dq_k\, dq_l = g^{KL}\, d\theta_K\, d\theta_L$ — see below): *positive definite in an R_n-region,*

where

$$g_{KL} = \mathbf{e}_K \cdot \mathbf{e}_L \ (= g_{LK}) \quad \text{and} \quad g^{KL} = \mathbf{e}^K \cdot \mathbf{e}^L \ (= g^{LK}), \qquad (3.12.8b)$$

are, respectively, the *covariant* and *contravariant NH components of the metric tensor.**
Hence, the following metric tensor and coordinate differential transformations:

$$g_{KL} = A^k{}_K\, A^l{}_L\, g_{kl} \Leftrightarrow g_{kl} = A^K{}_k\, A^L{}_l\, g_{KL},$$

$$g^{KL} = A^K{}_k\, A^L{}_l\, g^{kl} \Leftrightarrow g^{kl} = A^k{}_K\, A^l{}_L\, g^{KL},$$

$$g_{KL}\, g^{LR} = \delta^R{}_K (= g^R{}_K), \quad g_{KL}\, g^{KR} = \delta^R{}_L (= g^R{}_L); \qquad (3.12.9a)$$

*** Since the θ^K do not really exist, some authors denote them by $(dq)^K \equiv (d\theta)^K$ or $d(q)^K \equiv d(\theta)^K$: contravariant differentials of quasi-coordinates. To those suspicious of such nonexistent quantities, we offer the following reassuring remarks by Synge (1936, p. 29): "In the theory of quasi-coordinates in dynamics, however, it pays to live dangerously and to use the notation (our) $(3.12.3)_1$ although the latter, illicitly, suggests the existence of the θs which is not the case, unless the $A^K{}_k dq^k$ are perfect differentials." However, as soon as the q^k become known functions of some motion parameter u (e.g., time or arc-length), i.e., *after* the particular physical problem has been solved, then

$$d\theta^K/du = A^K{}_k[q(u)]\,(dq^k/du) \equiv g^K(u): \textit{known} \text{ functions of } u, \qquad (a)$$

and the θ^K can be found (starting from some "initial" value of u, say, $u_i = 0$) by integration of Equation a:

$$\theta^K(u) - \theta^K(0) = \int_0^u g^K(x)dx, \qquad (b)$$

i.e., the current (or final) values $\theta^K(u)$ depend on the particular integration path (just like the work of *nonpotential* forces, which is expressed as Equation b-like line integral), while genuine coordinates, like the q^ks, are path independent.

* We recall (Section 2.12) that in this case $T_n(P)$ and $\mathbf{T_n(P)}$ are (properly) Euclidean spaces.

$$d\theta^K = A^K{}_k\, dq^k = A^K{}_k(g^{kl}\, dq_l) \equiv A^{Kk}\, dq_k$$

$$\Leftrightarrow dq^k = A^k{}_K\, d\theta^K = A^k{}_K(g^{KL}\, d\theta_L) \equiv A^{kK}\, d\theta_K,$$

$$d\theta_K \equiv g_{KL}\, d\theta^L = g_{KL}(A^L{}_l\, dq^l) \equiv A_{Kk}\, dq^k \Leftrightarrow dq^k = A^{kK}\, d\theta_K; \quad (3.12.9b)$$

i.e.,

$$A^{Kk} = A^{kK} \equiv \mathbf{e}^K \cdot \mathbf{e}^k = g^{kl}\, A^K{}_l = g^{KL}\, A^k{}_L, \quad (3.12.10a)$$

$$A_{Kk} = A_{kK} \equiv \mathbf{e}_K \cdot \mathbf{e}_k = g_{kl}\, A^l{}_K = g_{KL}\, A^L{}_k, \quad (3.12.10b)$$

$$A^K{}_k = A_k{}^K \equiv \mathbf{e}^K \cdot \mathbf{e}_k = g^{KL}\, A_{Lk} = g_{kl}\, A^{Kl}; \quad (3.12.10c)$$

and so we have, in addition to Equations 3.12.4a and b,

$$\mathbf{e}^K = A^{Kk}\, \mathbf{e}_k\, (\Leftrightarrow \mathbf{e}_k = A_{kK}\, \mathbf{e}^K) \quad \text{and} \quad \mathbf{e}_K = A_{Kk}\, \mathbf{e}^k\, (\Leftrightarrow \mathbf{e}^k = A^{kK}\, \mathbf{e}_K). \quad (3.12.11)$$

3.12.5 NH Vectors and Tensors

The contravariant and covariant NH components of a general L_n-vector **V** are defined by

$$\mathbf{V} = V_k\, \mathbf{e}^k = V^k\, \mathbf{e}_k = V_K\, \mathbf{e}^K = V^K\, \mathbf{e}_K, \quad (3.12.12a)$$

where

$$\text{H } \textit{covariant} \text{ components: } V_k \equiv \mathbf{V} \cdot \mathbf{e}_k,$$

$$\text{H } \textit{contravariant} \text{ components: } V^k \equiv \mathbf{V} \cdot \mathbf{e}^k, \quad (3.12.12b)$$

$$\text{NH } \textit{covariant} \text{ components: } V_K \equiv \mathbf{V} \cdot \mathbf{e}_K,$$

$$\text{NH } \textit{contravariant} \text{ components: } V^K \equiv \mathbf{V} \cdot \mathbf{e}^K; \quad (3.12.12c)$$

and in view of Equations 3.12.10a through c, are interrelated as follows:

$$V_K = A^k{}_K\, V_k = A_{Kk}\, V^k \Leftrightarrow V_k = A^K{}_k\, V_K, \quad V^k = A^{kK}\, V_K, \quad (3.12.12d)$$

$$V^K = A^K{}_k\, V^k = A^{Kk}\, V_k \Leftrightarrow V_k = A_{kK}\, V^K, \quad V^k = A^k{}_K\, V^K, \quad (3.12.12e)$$

and, just as in the H case, their indices can be raised (lowered) with g^{KL} (g_{KL}); for example,

$$V_K = A^k{}_K\, V_k = A^k{}_K(g_{kl}\, V^l) = A^k{}_K\, g_{kl}(A^l{}_L\, V^L)$$

$$[= A^k{}_K(A_{kL}V^L) = (A^k{}_K\, A_{kL})V^L\,] = g_{KL}\, V^L \text{ and } V^K = \ldots = g^{KL}\, V_L. \quad (3.12.13)^*$$

Here, too, in general, $V_K \neq V^K$, unless $\mathbf{e}_K = \mathbf{e}^K$ (i.e., orthonormal) $\Rightarrow d\theta^K$: Cartesian. Then,

$$A^K{}_k \equiv \partial(d\theta^K) \,/\, \partial(dq^k) = \cos(d\theta^K, dq^k)$$

$$= \cos(dq^k, d\theta^K) = \partial(dq^k) \,/\, \partial(d\theta^K) \equiv A^k{}_K. \quad (3.12.14)$$

Next, generalizing Equations 3.12.12a ff, we define the NH components of an L_n-tensor $\mathbf{T}(P) \equiv \mathbf{T}$, say, for concreteness, of the mixed third-rank (absolute) tensor $T^{kl}{}_m$, relative to the NH bases $\{P, \mathbf{e}_K\}$ and $\{P, \mathbf{e}^K\}$, as follows:

$$T^{KL}{}_M = A^K{}_k\, A^L{}_l\, A^m{}_M\, T^{kl}{}_m \Leftrightarrow T^{kl}{}_m = A^k{}_K\, A^l{}_L\, A^M{}_m\, T^{KL}{}_M, \quad (3.12.15)$$

where $T^{kl}{}_m$ are the familiar H components of $\mathbf{T}$ relative to the H bases $\{P, \mathbf{e}_k\}$ and $\{P, \mathbf{e}^k\}$.** From all these transformations we conclude that:

> The NH algebra (i.e., no differentiations) is formally identical to the H algebra; hence, if a tensor vanishes in an NH basis, at an L_n-point, it vanishes in every other H or NH basis there; i.e., we can write $\mathbf{T} = \mathbf{0}$.

3.12.6 Mixed H–NH Transformations

Under a $q \leftrightarrow q'$ transformation: $dq^k = A^k{}_{k'}\, dq^{k'} \leftrightarrow dq^{k'} = A^{k'}{}_k\, dq^k$, and under a $d\theta \leftrightarrow d\theta'$ transformation: $d\theta^K = A^K{}_{K'}\, d\theta^{K'} \leftrightarrow d\theta^{K'} = A^{K'}{}_K\, d\theta^K$. Therefore, $d\theta^K = A^K{}_k\, dq^k = A^K{}_k\,(A^k{}_{k'}\, dq^{k'}) \equiv A^K{}_{k'}\, dq^{k'}$, where (symbolically):

$$A^K{}_{k'} \equiv [\partial(d\theta^K) \,/\, \partial(dq^k)]\, [\partial(dq^k) \,/\, \partial(dq^{k'})] \equiv \partial(d\theta^K) \,/\, \partial(dq^{k'}) \quad (3.12.16a)$$

(see next section); i.e., under $q \leftrightarrow q'$, $A^K{}_k$ is a covariant vector in k. Similarly, since

$$d\theta^{K'} = A^{K'}{}_K\, d\theta^K = A^{K'}{}_K(A^K{}_k\, dq^k) \equiv A^{K'}{}_k\, dq^k,$$

$$\text{where} \quad A^{K'}{}_k \equiv \partial(d\theta^{K'}) \,/\, \partial(dq^k), \quad (3.12.16b)$$

---

* One could, conceivably, define *relative* NH tensors; for example,

$$T^K = |\partial(dq) \,/\, \partial(d\theta)|^w A^K{}_k\, T^k \equiv |A^l{}_L|^w A^K{}_k\, T^k\text{: relative NH contravariant vector, of weight } w,$$

but we are unaware of any applications of such tensors to mechanics. Therefore, from now on, our NH tensors will be understood to be absolute NH tensors, i.e., $w = 0$, like Equations 3.12.13 and 3.12.15.

** Just as in the H case (Section 2.11), invariant, or direct, representations are easily available. For example, for the second-rank tensor $\mathbf{T} = T_{kl}\,\mathbf{e}^k\,\mathbf{e}^l = T^{kl}\,\mathbf{e}_k\,\mathbf{e}_l = T^k{}_l\,\mathbf{e}_k\,\mathbf{e}^l = T_l{}^k\,\mathbf{e}_l\,\mathbf{e}^k$, we shall also have $\mathbf{T} = T_{KL}\,\mathbf{e}^K\,\mathbf{e}^L = T^{KL}\,\mathbf{e}_K\,\mathbf{e}_L = T^K{}_L\,\mathbf{e}_K\,\mathbf{e}^L = T_L{}^K\,\mathbf{e}_L\,\mathbf{e}^K$.

$A^K{}_k$ is a contravariant vector in K, relative to $d\theta \leftrightarrow d\theta'$, etc. In sum, $A^K{}_k$ and $A^k{}_K$ have the tensor character indicated by their indices, relative to transformations $q \leftrightarrow q'$ and $d\theta \leftrightarrow d\theta'$. Generally, if $T_k, \ldots$ are H tensors under $q \leftrightarrow q'$, then, since $T^K = A^K{}_k T^k, \ldots$, *the* $T^K, \ldots$ *are invariant under* $q \leftrightarrow q'$ *and tensorial under* $d\theta \leftrightarrow d\theta'$, *as indicated by their NH indices.*

Problem 3.12.1

By $\partial(\ldots)/\partial q^s \equiv (\ldots)_{,s}$-differentiation of Equations 3.12.3b: $A^k{}_K A^K{}_l = \delta^k{}_l$ and $A^K{}_k A^k{}_L = \delta^K{}_L$, and further use of them, show that

$$A^k{}_{L,s} = -A^k{}_K A^l{}_L A^K{}_{l,s} \quad \text{and} \quad A^K{}_{l,s} = -A^L{}_l A^K{}_k A^k{}_{L,s}. \tag{a}$$

Such calculations are needed in the following sections on NH tensor analysis.

3.13 NH TENSOR DIFFERENTIATION: OBJECT OF ANHOLONOMICITY

3.13.1 NH Differentiation

Now, with an eye toward the testing of the holonomicity of the basis $\{P, \mathbf{e}_K\}$ (to be carried out in Section 3.14) and, recalling Equations 3.12.2a and b, we introduce the following definition.

3.13.1.1. Definition

Since the θ^K do not exist, by $\partial(\ldots)/\partial\theta^K \equiv (\ldots)_{,K}$ we shall understand the *symbolic* "quasi-chain rule":

$$\partial(\ldots) / \partial\theta^K \equiv [\partial(\ldots) / \partial q^k]\, [\partial(dq^k) / \partial(d\theta^K)] \equiv A^k{}_K[\partial(\ldots) / \partial q^k],$$

or, compactly,

$$(\ldots)_{,K} \equiv A^k{}_K\, (\ldots)_{,k}, \tag{3.13.1a}$$

and, inversely,

$$\partial(\ldots) / \partial q^k \equiv [\partial(\ldots) / \partial\theta^K]\, [\partial(d\theta^K) / \partial(dq^k)] \equiv A^K{}_k[\partial(\ldots) / \partial q^k],$$

or, compactly,

$$(\ldots)_{,k} \equiv A^K{}_k(\ldots)_{,K}. \tag{3.13.1b}$$

This definition preserves the $d(\ldots)$-invariance in both H and NH coordinates:

$$d(\ldots) = [\partial(\ldots) / \partial q^k]dq^k = [\partial(\ldots) / \partial q^k]\, (A^k{}_K\, d\theta^K)$$

$$= [\partial(\ldots) / \partial\theta^K]d\theta^K. \tag{3.13.1c}$$

Then, the *gradient* of a scalar field $T = T(q)$ is the contravariant NH vector:

$$\partial T/\partial\theta^K = A^k{}_K(\partial T/\partial q^k); \quad \text{or, compactly, } T_{,K} = A^k{}_K\, T_{,k}; \tag{3.13.2a}$$

while the gradient of a, say, contravariant vector (field) $T^K = T^K(q)$ is

$$\partial T^K/\partial\theta^L = A^l{}_L(\partial T^K/\partial q^l); \quad \text{or, compactly, } T^K{}_{,L} = A^l{}_L\, T^K{}_{,l} \tag{3.13.2b}$$

(in general, not a tensor; but used in the definition of the *divergence* of a vector).

3.13.2 The Nonholonomicity (or Anholonomicity) Object

NH operations like Equations 3.13.2a and b do *not* differ formally from their H counterparts; but the *rotation/curl(-ing)* of a vector V_K, $V_{[K,L]} \equiv (V_{K,L} - V_{L,K})/2$ *does*! Indeed, we find, successively,

$$2V_{[K,L]} = (A^k{}_K\, V_k)_{,L} - (A^l{}_L\, V_l)_{,K} = A^l{}_L(A^k{}_K\, V_k)_{,l} - A^k{}_K(A^l{}_L\, V_l)_{,k}$$

$$= A^l{}_L(A^k{}_{K,l}\, V_k + A^k{}_K\, V_{k,l}) - A^k{}_K(A^l{}_{L,k}\, V_l + A^l{}_L\, V_{l,k})$$

$$= A^k{}_K\, A^l{}_L(V_{k,l} - V_{l,k}) + (A^l{}_L\, A^k{}_{K,l} - A^l{}_K\, A^k{}_{L,l})V_k,$$

or, finally,

$$V_{[K,L]} = A^k{}_K\, A^l{}_L\, V_{[k,l]} - \Omega^R{}_{KL}\, V_R, \tag{3.13.3a}$$

where

$$-\Omega^R{}_{KL}\, V_R \equiv (1/2)(A^l{}_L\, A^k{}_{K,l} - A^l{}_K\, A^k{}_{L,l})V_k. \tag{3.13.3b}$$

Let us examine this result carefully: the *first*, tensorlike, term $A^k{}_K\, A^l{}_L\, V_{[k,l]}$ is what we would obtain if the θ^K were new H coordinates; but the *second* term $-\Omega^R{}_{KL}\, V_R$ is new and *is due entirely to the nonholonomicity of the* θ^K. Let us see why: substituting into it $V_k = A^R{}_k\, V_R$ and

$$A^k{}_K\, A^R{}_k = \delta^R{}_K \Rightarrow A^k{}_{K,l}\, A^R{}_k = -A^k{}_K\, A^R{}_{k,l}, \quad A^k{}_L\, A^R{}_k = \delta^R{}_L \Rightarrow A^k{}_{L,l}\, A^R{}_k = -A^k{}_L\, A^R{}_{k,l},$$

we get

$$-\Omega^R{}_{KL}\, V_R = (1/2)[A^k{}_L(-A^l{}_K\, A^R{}_{l,k}) - A^l{}_K(-A^k{}_L\, A^R{}_{k,l})]V_R$$

$$= [(1/2)(A^R{}_{l,k} - A^R{}_{k,l})A^k{}_K\, A^l{}_L]V_R \quad \text{(after some dummy index changes)}$$

or, finally,

$$\Omega^R{}_{KL} \equiv (1/2)(A^R{}_{k,l} - A^R{}_{l,k})A^k{}_K\, A^l{}_L \equiv A^k{}_K\, A^l{}_L\, A^R{}_{[k,l]}$$

$$\equiv (1/2)(A^R{}_{k,L}\, A^k{}_K - A^R{}_{k,K}\, A^k{}_L) \equiv A^R{}_{k,[L}\, A^k{}_{K]}$$ (recalling Equations 3.13.1a and b):

Object of Nonholonomicity, or Anholonomicity (AO) (3.13.4a)*

of the NH "coordinates" θ^K, measured along these coordinates.

3.13.2.1 Remarks on AO

i. In terms of its "intermediate" (holonomic) components $\Omega^R{}_{kl} \equiv A^R{}_{[k,l]}$, Equation 3.13.4a can be rewritten in the tensorlike form:

$$\Omega^R{}_{KL} = A^k{}_K\, A^l{}_L\, \Omega^R{}_{kl} \quad (\Leftrightarrow \Omega^R{}_{kl} = A^K{}_k\, A^L{}_l\, \Omega^R{}_{KL}). \tag{3.13.4b}$$

ii. We notice that (by the NH counterpart of Equations 3.12.1, or by Equations 3.12.4g): $A^R{}_{[K,L]} = \delta^R{}_{[K,L]} = 0$; i.e., in general, $\Omega^R{}_{KL} \neq A^R{}_{[K,L]}$, in spite of notational temptations!

iii. Clearly, *the AO is antisymmetric in its two lower indices:*

$$\Omega^R{}_{KL} = -\Omega^R{}_{LK} \quad (\text{also: } \Omega^R{}_{kl} = -\Omega^R{}_{lk}). \tag{3.13.4c}$$

It can be shown that, as in the H case, this is an invariant property of the AO; i.e., it holds in any NH coordinates θ^K, $\theta^{K'}$, ... (Due to Equation 3.13.4c one has to be extra careful when comparing references.)

iv. That the $n^2(n-1)/2$ $\{\Omega^R{}_{KL}\}$ do *not* constitute a tensor, but a *geometrical object,* is clear from the fact that in some coordinates (H) they vanish, while in others (NH) they do not (like the H Christoffels; but unlike them, the AO do not involve neither the NH metric tensor, nor its gradients).

v. Equations 3.13.3a ff embody the following "rule":

When an expression (say, a tensorial differential equation) is transformed from H to NH coordinates, the result is an expression just like the H one in form, but, in general, with an additional "correction term" that contains the AO.

vi. The AO, in a particular NH system θ^K, is independent of the particular (original) H system q^k used for its derivation; i.e., if $dq^k \to d\theta^K$ yields $\Omega^R{}_{KL}$, then $dq^{k'} \to d\theta^K$ yields $\Omega^{R'}{}_{K'L'} = \Omega^R{}_{KL}$.**

---

* $\Omega^R{}_{KL}$ seems to have been introduced to tensor calculus by Hessenberg (1916, publ. 1918; p. 211–a classic article, still worth reading today); also by Lagrange (1926, pp. 17 ff) and Vranceanu [1926, see Vranceanu, (1936)]. In the early 1930s, J. A. Schouten and D. v. Dantzig studied it systematically, and dubbed it *anholonomity* object; but we, for the sake of grammatical consistency, shall be calling it anholonomicity object (see also Golab, 1974, pp. 140ff), In mechanics, however, these quantities had been introduced earlier: Volterra (1898) and Hamel (1903–4, in complete generality); see Equation 3.13.5 below.

** The proof is straightforward, and is left as an exercise [see, e.g., Golab (1974, pp. 140–141)].

vii. In NH system mechanics (Chapters 6 and 7), we use, instead of $\Omega^R{}_{KL}$, its double, which are called *the Hamel (–Volterra) transitivity coefficients*;* i.e.,

$$\gamma^R{}_{KL} \equiv 2\Omega^R{}_{KL} = (A^R{}_{k,l} - A^R{}_{l,k})A^k{}_K\, A^l{}_L \equiv 2A^k{}_K\, A^l{}_L\, A^R{}_{[k,l]} \equiv A^k{}_K\, A^l{}_L\, \gamma^R{}_{kl}; \quad (3.13.5)$$

to stress their intimate connection with the NH Christoffels (detailed later).

3.13.2.2 Additional Uses of the AO

Rearranging slightly Equations 3.13.3a and b, we obtain

$$A^k{}_K\, A^l{}_L\, V_{[k,l]} = V_{[K,L]} + \Omega^R{}_{KL}\, V_R, \quad (3.13.6a)$$

and dot multiplying this with $A^K{}_r$, $A^L{}_s$, etc. (so as to isolate $V_{[k,l]}$), we get the transformation equation:

$$V_{[k,l]} = A^K{}_k\, A^L{}_l[V_{[K,L]} + \Omega^R{}_{KL}\, V_R]; \quad (3.13.6b)$$

from which we readily conclude that the necessary and sufficient conditions for V_K to be the gradient of a scalar function $f = f(q)$, i.e., $V_K = \partial f/\partial\theta^K \equiv f_{,K} \equiv A^k{}_K\, f_{,k}$, are:

$$V_{[K,L]} = -\Omega^R{}_{KL}\, V_R = \Omega^R{}_{LK}\, V_R. \quad (3.13.6c)$$

With $V_K \to f_{,K}$ this yields the following *fundamental noncommutativity equation for NH mixed (symbolic!) partial derivatives:*

$$\partial/\partial\theta^L(\partial f/\partial\theta^K) - \partial/\partial\theta^K(\partial f/\partial\theta^L) = 2\Omega^R{}_{LK}(\partial f/\partial\theta^R) \equiv \gamma^R{}_{LK}(\partial f/\partial\theta^R). \quad (3.13.6d)$$

3.13.2.3 Other Expressions for (AO)

$(\ldots)_{,r}$-differentiating Equation 3.12.3b$_2$ yields the earlier: $A^R{}_{k,l}\, A^k{}_K = -A^R{}_k\, A^k{}_{K,l}$, and dotting this with $A^K{}_r$ we obtain: $A^R{}_{k,l} = -A^r{}_{K,l}\, A^K{}_k\, A^R{}_r$ (recall Problem 3.12.1), and accordingly,

$$A^R{}_{k,l}\, A^k{}_K\, A^l{}_L = (-A^r{}_{M,l}\, A^M{}_k\, A^R{}_r)\, A^k{}_K\, A^l{}_L = -A^r{}_{M,l}\, \delta^M{}_K\, A^R{}_r\, A^l{}_L \quad (3.13.7a)$$

$$= -A^r{}_{K,l}\, A^R{}_r\, A^l{}_L = -A^r{}_{K,L}\, A^R{}_r \quad \text{(recalling Equation 3.13.1a).}$$

Hence, successively,

$$\gamma^R{}_{KL} = A^R{}_{k,l}\, A^k{}_K\, A^l{}_L - A^R{}_{l,k}\, A^k{}_K\, A^l{}_L = A^R{}_{k,l}(A^k{}_K\, A^l{}_L - A^l{}_K\, A^k{}_L)$$

$$= -A^r{}_{K,L}\, A^R{}_r - (-A^r{}_{L,K}\, A^R{}_r) = (A^r{}_{L,K} - A^r{}_{K,L})A^R{}_r. \quad (3.13.7b)$$

---

* Or (not quite correctly) "*Ricci–Boltzmann–Hamel three-index symbols.*"

In sum, we have found the following three γ-expressions:

$$\gamma^R{}_{KL} = A^k{}_K\, A^l{}_L(A^R{}_{k,l} - A^R{}_{l,k}) \equiv 2A^k{}_K\, A^l{}_L\, A^R{}_{[k,l]}, \tag{3.13.7c}$$

$$= A^R{}_{k,l}\,(A^k{}_K\, A^l{}_L - A^l{}_K\, A^k{}_L) \equiv 2A^R{}_{k,l}\, A^{[k}{}_K\, A^{l]}{}_L, \tag{3.13.7d}$$

$$= A^R{}_l\,(A^l{}_{L,k}\, A^k{}_K - A^l{}_{K,k}\, A^k{}_L) \equiv A^R{}_r\,(A^r{}_{L,K} - A^r{}_{K,L}) \equiv 2A^R{}_r\, A^r{}_{[L,K]}. \tag{3.13.7e}$$

Problem 3.13.1

Show, by direct calculation, that:

$$\Omega^R{}_{K[L,S]} = 2\Omega^M{}_{[LS}\Omega^R{}_{M]K}, \tag{a}$$

or

$$\gamma^R{}_{KL,S} - \gamma^R{}_{KS,L} = 2(\gamma^M{}_{LS}\,\gamma^R{}_{MK} - \gamma^M{}_{MS}\,\gamma^R{}_{LK}). \tag{b}$$

Problem 3.13.2

By differentiating Equations 3.13.7c through e, symbolically, relative to S and then permuting K, L, and S cyclically, etc., prove the Jacobi-like identity:

$$\gamma^R{}_{SM}\,\gamma^M{}_{KL} + \gamma^R{}_{LM}\,\gamma^M{}_{SK} + \gamma^R{}_{KM}\,\gamma^M{}_{LS} + \gamma^R{}_{KL,S} + \gamma^R{}_{SK,L} + \gamma^R{}_{LS,K} = 0. \tag{a}$$

This extends a well-known result from the theory of Lie groups — from constant "structure constants" (i.e., Equation a with constant γs) to variable ones.

3.13.2.4 The AO in Three-Dimensional Torsionless Space

Let us calculate the contravariant H components of the rotation/curling of the NH contravariant basis vectors $\mathbf{e}^R$, i.e., $(\text{curl } \mathbf{e}^R)^b$; say, in an E_3. Recalling the curl(…) definition (Equations 2.9.19a through c and Equation 3.6.4) we find, successively,

$$(\text{curl } \mathbf{e}^R)^b = e^{blk}\, A^R{}_{k,l} = (e^{blk}/2)(A^R{}_{k,l} - A^R{}_{l,k}) \equiv e^{blk}\, \Omega^R{}_{kl}$$

$$\equiv (e^{blk}/2)\gamma^R{}_{kl} = (e^{BLK}\, A^b{}_B\, A^l{}_L\, A^k{}_K)\Omega^R{}_{kl} = (e^{BLK}\, \Omega^R{}_{KL})A^b{}_B, \tag{3.13.8a}$$

i.e.,

$$(\text{curl } \mathbf{e}^R)^B = e^{BLK}\, \Omega^R{}_{KL} = -e^{BKL}\, \Omega^R{}_{KL} = -e^{BKL}\, \gamma^R{}_{KL}\, /\, 2$$

$$\Rightarrow \text{curl } \mathbf{e}^R = e^{BLK}\, \Omega^R{}_{KL}\, \mathbf{e}_B = -(1/2)e^{BKL}\, \gamma^R{}_{KL}\, \mathbf{e}_B\, (\equiv -\gamma^{RB}\, \mathbf{e}_B \equiv -\gamma^R{}_B\, \mathbf{e}^B)$$

and

$$\mathbf{e}^B \cdot \text{curl } \mathbf{e}^R = (1/2)e^{BLK}\, \gamma^R{}_{KL}, \tag{3.13.8b}$$

and solving for $\gamma^R{}_{KL}$ (see footnote below):

$$\gamma^R{}_{KL} = -e_{BKL}(\text{curl } \mathbf{e}^R)^B = e_{BLK}(\text{curl } \mathbf{e}^R)^B; \tag{3.13.8c}$$

and, similarly,

$$\gamma^R{}_{kl} \equiv 2\Omega^R{}_{kl} = e_{blk}(\text{curl } \mathbf{e}^R)^b. \tag{3.13.8d}$$

But, further,

$$\mathbf{e}_K \times \mathbf{e}_L = (A^k{}_K\, \mathbf{e}_k)(A^l{}_L\, \mathbf{e}_l) = A^k{}_K\, A^l{}_L(\mathbf{e}_k \times \mathbf{e}_l),$$

and, therefore*

$$\begin{aligned}(\mathbf{e}_K \times \mathbf{e}_L) \cdot \text{curl } \mathbf{e}^R &= (e^{blk}\, A^R{}_{[k,l]}\, A^p{}_K\, A^q{}_L)\ [\mathbf{e}_b \cdot (\mathbf{e}_p \times \mathbf{e}_q)] \\ &= (A^p{}_K\, A^q{}_L\, A^R{}_{[k,l]})(e^{blk}\, e_{bpq}) \\ &= A^R{}_{[k,l]}\, A^p{}_K\, A^q{}_L\, \delta^l{}_p \delta^k{}_q - A^R{}_{[k,l]}\, A^p{}_K\, A^q{}_L\, \delta^k{}_p\, \delta^l{}_q \\ &= A^R{}_{[k,l]}\, A^l{}_K\, A^k{}_L - A^R{}_{[k,l]}\, A^k{}_K\, A^l{}_L = -2A^R{}_{[k,l]}\, A^k{}_K\, A^l{}_L\end{aligned}$$

(with some dummy index changes)

i.e.,

$$(\mathbf{e}_K \times \mathbf{e}_L) \cdot \text{curl } \mathbf{e}^R = -\gamma^R{}_{KL} = \gamma^R{}_{LK}; \tag{3.13.8e}$$

also, recalling Equations 2.9.4 ff,

$$\mathbf{e}_K \times \mathbf{e}_L = v\ \varepsilon_{KLR}\ \mathbf{e}^R = e_{KLR}\ \mathbf{e}^R,$$

where

$$v \equiv \mathbf{e}_A \cdot (\mathbf{e}_B \times \mathbf{e}_C) = 1/\ v^*, \quad v^* \equiv \mathbf{e}^A \cdot (\mathbf{e}^B \times \mathbf{e}^C) = 1/v$$

$$\text{(with } A, B, C = 1, 2, 3). \tag{3.13.8f}$$

---

* By recalling Example 2.5.4 and Equation 1.2.6c, $e^{blk} e_{bpq} = \left(\varepsilon^{blk}/\sqrt{g}\right)\left(\sqrt{g}\varepsilon_{bpq}\right) = \varepsilon^{blk}\, \varepsilon_{bpq} = \delta^l{}_p\, \delta^k{}_q - \delta^k{}_p\, \delta^l{}_q$.

Hence, finally,

$$\mathbf{e}^B \cdot \text{curl } \mathbf{e}^R = (1/2)e^{BLK}\, \gamma^R{}_{KL} \quad \text{and}$$

$$e_{KLB}(\mathbf{e}^B \cdot \text{curl } \mathbf{e}^R) = \nu\, \varepsilon_{KLB}\, (\mathbf{e}^B \cdot \text{curl } \mathbf{e}^R) = -\gamma^R{}_{KL}; \qquad (3.13.8g)$$

or, *in extenso,*

$$\nu\, \mathbf{e}^1 \cdot \text{curl } \mathbf{e}^1 = -\gamma^1{}_{23} = \gamma^1{}_{32}, \qquad (3.13.8h)$$

$$\nu\, \mathbf{e}^2 \cdot \text{curl } \mathbf{e}^2 = -\gamma^2{}_{31} = \gamma^2{}_{13}, \qquad (3.13.8i)$$

$$\nu\, \mathbf{e}^3 \cdot \text{curl } \mathbf{e}^3 = -\gamma^3{}_{12} = \gamma^3{}_{21}, \qquad (3.13.8j)$$

$$\nu\, \mathbf{e}^1 \cdot \text{curl } \mathbf{e}^2 = -\gamma^2{}_{23} = \gamma^2{}_{32}, \qquad (3.13.8k)$$

$$\nu\, \mathbf{e}^1 \cdot \text{curl } \mathbf{e}^3 = -\gamma^3{}_{23} = \gamma^3{}_{32}, \qquad (3.13.8l)$$

$$\nu\, \mathbf{e}^2 \cdot \text{curl } \mathbf{e}^1 = -\gamma^1{}_{31} = \gamma^1{}_{13}, \qquad (3.13.8m)$$

$$\nu\, \mathbf{e}^2 \cdot \text{curl } \mathbf{e}^3 = -\gamma^3{}_{31} = \gamma^3{}_{13}, \qquad (3.13.8n)$$

$$\nu\, \mathbf{e}^3 \cdot \text{curl } \mathbf{e}^1 = -\gamma^1{}_{12} = \gamma^1{}_{21}, \qquad (3.13.8o)$$

$$\nu\, \mathbf{e}^3 \cdot \text{curl } \mathbf{e}^2 = -\gamma^2{}_{12} = \gamma^2{}_{21}. \qquad (3.13.8p)$$

From the above it follows that *the vanishing of* $\gamma^K{}_{\ldots}$ *is a necessary and sufficient condition for curl* $\mathbf{e}^K = \mathbf{0}$ *(*$\Rightarrow \mathbf{e}^K$*: irrotational, or gradient of a scalar field); and hence, for the existence of a one-parameter family of surfaces everywhere normal to the field* $\mathbf{e}^K(P)$, i.e., for its line to intersect these surfaces at right angles. (See also Equation 3.15.3e and the theorem following it; and books on vector field theory.) In an *orthonormal* basis, $\mathbf{e}_K = \mathbf{e}^K \Rightarrow \nu = \nu^* = 1$, and so the above combine to (no sum on K):

$$\mathbf{e}_K \cdot \text{curl } \mathbf{e}_K = -\gamma^K{}_{K+1,K+2} \equiv -\gamma_{K,K+1,K+2}$$

$$\text{(if an index exceeds 3, we subtract from it 3)} \qquad (3.13.8q)$$

For extensions of the above to n-dimensions, etc., see the Examples and Problems of Section 3.16.

Problem 3.13.3

Verify that, in an orthonormal NH basis:

$$\mathbf{e}_K \cdot \text{curl } \mathbf{e}_K = -\gamma^1{}_{23} - \gamma^2{}_{31} - \gamma^3{}_{12} = \gamma^1{}_{32} + \gamma^2{}_{13} + \gamma^3{}_{21}. \qquad (a)$$

Problem 3.13.4

Show that in three-dimensional orthogonal curvilinear coordinates:

$$\text{curl } \mathbf{e}_1 = \left(1/\sqrt{g}\right)[(\partial g_{11}/\partial q^3)\mathbf{e}_2 - (\partial g_{11}/\partial q^2)\mathbf{e}_3], \quad \text{etc., cyclically.} \tag{a}$$

Hint: There: grad $g_{11} = (g_{11,1})(\mathbf{e}_1/g_{11}) + (g_{11,2})(\mathbf{e}_2/g_{22}) + (g_{11,3})(\mathbf{e}_3/g_{33})$,

grad $q^1 = \mathbf{e}_1 / g_{11}$, etc., cyclically (commas denote q^k-derivatives).

3.14 NH TENSOR ANALYSIS: THE TRANSITIVITY EQUATIONS

3.14.1 Basic Results

The preceding AO definitions, Equations 3.13.4a and 3.13.7c through e, are not the only, or most natural, way to calculate them in concrete problems, or to use them in theoretical arguments; for example, they do not show clearly how these quantities vary under NH transformations $d\theta \to d\theta' \to d\theta'' \to \ldots$. Below we present such simpler and invariant definitions.

We begin with the NH transformation $d_1q \to d_1\theta$: $d_1\theta^R = A^R{}_k\, d_1q^k$, and then apply to it a second differentiation $d_2(\ldots)$, *not* to be confused with the second differential $d^2(\ldots)$:

$$\begin{aligned} d_2(d_1\theta^R) &= d_2A^R{}_k\, d_1q^k + A^R{}_k\, d_2(d_1q^k) = (A^R{}_{k,l}\, d_2q^l)d_1q^k + A^R{}_k\, d_2(d_1q^k) \\ &= A^R{}_{k,l}(A^l{}_L\, d_2\theta^L)(A^k{}_K\, d_1\theta^K) + A^R{}_k\, d_2(d_1q^k) \\ &= (A^R{}_{k,l}\, A^l{}_L\, A^k{}_K)d_1\theta^K\, d_2\theta^L + A^R{}_r\, d_2(d_1q^r). \end{aligned} \tag{3.14.1a}$$

Reversing the order of differentiations, and operating in an entirely analogous fashion, we find

$$d_1(d_2\theta^R) = \ldots = (A^R{}_{l,k}\, A^k{}_K\, A^l{}_L)d_1\theta^K\, d_2\theta^L + A^R{}_r\, d_1(d_2q^r); \tag{3.14.1b}$$

and, therefore, subtracting Equations 3.14.1a and b side by side, we get

$$\begin{aligned} d_2(d_1\theta^R) - d_1(d_2\theta^R) &= [(A^R{}_{k,l} - A^R{}_{l,k})A^k{}_K\, A^l{}_L]d_1\theta^K\, d_2\theta^L \\ &\quad + A^R{}_r[(d_2(d_1q^r) - d_1(d_2q^r)], \end{aligned} \tag{3.14.1c}$$

or, recalling Equations 3.13.7c through e, we finally obtain the following fundamental *transitivity equations*:

$$d_2(d_1\theta^R) - d_1(d_2\theta^R) = \gamma^R{}_{KL}\, d_1\theta^K\, d_2\theta^L + A^R{}_r[d_2(d_1q^r) - d_1(d_2q^r)]. \tag{3.14.1d}$$

The above can serve as a *definition* of the γs: the latter may be simply read off as the coefficients of the *bilinear* form $(\ldots)\, d_1\theta^K\, d_2\theta^L$.*

Inverting Equation 3.14.1d, by appropriate dot multiplication with $A^r{}_R$, etc., yields

$$d_2(d_1 q^r) - d_1(d_2 q^r) = A^r{}_R\{[d_2(d_1\theta^R) - d_1(d_2\theta^R)] - \gamma^R{}_{KL}\, d_1\theta^K\, d_2\theta^L\}. \quad (3.14.1e)$$

Normally, since the dq are H coordinates, $d_2(d_1 q^r) = d_1(d_2 q^r)$; or, simply: $d_2(d_1 q) = d_1(d_2 q)$. However, in some dynamic derivations (Section 7.5) we keep those terms for extra generality of interpretation of the results. Enforcing that assumption into the *general transitivity equations* (Equations 3.14.1d or e), we obtain the *simpler transitivity equations:*

$$d_2(d_1\theta^R) - d_1(d_2\theta^R) = \gamma^R{}_{KL}\, d_1\theta^K\, d_2\theta^L, \quad \gamma^R{}_{KL} \equiv A^k{}_K\, A^l{}_L(A^R{}_{k,l} - A^R{}_{l,k}). \quad (3.14.1f)$$

3.14.2 Transformation of the γ Terms

The above allow us to find how the γs vary under a $d\theta^{K'} = A^{K'}{}_K\, d\theta^K \Leftrightarrow d\theta^K = A^K{}_{K'}\, d\theta^{K'}$ transformation. Indeed, applying Equation 3.14.1c between the NH $d\theta^{K'}$ and $d\theta^K$ (which is legitimate application of that formula) yields:

$$\begin{aligned} d_2(d_1\theta^{R'}) - d_1(d_2\theta^{R'}) &= A^{R'}{}_R[(d_2(d_1\theta^R) - d_1(d_2\theta^R)] \\ &\quad + (A^{R'}{}_{K,L} - A^{R'}{}_{L,K}) d_1\theta^K\, d_2\theta^L \\ &= A^{R'}{}_R(\gamma^R{}_{KL}\, d_1\theta^K\, d_2\theta^L) + (A^{R'}{}_{K,L} - A^{R'}{}_{L,K}) d_1\theta^K\, d_2\theta^L \\ &\quad \text{(by Equation 3.14.1f)} \end{aligned} \quad (3.14.2a)$$

But from Equation 3.14.1f, applied between the NH $d\theta^{R'}$ and the H $q^k \to dq^k$, we have

$$\begin{aligned} d_2(d_1\theta^{R'}) - d_1(d_2\theta^{R'}) &= \gamma^{R'}{}_{K'L'}\, d_1\theta^{K'}\, d_2\theta^{L'} \\ &= \gamma^{R'}{}_{K'L'}\, (A^{K'}{}_K\, d_1\theta^K)(A^{L'}{}_L\, d_2\theta^L). \end{aligned} \quad (3.14.2b)$$

Hence, equating the right sides of Equations 3.14.2a and b, we find the $\gamma \to \gamma'$ *transformation equations:*

$$A^{R'}{}_{K,L} - A^{R'}{}_{L,K} = A^{K'}{}_K\, A^{L'}{}_L\, \gamma^{R'}{}_{K'L'} - A^{R'}{}_R\, \gamma^R{}_{KL}; \quad (3.14.2c)$$

---

* *Historical*: The earliest transitivity equations, a special case of Equation 3.14.1f applied to rigid-body kinematics, was given by Lagrange (2nd edition of his *Mécanique Analytique,* 1811–1815). Other special cases of them were given in the late 19th century (C. Neumann, V. Volterra, et al.). Their general formulation, shown here, along with detailed applications to mechanics, is due to K. Heun and G. Hamel (1902–6). To them is also due the term transitivity equations, because they afford a *transition* from Lagrangean to Eulerian (rigid-body) mechanics. See also Chapters 6 and 7.

or, isolating $\gamma^{R'}{}_{K'L'}$ by simple dot multiplications,

$$\gamma^{R'}{}_{K'L'} = A^{R'}{}_R\, A^K{}_{K'}\, A^L{}_{L'}\, \gamma^R{}_{KL} + A^K{}_{K'}\, A^L{}_{L'}(A^{R'}{}_{K,L} - A^{R'}{}_{L,K})$$

$$\text{(Tensor–like part + Nontensorial part).} \qquad (3.14.2d)^*$$

The above shows immediately that if $A^{R'}{}_{K,L} - A^{R'}{}_{L,K} \equiv A^{R'}{}_{K,l}\, A^l{}_L - A^{R'}{}_{L,k}\, A^k{}_K = 0$, in which case $d\theta$ and $d\theta'$ are called *relatively holonomic*, then the γs transform as tensors.

Problem 3.14.1

Show by (symbolic) differentiation of the $d\theta \leftrightarrow d\theta'$ compatibility equations $A^{R'}{}_K\, A^K{}_{K'} = \delta^{R'}{}_{K'}$, that

$$\Delta^{R'}{}_{K'L'} \equiv A^K{}_{K'}\, A^L{}_{L'}(A^{R'}{}_{K,L} - A^{R'}{}_{L,K}) = (A^K{}_{K'}\, A^L{}_{L'} - A^K{}_{L'}\, A^L{}_{K'})A^{R'}{}_{K,L}$$

$$= \ldots = -A^{R'}{}_K(A^K{}_{K',L'} - A^K{}_{L',K'}); \qquad \text{(a)}$$

where $(\ldots)_{,K} \equiv A^k{}_K\, (\ldots)_{,k}$, $(\ldots)_{,K'} \equiv A^k{}_{K'}\, (\ldots)_{,k}$, and $(\ldots)_{,k} \equiv A^K{}_k\, (\ldots)_{,K} = A^{K'}{}_k(\ldots)_{,K'}$, so that $(\ldots)_{,K'} = A^K{}_{K'}\, (\ldots)_{,K}$.

Problem 3.14.2

A *second* transitivity coefficient (or AO) is defined by (as with the H first- and second-kind Christoffels):

$$\gamma_{S,KL} \equiv g_{SR}\, \gamma^R{}_{KL} \leftrightarrow \gamma^R{}_{KL} = g^{RS}\, \gamma_{S,KL}; \qquad \text{(a)}$$

this and the first-kind Christoffels being the only cases where a comma does *not* signify partial differentiation. Show that its law of transformation under $d\theta \leftrightarrow d\theta'$ is

$$\gamma_{S',K'L'} \equiv A^S{}_{S'}\, A^K{}_{K'}\, A^L{}_{L'}\, \gamma_{S,KL} + g_{S'R'}\, \Delta^{R'}{}_{K'L'}$$

$$\text{(Recall Equation a of preceding Problem 3.14.1).} \qquad \text{(b)}$$

---

* Alternative derivation of Equation 3.14.2d: Direct application of Equation 3.13.4a yields, successively,

$$\Omega^{R'}{}_{K'L'} = A^k{}_{K'}\, A^l{}_{L'}\, A^{R'}{}_{[k,l]} = (A^k{}_K\, A^K{}_{K'})(A^l{}_L\, A^L{}_{L'})[A^R{}_{[k,l]}\, A^{R'}{}_R + A^R{}_{[k}\, A^{R'}{}_{R,l]}]$$

$$= A^K{}_{K'}\, A^L{}_{L'}\, A^{R'}{}_R\, \Omega^R{}_{KL} + A^K{}_{K'}\, A^L{}_{L'}\, A^{R'}{}_{[K,L]}$$

$$= A^K{}_{K'}\, A^L{}_{L'}\, A^{R'}{}_R\, \Omega^R{}_{KL} - A^{R'}{}_K\, A^K{}_{[K',L']}$$

$$= A^K{}_{K'}\, A^L{}_{L'}\, A^{R'}{}_R\, \Omega^R{}_{KL} + A^L{}_{[L'}\, A^K{}_{K']}\, A^{R'}{}_{K,L}. \qquad (3.14.2e)$$

Problem 3.14.3

Let us consider an arbitrary scalar function $f(q)$. We have, successively,

$$d_1 f = f_{,k}\, d_1 q^k = f_{,k}(A^k{}_K\, d_1\theta^K) = (f_{,k}\, A^k{}_K) d_1\theta^K = f_{,K}\, d_1\theta^K, \tag{a}$$

and, therefore,

$$d_2(d_1 f) = d_2(f_{,K})\, d_1\theta^K + f_{,K}\, d_2(d_1\theta^K) = f_{,KL}\, d_1\theta^K\, d_2\theta^L + f_{,K}\, d_2(d_1\theta^K); \tag{b}$$

and, similarly, swapping the roles of 1 and 2,

$$d_1(d_2 f) = f_{,LK}\, d_1\theta^K\, d_2\theta^L + f_{,K}\, d_1(d_2\theta^K). \tag{c}$$

Now, subtracting the above side by side and then invoking Equations 3.13.6d, 3.14.1d, etc., verify that

$$d_2(d_1 f) - d_1(d_2 f) = f_{,k}[d_2(d_1 q^k) - d_1(d_2 q^k)] = 0. \tag{d}$$

3.14.3 Geometrical Interpretation of the Transitivity Equations

In a manifold (say, an L_n or an R_n) let us consider the point $P(q)$ and its neighboring points $Q(q + d_1\theta^R)$ and $R(q + d_2\theta^R)$, obtained by applying $d_1(\ldots)$ and $d_2(\ldots)$, respectively, to the coordinates of P. Further application of $d_1(\ldots)$ to the coordinates of R produces $S[q + d_2\theta^R + d_1(q + d_2\theta^R)]$ and of $d_2(\ldots)$ to those of Q produces $T[q + d_1\theta^R + d_2(q + d_1\theta^R)]$. (See the resulting two families of curves in Figure 3.6; and note slight difference from Figure 3.3 of Section 3.9.)

However, here, and contrary to the H case, as the transitivity equations show,

$$d_2(d_1\theta^R) - d_1(d_2\theta^R) = \gamma^R{}_{KL}\, d_1\theta^K\, d_2\theta^L = \gamma^R{}_{KL}(d_1\theta^K\, d_2\theta^L - d_1\theta^L\, d_2\theta^K)/2$$

$$\equiv d_2(d_1 P) - d_1(d_2 P) = T - S \neq 0 \tag{3.14.3}$$

: *gap ST* (i.e., a measure of the *misclosure* of the infinitesimal parallelogram/pentagon *PQTSR*).

That is, S and T do not coincide; or, more precisely, if $d_1\theta$ and $d_2\theta$ are of the *first* order, then the gap $d_2(d_1\theta^R) - d_1(d_2\theta^R)$ is *bilinear*, that is, of the *second* order; i.e., closing it means that it should have been of the *third* order. Also, since $d_1\theta\, d_2\theta$ is of the order of the area of the elementary figure *PQTSR*, the γs, generally, have dimensions of $(\text{length})^{-1}$.*

* See also Misner et al. (1973, pp. 235–240).

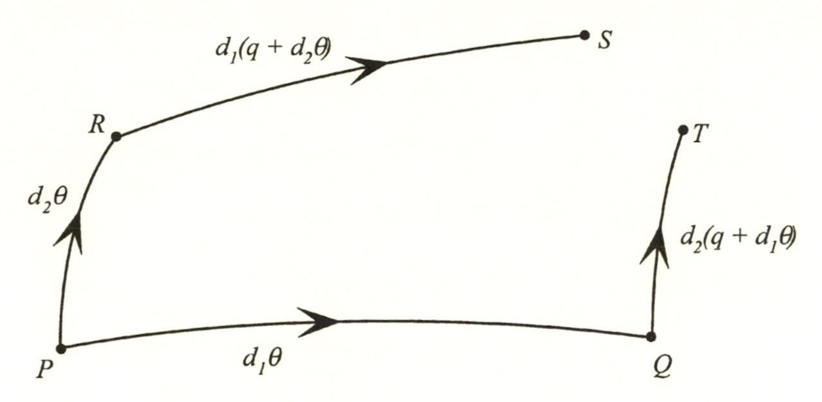

FIGURE 3.6 Gap *ST* of four-legged "quadrilateral" *PQTSR*.

$$P(q), \quad Q(q + d_1\theta) \equiv d_1P, \quad R(q + d_2\theta) \equiv d_2P,$$

$$T[q + d_1\theta^R + d_2(q + d_1\theta^R)] \equiv d_2Q = d_2(d_1P),$$

$$S[q + d_2\theta^R + d_1(q + d_2\theta^R)] \equiv d_1R = d_1(d_2P).$$

3.15 NH TENSOR ANALYSIS: NH AFFINITIES AND CHRISTOFFELS

3.15.1 NH Basis Gradients and Affinities

Let us now find the NH affinities and study their properties; and, in the process, check the holonomicity of the covariant basis $\{P, \mathbf{e}_K\}$. For the NH gradients of a general NH covariant basis $\{\mathbf{e}_K\}$ (assuming an L_n, for extra generality, recalling Section 3.8), we find, successively,

$$\begin{aligned}
\partial\mathbf{e}_K/\partial\theta^L &\equiv \mathbf{e}_{K,L} \equiv (\partial\mathbf{e}_K/\partial q^l)(\partial q^l/\partial\theta^L) = A^l{}_L\,\mathbf{e}_{K,l} = A^l{}_L(A^k{}_K\,\mathbf{e}_k)_{,l} \\
&= A^l{}_L(A^k{}_{K,l}\,\mathbf{e}_k + A^k{}_K\,\mathbf{e}_{k,l}) \\
&= A^l{}_L[A^k{}_{K,l}\,\mathbf{e}_k + A^k{}_K\,(\Lambda^r{}_{kl}\,\mathbf{e}_r)] \\
&= A^l{}_L(A^k{}_{K,l} + A^r{}_K\,\Lambda^k{}_{rl})\mathbf{e}_k \equiv A^l{}_L(A^k{}_K|_l\,\mathbf{e}_k) \equiv A^l{}_L(A_{Kk}|_l\,\mathbf{e}^k) \\
&\equiv A^k{}_K|_L\,\mathbf{e}_k \equiv A_{Kk}|_L\,\mathbf{e}^k \\
&\equiv (A^R{}_k\,A^k{}_K|_L)\mathbf{e}_R \equiv (A^k{}_R\,A_{Kk}|_L)\mathbf{e}^R, \qquad (3.15.1\text{a})
\end{aligned}$$

or

$$\mathbf{e}_{K,L} = \Lambda^R{}_{KL}\,\mathbf{e}_R = \Lambda_{R,KL}\,\mathbf{e}^R \qquad (3.15.1\text{b})$$

$$[\text{and thus, } \mathbf{e}_{K,k} = A^L{}_k\,\mathbf{e}_{K,L} = A^L{}_k(\Lambda^R{}_{KL}\,\mathbf{e}_R) = A^L{}_k(\Lambda_{R,KL}\,\mathbf{e}^R)] \qquad (3.15.1\text{c})$$

where

$$\Lambda^R{}_{KL} = \mathbf{e}_{K,L} \cdot \mathbf{e}^R \equiv A^R{}_K|_L \equiv A^R{}_k\, A^k{}_K|_L \equiv A^R{}_k\, A^l{}_L\, A^k{}_K|_l, \tag{3.15.1d}$$

$$\Lambda_{R,KL} = \mathbf{e}_{K,L} \cdot \mathbf{e}_R \equiv A_{KR}|_L \equiv A^k{}_R\, A_{Kk}|_L \equiv A^k{}_R\, A^l{}_L\, A_{Kk}|_l, \tag{3.15.1e}$$

$$[\Lambda^R{}_{KL} = g^{RS}\, \Lambda_{S,KL} \leftrightarrow \Lambda_{S,KL} = g_{SR}\, \Lambda^R{}_{KL}] \tag{3.15.1f}$$

are, respectively, the *nontensorial and nonsymmetrical NH counterparts of the second and first kind affinities,* (or generalized Ricci's coefficients of rotation, see Section 3.17); and, as shown below, play the same role in NH covariant differentiation as the H Λs, Γs do in the H case (Sections 3.3, 3.4, 3.8).

3.15.2 Properties of the NH Affinities

i. $(\partial \ldots / \partial\theta^L)$ — differentiating $\mathbf{e}_K \cdot \mathbf{e}^R = \delta^R{}_K\ (= g^R{}_K)$, yields $\mathbf{e}_{K,L} \cdot \mathbf{e}^R + \mathbf{e}_K \cdot \mathbf{e}^R{}_{,L} = 0$, and, therefore,

$$\Lambda^R{}_{KL} = \mathbf{e}_{K,L} \cdot \mathbf{e}^R = -\,\mathbf{e}^R{}_{,L} \cdot \mathbf{e}_K\ (\Rightarrow \mathbf{e}^R{}_{,L} = -\Lambda^R{}_{KL}\, \mathbf{e}^K). \tag{3.15.2a}$$

ii. From Equation 3.15.1e, we find, sucessively,

$$\Lambda_{R,KL} = \mathbf{e}_{K,L} \cdot \mathbf{e}_R = (\mathbf{e}_K \cdot \mathbf{e}_R)_{,L} - \mathbf{e}_{R,L} \cdot \mathbf{e}_K = g_{KR,L} - \Lambda_{K,RL},$$

$$\Rightarrow \partial g_{KR}/\partial\theta^L \equiv g_{KR,L} = \Lambda_{R,KL} + \Lambda_{K,RL} = 2\Lambda_{(R,K)L}. \tag{3.15.2b}$$*

3.15.3 NH Affinities in a Riemannian Space (i.e., $L_n \rightarrow R_n$)

There (recalling Section 3.8), the Λs are *symmetric* and equal the Christoffels: $\Lambda^k{}_{rl} = \Gamma^k{}_{rl}$; and as a result $\mathbf{e}_{k,l} - \mathbf{e}_{l,k} \equiv 2\mathbf{e}_{[k,l]} = \mathbf{0}$, for all k, l; i.e., the basis $\{P, \mathbf{e}_k\}$ is holonomic (or gradient). Let us see if that also holds for the basis $\{P, \mathbf{e}_K\}$. From $\mathbf{e}_k = A^K{}_k\, \mathbf{e}_K$, we find, successively,

$$\begin{aligned} \mathbf{e}_{k,l} &= (A^K{}_k\, \mathbf{e}_K)_{,l} \\ &= A^K{}_{k,l}\, \mathbf{e}_K + A^K{}_k\, \mathbf{e}_{K,l} = A^K{}_{k,l}\, \mathbf{e}_K + A^K{}_k(\mathbf{e}_{K,R}\, A^R{}_l) \\ &= A^K{}_{k,l}\, \mathbf{e}_K + A^K{}_k(A^R{}_l\, \Lambda^S{}_{KR}\, \mathbf{e}_S) = \ldots = (A^K{}_{k,l} + \Lambda^K{}_{SR}\, A^S{}_k\, A^R{}_l)\mathbf{e}_K; \end{aligned} \tag{3.15.3a}$$

similarly,

$$\mathbf{e}_{l,k} = (A^L{}_l\, \mathbf{e}_L)_{,k} = \ldots = (A^K{}_{l,k} + \Lambda^K{}_{RS}\, A^S{}_k\, A^R{}_l)\mathbf{e}_K; \tag{3.15.3b}$$

* This can also be obtained from the NH Ricci's lemma (Section 3.16):

$$0 = g_{KL}|_R = g_{KL,R} - \Lambda^S{}_{KR}\, g_{SL} - \Lambda^S{}_{LR}\, g_{SK} = g_{KL,R} - (\Lambda_{L,KR} + \Lambda_{K,LR}), \quad \text{q.e.d.}$$

and, therefore, subtracting Equations 3.15.3a and b side by side, we find

$$\mathbf{0} = \mathbf{e}_{k,l} - \mathbf{e}_{l,k} = [(\Lambda^K{}_{SR} - \Lambda^K{}_{RS})A^S{}_k\, A^R{}_l + (A^K{}_{k,l} - A^K{}_{l,k})]\mathbf{e}_K,$$

and dotting this with $A^k{}_C\, A^l{}_D$, etc., to isolate (the antisymmetric part) $\Lambda^K{}_{SR} - \Lambda^K{}_{RS}$, we get

$$0 = \Lambda^K{}_{CD} - \Lambda^K{}_{DC} + A^k{}_C\, A^l{}_D(A^K{}_{k,l} - A^K{}_{l,k})$$

(then recalling Equations 3.13.7c through e)

$$\Rightarrow \gamma^R{}_{KL} = \Lambda^R{}_{LK} - \Lambda^R{}_{KL} \equiv 2\Lambda^R{}_{[L,K]} = -2\Lambda^R{}_{[KL]}, \tag{3.15.3c}$$

$$[\text{also, } \gamma_{R,KL} = \Lambda_{R,LK} - \Lambda_{R,KL} \equiv 2\Lambda_{R,[L,K]} = -2\Lambda_{R,[KL]}]; \tag{3.15.3d}$$

i.e., even in an R_n, the NH Λs, *as defined by Equations 3.15.1d through f*, are nonsymmetric.* Utilizing this basic result in Equation 3.15.1d leads to the following fundamental *nonintegrability/noncommutativity* relations

$$(\mathbf{e}_{K,L} - \mathbf{e}_{L,K}) \cdot \mathbf{e}^R = \Lambda^R{}_{KL} - \Lambda^R{}_{LK} = \gamma^R{}_{LK}$$

$$\Rightarrow \partial\mathbf{e}_K/\partial\theta^L - \partial\mathbf{e}_L/\partial\theta^K = \gamma^R{}_{LK}\,\mathbf{e}_R = \gamma_{R,LK}\,\mathbf{e}^R; \tag{3.15.3e}$$

which can be viewed as *the vectorial transitivity* parallel to Equation 3.14.1d.

(Also, compare with its holonomic counterpart, Equations 3.12.2a and b.) Hence, the

Theorem: If (in an R_n) *all* the transitivity coefficients vanish identically, the basis $\{P, \mathbf{e}_K = (A_{Kk}) = (A^k{}_K)\}$ is holonomic, or gradient, i.e., the $\{\mathbf{e}_K\}$ are the tangent vectors of a global coordinate system. But if even one of these γs does not vanish, the basis $\{P, \mathbf{e}_K\}$ is nonholonomic (or noncoordinate). And conversely, if $\{P, \mathbf{e}_K\}$ is holonomic, all its γs vanish.**

Notation: From now on, we shall denote the NH affinities of a Riemannian space R_n as $\Gamma^R{}_{LK}$, $\Gamma_{R,LK}$; and reserve the symbols $\Lambda^R{}_{LK}$, $\Lambda_{R,LK}$ (if needed) for the NH affinities of a general linearly connected manifold L_n.

Problem 3.15.1. Alternative Derivations of Equation 3.15.3c

Take $d_1\mathbf{P} = \mathbf{e}_K\, d_1\theta^K$. Then, $d_2(d_1\mathbf{P}) = d_2\mathbf{e}_K\, d_1\theta^K + \mathbf{e}_K\, d_2(d_1\theta^K) = (\Lambda^R{}_{KL}\, d_2\theta^L)\,\mathbf{e}_R\, d_1\theta^K + \mathbf{e}_R\, d_2(d_1\theta^R)$ (by Equations 3.15.1a ff), and, similarly, $d_1(d_2\mathbf{P}) = \ldots$. By applying the integrability condition: $d_2(d_1\mathbf{P}) = d_1(d_2\mathbf{P})$, while invoking Equations 3.14.1d ff, derive Equation 3.15.3c.

---

* See also transformation of the NH counterpart of the Λs below, Equations 3.15.4a ff.

** (See following page for footnote.)

Example 3.15.1. Alternative Derivation of the Transitivity Equations 3.14.1f

By direct differentiations, we find

$$d_1(d_2\mathbf{P}) - d_2(d_1\mathbf{P}) = [d_1(d_2\theta^K) - d_2(d_1\theta^K)]\mathbf{e}_K + (d_2\theta^K\, d_1\mathbf{e}_K - d_1\theta^K\, d_2\mathbf{e}_K).$$

But,

$$d_1\mathbf{e}_K = d_1(A^k{}_K\, \mathbf{e}_k) = d_1A^k{}_K\, \mathbf{e}_k + A^k{}_K\, d_1\mathbf{e}_k = A^k{}_{K,l}(A^l{}_L\, d_1\theta^L)\mathbf{e}_k + A^k{}_K(\mathbf{e}_{k,l}\, d_1q^l).$$

Therefore, and recalling that here $\mathbf{e}_{k,l} = \mathbf{e}_{l,k}$ as well as the γ-definitions (Equations 3.13.7c through e), we find

$$\mathbf{0} = d_1(d_2\mathbf{P}) - d_2(d_1\mathbf{P}) = \ldots = [d_1(d_2\theta^R) - d_2(d_1\theta^R)]\mathbf{e}_R$$

$$+ [(A^k{}_{K,l}\, A^l{}_L - A^k{}_{L,l}\, A^l{}_K)A^R{}_k]d_1\theta^L\, d_2\theta^K\, \mathbf{e}_R + A^k{}_K\, A^l{}_L(\mathbf{e}_{k,l} - \mathbf{e}_{l,k})d_1\theta^L\, d_2\theta^K$$

$$= [d_1(d_2\theta^R) - d_2(d_1\theta^R) + \gamma^R{}_{LK}\, d_1\theta^L\, d_2\theta^K]\, \mathbf{e}_R, \quad \text{q.e.d.}$$

3.15.4 Transformation of the NH Affinities

Reasoning as in Sections 3.3 and 3.8 (also, see below), we easily see that under a transformation $(dq^k, \mathbf{e}_k) \rightarrow (d\theta^K, \mathbf{e}_K)$ *the H affinities transform to the NH affinities in exactly the same way as H Christoffels transform to other also H Christoffels* (but, here, note subscript order!):

$$\Lambda^R{}_{KL} = A^R{}_r\, A^k{}_K\, A^l{}_L\, \Lambda^r{}_{kl} + A^R{}_r\, A^r{}_{K,L}$$

$$= A^R{}_r\, A^k{}_K\, A^l{}_L\, \Lambda^r{}_{kl} + A^R{}_r\, A^l{}_L\, A^r{}_{K,l}$$

** *Alternative Derivation of Integrability Conditions:* By Stokes' theorem (Example 3.6.1), with $T_r \rightarrow A^R{}_r$, we obtain successively:

$$J^R \equiv \int_C d\theta^R \equiv \int_C A^R{}_r\, dq^r = \int_C \mathbf{e}^R \cdot d\mathbf{R} = (1/2)\iint_A (A^R{}_{k,l} - A^R{}_{l,k})dA^{lk}, \tag{3.15.3f}$$

or, since: $dA^{lk} \equiv d_1q^l\, d_2q^k - d_1q^k\, d_2q^l = (A^l{}_L\, d_1\theta^L)(A^k{}_K\, d_2\theta^K) - (A^k{}_K\, d_1\theta^K)(A^l{}_L\, d_2\theta^L)$

$$= A^k{}_K A^l{}_L(d_1\theta^L\, d_2\theta^K - d_1\theta^K\, d_2\theta^L) \equiv A^k{}_K A^l{}_L\, dA^{LK},$$

$$J^R = \iint_A (A^R{}_{[k,l]}\, A^k{}_K\, A^l{}_L)dA^{LK} = \iint_A \Omega^R{}_{KL}\, dA^{LK} = \iint_A (1/2)\gamma^R{}_{KL}\, dA^{LK}. \tag{3.15.3g}$$

Hence, if $J^R = 0$ for *any* (size) boundary, then $\gamma^R{}_{K,L} = 0$; and vice versa.

$$= A^R{}_r\, A^k{}_K\, A^l{}_L\, \Lambda^r{}_{kl} - A^k{}_K\, A^l{}_L\, A^R{}_{k,l}$$

$$= A^R{}_r\, A^k{}_K\, A^l{}_L\, \Lambda^r{}_{kl} - A^k{}_K\, A^R{}_{k,L}; \qquad (3.15.4a)$$

where, in the last two, we used: $(A^R{}_r\, A^r{}_K)_{,l} = (\delta^R{}_K)_{,l} = 0 \Rightarrow A^R{}_r\, A^r{}_{K,l} = -A^r{}_K\, A^R{}_{r,l}$.* Similarly, under a transformation $d\theta \to d\theta'$ (whether both H, or both NH, or one of each):

$$\Lambda^{R'}{}_{K'L'} = A^{R'}{}_R\, A^K{}_{K'}\, A^L{}_{L'}\, \Lambda^R{}_{KL} + A^{R'}{}_R\, A^R{}_{K',L'}, \quad \text{etc.} \qquad (3.15.4b)$$

Next, to the antisymmetric part of the NH Λs. From Equation 3.15.4a, with $K \to L$ and $L \to K$, we get

$$\Lambda^R{}_{LK} = A^R{}_r\, A^k{}_K\, A^l{}_L\, \Lambda^r{}_{lk} - A^k{}_K\, A^l{}_L\, A^R{}_{l,k},$$

and, therefore,

$$\Lambda^R{}_{KL} - \Lambda^R{}_{LK} = A^R{}_r\, A^k{}_K\, A^l{}_L(\Lambda^r{}_{kl} - \Lambda^r{}_{lk}) + A^k{}_K\, A^l{}_L(A^R{}_{l,k} - A^R{}_{k,l}),$$

i.e.,

$$\Lambda^R{}_{[KL]} = A^R{}_r\, A^k{}_K\, A^l{}_L\, \Lambda^r{}_{[kl]} + (1/2)\gamma^R{}_{LK}; \qquad (3.15.5a)$$

or, finally, in terms of the *torsion* tensor: $S^R{}_{KL} = A^R{}_r\, A^k{}_K\, A^l{}_L\, S^r{}_{kl} = A^R{}_r\, A^k{}_K\, A^l{}_L\, \Lambda^r{}_{[kl]}$ (Sections 3.8 and 3.9),

$$\Lambda^R{}_{[KL]} = S^R{}_{KL} + (1/2)\gamma^R{}_{LK} = S^R{}_{KL} + \Omega^R{}_{LK}. \qquad (3.15.5b)^{**}$$

In a *torsion-free* space, like an R_n, $\Lambda^r{}_{kl} \to \Gamma^r{}_{kl} \Rightarrow \Gamma^r{}_{[kl]} = S^r{}_{kl} = 0 \Rightarrow S^R{}_{KL} = 0$; i.e., since there $\Lambda^R{}_{KL} \to \Gamma^R{}_{KL}$, we recover Equation 3.15.3c.

Similarly, for the symmetric part of $\Lambda^R{}_{KL}$, we find

$$\Lambda^R{}_{(KL)} = A^R{}_r\, A^k{}_K\, A^l{}_L\, \Lambda^r{}_{(kl)} - (1/2)A^k{}_K\, A^l{}_L(A^R{}_{k,l} + A^R{}_{l,k}); \qquad (3.15.6a)$$

and, therefore, in a torsion-free space:

$$\Lambda^R{}_{(KL)} \to \Gamma^R{}_{(KL)} = A^R{}_r\, A^k{}_K\, A^l{}_L\, \Gamma^r{}_{kl} - A^k{}_K\, A^l{}_L\, A^R{}_{(k,l)}. \qquad (3.15.6b)$$

---

* *Proof of Equation 3.15.4a:* We have $\Lambda^R{}_{KL} = \mathbf{e}_{K,L} \cdot \mathbf{e}^R = (A^l{}_L\, \mathbf{e}_{K,l}) \cdot (A^R{}_r\, \mathbf{e}^r) = A^l{}_L\, A^R{}_r\, (\mathbf{e}_{K,l} \cdot \mathbf{e}^r)$. But, $\mathbf{e}_{K,l} = (A^k{}_K\, \mathbf{e}_k)_{,l} = A^k{}_{K,l}\, \mathbf{e}_k + A^k{}_K \mathbf{e}_{k,l} = \ldots = (A^s{}_{K,l} + \Lambda^s{}_{kl} A^k{}_K)\, \mathbf{e}_s$ (by Section 3.8). Hence, $\mathbf{e}_{K,l} \cdot \mathbf{e}^r = A^r{}_{K,l} + \Lambda^r{}_{kl} A^k{}_K$, from which Equation 3.15.4a$_2$ follows easily.

** Some authors call $\Lambda^R{}_{(KL)} \equiv S^R{}_{KL}$. Then, instead of Equations 3.15.5a and b:

$$S^R{}_{KL} = A^R{}_r A^k{}_K A^l{}_L \Lambda^r{}_{(kl)} + (1/2)\gamma^R{}_{LK}.$$

The transformation of the first-kind NH affinities follows readily from $\Gamma_{R,KL} \equiv g_{RS}\,\Gamma^S{}_{KL}$:

$$\Lambda_{R,KL} = A^r{}_R\, A^k{}_K\, A^l{}_L\, \Lambda_{r,kl} + A_{Rr}\, A^r{}_{K,L}$$

$$= A^r{}_R\, A^k{}_K\, A^l{}_L\, \Lambda_{r,kl} + A_{Rr}\, A^l{}_L\, A^r{}_{K,l}$$

$$= A^r{}_R\, A^k{}_K\, A^l{}_L\, \Lambda_{r,kl} - A^k{}_K\, A^l{}_L\, A_{Rk,l}$$

$$= A^r{}_R\, A^k{}_K\, A^l{}_L\, \Lambda_{r,kl} - A^k{}_K\, A_{Rk,L}. \tag{3.15.7a}$$

Similarly, swapping K and L in Equation 3.15.7a: $\Lambda_{R,LK} = A^r{}_R\, A^k{}_K\, A^l{}_L\, \Lambda_{r,lk} - A^k{}_K A^l{}_L A_{Rl,k}$, and so:

$$\Lambda_{R,KL} - \Lambda_{R,LK} = A^r{}_R\, A^k{}_K\, A^l{}_L(\Lambda_{r,kl} - \Lambda_{r,lk}) + A^k{}_K\, A^l{}_L(A_{Rl,k} - A_{Rk,l}),$$

i.e.,

$$\Lambda_{R,[KL]} = A^r{}_R\, A^k{}_K\, A^l{}_L\, S_{r,kl} + (1/2)\gamma_{R,LK} = S_{R,KL} + \Omega_{R,LK}; \tag{3.15.7b}$$

where

$$\gamma_{R,LK} \equiv g_{RS}\, \gamma^S{}_{LK}\,(= -\gamma_{R,KL}). \tag{3.15.7c}$$

From the above, it follows readily that, in a torsionless manifold, the *first* kind integrability/transitivity equations (if we call Equation 3.15.3e transitivity equations of the *second* kind) are

$$\mathbf{e}_{K,L} \cdot \mathbf{e}_R - \mathbf{e}_{L,K} \cdot \mathbf{e}_R = \Gamma_{R,KL} - \Gamma_{R,LK} \equiv 2\Gamma_{R,[KL]} = \gamma_{R,LK}, \tag{3.15.7d}$$

or

$$\mathbf{e}_{K,L} - \mathbf{e}_{L,K} = \gamma_{R,LK}\, \mathbf{e}^R. \tag{3.15.7e}$$

3.15.5 The First-Kind NH Christoffel-Like Symbols and Their Properties

Let us, next, examine the symmetric NH Christoffel-like symbols of the first kind (formally identical to the H first-kind Christoffels, $\Gamma_{r,kl}$, of Section 3.3):

$$C_{R,KL} = C_{R,LK} \equiv (1/2)(g_{RL,K} + g_{RK,L} - g_{KL,R}) \neq \Gamma_{R,KL}. \tag{3.15.8a}$$

To relate these with, say (with no real loss in generality), the Riemannian $\Gamma_{R,KL}$, we begin by $(\ldots)_{,S}$–differentiating $g_{RL} = A^r{}_R\, A^l{}_L\, g_{rl}$:

$$g_{RL,K} = A^r{}_R\, A^l{}_L\, g_{rl,K} + A^r{}_{R,K}\, A^l{}_L\, g_{rl} + A^l{}_{L,K}\, A^r{}_R\, g_{rl} = A^r{}_R\, A^l{}_L\, A^k{}_K\, g_{rl,k} + \ldots,$$

and, similarly, $g_{KR,L} = \ldots$, $g_{KL,R} = \ldots$. Hence,

$$C_{R,KL} = (1/2)(g_{rl,k} + g_{kr,l} - g_{kl,r})A^r{}_R\, A^k{}_K\, A^l{}_L + \ldots,$$

and due to $(A^k{}_K\, A^L{}_k)_{,r} = (\delta^L{}_K)_{,r} = 0 \Rightarrow A^L{}_k\, A^k{}_{K,r} = -A^k{}_K\, A^L{}_{k,r}$ and Equation 3.15.7c:

$$\gamma_{R,KL} \equiv g_{RS}\, \gamma^S{}_{KL} = g_{RS}\, [A^k{}_K\, A^l{}_L(A^S{}_{k,l} - A^S{}_{l,k})],$$

after a long but straightforward algebra, we find

$$C_{R,KL} = \Gamma_{R,KL} - (1/2)(\gamma_{K,RL} + \gamma_{L,RK} - \gamma_{R,KL}), \tag{3.15.8b)*}$$

or

$$\begin{aligned} \Gamma_{R,KL} &= C_{R,KL} + (1/2)(\gamma_{K,RL} + \gamma_{L,RK} - \gamma_{R,KL}), \\ &= A^r{}_R\, A^k{}_K\, A^l{}_L\, \Gamma_{r,kl} + g_{kr}\, A^r{}_R\, A^k{}_{K,L}; \end{aligned} \tag{3.15.8c}$$

which shows that in H coordinates, where all the γs vanish, $\Gamma_{r,kl} = C_{r,kl}$; and, also, why we were right in *not* denoting $A^r{}_R\, A^k{}_K\, A^l{}_L\, \Gamma_{r,kl}$ as $\Gamma_{R,KL}$.

Problem 3.15.2

Using Equation 3.15.8c verify that $\Gamma_{R,KL} - \Gamma_{R,LK} = \ldots = \gamma_{R,LK}$, i.e., Equations 3.15.7d and e.

Example 3.15.2. Alternative Derivations of Equations 3.15.8b and c

i. Due to Equation 3.15.2b, Equation 3.15.8a can be rewritten as

$$\begin{aligned} C_{R,KL} &= \Gamma_{(R,L)K} + \Gamma_{(K,R)L} - \Gamma_{(K,L)R} \\ &= (1/2)(\Gamma_{R,LK} + \Gamma_{L,RK} + \Gamma_{K,RL} + \Gamma_{R,KL} - \Gamma_{K,LR} - \Gamma_{L,KR}) \\ &= (1/2)(\Gamma_{R,LK} + 2\Gamma_{L,[RK]} + 2\Gamma_{K,[RL]} + \Gamma_{R,KL}) \quad \text{(adding and subtracting } \Gamma_{R,LK}/2) \\ &= \Gamma_{R,LK} + (\Gamma_{L,[RK]} + \Gamma_{K,[RL]} - \Gamma_{R,[LK]}), \end{aligned} \tag{a}$$

and since by Equation 3.15.7d, cyclically,

---

* See for instance, Bewley (1961, pp. 127ff) and Example 3.15.2. As shown later (Chapter 7), because of Equations 3.15.1a through f, *it is the nonsymmetric* $\Gamma_{R,KL}$ *that are important to NH geometry and mechanics; not the* $C_{R,KL}$.

$$\Gamma_{K,LR} - \Gamma_{K,RL} = \gamma_{K,RL}, \quad \Gamma_{L,RK} - \Gamma_{L,KR} = \gamma_{L,KR}, \quad \Gamma_{R,KL} - \Gamma_{R,LK} = \gamma_{R,LK},$$

we obtain Equation 3.15.8b (with $L \to K$ and $K \to L$):

$$C_{R,KL} = C_{R,LK} = \Gamma_{R,LK} + (1/2)(\gamma_{L,KR} + \gamma_{K,LR} - \gamma_{R,KL}). \tag{b}$$

ii. By the NH Ricci's lemma (Section 3.16), cyclically (and, temporarily, using for extra clarity lower case Greek indices for NH components, etc., instead of the customary uppercase Latin):

$$g_{\alpha\beta}|_\gamma = g_{\alpha\beta,\gamma} - \Gamma^\delta{}_{\alpha\gamma}\, g_{\delta\beta} - \Gamma^\delta{}_{\beta\gamma}\, g_{\alpha\delta} = 0 \Rightarrow g_{\alpha\beta,\gamma} = \Gamma_{\beta,\alpha\gamma} + \Gamma_{\alpha,\beta\gamma}, \tag{c1}$$

$$g_{\beta\gamma}|_\alpha = g_{\beta\gamma,\alpha} - \Gamma^\delta{}_{\beta\alpha}\, g_{\delta\gamma} - \Gamma^\delta{}_{\gamma\alpha}\, g_{\beta\delta} = 0 \Rightarrow g_{\beta\gamma,\alpha} = \Gamma_{\gamma,\beta\alpha} + \Gamma_{\beta,\gamma\alpha}, \tag{c2}$$

$$g_{\gamma\alpha}|_\beta = g_{\gamma\alpha,\beta} - \Gamma^\delta{}_{\gamma\beta}\, g_{\delta\alpha} - \Gamma^\delta{}_{\alpha\beta}\, g_{\gamma\delta} = 0 \Rightarrow g_{\gamma\alpha,\beta} = \Gamma_{\alpha,\gamma\beta} + \Gamma_{\gamma,\alpha\beta}. \tag{c3}$$

Now, adding Equations c1 and c3 and subtracting from the result Equation c2 yields

$$2\Gamma_{\alpha,\beta\gamma} \equiv g_{\gamma\alpha,\beta} + g_{\alpha\beta,\gamma} - g_{\beta\gamma,\alpha} = \ldots = 2\Gamma_{\alpha,(\beta\gamma)} + 2\Gamma_{\beta,[\alpha\gamma]} + 2\Gamma_{\gamma,[\alpha\beta]},$$

or, after adding and subtracting $\Gamma_{\alpha,\beta\gamma}$,

$$C_{\alpha,\beta\gamma} = \Gamma_{\alpha,\beta\gamma} + (\Gamma_{\beta,[\alpha\gamma]} + \Gamma_{\gamma,[\alpha\beta]} - \Gamma_{\alpha,[\beta\gamma]}) \quad \text{(invoking Equation 3.15.7d)}$$

$$= \Gamma_{\alpha,\beta\gamma} - (1/2)(\gamma_{\beta,\alpha\gamma} + \gamma_{\gamma,\alpha\beta} - \gamma_{\alpha,\beta\gamma}),$$

i.e., Equation 3.15.8b, with $\alpha \to R$, $\beta \to K$, $\gamma \to L$.

Similarly, we can define the *second*-kind Christoffel-like symbols: $C^R{}_{KL} = g^{RS}\,\Gamma_{S,KL}$, and obtain equations analogous to Equation 3.15.8b, etc.

Problem 3.15.3

Show that $\Gamma_{R,(KL)} = C_{R,KL} + (1/2)(\gamma_{K,RL} + \gamma_{L,RK})$.

3.16 NH TENSOR ANALYSIS: NH COVARIANT DERIVATIVE

Let us define the NH covariant derivative of a general (say absolute) NH vector V^K by

$$V^K|_L \equiv V^K{}_{,L} + \Lambda^K{}_{RL}\, V^R; \tag{3.16.1a}$$

i.e., formwise identical to the H case (Section 3.4). Then, we have, successively,

$$V^K|_L = A^l{}_L(A^K{}_k\ V^k)_{,l} + \Lambda^K{}_{RL}\ V^R \quad \text{(then invoking the } \textit{third} \text{ of Equations 3.15.4a)}$$

$$= A^l{}_L\ A^K{}_{k,l}\ V^k + A^l{}_L\ A^K{}_k\ V^k{}_{,l} + (A^K{}_k\ A^r{}_R\ A^l{}_L\ \Lambda^k{}_{rl} - A^k{}_R\ A^l{}_L\ A^K{}_{k,l})(A^R{}_r\ V^r)$$

$$= A^K{}_k\ A^l{}_L(V^k{}_{,l} + \Lambda^k{}_{rl}\ V^r) \quad \text{(after some dummy index changes and use of Equations 3.12.3b)}$$

i.e.,

$$V^K|_L = A^K{}_k\ A^l{}_L\ V^k|_l\text{: a NH tensor.} \tag{3.16.1b}$$

Similarly, we can show that

$$V_K|_L \equiv V_{K,L} - \Lambda^R{}_{KL}\ V_R = \ldots = A^k{}_K\ A^l{}_L\ V_k|_l; \tag{3.16.1c}$$

and analogously for higher-rank NH tensors; for, e.g.,

$$T^K{}_L|_R \equiv T^K{}_{L,R} + \Lambda^K{}_{SR}\ T^S{}_L - \Lambda^S{}_{LR}\ T^K{}_S = A^K{}_k\ A^l{}_L\ A^r{}_R\ T^k{}_l|_r. \tag{3.16.1d}$$

Now we can define the NH gradient and differential of a vector **V**, and the NH absolute differentials and derivatives of its NH components, respectively, in complete analogy with the H case (Section 3.5), as follows

- $\partial\mathbf{V}/\partial\theta^L = (V^K\ \mathbf{e}_K)_{,L} = V^K{}_{,L}\ \mathbf{e}_K + V^K\ \mathbf{e}_{K,L} = (V^K{}_{,L} + \Lambda^K{}_{RL}\ V^R)\mathbf{e}_K$ (invoking Equation 3.15.1b) or, finally,

$$\partial\mathbf{V}/\partial\theta^L = V^K|_L\ \mathbf{e}_K = V_K|_L\ \mathbf{e}^K; \tag{3.16.2a}$$

- $d\mathbf{V} = (\partial\mathbf{V}/\partial\theta^L)d\theta^L = (V^K|_L\ \mathbf{e}_K)d\theta^L \equiv DV^K\ \mathbf{e}_K$
  $= (V_K|_L\ \mathbf{e}^K)d\theta^L \equiv DV_K\ \mathbf{e}^K;$ (3.16.2b)

- $DV^K \equiv V^K|_L\ d\theta^L = dV^K + \Lambda^K{}_{RL}\ V^R\ d\theta^L,$
  $DV_K \equiv V_K|_L\ d\theta^L = dV_K - \Lambda^R{}_{KL}\ V_R\ d\theta^L,$ (3.16.2c)

- $DV^K/Du \equiv V^K|_L\ (d\theta^L/du),$
  $DV_K/Du \equiv V_K|_L\ (d\theta^L/du)$ (u: curve parameter). (3.16.2d)

Hence, the *parallel transport* conditions for V^K and V_K are

$$DV^K = 0 \Rightarrow d^*V^K = -\Lambda^K{}_{RL}\ V^R\ d\theta^L, \tag{3.16.3a}$$

$$DV_K = 0 \Rightarrow d^*V_K = \Lambda^R{}_{KL}\ V_R\ d\theta^L; \tag{3.16.3b}$$

while the NH counterpart of Ricci's theorem (Equation 3.5.3 and Problem 3.8.1) is $g_{KL}|_R = 0$, and

$$\mathbf{e}_K|_L = \mathbf{e}_{K,L} - \Lambda^R{}_{KL}\, \mathbf{e}_R = \mathbf{0} \Rightarrow \mathbf{e}_{K,L} = \Lambda^R{}_{KL}\, \mathbf{e}_R, \tag{3.16.4a}$$

$$\mathbf{e}^K|_L = \mathbf{e}^K{}_{,L} + \Lambda^K{}_{RL}\, \mathbf{e}^R = \mathbf{0} \Rightarrow \mathbf{e}^K{}_{,L} = -\Lambda^K{}_{RL}\, \mathbf{e}^R; \tag{3.16.4b}$$

i.e., Equations 3.15.1b and 3.15.2a.

3.16.1 Nonholonomic Riemann–Christoffel Tensor

Finally, building $(V_K|_L)|_S \equiv V_K|_{LS}$, with the help of Equations 3.16.1b and c, and then $V_K|_{LS} - V_K|_{SL} = \ldots$, à la Equations 3.11.1ff, we can obtain R–C in nonholonomic coordinates, $R^P{}_{QLS}$, and find its relation to its holonomic counterpart (Equations 3.11.3b and c). For details on this topic, see, e.g., (alphabetically): Golab (1974, pp. 214 ff), Schmutzer (1968, pp. 103ff), Schouten (1954b, pp. 102–103, 123).

Example 3.16.1. Kinematical Interpretation of the NH Affinities

Proceeding as in Example 3.8.3, we obtain, successively,

$$d\mathbf{e}_K = \mathbf{e}_{K,L}\, d\theta^L = (\Lambda^R{}_{KL}\, \mathbf{e}_R) d\theta^L = (\Lambda^R{}_{KL}\, d\theta^L)\mathbf{e}_R \equiv d\phi_K{}^R\, \mathbf{e}_R \tag{a1}$$

$$= (\Lambda_{R,KL}\, \mathbf{e}^R) d\theta^L = (\Lambda_{R,KL}\, d\theta^L)\mathbf{e}^R \equiv d\phi_{KR}\, \mathbf{e}^R; \tag{a2}$$

i.e.,

$$d\phi_K{}^R = d\mathbf{e}_K \cdot \mathbf{e}^R = \Lambda^R{}_{KL}\, d\theta^L, \quad d\phi_{KR} = d\mathbf{e}_K \cdot \mathbf{e}_R = \Lambda_{R,KL}\, d\theta^L. \tag{a3}$$

These equations show that *the tensorial $d\phi_{\ldots}$ (by the Quotient Rule) are the NH components of the elementary, or first-order, change of the NH basis vectors, along that basis;* and similarly for the (nontensorial) affinities.

Next, we show that if the basis $\{\mathbf{e}_K\}$ undergoes *a rigid-body motion* (i.e., translation and rotation without deformation/strain) the $d\phi_{KR}$ are *antisymmetric*. Indeed, then

$$\mathbf{e}_K \cdot \mathbf{e}_R = g_{KR} = \text{constant} \quad (= \delta_{KL}, \text{ if the basis is orthonormal}),$$

and, therefore,

$$0 = dg_{KR} = d(\mathbf{e}_K \cdot \mathbf{e}_R) = d\mathbf{e}_K \cdot \mathbf{e}_R + d\mathbf{e}_R \cdot \mathbf{e}_K$$

$$\Rightarrow d\phi_{KR} \equiv d\mathbf{e}_K \cdot \mathbf{e}_R = -\, d\mathbf{e}_R \cdot \mathbf{e}_K \equiv -d\phi_{RK}, \quad \text{q.e.d.}; \tag{b1}$$

and, similarly,

$$0 = dg_K{}^R = d\delta_K{}^R = d(\mathbf{e}_K \cdot \mathbf{e}^R)$$

$$\Rightarrow d\phi_K{}^R \equiv d\mathbf{e}_K \cdot \mathbf{e}^R = -\, d\mathbf{e}^R \cdot \mathbf{e}_K \equiv -\, d\phi^R{}_K. \tag{b2}$$

For an orthonormal basis in three dimensions (e.g., the Frenet–Serret triad, Section 5.3), $d\phi_{KR} = d\phi_K{}^R$ can be replaced by its *axial vector* (Equations 2.9.22a ff)

$$d\phi_L \equiv (1/2)e_{LKR}\, d\phi_{KR} = (1/2)\varepsilon_{LKR}\, d\phi_{KR} \quad \text{(since } g = 1\text{)}; \tag{c)*}$$

explicitly, in an orthonormal basis:

$$d\phi_1 = d\phi_{23} = -\, d\phi_{32}, \quad d\phi_2 = d\phi_{31} = -\, d\phi_{13}, \quad d\phi_3 = d\phi_{12} = -\, d\phi_{21}. \tag{c1}$$

Example 3.16.2. The Ricci Coefficients of Rotation (c. 1895)

These are *the affinities of an orthonormal basis undergoing rigid-body motion,* say, in an R_n; i.e., they are *the physical components of the Christoffels* of the associated orthogonal curvilinear coordinates.** These nontensorial quantities are usually denoted by $\lambda_{R,KL}$ (where the comma is just convenient notation, not some differentiation; i.e., as in the first-kind affinities, Christoffels) and, as shown in the preceding Example 3.16.1, they are *antisymmetric* in R and K:

$$\Lambda_{R,KL} \to \Gamma_{R,KL} \to \lambda_{R,KL} = \mathbf{e}_{K,L} \cdot \mathbf{e}_R = -\lambda_{K,RL} = -\mathbf{e}_{R,L} \cdot \mathbf{e}_K, \tag{a1}$$

where

$$\mathbf{e}_{K,L} = \lambda_{R,KL}\, \mathbf{e}_R, \ \{\mathbf{e}_K\}\text{: orthonormal (but, generally, } \textit{nongradient}\text{) basis.} \tag{a2}$$

This can also be derived nonvectorially as follows. Since the basis is orthonormal,

$$\mathbf{e}_L \cdot \mathbf{e}_K = A_{Lk}\, A^k{}_K = \delta_{LK} \Rightarrow A_{Lk}|_l\, A^k{}_K = -A_{Lk}\, A^k{}_K|_l = -A_{Kk}|_l\, A^k{}_L, \tag{b1}$$

$$\mathbf{e}_R \cdot \mathbf{e}_K = A_{Rk}\, A^k{}_K = \delta_{RK} \Rightarrow A_{Rk}|_l\, A^k{}_K = -A_{Rk}\, A^k{}_K|_l = -A_{Kk}|_l\, A^k{}_R; \tag{b2}$$

and, therefore, recalling Equations 3.15.1a ff, we have

$$\lambda_{R,KL} = \mathbf{e}_R \cdot \mathbf{e}_{K,L} = (A^r{}_R\, \mathbf{e}_r) \cdot [(A_{Kk}|_l\, \mathbf{e}^k)A^l{}_L]$$

$$= A_{Kk}|_l\, A^k{}_R\, A^l{}_L = -A_{Rk}|_l\, A^k{}_K\, A^l{}_L = -\lambda_{K,RL}, \quad \text{q.e.d.} \tag{b3}$$

Next, let us relate the Hamel coefficients of that orthonormal basis with its Ricci coefficients.

---

* Or, sometimes, as $d\phi_L \equiv -(1/2)e_{LKR}\, d\phi_{KR} = -(1/2)\varepsilon_{LKR}\, d\phi_{KR}$.

** Continuum mechanics authors call them *wryness coefficients* of these coordinates. For further details on the Ricci coefficients, see, e.g.,(alphabetically): Cisotti (1928, pp. 83–86), Jaunzemis (1967, pp. 110–120), Levi-Civita (1926, pp. 268–272), Warsi (1996), Weatherburn (1963, pp. 98–109), Wright (1908, pp. 68–71).

i. *Vectorially:* Guided by Equations 3.15.3c and d, and using Equations a1 and a2, we find

$$\lambda_{R,KL} - \lambda_{R,LK} = \mathbf{e}_{K,L} \cdot \mathbf{e}_R - \mathbf{e}_{L,K} \cdot \mathbf{e}_R = (\mathbf{e}_{K,L} - \mathbf{e}_{L,K}) \cdot \mathbf{e}_R$$

$$\Rightarrow \lambda_{R,[KL]} = \gamma_{R,LK}/2 \quad \text{(recalling Equation 3.15.3e);} \qquad \text{(c1)}$$

ii. *Nonvectorially:* Using expression b3 and Equations b1 and b2, we obtain, successively,

$$\lambda_{R,KL} - \lambda_{R,LK} = A_{Kk}|_l\, A^k{}_R\, A^l{}_L - A_{Lk}|_l\, A^k{}_R\, A^l{}_K = A_{Kl}|_k\, A^l{}_R\, A^k{}_L - A_{Lk}|_l\, A^k{}_R\, A^l{}_K$$

$$= -A_{Rl}|_k\, A^l{}_K\, A^k{}_L - (-A_{Rk}|_l\, A^k{}_L\, A^l{}_K)$$

$$= (A_{Rk}|_l - A_{Rl}|_k) A^k{}_L\, A^l{}_K = (A_{Rl}|_k - A_{Rk}|_l) A^k{}_K\, A^l{}_L$$

$$= (A_{Rl,k} - A_{Rk,l}) A^l{}_L\, A^k{}_K \quad \text{(since the H Christoffels are symmetric in } l \text{ and } k\text{)}$$

i.e.,

$$\lambda_{R,[KL]} = A^l{}_L\, A^k{}_K\, A_{R[l,k]} = \gamma_{R,LK}/2. \qquad \text{(c2)}$$

Problem 3.16.1. Ricci Coefficients in Terms of the Hamel Coefficients

Show that for an orthonormal basis undergoing rigid-body motion (i.e., case of Ricci coefficients), Equations a1 and 2 of the preceding Example 3.16.2 applied to Equations 3.15.8b and c yields $C_{R,KL} + C_{K,RL} = 0$. Next, by combining this result with $C_{R,KL} - C_{R,LK} = 0$ show that, in this case, $C_{R,KL} = 0$, and, therefore,

$$2\lambda_{R,KL} = \gamma_{K,RL} + \gamma_{L,RK} - \gamma_{R,KL} \quad \text{(i.e., no metric needed).} \qquad \text{(a)}$$

Example 3.16.3

Continuing from the preceding example and problem, let us express the covariant derivatives of the basis vectors components, i.e., $A_{Kr}|_s$, in terms of the Ricci coefficients. Dotting the λ-definition: $\lambda_{R,KL} = A_{Kr}|_l\, A^r{}_R\, A^l{}_L$ with $A^R{}_p\, A^L{}_q$ yields

$$A^R{}_p\, A^L{}_q\, \lambda_{R,KL} = (A^R{}_p\, A^r{}_R)(A^L{}_q\, A^l{}_L) A_{Kr}|_l = \delta^r{}_p\, \delta^l{}_q\, A_{Kr}|_l = A_{Kp}|_q,$$

i.e., finally,

$$A_{Kr}|_l = A^R{}_r\, A^L{}_l\, \lambda_{R,KL}; \qquad \text{(a)}$$

from which (since the H affinities are again assumed symmetric):

$A_{Kr}|_s - A_{Ks}|_r = A_{Kr,s} - A_{Ks,r}$ (then invoking Equations c1 and c2 of Example 3.16.2)

$= A^R{}_r\, A^S{}_s(\lambda_{K,SR} - \lambda_{K,RS})$

$= A^R{}_r\, A^S{}_s(\lambda_{R,KS} - \lambda_{S,KR})$ (by Equations a1 and b3 of Example 3.16.2)

$\equiv A^R{}_r\, A^S{}_s\, \gamma_{K,RS} \equiv \gamma_{K,rs}$, (as expected, Equations 3.13.4b and 3.13.5). (b)

Problem 3.16.2

From Equations a1 and a2 of Example 3.16.2 and the kinematic type representation: $\mathbf{e}_{K,L} = \mathbf{D}_L \times \mathbf{e}_K$, where $\mathbf{D}_L \equiv D_{LS}\, \mathbf{e}_S$, deduce that

$$\lambda_{R,KL} = e_{RSK}\, D_{LS} = \varepsilon_{RSK}\, D_{LS}, \tag{a}$$

and, inversely,

$$2D_{LS} = e_{SKR}\, \lambda_{R,KL} = \varepsilon_{SKR}\, \lambda_{R,KL}. \tag{b}$$

Then, Equation a of the preceding Example 3.16.3 transforms to

$$A_{Kr}|_l = e_{RQK}\, A^R{}_r\, A^L{}_l\, D_{LQ} = \varepsilon_{RQK}\, A^R{}_r\, A^L{}_l\, D_{LQ}. \tag{c}$$

Problem 3.16.3

Verify that, in direct notation (recall Section 2.11),

i. The covariant derivative of a general n-dimensional vector $\mathbf{V}$ is

$$[(\ldots)_{,K}\, \mathbf{e}^K] \otimes (V_L\, \mathbf{e}^L) = \ldots = V_K|_L\, \mathbf{e}^L \otimes \mathbf{e}^K = V^K|_L\, \mathbf{e}^L \otimes \mathbf{e}_K, \tag{a}$$

whether the coordinates are holonomic or nonholonomic.

ii. The curl/rotation tensor of $\mathbf{V}$, Curl $\mathbf{V}$, is (assuming symmetric holonomic affinities):

$$\text{Curl}\, \mathbf{V} \equiv \nabla \otimes \mathbf{V} - \mathbf{V} \otimes \nabla \equiv [(\ldots)_{,k}\, \mathbf{e}^k] \otimes (V_l\, \mathbf{e}^l) - (V_l\, \mathbf{e}^l) \otimes [(\ldots)_{,k}\, \mathbf{e}^k]$$

$$= \ldots = (V_{k,l} - V_{l,k})\mathbf{e}^l \otimes \mathbf{e}^k \equiv (\partial_l V_k - \partial_k V_l)\mathbf{e}^l \otimes \mathbf{e}^k \equiv V_{lk}\, \mathbf{e}^l \otimes \mathbf{e}^k,$$

$$\Rightarrow V_{lk} = (\text{Curl}\, \mathbf{V})_{lk} = \mathbf{e}_l \cdot \text{Curl}\, \mathbf{V} \cdot \mathbf{e}_k, \tag{b}^*$$

---

* Of which, $n!/2!\,(n-2)! = n(n-1)/2$ are independent.

$$\text{Curl*}\mathbf{V} \equiv [(\ldots)_{,K}\, \mathbf{e}^K] \otimes (V_L\, \mathbf{e}^L) - (V_L\, \mathbf{e}^L) \otimes [(\ldots)_{,K}\, \mathbf{e}^K]$$

$$= \ldots = (V_{K,L} - V_{L,K})\mathbf{e}^L \otimes \mathbf{e}^K \equiv (\partial_L V_K - \partial_K V_L)\mathbf{e}^L \otimes \mathbf{e}^K \equiv V_{LK}\, \mathbf{e}^L \otimes \mathbf{e}^K,$$

$$\Rightarrow V_{LK} = (\text{Curl*}\mathbf{V})_{LK} = \mathbf{e}_L \cdot \text{Curl*}\mathbf{V} \cdot \mathbf{e}_K; \tag{c}$$

and so, the earlier basic transformation formula (Equations 3.13.3a and 3.13.5):

$$V_{LK} = A^k{}_K\, A^l{}_L\, V_{lk} + \gamma^R{}_{LK}\, V_R, \tag{d}$$

can be rewritten as

$$\text{Curl*}\mathbf{V} = [A^k{}_K\, A^l{}_L(V_{k,l} - V_{l,k}) + \gamma^R{}_{LK}\, V_R]\mathbf{e}^L \otimes \mathbf{e}^K, \tag{d1}$$

or

$$(\text{Curl*}\mathbf{V})_{LK} = A^k{}_K\, A^l{}_L(\text{Curl }\mathbf{V})_{lk} + \gamma^R{}_{LK}\, V_R, \tag{d2}$$

or

$$\mathbf{e}_L \cdot \text{Curl*}\mathbf{V} \cdot \mathbf{e}_K = A^l{}_L(\mathbf{e}_l \cdot \text{Curl }\mathbf{V} \cdot \mathbf{e}_k)A^k{}_K + \gamma^R{}_{LK}\, V_R$$

$$= \mathbf{e}_L \cdot \text{Curl }\mathbf{V} \cdot \mathbf{e}_K + \gamma^R{}_{LK}\, V_R. \tag{d3}$$

These formulae extend the concept of the Curl(…) of a vector (Equation 3.6.4) to n-dimensions and general nonholonomic coordinates.

iii. Show that, for the special case: $\mathbf{V} \to \mathbf{e}^S = \delta^S{}_K\, \mathbf{e}^K = A^S{}_k\, \mathbf{e}^k$, i.e., $V_K \to \delta^S{}_K$ (constant) and $V_k \to A^S{}_k$, the above yield the following invariant representation of the Hamel coefficients:

$$\gamma^R{}_{LK} = \mathbf{e}_K \cdot \text{Curl }\mathbf{e}^R \cdot \mathbf{e}_L = A^k{}_K(\text{Curl }\mathbf{e}^R)_{kl}\, A^l{}_L$$

$$[= A^k{}_K(\mathbf{e}_k \cdot \text{Curl }\mathbf{e}^R \cdot \mathbf{e}_l)A^l{}_L = A^k{}_K(\partial_k A^R{}_l - \partial_l A^R{}_k)A^l{}_L \equiv A^l{}_L\, A^k{}_K\, \gamma^R{}_{lk}]. \tag{e}$$

This form is useful in the geometrical interpretation of Frobenius' integrability conditions (Section 6.9).

For further details on the Curl(…) *of* general n-dimensional vectors and tensors, see, e.g., (alphabetically): Brand (1947, pp. 372–376), Eringen (1971, pp. 75–76), Schmutzer (1968, pp. 86 ff), Veblen (1927, pp. 64–66).

Part II

Analytical Dynamics

4 Introduction to Analytical Dynamics

The following is a very compact, selective, intuitive, and unavoidably incomplete (but adequate for our purposes) summary of some aspects of the fundamental *concepts* and *equations of motion* of classical (nonrelativistic and nonquantum) dynamics. The problem and method of analytical dynamics (AD) are also outlined.*

4.1 FUNDAMENTAL CONCEPTS

> I would mention the experience that it is exceedingly difficult to expound to thoughtful hearers the very introduction to mechanics without being occasionally embarrassed, without feeling tempted now and again to apologize, without wishing to get as quickly as possible over the rudiments and on to the examples which speak for themselves.
>
> H. Hertz (*The Principles of Mechanics,* originally in German, 1894)

The mathematical model, or map, of nature, to be studied here in the language of tensors, is the fundamental branch of classical theoretical dynamics called Lagrangean, or *analytical dynamics* (AD). Here analytical or *deductive* mechanics (as contrasted to synthetic, or inductive, mechanics) means the particular *energetic* form of mechanics invented by Lagrange (second half of 18th century), and further elaborated and perfected by Poisson, Hamilton, Jacobi, W. Thomson, Tait, Gibbs, Helmholtz, Routh, Boltzmann, Hertz, Maggi, Appell, Voronets, Whittaker, Heun, Hamel, et al. (19th and early 20th century); as contrasted to the other, also deductive but *momentum* mechanics invented by Euler (18th century) and elaborated by Poinsot, Coriolis, Minding, Möbius, Schell, Somoff, Ball, Study, A. Föppl, et al. (19th century).**

As in the rest of physics, the relation of AD to nature is based on certain basic mathematical and physical mental building blocks called *categories*, or *concepts*. Here are some of the most important ones.

Event. A physical occurrence, or material process, that is sharply localized in both space and time; e.g., the arrival of a train at a certain station on a certain time. A "point" in the four-dimensional *space–time*, or *event–space*, has coordinates $(q^1, q^2, q^3, q^4 \equiv t) \equiv (q^k, t) \equiv (q^\alpha)$, $\alpha = 1, 2, 3, 4$. The correspondence between all possible

---

* For complementary reading, we recommend (alphabetically): Bergmann (1942, 1962), Hamel (1912, 1949), Langhaar (1962), Lindsay and Margenau (1936), Mittelstaedt (1970), Synge and Griffith (1959), Synge (1960), Truesdell and Toupin (1960); also Papastavridis (1998) and references given there.

** Both these mechanics are *vectorial*; the latter in ordinary three-dimensional Euclidean space, and the former in a, generally curved (e.g., Riemannian) n-dimensional metric manifold.

events and the totality of ordered quadruples (q^k, t), where all variables range from $-\infty$ to $+\infty$, is one-to-one. Event is the fundamental physical phenomenon of macroscopic physics.

Space–Time. The totality of all physically possible events, or "world points," in the universe is a four-dimensional differentiable manifold (with torsionless linear connection), called *Newtonian*, or *absolute*, space–time $A_{S/T}$; it is the *union* of absolute space A_S and absolute (or "universal") time A_T. Symbolically, $A_{S/T} = A_S + A_T$. A_S is the totality of all possible *positions* of all events, whereas A_T is that of all possible *time instants*. Two positions define a *distance* in A_S; while two instants define an *interval* in A_T. If the qs are close, so are the events. The examination of the primary concepts of space and time is far outside the borders of this book ($\rightarrow$ Relativity, Cosmology, etc.). For our purposes, it will suffice to say this: we start with the concepts of spatiotemporal relations among events, and then by theoretical abstraction we arrive at the concepts of space and time; these in turn, are manifestations (or forms, or modes of existence) of that objective reality called *matter.*

Frames of Reference. Clearly, the association between events, or $A_{S/T}$-points, and ordered quadruples (q^α; $\alpha = 1, \ldots, 4$) is *nonunique*: any convention, or set of measurements, that associates with every event a set of four real numbers, q^k: spatial coordinates, q^4: temporal coordinate (see Time, below) is called a *frame of reference* (F) in $A_{S/T}$, or simply *frame*, i.e., an F is a *nonunique representation* of $A_{S/T}$; and, conversely, the totality of all oriented Euclidean frames E_3 (space) + E_1 (time) $\equiv E_{3+1}$ defines, or constitutes, $A_{S/T}$. This "coordinate-ization" of events must be made relative to *material* bodies; i.e., a frame of reference is *not* an inherent property of $A_{S/T}$ independent of matter: the *coordinate system* can be conceptualized as a net, or framework, of material, straight or curved, wires invariably welded to an extended rigid body, and participating completely in the motion of the latter (or, like a network of streets and avenues in a city), while time is universal and absolute, that is, independent of space and motion, and therefore common to all conceivable rigid bodies (and associated co–moving observers).

Units and Measurements. In classical mechanics, the units of measuring *distances* and *time intervals*, that is, length and time, are completely arbitrary. These measurements are carried out by a team of equivalent *observers*, one for each point of the frame used, and equipped with calibrated *yardsticks* (length) and *clocks* (time), all in continuous communication with each other, say, with walkie-talkies or flashlights, and with some "chief" observer situated at the "origin" of the frame and being the representative of the entire team of that frame. Therefore, sometimes we use the terms frame and observer synonymously.*

Frame of Reference Transformations. In another frame of reference F', the *same* event of $A_{S/T}$ may be described by the new coordinates $q^{\alpha'} = q^{k'}$, $q^{4'} \equiv t'$. The *explicitly time-dependent transformation* between the corresponding measurements of F and F':

---

* The above bring out the importance of the *properties of light* (since the team members communicate with light signals) and of *rigid bodies* in defining frames of reference and in determining the nature of the geometry of our physical space. It is no accident that the relativistic critique of classical mechanics began with a close scrutiny of these two concepts.

$$q^{k'} = q^{k'}(q^k, t) \quad (k, k': 1, 2, 3) \tag{4.1.1}$$

$t' = t'(t) = t$ (*absolute* time; always possible with proper time *scale* and *origin* choices), is the mathematical form of an admissible *frame of reference transformation* in the three-dimensional Euclidean space of classical mechanics. On the other hand, the *explicitly time-independent* equations $q^{k'} = q^{k'}(q^k)$ denote a transformation between two coordinate systems both invariably embedded in the *same* frame, say, F; i.e., the nets of "wires" q^k and $q^{k'}$ do *not* move relative to each other and to F. A coordinate transformation is a purely mathematical (*geometrical*) affair, while a frame of reference transformation is both a geometrical and, most importantly, a *physical* one; i.e., it involves new and nontrivial experimental and theoretical considerations!

Inertial Frames of Reference. In $A_{S/T}$ we assume the existence of a particular frame, called *inertial*, in which the laws of dynamics take their simplest form; that is, one where Newton's *first* law holds: *in the absence of interaction with other bodies (i.e., no forces, or if the body in question is sufficiently far removed from all other bodies) a body preserves its state of rest or uniform linear motion in that frame;* in which case its Cartesian spatial coordinates are, at most, *linear* functions of time. Further, as Newton's *second* law shows, once an inertial frame has been found, any other frame moving relative to it uniformly and in a straight line, that is, with *zero* acceleration (vectorially), is also inertial; i.e., we have an entire family, or *group*, of inertial frames in accelerationless motion relative to each other (Galilean Principle of Relativity). The rectangular Cartesian coordinates and times of any two inertial frames, (y^k, t) and $(y^{k'}, t)$, are related by the *Galilean transformation* (recall Equation 2.5.10a and following):

$$y^{k'} = G^{k'}{}_k\, y^k + G^{k'}\, t + g^{k'}, \quad t' = t + t_o \quad (k, k' = 1, 2, 3) \tag{4.1.2}$$

where the *tengroup parameters:* t_o (frequently chosen to be zero), the three $g^{k'}$, the three $G^{k'}$, and the nine $G^{k'}{}_k \equiv cos(y^{k'}, y^k) = cos(y^k, y^{k'}) \equiv G^k{}_{k'}$ (elements of a *proper orthogonal* matrix $\Rightarrow$ only three are independent) are constant. As a result of Equation 4.1.2:

$$dy^{k'}/dt = G^{k'}{}_k(dy^k/dt) + G^{k'} \quad \text{and} \quad d^2y^{k'}/dt^2 = G^{k'}{}_k(d^2y^k/dt^2); \tag{4.1.2a}$$

in words, apart from physically inessential axes orientation $(G^{k'}{}_k)$, the velocities differ by the relative velocity between the two frames $(G^{k'})$, while the accelerations are the same (which makes the basic dynamic equations *form invariant* under Equation 4.1.2). To the best approximation available to us today, the primary inertial frame is the *astronomical* frame, that is, the frame determined by the center of mass of our solar system (origin) and the directions of the "fixed stars" ("Copernican axes"). Every other frame moving relative to it with vectorially constant translational velocity is called *secondary* inertial frame. In many engineering problems, an *earthbound* frame is, to an excellent approximation, inertial.

Mass. A positive number, m. The "price," or quantity, of matter in a body; a measure of its inertia, i.e., of its disinclination to *change its velocity* (Newton's laws of motion), or of its capacity to attract another body (Newton's law of gravitation).

- Mass is *additive,* the mass of a system equals the sum of the masses of its parts. In mathematical terms, mass is a measure.
- The mass of a body is not affected by its motion and/or deformation; i.e., mass is an *invariant* body characteristic.
- Mass has its own physical dimension, independent of those of length and time.

Mass, and other physical concepts, can be modeled either as *discrete* or as *continuous*. In the following we shall use both concepts interchangeably and equivalently. In particular, we shall use frequently *discrete language and continuous mathematics.*

Particle or Material Point. An idealized physical system having position and inertia but no spatial extension; a body whose *rotation* and *deformation* can be neglected. According to the discrete viewpoint, particles are the fundamental building blocks of material systems.

Rigid Body. (i) Under the *discrete* model of matter, a rigid body is a system of particles (at least three of them, and non-collinear, assumed present) whose mutual distances are invariable, i.e., independent of the body's position and orientation; or, a system of particles so interconnected that, throughout their motion, the distance between *any* two of them, i.e., of each one from every other, remains constant. (ii) Under the *continuum* model of matter, a rigid body is a material system that strongly resists volume and/or shape changes, although it may change its position relative to its environment; or, *a body whose deformation, but not rotation, can be neglected.*

Time (t). The real number we associate, or order, with each position, or configuration, of a moving particle relative to a frame of reference; that is, to define time we must have at least one particle changing position in some frame – if nothing moved, time would freeze too, i.e., no moving matter, no "flowing" time. Time arises in connection with assumptions of *causal* connections between physical events (see past/future, below). After such a choice of chronology, we can build a *clock*, i.e., a perfectly *periodic* (repeatable) physical system whose nature depends on the dynamical equations of our physical theory. Then the duration between two successive repetitions in the regular sequence of such events, generated by our designated clock (e.g., the to → fro → back time interval of the bob of a mathematical pendulum), can be chosen as the time unit. Essentially, clocks interpret time measurements as changes of position, e.g., rotations of clock hands, and such spatial measurements are the ultimate experimentally observables in physics. It should be stressed that, in spite of formal appearances (e.g., four-dimensional space time), time is not just another "coordinate," but is qualitatively very different from space, even for one-dimensional problems: a particle may occupy the same position at two or more instants of time, but it cannot occupy more than one position at an instant of time; and, during its motion its velocity dq/dt may vanish, but dt/dq may not!*

Displacement, Velocity, Acceleration of a Particle P. If its position at time t, relative to a frame F, is given by the vector $\mathbf{r} = (x, y, z)$ and at time $t + \Delta t$ by $\mathbf{r} + \Delta\mathbf{r}$, then its displacement during Δt is $\Delta\mathbf{r}$. The *path* between $\mathbf{r}$ and $\mathbf{r} + \Delta\mathbf{r}$ is assumed traversed continuously, and expressible as the sum of the elementary displacements $d\mathbf{r} = (dx, dy, dz)$. The curve $\mathbf{r}(t) = [x(t), y(t), z(t)]$, in the event space/space time (x, y, z, t), is called *history* or *world-line* of the particle; and it's projection on its Euclidean subspace E_3 is called *trajectory* or *path* of P. The instantaneous *velocity* and *acceleration* of P in F are the derivatives:

$$\mathbf{v} \equiv d\mathbf{r}/dt = (dx/dt,\ dy/dt,\ dz/dt),$$

$$\mathbf{a} \equiv d\mathbf{v}/dt = d^2\mathbf{r}/dt^2 = (d^2x/dt^2,\ d^2y/dt^2,\ d^2z/dt^2), \qquad (4.1.3)$$

respectively. In classical mechanics, the $dx/dt, dy/dt, dz/dt$ are unrestricted.

Force. The cause of motion and/or deformation of bodies. An objective, i.e., frame-independent quantity. Forces are classified multiply as external or internal (or mutual); contact or body (or field); concentrated or distributed; potential or nonpotential; active (or impressed, or physical) or constraint reactions (geometrical, or kinematical); finite or impulsive, particle or system etc. Force, along with space, time, and mass are the fundamental *primitive*, or (preferably) undefined concepts or building blocks of mechanics (like the famous "God-given" integers of Kronecker). Some continuum mechanics add to this list the rigid body.**

Newton–Euler Law. The equation of motion of a particle P of mass dm is

$$dm\ \mathbf{a} = d\mathbf{f} \quad \text{(field equation)}, \quad d\mathbf{f} = d\mathbf{f}(t,\ \mathbf{r},\ \mathbf{v}) \quad \text{(constitutive equation)}, \qquad (4.1.4)$$

where $\mathbf{r}$: position of P relative to some origin O, fixed in an inertial frame, $\mathbf{v} \equiv d\mathbf{r}/dt$: inertial velocity of P, $\mathbf{a} = d^2\mathbf{r}/dt^2$: inertial acceleration of P; and $d\mathbf{f}$: *total* force on P.

4.2 CONFIGURATION SPACE

> Tensorial methods are not applied to dynamics primarily for the purpose of solving definite dynamical problems. The purpose of these methods is rather to permit an adequate treatment of the so-called "general problem of dynamics," that is to permit the ideas of Riemannian or more general geometries to impinge on dynamical theory. Here the results are unexpectedly beautiful. We find that the behavior of a general

* And, of course, there is the fundamental issue of time *directedness*, or *unilateralness*; the famous "time arrow" of A. S. Eddington, which shows clearly that time is not just another one–dimensional space: a traveling particle separates past from future; but, in general, not left from right or up from down. In such *irreversible* (nonperiodic) phenomena we distinguish between past and future by the following *principle of determinism:* causes (or inputs, or totality of initial conditions) precede effects (or outputs, or responses); or, the future cannot affect the past! Analytically, irreversibility is detected as follows: if the equations of motion remain invariant under $dt \to -dt$, then the phenomena described by them are reversible; if not, they are irreversible. Finally, this irreversibility is not to be found in the fundamental (or field, or ponderomotive) equations of mechanics and physics (such as those of Newton–Euler, or Maxwell, or Schrödinger, etc.), but in their *constitutive* equations (such as those of friction, diffusion, etc.).

** Overcoming the difficulties contained in the above definitions, led to the theory of *relativity*; or, better, the theory of *space, time, and gravitation* of A. Einstein (mid-1910s).

dynamical system is precisely that which we would naturally assign to a particle in a space of N dimensions. *Thus we resuscitate the geometrical spirit which the work of Lagrange and Hamilton did so much to destroy, and see the system moving, not as a complicated set of particles in Euclidean* 3*-space, but as a single particle in Riemannian N-space*.... It is the least important systems dynamically — namely, *nonholonomic* systems and systems with *moving constraints* — that are most stimulating to the geometer ... [italics added]

J. L. Synge (1936, pp. 7–8)

4.2.1 Kinematics

We consider a mechanical system S consisting of N particles moving in ordinary three-dimensional Euclidean space E_3. Their simultaneous instantaneous positions relative to a (not necessarily inertial) E_3-frame F are the *configuration* of the system; and the set of all configurations that S can take is its *configuration space*. One of the key problems of dynamics is to *express the configurations* of a given S relative to an F, under the action of a group of known and unknown forces (and initial conditions), as *functions of time*. Quantitatively, this configuration is defined by an *F-system of coordinates* (CS). Since E_3 is flat, let that CS be, with no loss in generality, a *rectangular Cartesian* one, and an associated *local ortho-normal–dextral* (*OND*) *basis* $\{O;\ \mathbf{i},\ \mathbf{j},\ \mathbf{k}\}$. Then, a configuration of S, in F, is given by the N particle *position* vectors:

$$\mathbf{r}_P = x_P\,\mathbf{i} + y_P\,\mathbf{j} + z_P\,\mathbf{k} \equiv r_{P,\alpha}\,\mathbf{u}_\alpha \equiv r_{P\alpha}\,\mathbf{u}_\alpha \quad (P = 1, \ldots, N;\ \alpha = 1, 2, 3); \quad (4.2.1)$$

i.e., by the $3N$ scalar functions of time $\{x_P(t), y_P(t), z_P(t)\}$ or $\{r_{P\alpha}(t)\}$. Now, let us push the geometrical description one step further: let us view these $3N$ functions as the rectangular Cartesian coordinates of the position vector ξ of a fictitious, or *figurative*, particle representing the *entire* system, in an OND basis $\{O,\ \mathbf{u}_S;\ S = 1, \ldots, 3N\}$ in an N-dimensional Euclidean space E_N:

$$\xi = \xi^1\,\mathbf{u}_1 + \xi^2\,\mathbf{u}_2 + \ldots + \xi^{3N-2}\,\mathbf{u}_{3N-2} + \xi^{3N-1}\,\mathbf{u}_{3N-1} + \xi^{3N}\,\mathbf{u}_{3N}, \quad (4.2.2)$$

where

$$\mathbf{u}_S = \partial\xi/\partial\xi^S \quad \text{(unit vectors);} \quad (4.2.2a)$$

$$\xi^1 \equiv x_1 = r_{11}, \quad \xi^2 \equiv y_1 = r_{12}, \quad \xi^1 \equiv z_1 = r_{13},$$

$$\ldots\ldots\ldots\ldots\ldots\ldots\ldots\ldots\ldots\ldots\ldots\ldots$$

$$\xi^{3N-2} \equiv x_N = r_{N1}, \quad \xi^{3N-1} \equiv y_N = r_{N2}, \quad \xi^{3N} \equiv z_N = r_{N3}. \quad (4.2.2b)$$

We shall call E_{3N} the *free*, or *unconstrained*, configuration space of the system; and if we add to it the one-dimensional "space" of time, we obtain the unconstrained *extended* configuration space, or *configuration space–time* of the system E_{3N+1}. Conversely, E_{3N} (E_{3N+1}) can be viewed as the totality of all configurations (configurations plus time) that our (hitherto assumed unconstrained) system S can attain; that is, the locus of a figurative point, representing the instantaneous system configuration, as it roams over E_{3N}.

4.2.2 Kinetics

Relative to an inertial E_3-frame, the equations of motion of a typical S-particle are

$$m_P(d^2x_P/dt^2) = f_{P,x} \equiv f_{Px}, \quad m_P(d^2y_P/dt^2) = f_{P,y} \equiv f_{Py},$$

$$m_P(d^2z_P/dt^2) = f_{P,z} \equiv f_{Pz}, \tag{4.2.3a}$$

where

$$\mathbf{f}_P = f_{Px}\,\mathbf{i} + f_{Py}\,\mathbf{j} + f_{Pz}\,\mathbf{k}\text{: } \textit{total} \text{ force on } P, \tag{4.2.3b}$$

$$m_P\text{: mass of } P, \tag{4.2.3c}$$

$$\mathbf{a}_P = (d^2x_P/dt^2)\,\mathbf{i} + (d^2y_P/dt^2)\,\mathbf{j} + (d^2z_P/dt^2)\,\mathbf{k}\text{:}$$

inertial acceleration of P;(4.2.3d)

or, compactly (no sum on P):

$$m_P(d^2r_{P\alpha}/dt^2) = f_{P\alpha} \quad (P = 1, \ldots, N;\ \alpha = 1, 2, 3). \tag{4.2.3e}$$

Let us, next, transform the above into equations of motion of the single figurative particle. With the help of the earlier *system position vector* ξ, the *system force vector*

$$\Xi = \sum \Xi^S\,\mathbf{u}_S \quad (S = 1, \ldots, 3N), \tag{4.2.4a}$$

where

$$\Xi^1 = f_{1x},\quad \Xi^2 = f_{1y},\quad \Xi^3 = f_{1z}, \ldots,\quad \Xi^{3N-2} = f_{Nx},\quad \Xi^{3N-1} = f_{Ny},\quad \Xi^{3N} = f_{Nz}, \tag{4.2.4b}$$

and the mass notation:

$$M_1 = M_2 = M_3 = m_1, \ldots, M_{3N-2} = M_{3N-1} = M_{3N} = m_N, \tag{4.2.4c}$$

we can rewrite the $3N$ equations 4.2.3a (in a so-obtained inertial E_{3N}–frame) as

$$M_S(d^2\xi^S/dt^2) = \Xi^S \quad (S = 1, \ldots, 3N; \text{ no sum on } S). \tag{4.2.5a}^*$$

A final simplification of Equation 4.2.5a is obtained if we rewrite it as

$$d^2/dt^2\left(\sqrt{M_S}\xi^S\right) = \Xi^S/\sqrt{M_S}; \tag{4.2.6a}$$

or

$$d^2y^S/dt^2 = Y^S, \quad \text{where } y^S \equiv \sqrt{M_S}\,\xi^S, \quad Y^S \equiv \Xi^S/\sqrt{M_S}; \tag{4.2.6b}$$

or, vectorially,

$$d^2\mathbf{R}/dt^2 = \mathbf{Q}, \tag{4.2.6c}$$

where

$$\mathbf{R} = \sum y^S\,\mathbf{u}_S = \sum\left(\sqrt{M_S}\xi^S\right)\mathbf{u}_S \quad (\neq \rho), \tag{4.2.6d}$$

$$\mathbf{Q} = \sum Y^S\,\mathbf{u}_S = \sum\left(\Xi^S/\sqrt{M_S}\right)\mathbf{u}_S \quad (\neq \Xi). \tag{4.2.6e}$$

In sum, the original Equations 4.2.3a ff are first rewritten as $\sqrt{m_P}\,(d^2x_P/dt^2) = f_{Px}/\sqrt{m_P}$, etc. (no sum on P) and then as: $d^2y^{3P-2}/dt^2 = Y^{3P-2}$, $d^2y^{3P-1}/dt^2 = Y^{3P-1}$, $d^2y^{3P}/dt^2 = Y^{3P}$; so that the motion of the system is given by the motion of the tip of the E_{3N}-vector $\mathbf{y}$.** Finally, we could have chosen *curvilinear* coordinates in E_3 (to describe $\mathbf{r}_P$, etc.) and E_{3N} (to describe $\mathbf{y}$, etc.), but at this point, i.e., in the absence of constraints, no particular advantage would have resulted from such generality.

4.3 INTRODUCTION TO CONSTRAINTS — PURPOSE OF ANALYTICAL MECHANICS (AM)

If no functional relation(s) exist among the N $\{\mathbf{r}_P\}$, independently of any kinetic considerations, the system S is called *free,* or *unconstrained.* Then, the motion of

* Or, vectorially,

$$d^2\rho/dt^2 = \Xi, \tag{4.2.5b}$$

where

$$\rho \equiv (M_1\,\xi^1)\mathbf{u}_1 + \ldots + (M_{3N}\,\xi^{3N})\mathbf{u}_{3N}. \tag{4.2.5c}$$

its particles is found by solving its (generally nonlinear and coupled) differential equations of motion (Equations 4.2.6b and c): $d^2y^S/dt^2 = Y^S(t, \mathbf{y}, d\mathbf{y}/dt)$: *known function of its arguments,* plus appropriate initial (and/or boundary) conditions. The preeminent, or prototypical, areas of application of this unconstrained dynamics are in celestial mechanics, ballistics (e.g., the famous three-body problem). There, for an N-particle system, we have $3N$ equations of motion for the $3N$ unknown time functions $\{x_P, y_P, z_P;\ P = 1, \ldots, N\}$ or $\{y^S;\ S = 1, \ldots, 3N\}$; i.e., a mathematically determinate problem.

However, most engineering and earthly problems are of the *constrained* kind; i.e., during its motion, our system S comes, sooner or later, into contact with other bodies or systems, $S_1, S_2, \ldots$ and the latter impose restrictions, or *constraints*, on its external and/or internal mobility; i.e., on the displacements and/or velocities of its particles. Here are examples of such constraints:

Geometrical Constraints:

External: Particle forced to move on the surface of a sphere; e.g., mathematical pendulum;

Internal: Rigid body, incompressible fluid;

Kinematical Constraints:

External: Sphere rolling on a rough plane;

Internal: Sphere rolling inside an ellipsoid that is itself rolling on a rough plane.

This is a first classification of constraints; a finer one will be given later (Chapter 6).

From the kinetic viewpoint, these constraints mean that the total force $d\mathbf{f}$ is *not* completely known. Instead of Equation 4.1.4, we now have (the modern version of) *d'Alembert's physical decomposition:*

$$d\mathbf{f} = d\mathbf{F} + d\mathbf{R} \Rightarrow dm\,\mathbf{a} = d\mathbf{F} + d\mathbf{R}, \tag{4.3.1}$$

** Some authors (e.g., Poliahov et al., 1985, Ch. 2) introduce the following variables, instead of ours (no sum on repeated indices, unless shown explicitly):

$$M \equiv \sum m_P = \sum (M_S/3) \quad \text{(total mass)},$$

$$\eta^S \equiv \xi^S (M_S/M)^{1/2},\ \eta = \sum \eta^S \mathbf{u}_S; \quad H^S \equiv \Xi^S/(M_S/M)^{1/2}, \quad \mathbf{H} = \sum H^S \mathbf{u}_S,$$

in which case the system equation of motion (for a fictitious particle of mass M) is

$$M\,(d^2\eta^S/dt^2) = H^S; \qquad \text{or, vectorially,} \quad M\,(d^2\eta/dt^2) = \mathbf{H}.$$

In *continuum* mechanics notation, the above read simply (with some easily understood notation):

$$dm\,\mathbf{a} = d\mathbf{f} \Rightarrow \sqrt{dm}\,\mathbf{a} = d\mathbf{f}/\sqrt{dm}, \text{ etc.}$$

where $d\mathbf{F} = d\mathbf{F}(t, \mathbf{r}, \mathbf{v})$: nonconstraint, *physical*, or *impressed*, force; given by a constitutive equation, i.e., *containing physical/material constants, coefficients*, etc. (e.g., gravity, elastic stress, sliding/slipping friction)
$d\mathbf{R} = d\mathbf{R}(t, \mathbf{r})$: *constraint reaction*, unknown (e.g., rolling friction, tension in inextensible cable), *not containing material constants.*

4.3.1 Whence the Need for AM

The above show clearly that the unknowns of this constrained dynamics problem are both the *motion* and *reactions*, coupled together in a, generally, complicated fashion. And here is where AM comes in, and draws its theoretical and practical significance: with the help of a small number of strategically chosen, simple but powerful, concepts and postulates it succeeds in completely *uncoupling* the calculation of motion from that of the reactions, by formulating two kinds of equations:

- Reactionless, or *kinetic*, equations, that solve the problem of motion; and
- Reaction-containing, or *kinetostatic*, equations that solve the problem of reactions, *after* the motion has been determined.

The solution of the kinetic equations is the lion's share of the problem; once the motion has been found, the solution of the kinetostatic equations follows easily. Mixed formulations, combining both motion and reactions, are also available.*

Here is a summary (bird's eye preview) of the philosophy, method, and features of AM (Chapters 6 and 7):

- The transition away from the particle(s) and rigid body(ies) of the "elementary"Newton–Euler approach to the *system* (and its appropriate variables) necessitates the passage from the physical space E_3 to the configuration space E_{3N}, while *the imposition of constraints, generally, changes the flat space E_{3N} to a curved one* (e.g., Riemannian).
- The uniform transition from Eulerian rigid-body mechanics to Lagrangean system mechanics, along with the earlier-described complete decoupling of the equations of constrained motion into kinetic and kinetostatic, necessitates the introduction of *nonholonomic* variables, and eventually leads to the fundamental equations of motion of Hamel (or Lagrange–Euler, as he called them).
- AM is based on two related postulates: *Lagrange's principle,* for eliminating constraint reactions, and *Lagrange's principle of relaxation of the constraints* (method of multipliers), for retrieving them, if needed; and both these principles involve the fundamental concepts of *virtual displacement* and *virtual work.* The first principle yields the *kinetic* equations

* It should be stressed that AM can prove quite useful even in unconstrained problems, e.g., in discovering conservation/energy theorems, integral invariants in phase space, let alone the indispensable background for quantum mechanics.

(motion), while the second supplies the *kinetostatic* equations (constraint reactions) (Section 7.4).

- Finally, the earlier-mentioned nonflatness of the constrained configuration space, the study of its *geodesics* and their stability, and the analytical implications and geometrical interpretation of the *nonintegrability* of existing system constraints, makes tensors the natural "language." In sum, AM is not just another abstract and formalistic version of good old Newton–Euler mechanics. Even for engineers, it is not a dispensable luxury that duplicates the work of other disciplines, but a powerful, fertile, and yet simple and economical tool that grew out of the needs of constrained system dynamics.

[Such persuading is hardly needed for physicists and/or mathematicians: AM was necessary in the invention of both the Theory of Relativity (Configuration Space → Riemannian geometry → Tensor Calculus) and Quantum Mechanics (Phase Space → Hamiltonian Mechanics, etc.); and also of key branches of classical physics, e.g., statistical mechanics.]

Arguably, AM is the greatest intellectual achievement of classical physics!

5 Particle on a Curve and on a Surface

5.1 INTRODUCTION

In this chapter, using the hitherto developed tensorial machinery, and since analytical dynamics (AD) models the motions of the most general mechanical systems as motions of a single fictitious, or figurative, particle in configuration space (Section 4.2), we begin our study of mechanics by discussing the dynamics of a particle in ordinary three-dimensional Euclidean (flat) space E_3: in general and in intrinsic variables; on a general space, or skew, or twisted, curve; and on a general curved surface lying there. Then, we extend these results to n-dimensional (Riemannian) surfaces and, finally, discuss the theory of perturbation of (actual or figurative) particle trajectories in configuration space, and their stability.*

5.2 PARTICLE IN ORDINARY SPACE: GENERAL COORDINATES

We consider a particle P of mass m and position vector, relative to, say, inertial rectangular Cartesian axes x, y, z of fixed origin O and corresponding basis $\{O; \mathbf{i}, \mathbf{j}, \mathbf{k}\}$,

$$\mathbf{r} = x\mathbf{i} + y\mathbf{j} + z\mathbf{k}. \tag{5.2.1a}$$

To simplify matters, i.e., get rid of m, we introduce the following coordinate transformation:

$$\mathbf{r} \rightarrow \mathbf{R} = \sqrt{m}\,\mathbf{r} \text{ (normalized position), or in components:}$$

$$(X, Y, Z) \equiv \sqrt{m}(x, y, z); \tag{5.2.1b}$$

also, we introduce *curvilinear* coordinates $X = X(q^k)$, $Y = Y(q^k)$, $Z = Z(q^k)$ ($k = 1, 2, 3$; and similarly for all other Latin indices).

Next, we define the following fundamental mechanical quantities:

---

* For complementary reading, we recommend (alphabetically): Ferrarese (1980), Korenev (1967), Lur'e (1968), McConnell (1931), Synge (1927, 1936, 1949, 1959, 1960). Also, for ordinary and Riemannian differential geometry: Blaschke and Leichtweiss (1973), Dubrovin et al. (1984), Eisenhart (1926, 1947), Favard (1957), Gerretsen (1962), Kreyszig (1959), Lipschutz (1969), Mishchenko and Fomenko (1988), Sokolnikoff (1951), Thomas (1965), Willmore (1959).

- The (inertial) *kinetic energy* of the particle, T, is

$$2T \equiv m(d\mathbf{r}/dt)\cdot(d\mathbf{r}/dt) \equiv m\mathbf{v}\cdot\mathbf{v} = \ldots = m\, g_{kl}(dq^k/dt)(dq^l/dt)$$

$$= (d\mathbf{R}/dt)\cdot(d\mathbf{R}/dt) \equiv \mathbf{V}\cdot\mathbf{V} = \ldots = M_{kl}(dq^k/dt)(dq^l/dt), \quad (5.2.2a)$$

where, clearly,

$$\mathbf{v} \equiv d\mathbf{r}/dt = (\partial\mathbf{r}/\partial q^k)(dq^k/dt) \equiv v^k\,\mathbf{e}_k \quad \text{(inertial } \textit{velocity} \text{ of } P) \quad (5.2.2b)$$

$$\mathbf{V} \equiv d\mathbf{R}/dt = (\partial\mathbf{R}/\partial q^k)(dq^k/dt) \equiv V^k\,\mathbf{E}_k = \sqrt{m}(dq^k/dt)\mathbf{e}_k = \sqrt{m}\mathbf{v}, \quad (5.2.2c)$$

$$g_{kl} \equiv \mathbf{e}_k\cdot\mathbf{e}_l, \quad M_{kl} \equiv \mathbf{E}_k\cdot\mathbf{E}_l = m\, g_{kl}, \quad dq^k/dt \equiv v^k = V^k; \quad (5.2.2d)$$

and the corresponding *kinematical elements, ds* and *dS*, are

$$(dS)^2 = m(ds)^2 = (2T)(dt)^2 = M_{kl}\,dq^k dq^l = mg_{kl}\,dq^k dq^l. \quad (5.2.2e)$$

- The (inertial) *acceleration* of P, $\mathbf{a}$, is

$$\mathbf{a} = d\mathbf{v}/dt = d/dt(v^k\mathbf{e}_k) = (dv^k/dt)\mathbf{e}_k + v^k[\mathbf{e}_{k,s}(dq^s/dt)]$$

$$= (d^2q^k/dt^2)\mathbf{e}_k + (dq^k/dt)[\lambda^r{}_{ks}\,\mathbf{e}_r(dq^s/dt)]$$

$$= [d^2q^k/dt^2 + \lambda^k{}_{rs}(dq^r/dt)(dq^s/dt)]\mathbf{e}_k \equiv a^k\,\mathbf{e}_k = a_k\,\mathbf{e}^k, \quad (5.2.3a)$$

i.e.,

$$a_k = g_{kl}\,a^l = \ldots = g_{kl}(d^2q^l/dt^2) + \lambda_{k,rs}(dq^r/dt)(dq^s/dt), \quad (5.2.3b)$$

where the $\lambda^k{}_{rs}$ and $\lambda_{k,rs} \equiv g_{kl}\,\lambda^l{}_{rs}$ are, respectively, the second- and first-kind Christoffels *based on g_{kl}*. Therefore,

$$\mathbf{a} = d/dt\left[(dq^k/dt)\left(\mathbf{E}_k/\sqrt{m}\right)\right] = \left(1/\sqrt{m}\right)d/dt[(dq^k/dt)\mathbf{E}_k]$$

$$= \ldots = \left(1/\sqrt{m}\right)(A^k\,\mathbf{E}_k) = \left(1/\sqrt{m}\right)(A_k\,\mathbf{E}^k) = \left(1/\sqrt{m}\right)\mathbf{A}, \quad (5.2.3c)$$

where

$$A^k = d^2q^k/dt^2 + \Gamma^k{}_{rs}(dq^r/dt)(dq^s/dt),$$

$$A_k = M_{kl}\,A^l = \ldots = M_{kl}(d^2q^l/dt^2) + \Gamma_{k,rs}(dq^r/dt)(dq^s/dt) = m\,a_k, \quad (5.2.3d)$$

and the $\Gamma^k{}_{rs}$ are the second-kind Christoffels *based on* M_{kl}. Also, it is not hard to verify that

$$A_k = E_k(T) = E_k(M_{rs}V^rV^s/2) \equiv d/dt(\partial T/\partial V^k) - \partial T/\partial q^k:$$

Lagrange's kinematico-inertial identity. (5.2.3e)

- The (first-order) *virtual work*, $\delta' W$, of the total impressed force on P (Section 4.3), $\mathbf{F}$, is

$$\delta' W \equiv \mathbf{F} \cdot \delta\mathbf{r} = \left(\mathbf{F}/\sqrt{m}\right)\cdot\left(\sqrt{m}\,\delta\mathbf{r}\right) \equiv \mathbf{Q} \cdot \delta\mathbf{R} = Q_k\, \delta q^k = Q^k\, \delta q_k, \qquad (5.2.4a)$$

where

$$\mathbf{Q} \equiv \mathbf{F}/\sqrt{m} = Q^k\, \mathbf{E}_k = Q_k\, \mathbf{E}^k \quad (Q_k \equiv M_{kl}\, Q^l), \qquad (5.2.4b)$$

$$\delta\mathbf{R} \equiv (\partial\mathbf{R}/\partial q^k)\delta q^k = \mathbf{E}_k\, \delta q^k = \mathbf{E}^k\, \delta q_k \qquad (5.2.4c)$$

(normalized) *virtual displacement* of P. With the help of the above, the Newton–Euler law (Equation 4.3.1) for this, *hitherto unconstrained,* system (i.e., $\mathbf{f} = \mathbf{F}$) transforms successively to

$$m(d^2\mathbf{r}/dt^2) = m\,\mathbf{a} = \mathbf{F} \Rightarrow d^2\left(\sqrt{m}\,\mathbf{r}\right)/dt^2 = \mathbf{F}/\sqrt{m}, \qquad (5.2.5a)$$

or

$$\mathbf{A} = \mathbf{Q} \Rightarrow A^k = Q^k, \quad A_k = Q_k; \qquad (5.2.5b)$$

where, *in extenso* (with $V^k \equiv dq^k/dt$),

$$A^k \equiv DV^k/Dt = dV^k/dt + \Gamma^k{}_{rs}\, V^r(dq^s/dt)$$

$$\equiv d^2q^k/dt^2 + \Gamma^k{}_{rs}(dq^r/dt)(dq^s/dt) \qquad (5.2.5c)$$

$$A_k \equiv DV_k/Dt = M_{kl}(DV^l/Dt) = dV_k/dt - \Gamma^r{}_{ks}\, V_r(dq^s/dt)$$

$$\equiv M_{kl}(d^2q^l/dt^2) + \Gamma_{k,rs}(dq^r/dt)(dq^s/dt). \qquad (5.2.5d)$$

Example 5.2.1

i. *Cylindrical Coordinates.* Here, with the usual notations [i.e., ϕ = angle(Ox, Or)], $x = r\cos\phi$, $y = r\sin\phi$, $z = z$ and $q^1 = r$, $q^2 = \phi$, $q^3 = z$, and (with $m = 1$):

$$dS^2 = dx^2 + dy^2 + dz^2 = (1)dr^2 + (r^2)d\phi^2 + (1)dz^2, \tag{a}$$

i.e., $M_{11} = 1$, $M_{22} = r^2$, $M_{33} = 1$; hence, the nonvanishing Christoffels are

$$\Gamma^1{}_{22} \equiv \Gamma^r{}_{\phi\phi} = -r, \quad \Gamma^2{}_{12} = \Gamma^2{}_{21} \equiv \Gamma^\phi{}_{r\phi} = \Gamma^\phi{}_{\phi r} = 1/r, \tag{b}$$

the velocities are

$$V^1 = dr/dt, \quad V^2 = d\phi/dt, \quad V^3 = dz/dt, \tag{c}$$

$$V_{<1>} \equiv V_{<r>} = dr/dt, \quad V_{<2>} \equiv V_{<\phi>} = r(d\phi/dt), \quad V_{<3>} \equiv V_{<z>} = dz/dt; \tag{d}$$

and therefore the equations of motion are

$$A^1 = A^r = d^2r/dt^2 - r(d\phi/dt)^2 = Q^1 = Q^r, \tag{e1}$$

$$A^2 = A^\phi = d^2\phi/dt^2 + (2/r)(dr/dt)(d\phi/dt) = Q^2 = Q^\phi, \tag{e2}$$

$$A^3 = A^z = d^2z/dt^2 = Q^3 = Q^z. \tag{e3}$$

ii. *Spherical Coordinates.* Here $q^1 = r$, $q^2 = \theta$ [= angle(Oz, Or)], $q^3 = \phi$ [= angle(Ox, projection of Or on plane O–xy)] and (with $m = 1 \Rightarrow ds = dS$):

$$dS^2 = dx^2 + dy^2 + dz^2 = (1)dr^2 + (r^2)d\theta^2 + (r^2 \sin^2\theta)d\phi^2, \tag{f}$$

i.e.,

$$M_{11} = 1, \quad M_{22} = r^2, \quad M_{33} = r^2 \sin^2\theta;$$

and, therefore,

$$2T = (dS/dt)^2 = (dr/dt)^2 + r^2(d\theta/dt)^2 + (r^2 \sin^2\theta)(d\phi/dt)^2. \tag{g}$$

In this example, instead of finding the Christoffels and then the equations of motion, we shall, instead, calculate the accelerations from $A_k = d/dt(\partial T/\partial V^k) - \partial T/\partial q^k$, and then from $A^k = M^{kl} A_l = d^2q^k/dt^2 + \Gamma^k{}_{rs}(dq^r/dt)(dq^s/dt)$ (here $M^{kk} = 1/M_{kk}$ – no sum over k), read off the Christoffels:

$$A_1 = d/dt(\partial T/\partial V^r) - \partial T/\partial r = d^2r/dt^2 - r(d\theta/dt)^2 - r \sin^2\theta(d\phi/dt)^2, \tag{h1}$$

$$A_2 = d/dt(\partial T/\partial V^\theta) - \partial T/\partial\theta = d^2\theta/dt^2 + (2/r)(dr/dt)(d\theta/dt) - (\sin\theta \cos\theta)(d\phi/dt)^2, \tag{h2}$$

$$A_3 = d/dt(\partial T/\partial V^{\phi}) - \partial T/\partial \phi$$

$$= d^2\phi/dt^2 + (2/r)(dr/dt)(d\phi/dt) + (2 \cot \theta)(d\theta/dt)(d\phi/dt); \qquad \text{(h3)}$$

and comparing with A^k (while recalling that, due to $\Gamma^k{}_{rs} = \Gamma^k{}_{sr}$, *mixed* terms in the (dq/dt)s occur twice) we find:

$$\Gamma^1{}_{22} = \Gamma^r{}_{\theta\theta} = -r, \quad \Gamma^1{}_{33} = \Gamma^r{}_{\phi\phi} = -r \sin^2\theta, \qquad \text{(i1)}$$

$$\Gamma^2{}_{12} = \Gamma^2{}_{21} = \Gamma^\theta{}_{r\theta} = \Gamma^\theta{}_{\theta r} = 1/r, \quad \Gamma^2{}_{33} = \Gamma^\theta{}_{\phi\phi} = -\sin\theta\cos\theta, \qquad \text{(i2)}$$

$$\Gamma^3{}_{13} = \Gamma^3{}_{31} = \Gamma^\phi{}_{r\phi} = \Gamma^\phi{}_{\phi r} = 1/r, \quad \Gamma^3{}_{23} = \Gamma^3{}_{32} = \Gamma^\phi{}_{\theta\phi} = \Gamma^\phi{}_{\phi\theta} = \cot\theta. \qquad \text{(i3)}$$

5.3 PARTICLE IN ORDINARY SPACE: NATURAL, OR INTRINSIC, VARIABLES

Here we treat the same problem, in terms of the *arc-length* s (or $S = \sqrt{m}\, s$) along the path of the particle P, C, measured from some origin on it (positive in one direction, negative in the other). We begin with some differential-geometric background. Let the curve equations be

$$C: \quad q^k = q^k(u) \quad [k = 1, 2, 3;\ u_1 \le u: \text{curve parameter} \le u_2]; \qquad (5.3.1)$$

where the $q^k(\ldots)$ are assumed to be as well behaved as needed. The square of the *elementary arc-length* of C, ds, between its neighboring points $P(q)$ and $P'(q + dq)$, is {with metric $g_{kl} = g_{lk}$ evaluated on C, i.e., $g_{kl}(q) = g_{kl}[q^r(u)] = g_{kl}(u)$}:

$$ds^2 = g_{kl}\, dq^k\, dq^l = [g_{kl}(dq^k/du)(dq^l/du)]du^2 \quad (\ge 0); \qquad (5.3.2)$$

and integrating Equation 5.3.2 between u_1 and u_2 we can find the (finite) arc-length $\mathbf{P_1P_2} \equiv s_{12} = s_{21}$, where $P_1 = P(u_1)$ and $P(u_2)$.

Tangent. Let $\mathbf{r}$ be the position vector of the typical C-point P,* relative to some fixed E_3-origin. Then the *unit tangent vector* to C, at P, in the sense of increasing arc-length s, τ, is defined by:

$$\tau \equiv d\mathbf{r}/ds = (d\mathbf{r}/dq^k)(dq^k/ds) \equiv \tau^k\, \mathbf{e}_k \quad (\tau^k = dq^k/ds); \qquad (5.3.3)$$

$$\tau \cdot \tau = g_{kl}\, \tau^k\, \tau^l = g_{kl}(dq^k/ds)(dq^l/ds) = 1 \qquad (5.3.4)$$

(by Equation 5.3.2), i.e., $|\tau| = 1$.

* We use P for our traveling particle as well as for the generic C-point with which it, instantaneously, coincides.

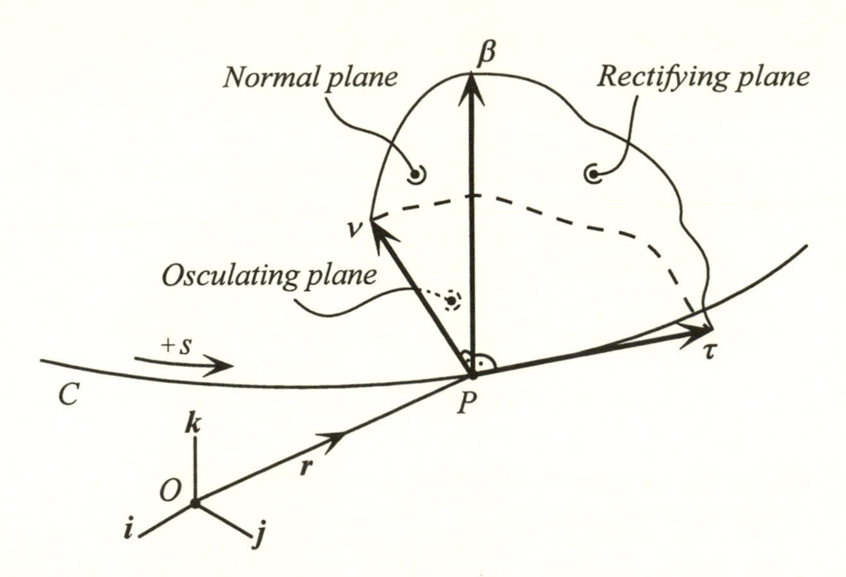

FIGURE 5.1 The Frenet (or Frenet–Serret) moving trihedron of a space curve.

Normal. D-differentiating the above along C, while recalling *Ricci's* theorem (Section 3.5: $Dg_{kl} = 0$), we find: $D(g_{kl}\,\tau^k\,\tau^l)/Ds = 0 \Rightarrow g_{kl}\,\tau^k(D\tau^l/Ds) = 0$, i.e., either $D\tau^l/Ds = 0$ ($C \rightarrow$ straight line) or $D\tau^l/Ds$ is perpendicular, or normal, to C at P. So, in general,

$$d^2\mathbf{r}/ds^2 = d\boldsymbol{\tau}/ds = (D\tau^l/Ds)\mathbf{e}_l \equiv \kappa_{(\nu)}\,\nu^l\,\mathbf{e}_l = \kappa_{(\nu)}\,\boldsymbol{\nu}; \tag{5.3.5}$$

i.e., $D\tau^l/Ds = \kappa_{(\nu)}\,\nu^l$, $\kappa_{(\nu)} \geq 0$ (zero for a straight line). The vector $\boldsymbol{\nu} \equiv \nu^l\,\mathbf{e}_l$ is the *first,* or principal, unit *normal* to C, at P; and $\kappa_{(\nu)}$ (frequently denoted by κ, and introduced to make $\boldsymbol{\nu}$ unit: $\boldsymbol{\nu}\cdot\boldsymbol{\nu} = g_{kl}\,\nu^k\,\nu^l = 1$) is the *first,* or principal, *curvature* there; while its inverse $\rho_{(\nu)} = 1/\kappa_{(\nu)}$ is the corresponding *radius of curvature.* The plane of $\boldsymbol{\tau}$, $\boldsymbol{\nu}$ is called *osculating plane* of C at P (Figure 5.1).

Binormal. Finally, the *binormal,* or *second* normal, to C at P, $\boldsymbol{\beta}$, is the unit vector defined so that $\{\boldsymbol{\tau}, \boldsymbol{\nu}, \boldsymbol{\beta}\}$ build an orthonormal and right-handed (or dextral) local basis, known as *Frenet's moving trihedron* there. The vectors $\boldsymbol{\nu}$, $\boldsymbol{\beta}$ define the *normal plane* of C at P, while $\boldsymbol{\tau}$, $\boldsymbol{\beta}$ define its *rectifying plane* there. The above allow us to write:

$$\boldsymbol{\beta} = \beta^k\,\mathbf{e}_k = \boldsymbol{\tau}\times\boldsymbol{\nu} = (e^{klr}\,\tau_l\,\nu_r)\mathbf{e}_k = \left(\varepsilon^{klr}\mathbf{y}\sqrt{g}\right)\tau_l\,\nu_r\,\mathbf{e}_k, \tag{5.3.6}$$

where $\tau_l = g_{lk}\,\tau^k$, $\nu_r = g_{rk}\,\nu^k$, and, therefore,

$$(\boldsymbol{\tau}, \boldsymbol{\nu}, \boldsymbol{\beta}) \equiv \boldsymbol{\tau}\cdot(\boldsymbol{\nu}\times\boldsymbol{\beta}) = \ldots = \sqrt{g}\,\varepsilon_{klr}\,\tau^k\,\nu^l\,\beta^r = +\,1. \tag{5.3.7}$$

Some Properties of the Frenet Triad in Tensor Form

i. Differentiating the normalization condition $\nu \cdot \nu = 1$ along C yields

$$(d\nu/ds) \cdot \nu = g_{kl}\, \nu^k (D\nu^k/Ds) = 0 \Rightarrow D\nu^k/Ds \text{ is normal to } \nu^k. \quad (5.3.8)$$

ii. $[d(\ldots)/ds]$-diferentiating the orthogonality condition $\tau \cdot \nu = g_{kl}\, \tau^k\, \nu^l = 0$ yields

$$\tau \cdot (d\nu/ds) + (d\tau/ds) \cdot \nu = g_{kl}\, \tau^k (D\nu^l/Ds) + g_{kl}\, \nu^l (D\tau^k/Ds) = 0, \quad (5.3.9)$$

from which (invoking Equation 5.3.5): $g_{kl}\, \tau^k(D\nu^l/Ds) = -g_{kl}\, \nu^l(D\tau^k/Ds) = -g_{kl}\, \nu^l(\kappa_{(\nu)}\, \nu^k) = -\kappa_{(\nu)}(g_{kl}\, \nu^k\, \nu^l) = -g_{kl}\, \tau^l\, \tau^k = -\kappa_{(\nu)}$, or finally,

$$g_{kl}\, \tau^k[(D\nu^l/Ds) + \kappa_{(\nu)}\, \tau^l] = 0 \Rightarrow \mu^l \equiv (D\nu^l/Ds) + \kappa_{(\nu)}\, \tau^l \quad (5.3.9a)$$

is normal to τ^l.
But also $g_{kl}\, \nu^k\, \nu^l = 1$, and therefore τ^l, ν^l, μ^l build a special orthogonal basis at P. Analytically, the binormal vector β is defined by

$$\beta = \beta^l\, \mathbf{e}_l, \quad \beta^l \equiv \mu^l/\kappa_{(\beta)} = (1/\kappa_{(\beta)})[(D\nu^l/Ds) + \kappa_{(\nu)}\, \tau^l], \quad (5.3.10)$$

where $\kappa_{(\beta)}$ (frequently denoted by τ, and introduced to make β unit: $\beta \cdot \beta = g_{kl}\, \beta^k\, \beta^l = 1$) is the *second* curvature, or *torsion* of C at P (unrelated to the torsion of a manifold, Sections 3.8 and 3.9); and its inverse $\rho_{(\beta)} = 1/\kappa_{(\beta)}$ is the corresponding *radius of second curvature*, or *radius of torsion*, there. Clearly, for a plane curve $\kappa_{(\beta)}$ vanishes; while for a straight line both $\kappa_{(\nu)}$ and $\kappa_{(\beta)}$ vanish.

iii. From Equation 5.3.5 and the above, we readily find

$$g_{dh}(D\tau^d/Ds)(D\tau^h/Ds) = g_{dh}(\kappa_{(\nu)}\, \nu^d)(\kappa_{(\nu)}\, \nu^h) = \ldots = \kappa_{(\nu)}{}^2,$$

i.e.,

$$\kappa_{(\nu)}{}^2 = (d\tau/ds) \cdot (d\tau/ds) = g_{dh}(D\tau^d/Ds)(D\tau^h/Ds). \quad (5.3.11)$$

If $d\tau = \mathbf{0}$ ($\Rightarrow D\tau^h = 0$; although, in general, $d\tau^h \neq 0$), then $\kappa_{(\nu)} = 0$ (straight line).

iv. With the help of Equation 5.3.10 we find successively: $e_{hld}\, \tau^h\, \nu^l(D\nu^d/Ds) = e_{hld}\, \tau^h\, \nu^l(\kappa_{(\beta)}\, \beta^d - \kappa_{(\nu)}\, \tau^d) = (e_{hld}\, \tau^h\, \nu^l\, \beta^d)\kappa_{(\beta)} - (e_{hld}\, \tau^h\, \nu^l\, \tau^d)\kappa_{(\nu)} = (1)\kappa_{(\beta)} - (0)\kappa_{(\nu)}$, since $e_{hld}\, \tau^h\, \tau^d = 0$, i.e.,

$$\kappa_{(\beta)} = e_{hld}\, \tau^h\, \nu^l(D\nu^d/Ds) = \tau \cdot [\nu \times (d\nu/ds)]. \quad (5.3.12)$$

The Frenet (or Frenet–Serret) Formulae. These are three remarkable differential equations for τ, ν, β which *contain all essential geometrical properties of C in the small.* We have already seen *two* of them: Equations 5.3.5 and 5.3.10:

$$d\tau/ds = \kappa_{(\nu)}\,\nu \quad \text{or} \quad D\tau^l/Ds = \kappa_{(\nu)}\,\nu^l, \tag{5.3.13}$$

$$d\nu/ds = -\kappa_{(\nu)}\,\tau + \kappa_{(\beta)}\,\beta \quad \text{or} \quad D\nu^l/Ds = -\kappa_{(\nu)}\,\tau^l + \kappa_{(\beta)}\,\beta^l. \tag{5.3.14}$$

The *third* one, proved below, is

$$d\beta/ds = -\kappa_{(\beta)}\,\nu \quad \text{or} \quad D\beta^l/Ds = -\kappa_{(\beta)}\,\nu^l. \tag{5.3.15}$$

Indeed, $D(\ldots)/DS$-differentiating the component form of the β-definition: $\beta = \tau \times \nu$, while recalling that $De^{\cdots} = 0$ (Equation 3.5.3b), we find

$$\begin{aligned} D\beta^l/Ds &= D(e^{ljh}\tau_j\,\nu_h)/Ds = e^{ljh}\nu_h(D\tau_j/Ds) + e^{ljh}\tau_j(D\nu_h/Ds) \\ &= e^{ljh}\nu_h(\kappa_{(\nu)}\,\nu_j) + e^{ljh}\tau_j(\kappa_{(\beta)}\,\beta_h - \kappa_{(\nu)}\,\tau_h) \\ &= \kappa_{(\nu)}(e^{ljh}\,\nu_h\,\nu_j) + \kappa_{(\beta)}(e^{ljh}\tau_j\,\beta_h) - \kappa_{(\nu)}(e^{ljh}\tau_j\,\tau_h) \\ &= \kappa_{(\nu)}(0) + \kappa_{(\beta)}(e^{ljh}\tau_j\,\beta_h) - \kappa_{(\nu)}\,(0) = \kappa_{(\beta)}(\tau \times \beta)^l \\ &= \kappa_{(\beta)}(-\nu)^l, \quad \text{q.e.d.} \end{aligned} \tag{5.3.15a}$$

In sum, the three Frenet equations are, in vector and tensor form:

$$d\tau/ds = (0)\tau + (\kappa_{(\nu)})\nu + (0)\beta, \text{or} \quad D\tau^l/Ds = \kappa_{(\nu)}\,\nu^l, \tag{5.3.16a}$$

$$d\nu/ds = (-\kappa_{(\nu)})\tau + (0)\nu + (\kappa_{(\beta)})\beta, \tag{5.3.16b}$$

$$\text{or} \quad D\nu^l/Ds = -\kappa_{(\nu)}\,\tau^l + \kappa_{(\beta)}\,\beta^l,$$

$$d\beta/ds = (0)\tau + (-\kappa_{(\beta)})\nu + (0)\beta, \text{or} \quad D\beta^l/Ds = -\kappa_{(\beta)}\,\nu^l; \tag{5.3.16c}$$

or in the (antisymmetric) *matrix* form:

$$\begin{pmatrix} d\boldsymbol{\tau}/ds \\ d\mathbf{v}/ds \\ d\boldsymbol{\beta}/ds \end{pmatrix} = \begin{pmatrix} 0 & \kappa_{(\nu)} & 0 \\ -\kappa_{(\nu)} & 0 & \kappa_{(\beta)} \\ 0 & -\kappa_{(\beta)} & 0 \end{pmatrix} \begin{pmatrix} \boldsymbol{\tau} \\ \mathbf{v} \\ \boldsymbol{\beta} \end{pmatrix}. \tag{5.3.17}$$

It can be shown that the natural, or intrinsic, equations: $\kappa_{(\nu)} = \kappa_{(\nu)}(s)$ and $\kappa_{(\beta)} = \kappa_{(\beta)}(s)$ determine *C to within a rigid-body motion.* A possible advantage of the tensor forms over the direct vector ones lies in the freedom of choice of the q's and their easy implementation for a specific system of such coordinates. For a straight line, i.e., $\kappa_{(\nu)} = 0$, and so Equation 5.3.16a assumes the *geodesic* form of our E_3 (with $\lambda^l{}_{kh}$ denoting the Christoffels based on the g_{kl}):

$$D\tau^l/Ds \equiv d\tau^l/ds + \lambda^l{}_{kh}\, \tau^k(dq^h/ds)$$

$$= d^2q^l/ds^2 + \lambda^l{}_{kh}\,(dq^k/ds)(dq^h/ds) = 0. \qquad (5.3.18)^*$$

Kinematic Interpretation of the Frenet Formulae, the Darboux Vector. In terms of the *Darboux vector*

$$\mathbf{D} \equiv \kappa_{(\beta)}\,\tau + \kappa_{(\nu)}\,\beta \equiv \mathbf{D}_\tau + \mathbf{D}_\beta$$

$$[\Rightarrow |\mathbf{D}| = [\kappa_{(\nu)}{}^2 + \kappa_{(\beta)}{}^2]^{1/2}\text{: } \textit{total curvature} \text{ of } C \text{ at } P] \qquad (5.3.19)$$

$$[\Rightarrow \kappa_{(\nu)} = \mathbf{D}\cdot\beta = \mathbf{D}_\beta\cdot\beta, \qquad (5.3.19a)$$

since > 0, **D** and β are always on the same side of the osculating plane]

$$[\Rightarrow \kappa_{(\beta)} = \mathbf{D}\cdot\tau = \mathbf{D}_\tau\cdot\tau\,, \qquad (5.3.19b)$$

if > 0, then **D** and τ make an *acute* angle; i.e., the osculating plane and its normal, that is the binormal, rotate in a right-hand sense around the positive tangent], the Frenet equations take the following *kinematic* (or *Poisson*) form, i.e., $d(\ldots)/dt = \omega \times (\ldots)$, where $(\ldots)$: any vector rigidly attached to axes rotating with angular velocity ω; here $\omega \to \mathbf{D}$ and $dt \to ds$ (recall Example 3.8.3):

$$d\tau/ds = \mathbf{D}\times\tau = \ldots = \kappa_{(\nu)}\,\nu, \qquad (5.3.20a)$$

$$d\nu/ds = \mathbf{D}\times\nu = \ldots = -\kappa_{(\nu)}\,\tau + \kappa_{(\beta)}\,\beta, \qquad (5.3.20b)$$

$$d\beta/ds = \mathbf{D}\times\beta = \ldots = -\kappa_{(\beta)}\,\nu. \qquad (5.3.20c)$$

From the above geometrical equations the following kinematical picture emerges (thinking of arc-length s as time t):

---

* Extensions of Frenet's equations to a general n-dimensional manifold are also available; see, e.g., Mishchenko and Fomenko (1988, pp. 154–160), Teichmann (1964, pp. 102–103).

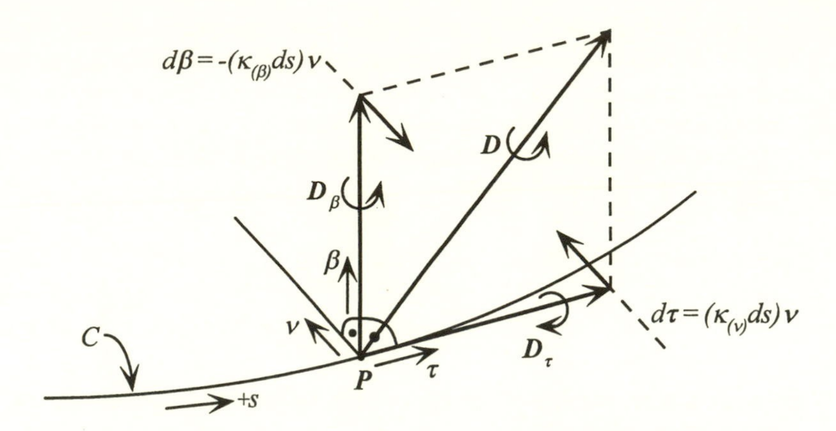

FIGURE 5.2 Kinematical interpretation of Frenet's moving trihedron and Darboux's vector: $\mathbf{D} \equiv \kappa_{(\beta)}\,\tau + \kappa_{(\nu)}\,\beta$; $d\tau/ds = \kappa_{(\nu)}\,\nu$, $d\nu/ds = -\kappa_{(\nu)}\,\tau + \kappa_{(\beta)}\,\beta$, $d\beta/ds = -\kappa_{(\beta)}\,\nu$.

- The tangent turns around the binormal, with (positive) "angular velocity" $\kappa_{(\nu)}$ (Equation 5.3.16a)
- The binormal turns around the tangent with "angular velocity" $\kappa_{(\beta)}$ (Equation 5.3.16c)
- The entire Frenet trihedron, or local frame, $(P\!: \tau, \nu, \beta)$ turns around the Darboux vector with (positive) angular velocity $|\mathbf{D}|$ (Equations 5.3.20a to c) (Figure 5.2).

Example 5.3.1

i. For a *circle C* in polar coordinates $q^1 = r = R$ (radius), $q^2 = \phi(s)$, $q^3 = 0$, we have

$$ds^2 = dR^2 + R^2\,d\phi^2 \Rightarrow g_{11} = 1, \quad g_{22} = R^2, \quad g_{33} = 0, \quad g_{kl} \neq 0 \quad (k \neq l); \quad \text{(a)}$$

the unit tangent vector has components $\tau = (\tau^k = dq^k/ds) = (\tau^1 = 0,\ \tau^2 = d\phi/ds,\ \tau^3 = 0)$, and so the normalization condition $1 = g_{kl}\,\tau^k\,\tau^l$ results in

$$R^2(d\phi/ds)^2 = 1 \Rightarrow d\phi/ds = 1/R = \kappa_{(\nu)} \equiv \kappa > 0; \quad \text{(b)}$$

and the nonvanishing Christoffels are: $\Gamma^1{}_{22} = -R$, $\Gamma^2{}_{12} = \Gamma^2{}_{21} = 1/R$. Hence, the *first* set of the Frenet equations reduces to

$$d\tau^1/ds + \Gamma^1{}_{kl}\,\tau^k(dq^l/ds) = \Gamma^1{}_{22}\,\tau^2(dq^2/ds) = -1/R = \kappa_{(\nu)}\,\nu^1, \quad \text{(c1)}$$

$$d\tau^2/ds + \Gamma^2{}_{kl}\,\tau^k(dq^l/ds) = \Gamma^2{}_{12}\,\tau^2(dq^1/ds) = 0 = \kappa_{(\nu)}\,\nu^2 \Rightarrow \nu^2 = 0, \quad \text{(c2)}$$

$$d\tau^3/ds + \Gamma^3{}_{kl}\,\tau^k(dq^l/ds) = 0 = \kappa_{(\nu)}\,\nu^3 \Rightarrow \nu^3 = 0; \quad \text{(c3)}$$

i.e., $\nu = (-1/R\kappa_{(1)}, 0, 0)$, and since $\kappa_{(\nu)}{}^2 = g_{sl}(\kappa_{(\nu)}\,\nu^s)(\kappa_{(\nu)}\,\nu^l) = 1/R^2 \Rightarrow \kappa_{(\nu)} = 1/R(> 0)$, finally $\nu = (-1, 0, 0)$. Similarly, we can show that $\beta = (0, 0, 1)$, i.e., $\kappa_{(\beta)} = 0$ (plane curve!).

ii. For a *straight* line, we have $D\tau^h/Ds = d^2q^h/ds^2 + \Gamma^h{}_{sl}(dq^s/ds)(dq^l/ds) = 0$. In rectangular Cartesian coordinates $\Gamma^h{}_{sl} = 0$, and so, there, the above reduces to $d^2q^h/ds^2 = 0 \Rightarrow q^h = c_1\,s + c_2$ ($c_{1,2}$: constants of integration).

Problem 5.3.1

Show that:

- $D^2\tau^h/Ds^2 = (d\kappa_{(\nu)}/ds)\nu^h + \kappa_{(\nu)}(\kappa_{(\beta)}\,\beta^h - \kappa_{(\nu)}\,\tau^h),$ (a)
- $D^2\nu^h/Ds^2 = (d\kappa_{(\nu)}/ds)\beta^h - (\kappa_{(\nu)}{}^2 + \kappa_{(\beta)}{}^2)\nu^h - (d\kappa_{(\nu)}/ds)\tau^h,$ (b)
- $D^2\beta^h/Ds^2 = \kappa_{(\beta)}(\kappa_{(\nu)}\,\tau^h - \kappa_{(\beta)}\,\beta^h) - (d\kappa_{(\beta)}/ds)\nu^h.$ (c)

Problem 5.3.2

Show that

- $\kappa_{(\beta)} = (1/\kappa_{(\nu)}{}^2)e_{hld}\,\tau^h(D\tau^l/Ds)(D\tau^d/Ds)$

$$= (1/\kappa_{(\nu)}{}^2)(d\mathbf{r}/ds,\ d^2\mathbf{r}/ds^2,\ d^3\mathbf{r}/ds^3)$$

$$= (d\mathbf{r}/ds,\ d^2\mathbf{r}/ds^2,\ d^3\mathbf{r}/ds^3)\ /\ [(d\mathbf{r}/ds)\times(d^2\mathbf{r}/ds^2)]^2; \tag{a}$$

- $e_{hld}(D\tau^h/Ds)(D^2\tau^l/Ds^2)(D^3\tau^d/Ds^3) = \kappa_{(\nu)}{}^5[d(\kappa_{(\beta)}/\kappa_{(\nu)})/ds];$ (b)
- $e_{hld}(D\beta^h/Ds)(D^2\beta^l/Ds^2)(D^3\beta^d/Ds^3) = \kappa_{(\beta)}{}^5[d(\kappa_{(\nu)}/\kappa_{(\beta)})/ds].$ (c)

Now let us resume our discussion of particle kinematics. Here we find, successively,

Velocity

$$\mathbf{v} = d\mathbf{r}/dt = (\partial\mathbf{r}/\partial q^k)(dq^k/dt) \equiv (dq^k/dt)\mathbf{e}_k = (dq^k/dt)\left(\mathbf{E}_k/\sqrt{m}\right) = \mathbf{V}/\sqrt{m}$$

$$= \mathbf{e}_k(dq^k/ds)(ds/dt) = (d\mathbf{r}/ds)(ds/dt) \equiv v_\tau\,\tau \equiv v\,\tau$$

$$= \left(\mathbf{E}_k/\sqrt{m}\right)(dq^k/dS)(dS/dt)$$

$$= \left(1/\sqrt{m}\right)(d\mathbf{R}/dS)(dS/dt) \equiv \left(1/\sqrt{m}\right)V\mathbf{u}, \tag{5.3.21}$$

where

$$\mathbf{u} \equiv d\mathbf{R}/dS = (dq^k/ds)\mathbf{E}_k = \sqrt{m}\,d\mathbf{r}\ /\ \sqrt{m}\,ds = d\mathbf{r}/ds = \tau. \tag{5.3.21a}$$

Acceleration

$$\mathbf{a} = d\mathbf{v}/dt = d/dt(v\ \tau)$$

$$= (dv/dt)\tau + v(d\tau/dt) = (dv/dt)(dq^k/ds)\mathbf{e}_k + v^2(d\tau/ds). \qquad (5.3.22a)$$

But

$$d\tau/ds = d/ds[(dq^k/ds)\mathbf{e}_k] = (d^2q^k/ds^2)\mathbf{e}_k + (d\mathbf{e}_k/ds)(dq^k/ds)$$

$$= (d^2q^k/ds^2)\mathbf{e}_k + [(\partial\mathbf{e}^k/\partial q^l)(dq^l/ds)](dq^k/ds)$$

$$= (d^2q^k/ds^2)\mathbf{e}_k + (\lambda^r{}_{kl}\ \mathbf{e}_r)(dq^l/ds)(dq^k/ds)$$

$$= [(d^2q^h/ds^2) + \lambda^h{}_{rs}(dq^r/ds)(dq^s/ds)]\mathbf{e}_h$$

$$\equiv \kappa_{(\mathrm{v})}{}^h\ \mathbf{e}_h = \kappa_{(\mathrm{v})h}\ \mathbf{e}^h = \kappa_{(\mathrm{v})}\ \mathrm{v} \text{ (by the first of Frenet's formulae).} \qquad (5.3.22b)$$

Thus, $\kappa_{(\mathrm{v})}{}^h = d^2q^h/ds^2 + \lambda^h{}_{rs}(dq^r/ds)(dq^s/ds)$ are the contravariant components of the first *curvature vector* $\kappa_{(\mathrm{v})} \equiv \kappa_{(\mathrm{v})}\ \mathrm{v}$; and $\kappa_{(\mathrm{v})}{}^2 = g_{hd}\ \kappa_{(\mathrm{v})}{}^h\ \kappa_{(\mathrm{v})}{}^d$ is the (square of the first) curvature of C at P. In view of Equation 5.3.22b, Equation 5.3.22a assumes the form:

$$\mathbf{a} = \mathbf{a}_\tau + \mathbf{a}_\mathrm{v} = (d^2s/dt^2)\tau + [(ds/dt)^2\kappa_{(\mathrm{v})}]\mathrm{v}, \qquad (5.3.22c)$$

$$\mathbf{a}_\mathrm{v} = (ds/dt)^2(\kappa_{(\mathrm{v})}{}^h\ \mathbf{e}_h)$$

$$= (ds/dt)^2\ [d^2q^h/ds^2 + \lambda^h{}_{rs}(dq^r/ds)(dq^s/ds)]\mathbf{e}_h:$$

normal acceleration, (5.3.22d)

$$\mathbf{a}_\tau = (d^2s/dt^2)(dq^k/ds)\mathbf{e}_k: \quad \textit{tangential}\ \text{acceleration.} \qquad (5.3.22e)$$

From the above, we readily conclude that the *normalized* acceleration of P, $\mathbf{A}$, equals

$$\mathbf{A} \equiv \sqrt{m}\,\mathbf{a} = d\mathbf{V}/dt = \sqrt{m}\,(d\mathbf{v}/dt)$$

$$= \mathbf{A}_\tau + \mathbf{A}_\mathrm{v} = (d^2S/dt^2)\tau + [(dS/dt)^2\mathrm{K}_{(\mathrm{v})}]\mathrm{v},$$

$$\mathrm{K}_{(\mathrm{v})} \equiv \kappa_{(\mathrm{v})}/\sqrt{m}, \qquad (5.3.22f)$$

$$\mathbf{A}_\mathrm{v} = (dS/dt)^2(\mathrm{K}_{(\mathrm{v})}{}^h\ \mathbf{E}_h)$$

$$= (dS/dt)^2[d^2q^h/dS^2 + \Gamma^h{}_{rs}(dq^r/dS)(dq^s/dS)]\mathbf{E}_h:$$

normal acceleration, (5.3.22g)

$$\mathbf{A}_\tau = (d^2S/dt^2)(dq^k/dS)\mathbf{E}_k\text{: } \textit{tangential} \text{ acceleration.} \tag{5.3.22h}$$

We notice that

$$d\tau/dS = K_{(\nu)}{}^h\,\mathbf{E}_h = [d^2q^h/dS^2 + \Gamma^h{}_{rs}(dq^r/dS)(dq^s/dS)]\mathbf{E}_h = K_{(\nu)}\,\nu \tag{5.3.22i}$$

$$\Rightarrow K_{(\nu)}{}^h = \left(1\mathbf{y}\sqrt{m}\right)\left(1\mathbf{y}\sqrt{m}\right)\kappa_{(\nu)}{}^h;\quad \mathbf{A}_\tau = \sqrt{m}\,\mathbf{a}_\tau,\quad \mathbf{A}_\nu\sqrt{m}\,\mathbf{a}_\nu. \tag{5.3.22j}$$

Force. The total impressed force $\mathbf{F}$ is decomposed conveniently into (i) a *tangential* component $\mathbf{F}\cdot\tau$, or

$$\mathbf{F}_\tau = (\mathbf{F}\cdot\tau)\tau = [(F_d\,\mathbf{e}^d)\cdot(dq^l/ds)\mathbf{e}_l](dq^k/ds)\mathbf{e}_k = F_{(\tau)}{}^k\,\mathbf{e}_k = F_\tau\,\tau,$$

i.e.,

$$F_{(\tau)}{}^k \equiv (dq^k/ds)(dq^l/ds)F_l; \tag{5.3.23a}$$

and (ii) a *normal* component $\mathbf{F}\cdot\nu$, or

$$\mathbf{F}_\nu = (\mathbf{F}\cdot\nu)\nu = \mathbf{F} - \mathbf{F}_\tau = [F^k - F_l(dq^l/ds)(dq^k/ds)]\mathbf{e}_k = F_{(\nu)}{}^k\,\mathbf{e}_k = F_\nu\,\nu,$$

i.e.,

$$F_{(\nu)}{}^k \equiv [g^{kl} - (dq^k/ds)(dq^l/ds)]F_l. \tag{5.3.23b}$$

Then, the Newton–Euler law transforms successively to:

$$m\,\mathbf{a} = \mathbf{F} \Rightarrow \sqrt{m}\,\mathbf{a} = \mathbf{F}/\sqrt{m} \Rightarrow \mathbf{A} = \mathbf{Q}$$

$$\Rightarrow \mathbf{A}_\tau = \mathbf{Q}_\tau \text{ (along tangent)},\quad \mathbf{A}_\nu = \mathbf{Q}_\nu \text{ (along normal)}; \tag{5.3.23c}$$

or, in components, to the *natural*, or *intrinsic*, equations of motion:

$$d^2S/dt^2 = Q_k(dq^k/dS) \qquad \text{(along tangent)}, \tag{5.3.23d}$$

$$(dS/dt)^2(D\tau^k/Dt) = (dS/dt)^2[d^2q^k/dS^2 + \Gamma^k{}_{rs}(dq^r/dS)(dq^s/dS)]$$

$$= [M^{kl} - (dq^k/dS)(dq^l/dS)]Q_l \quad \text{(along normal)}. \tag{5.3.23e}$$

Under wholly *potential* forces, i.e., $Q_k = -(\partial\Pi/\partial q^k)$, $\Pi = \Pi(q)$, Equation 5.3.23d transforms further:

$$d^2S/dt^2 = -(\partial\Pi/\partial q^k)(dq^k/dS) = -\partial\Pi/\partial S, \tag{5.3.23f}$$

or, multiplying with $dS = (dS/dt)dt$ and then, integrating once in time, and with $E \equiv$ total (constant) *energy* of particle,

$$(dS/dt)^2 = 2(E - \Pi). \tag{5.3.23g}$$

Finally, eliminating $(dS/dt)^2$ between Equations 5.3.23g and e we get the *equations of the trajectories of the particle under potential forces (or orbits):*

$$d^2q^k/dS^2 + \Gamma^k{}_{rs}(dq^r/dS)(dq^s/dS)$$

$$= -(\partial\Pi/\partial q^l)[M^{kl} - (dq^k/dS)(dq^l/dS)] \,/\, 2(E - \Pi); \tag{5.3.23h}$$

which is to be solved along with the "normalization" constraint

$$M_{kl}(dq^k/dS)(dq^l/dS) = g_{kl}(dq^k/ds)(dq^l/ds) = 1. \tag{5.3.23i}$$

If $\mathbf{Q} = \mathbf{0}$ (force-free case), then Equations 5.3.23g and h reduce, respectively, to

$$dS/dt = \text{constant}, \quad D(dq^k/dt)/DS = 0 \quad \text{(geodesic deviation);} \tag{5.3.23j}$$

i.e., P moves with constant magnitude velocity along a geodesic curve of the corresponding space — here in E_3 along a straight line. (The above extend directly to an n-dimensional configuration space, with all Latin indices running from 1 to n.)

5.4 PARTICLE ON A CURVE

Let us consider again the particle P, but now constrained to move on the specified *fixed* curve C, whose parametric representation is $\{q^k = q^k(s)$ or $q^k = q^k(S)$; $k = 1, 2, 3\}$. The *velocity* of P is

$$dq^k/dt = (dq^k/ds)(ds/dt) \equiv (dq^k/dt)\tau^k, \tag{5.4.1a}$$

$v_\tau = ds/dt$: tangential component of $\mathbf{v}$, $\tau^k \equiv dq^k/ds$: unit tangent vector; (5.4.1b)

while its *acceleration* is

$$a^k = D(dq^k dt)/Dt = D[(ds/dt)\tau^k]/Dt$$

$$= [d(ds/dt)/dt]\tau^k + (ds/dt)(D\tau^k/Ds)(ds/dt)$$

$$= (d^2s/dt^2)\tau^k + (ds/dt)^2(D\tau^k/Ds) = (d^2s/dt^2)\tau^k + [(ds/dt)^2\kappa_{(v)}]\nu^k, \tag{5.4.2}*$$

* Where, in the last step, we used the *first* of the Frenet formulae.

i.e., **a** *is coplanar with the tangent and (principal) normal, and lies wholly on the osculating plane* (of τ and ν) of C at P. The equation of (constrained) motion yields

$$m\,\mathbf{a} = \mathbf{F}_{(\text{total impressed force})} + \mathbf{Z}_{(\text{total constraint reaction})}$$

$$\Rightarrow \sqrt{m}\,\mathbf{a} = \mathbf{F}/\sqrt{m} + \mathbf{Z}/\sqrt{m}\,, \tag{5.4.3a}$$

or, with the customary renaming,

$$\mathbf{A} = \mathbf{Q}_{(\text{normalized total impressed force})} + \Lambda_{(\text{normalized total constraint reaction})}$$

$$\Rightarrow A^k = Q^k + \Lambda^k, \tag{5.4.3b}$$

or, further, with use of the normalized version of Equation 5.4.2, i.e., $ds/dt \to dS/dt \equiv V$, $\tau^k \to \mathrm{T}^k$, $\kappa_{(\nu)} \to \mathrm{K}_{(\nu)}$, $\nu^k \to \mathrm{N}^k$, $\beta^k \to \mathrm{B}^k$

$$(dV/dt)\mathrm{T}^k + V^2\,\mathrm{K}_{(\nu)}\,\mathrm{N}^k = Q^k + \Lambda^k \quad (2T = m\,v^2 = V^2) \tag{5.4.3c}$$

or, since $dV/dt = (dV/dS)(dS/dt) \equiv V(dV/dS) = d/dS(V^2/2) = dT/dS$,

$$(dT/dS)\mathrm{T}^k + (2T\mathrm{K}_{(\nu)})\mathrm{N}^k = Q^k + \Lambda^k. \tag{5.4.3d}$$

Under the *physical assumption* that Λ^k is *orthogonal* to the curve tangent, i.e., $\Lambda^k\,\mathrm{T}_k = 0$, projecting Equation 5.4.3d along the tangent, that is, dotting it with T_k, we get the *energy rate equation*:

$$Q^k\,\mathrm{T}_k = dT/dS; \tag{5.4.3e}$$

projecting it along the normal, that is, dotting it with N_k, we find

$$Q^k\,\mathrm{N}_k + \Lambda^k\,\mathrm{N}_k = 2T\,\mathrm{K}_{(\nu)}; \tag{5.4.3f}$$

and projecting it along the binormal, i.e., dotting it with B_k, we obtain the *force balance equation*:

$$Q^k\,\mathrm{B}_k + \Lambda^k\,\mathrm{B}_k = 0 \Rightarrow Q^k\,\mathrm{B}_k = -\Lambda^k\,\mathrm{B}_k, \tag{5.4.3g}$$

which shows that $Q_{(b)} \equiv Q^k\,\mathrm{B}_k \neq 0$, but instead the total binormal component $(Q^k + \Lambda^k)B_k$ vanishes. Integrating Equation 5.4.3e yields

$$T = \int (Q^k\,\mathrm{T}_k)dS = \int [Q^k(dq^k/dS)]dS, \tag{5.4.3h}$$

and if, further, $Q_k = -(\partial\Pi/\partial q^k)$, then

$$T = -\int (d\Pi/dS)dS \Rightarrow E \equiv T + \Pi = \text{constant.} \qquad (5.4.3\text{i})^*$$

If C is a *natural* trajectory, then (by definition) $\Lambda^k = 0$, and so Equation 5.4.3d reduces to

$$Q^k = (dT/dS)\mathrm{T}^k + (2T\,\mathrm{K}_{(v)})\mathrm{N}^k, \qquad (5.4.3\text{j})$$

i.e., *the impressed force lies on the osculating plane.* Equation 5.4.3g would have led us to the same result. If $Q^k = 0$, then Equation 5.4.3j leads immediately to $dT/dS = 0$ ($\Rightarrow \Pi$ = constant) and $\mathrm{K}_{(v)} = 0$ ($\Rightarrow$ straight-line path, i.e., "law" of inertia).

Example 5.4.1. Motion of a *Free* Particle in a *Uniform* Gravitiational Field Q^k

Since the field is *parallel* (Section 3.7), $DQ^k/DS = 0$, and so Equation 5.4.3j yields

$$D/DS[(dT/dS)\mathrm{T}^k + 2T\,\mathrm{K}_{(v)}\,\mathrm{N}^k] = 0, \qquad (\text{a})$$

or, since T, $\mathrm{K}_{(v)}$, and dT/dS are scalars,

$$(d^2T/dS^2)\mathrm{T}^k + (dT/dS)(D\mathrm{T}^k/DS) + (2\mathrm{K}_{(v)}\,\mathrm{N}^k)(dT/dS)$$

$$+ (2T\,\mathrm{N}^k)(d\mathrm{K}_{(v)}/dS) + (2T\,\mathrm{K}_{(v)})(D\mathrm{N}^k/dS) = 0, \qquad (\text{b})$$

or, using the Frenet equations in the second and last terms, and since the natural trajectory is a plane curve [$Q^k\,\mathrm{B}_k = 0 \Rightarrow (DQ^k/DS)\mathrm{B}_k + Q^k(D\mathrm{B}_k/DS) = 0 \Rightarrow Q^k(-\mathrm{K}_{(\beta)}\,\mathrm{N}_k) = 0 \Rightarrow \mathrm{K}_{(\beta)} = 0$]

$$[(d^2T/dS^2) - 2T\,\mathrm{K}_{(v)}{}^2]\mathrm{T}^k + [3\mathrm{K}_{(v)}(dT/dS) + 2T(d\mathrm{K}_{(v)}/dS)]\mathrm{N}^k$$

$$+ (2T\,\mathrm{K}_{(v)}\,\mathrm{K}_{(\beta)})B^k = 0 \quad \text{(the last term vanishes),} \qquad (\text{c})$$

from which we are immediately led to the following pair of differential equations:

$$d^2T/dS^2 = 2T\mathrm{K}_{(v)}{}^2 \quad \text{and} \quad dT/T = -(2/3)(d\mathrm{K}_{(v)}/\mathrm{K}_{(v)}). \qquad (\text{d})$$

The second of them integrates easily to $T = c\mathrm{K}_{(v)}{}^{-2/3}$ (c: integration constant), and substituting this result into the first (and with $\rho_{(v)} \equiv 1/\mathrm{K}_{(v)}$, normalized curvature) yields

$$d^2/dS^2[\rho_{(v)}{}^{2/3}] = 2\rho_{(v)}{}^{-4/3}, \qquad (\text{e})$$

---

* We use Π for the potential, instead of the customary V, to avoid confusion with our $dS/dt \equiv V$.

which, as the reader may verify, is the *intrinsic* equation of a *parabola.*

5.5 PARTICLE ON A SURFACE

5.5.1 INTRODUCTION TO SURFACES, VELOCITY

Here we consider the earlier particle P of mass m, but now forced to move on a *fixed* surface S. Let us begin by summarizing some relevant differential-geometric fundamentals. Certain properties of S can be expressed independently of the surrounding Euclidean space E_3, that is, in terms of quantities measured *without leaving the surface* (e.g., arc-length measurements carried out by ants who cannot get out of the surface); and the study of such intrinsic, or internal, properties is called the *intrinsic geometry* of S. Other properties of S that depend on the structure of E_3 as well, e.g., covariant differentiation of surface tensors or acceleration of P, we shall term extrinsic, or external, and their study *extrinsic geometry* of S.*

Now, let the position vector relative to a fixed E_3-origin O, $\mathbf{r}$, be expressed as

$$\mathbf{r} = x\,\mathbf{i} + y\,\mathbf{j} + z\,\mathbf{k} = \mathbf{r}[x(q^1, q^2), y(q^1, q^2), z(q^1, q^2)]$$

$$= \mathbf{r}(q^1, q^2) = \mathbf{r}(q^\alpha) \tag{5.5.1}$$

(in this section, *Greek* indices take the values 1, 2), where $q^\alpha \equiv (q^1, q^2)$ are the *Gaussian,* or *surface,* curvilinear coordinates of the S-point P. Similarly for its *normalized* image: $\mathbf{r} \to \mathbf{R} \equiv \sqrt{m}\,\mathbf{r} = \mathbf{R}(X, Y, Z)$, i.e., $X = \sqrt{m}\,x$ etc.

$$\mathbf{R} = \mathbf{R}[X(q^1, q^2), Y(q^1, q^2), Z(q^1, q^2)] = \mathbf{R}(q^1, q^2) \equiv \mathbf{R}(q^\alpha). \qquad (5.5.1\text{a})^{**}$$

Then, the actual and normalized *velocities* of P are, respectively,

$$\mathbf{v} \equiv d\mathbf{r}/dt = (\partial\mathbf{r}/\partial q^\alpha)(dq^\alpha/dt) \equiv v^\alpha\,\mathbf{e}_\alpha = v_\alpha\,\mathbf{e}^\alpha, \tag{5.5.2a}$$

$$\mathbf{V} \equiv d\mathbf{R}/dt = (\partial\mathbf{R}/\partial q^\alpha)(dq^\alpha/dt) \equiv V^\alpha\,\mathbf{E}_\alpha = V_\alpha\,\mathbf{E}^\alpha \quad \left(= \sqrt{m}\,\mathbf{v}\right). \tag{5.5.2b}$$

Here,

---

* It can be shown that *the intrinsic properties remain invariable under inextensible deformations of S* (e.g., those of a high-quality bedsheet), whereas the extrinsic ones do not. Analytical dynamics utilizes both kinds of properties; while relativity is primarily concerned with the intrinsic ones (of a four-dimensional surface).

** We assume that these parametrizations, and corresponding surface points, are ***regular***; i.e., are such that *the rank of the 3 × 2 Jacobian matrix* $J \equiv \partial(X, Y, Z)/\partial(q^1, q^2)$ *equals 2* (i.e., points for which S has a *unique tangent plane* ⇒ *unique normal* – see below). Points for which rank $J < 2$ (e.g., origin, in polar or spherical coordinates) are called *singular.* If, however, for *another* system of surface coordinates the same singular point becomes regular, the singularity is called *nonessential;* otherwise it is called *essential,* e.g., a cone vertex.

$$\mathbf{E}_\alpha \equiv \sqrt{m}\,\mathbf{e}_\alpha, \tag{5.5.3a}$$

$$\mathbf{E}^\alpha = M^{\alpha\beta}\,\mathbf{E}_\beta\ (\Leftrightarrow \mathbf{E}_\beta = M_{\beta\alpha}\,\mathbf{E}^\alpha), \quad \mathbf{e}^\alpha = g^{\alpha\beta}\,\mathbf{e}_\beta\ (\Leftrightarrow \mathbf{e}_\beta = g_{\beta\alpha}\,\mathbf{e}^\alpha), \tag{5.5.3b}$$

are the actual and normalized *fundamental surface vectors* [the covariant ones ($\mathbf{E}_\alpha$, $\mathbf{e}_\alpha$) tangent to the surface coordinate lines q^α, and constituting a natural *covariant* basis for the Euclidean *tangent vector space* there $\mathbf{T}_2(\mathbf{P})$, and the contravariant ones ($\mathbf{E}^\alpha$, $\mathbf{e}^\alpha$), defined as usual by $\mathbf{E}^\alpha \cdot \mathbf{E}_\beta = \mathbf{e}^\alpha \cdot \mathbf{e}_\beta = \delta^\alpha{}_\beta$, and constituting the corresponding *contravariant* basis for $\mathbf{T}_2(\mathbf{P})$]; so that the square of the actual (ds) and normalized (dS) elementary arc-lengths equal:

$$(dS)^2 \equiv 2T(dt)^2 = M_{\alpha\beta}\,dq^\alpha\,dq^\beta = m\,g_{\alpha\beta}\,dq^\alpha\,dq^\beta \equiv m(ds)^2, \tag{5.5.3c}$$

where $M_{\alpha\beta} \equiv \mathbf{E}_\alpha \cdot \mathbf{E}_\beta = m(\mathbf{e}_\alpha \cdot \mathbf{e}_\beta) \equiv m\,g_{\alpha\beta}$ are the associated covariant *surface metric tensors*; and the contravariant ones are defined, as usual, by $M_{\alpha\beta}\,M^{\beta\gamma} = \delta^\gamma{}_\alpha$, $g_{\alpha\beta}\,g^{\beta\gamma} = \delta^\gamma{}_\alpha$.*

5.5.2 Tensor Analysis on a Surface

The above show that the *algebra* of surface vectors and tensors (i.e., of vectors and tensors expressible as linear and homogeneous combinations of the $\mathbf{E}_\alpha$, $\mathbf{e}_\alpha$, etc. and therefore *lying wholly* on $\mathbf{T}_2(\mathbf{P})$, like the velocity) creates new surface vectors and tensors. However, as shown below, their differential calculus (analysis), does *not*: *the derivative of a surface vector, etc. is not a surface vector, but, in general, possesses a component normal to it.* (The reader is probably familiar from elementary mechanics of the fact that the time derivative of the velocity of a particle on a curve or surface has components both along the tangent to that curve/surface and normal to it.) To understand this, let $\mathbf{n}$ be the *unit* normal to S at P, defined so that $\mathbf{e}_1$, $\mathbf{e}_2$, and $\mathbf{n}$ build a *dextral* basis there:

---

* We notice that

$$|\mathbf{E}_1 \times \mathbf{E}_2|^2 = (\mathbf{E}_1)^2\,(\mathbf{E}_2)^2 - (\mathbf{E}_1 \cdot \mathbf{E}_2)^2 = M_{11}\,M_{22} - M_{12}{}^2 = \det(M_{\alpha\beta}) > 0; \tag{5.5.3d}$$

and the ($\Rightarrow$ positive definite) quantity

$$I \equiv d\mathbf{R} \cdot d\mathbf{R} = M_{\alpha\beta}\,dq^\alpha\,dq^\beta = m\,g_{\alpha\beta}\,dq^\alpha\,dq^\beta \equiv m\,(i) \tag{5.5.3e}$$

is called the *first fundamental* (*quadratic*) *surface form.* Further, and in analogy with results of Section 2.9, the *element of a surface area*, formed by $d\mathbf{R}_1 = \mathbf{E}_1\,dq^1$ and $d\mathbf{R}_2 = \mathbf{E}_2\,dq^2$ equals:

$$dA = |d\mathbf{R}_1 \times d\mathbf{R}_2| = |\mathbf{E}_1 \times \mathbf{E}_2|\,dq^1\,dq^2 = [\det(M_{\alpha\beta})]^{1/2}\,dq^1\,dq^2; \tag{5.5.3f}$$

while the angle ϕ between the coordinate lines q^1, q^2 is calculated from

$$\cos\phi = (d\mathbf{R}_1 \cdot d\mathbf{R}_2)\,/\,|d\mathbf{R}_1|\,|d\mathbf{R}_2| = M_{12}\,/\,(M_{11}\,M_{22})^{1/2}, \tag{5.5.3g}$$

a formula showing that if $M_{12} = 0$, the coordinate lines are orthogonal.

$$\mathbf{n} = (\mathbf{e}_1 \times \mathbf{e}_2) / |\mathbf{e}_1 \times \mathbf{e}_2| = (\mathbf{e}_1 \times \mathbf{e}_2) / [\det(g_{\alpha\beta})]^{1/2}, \quad \mathbf{e}_\alpha \cdot \mathbf{n} = 0; \quad (5.5.4a)$$

(recalling Equation 5.5.3d) and similarly for the normalized normal vector $\mathbf{N}$ (= $\mathbf{n}$). Now, $(\ldots)_{,\beta}$–differentiating Equation 5.5.4a$_2$ we obtain $\mathbf{e}_{\alpha,\beta} \cdot \mathbf{n} + \mathbf{e}_\alpha \cdot \mathbf{n}_{,\beta} = 0 \Rightarrow \mathbf{e}_{\alpha,\beta} \cdot \mathbf{n} = -\mathbf{e}_\alpha \cdot \mathbf{n}_{,\beta} \equiv b_{\alpha\beta}$ ($\neq 0$, in general); i.e., we can write the following basic $\mathbf{e}_{\alpha,\beta}$ representation:

$$\mathbf{e}_{\alpha,\beta} = A^\gamma{}_{\alpha\beta}\, \mathbf{e}_\gamma + b_{\alpha\beta}\, \mathbf{n}, \quad (5.5.4b)$$

from which, and the integrability condition for the basis $\{\mathbf{e}_\alpha\}$: $\mathbf{e}_{\alpha,\beta} = \mathbf{e}_{\beta,\alpha}$ (no torsion – recalling closing remarks in Section 3.11), we conclude that

$$A^\gamma{}_{\alpha\beta} = A^\gamma{}_{\beta\alpha} \equiv \mathbf{e}_{\alpha,\beta} \cdot \mathbf{e}^\gamma \quad \text{and} \quad b_{\alpha\beta} = b_{\beta\alpha} = \mathbf{e}_{\alpha,\beta} \cdot \mathbf{n}. \quad (5.5.4c)$$

To calculate the hitherto unknown coefficients $A^\gamma{}_{\alpha\beta}$ ($\rightarrow$ surface Christoffels – see below), we proceed as in Equations 3.8.14 ff: $(\ldots)_{,\gamma}$–differentiating $\mathbf{e}_\alpha \cdot \mathbf{e}_\beta \equiv g_{\alpha\beta}$ and invoking Equation 5.5.4b we find, successively,

$$\mathbf{e}_{\alpha,\gamma} \cdot \mathbf{e}_\beta + \mathbf{e}_\alpha \cdot \mathbf{e}_{\beta,\gamma} = g_{\alpha\beta,\gamma}$$

$$\Rightarrow (A^\delta{}_{\alpha\gamma}\, \mathbf{e}_\delta + b_{\alpha\gamma}\, \mathbf{n}) \cdot \mathbf{e}_\beta + \mathbf{e}_\alpha \cdot (A^\delta{}_{\beta\gamma}\, \mathbf{e}_\delta + b_{\beta\gamma}\, \mathbf{n}) = g_{\alpha\beta,\gamma}$$

$$\text{or} \quad A^\delta{}_{\alpha\gamma}\, g_{\delta\beta} + A^\delta{}_{\beta\gamma}\, g_{\delta\alpha} \equiv A_{\beta,\alpha\gamma} + A_{\alpha,\beta\gamma} = g_{\alpha\beta,\gamma} \quad (A_{\beta,\alpha\gamma} = A_{\beta,\gamma\alpha}). \quad (5.5.4d)$$

Repeating this cyclically for all α, β, γ, we obtain

$$A_{\gamma,\beta\alpha} + A_{\beta,\gamma\alpha} = g_{\beta\gamma,} \quad \text{and} \quad A_{\alpha,\gamma\beta} + A_{\gamma,\alpha\beta} = g_{\gamma\alpha,\beta}. \quad (5.5.4e)$$

From Equations 5.5.4d and e it readily follows that the (holonomic) *surface Christoffels* of the *first* kind ($A_{\gamma,\alpha\beta}$) and *second* kind ($A^\delta{}_{\alpha\beta}$) equal

$$A_{\gamma,\alpha\beta} \equiv \mathbf{e}_{\alpha,\beta} \cdot \mathbf{e}_\gamma = \mathbf{e}_{\beta,\alpha} \cdot \mathbf{e}_\gamma = (1/2)(g_{\gamma\beta,\alpha} + g_{\alpha\gamma,\beta} - g_{\alpha\beta,\gamma}) = A^\delta{}_{\alpha\beta}\, g_{\delta\gamma}, \quad (5.5.4f)$$

$$A^\delta{}_{\alpha\beta} \equiv A_{\gamma,\alpha\beta}\, g^{\delta\gamma} = (g^{\delta\gamma}/2)(g_{\gamma\beta,\alpha} + g_{\alpha\gamma,\beta} - g_{\alpha\beta,\gamma}). \quad (5.5.4g)$$

If $g_{\alpha\beta}$ = constant, then, clearly, the As vanish and (as discussed later) S admits a *global rectangular Cartesian* coordinate system. Finally, to express $\mathbf{e}^\alpha{}_{,\beta}$ à la Equation 5.5.4b, we proceed as follows: $(\ldots)_{,\beta}$–differentiating $\mathbf{e}^\alpha \cdot \mathbf{e}_\gamma = \delta^\alpha{}_\gamma$, we obtain

$$\mathbf{e}^\alpha{}_{,\beta} \cdot \mathbf{e}_\gamma + \mathbf{e}^\alpha \cdot \mathbf{e}_{\gamma,\beta} = 0$$

$$\Rightarrow \mathbf{e}^\alpha{}_{,\beta} \cdot \mathbf{e}_\gamma = -\mathbf{e}^\alpha \cdot \mathbf{e}_{\gamma,\beta} = -\mathbf{e}^\alpha \cdot (A^\varepsilon{}_{\gamma\beta}\, \mathbf{e}_\varepsilon + b_{\gamma\beta}\, \mathbf{n}) = \ldots = -A^\alpha{}_{\beta\gamma}, \quad (5.5.5a)$$

i.e.,

$$\mathbf{e}^{\alpha}{}_{,\beta} = -A^{\alpha}{}_{\beta\gamma}\,\mathbf{e}^{\gamma} + b^{\alpha}{}_{\beta}\,\mathbf{n}, \quad b^{\alpha}{}_{\beta} \equiv \mathbf{e}^{\alpha}{}_{,\beta}\cdot\mathbf{n}. \tag{5.5.5b}$$

The additional terms $b_{\alpha\beta}\,\mathbf{n}$ and $b^{\alpha}{}_{\beta}\,\mathbf{n}$ in Equations 5.5.4b and 5b, *in apparent disagreement with the corresponding formulae for* $\mathbf{e}_{k,l}$ *and* $\mathbf{e}^{k}{}_{,l}$ *of Sections 3.8ff,* arise from the fact that *our two-dimensional geometry is not intrinsic, but expressess the extrinsic properties of our two-dimensional space (surface) embedded in a three-dimensional space* (more on this in Section 5.6, and Problem 5.6.1). Let us examine the "out-of-S" coefficients/components $b_{\alpha\beta}$. From the above, we readily find

$$b_{\alpha\beta} = \mathbf{e}_{\alpha,\beta}\cdot\mathbf{n} = \mathbf{e}_{\beta,\alpha}\cdot\mathbf{n} = -\mathbf{e}_{\alpha}\cdot\mathbf{n}_{,\beta} = -\mathbf{e}_{\beta}\cdot\mathbf{n}_{,\alpha}$$

$$= \mathbf{e}_{\alpha,\beta}\cdot(\mathbf{e}_1\times\mathbf{e}_2)\,/\,|\mathbf{e}_1\times\mathbf{e}_2| = (\mathbf{e}_{\alpha,\beta}, \mathbf{e}_1, \mathbf{e}_2)\,/\,[\mathrm{Det}(g_{\alpha\beta})]^{1/2}. \tag{5.5.6}$$

The preceding results reveal that $\mathbf{n}_{,\alpha}$ is a *surface* vector. Indeed,

$$\mathbf{n}_{,\alpha} = -b_{\alpha\beta}\,\mathbf{e}^{\beta} = -b_{\alpha\beta}\,(g^{\beta\gamma}\,\mathbf{e}_{\gamma}) = -b^{\gamma}{}_{\alpha}\,\mathbf{e}_{\gamma}, \tag{5.5.7a}$$

where

$$b^{\beta}{}_{\alpha} = b_{\alpha\gamma}\,g^{\gamma\beta} = b^{\beta\gamma}\,g_{\gamma\alpha},$$

$$b^{\alpha\beta} = b^{\beta}{}_{\gamma}\,g^{\alpha\gamma} = b_{\gamma\delta}\,g^{\gamma\alpha}\,g^{\delta\beta}, \quad b_{\alpha\beta} = b^{\gamma}{}_{\alpha}\,g_{\gamma\beta} = b^{\gamma\delta}\,g_{\gamma\alpha}\,g_{\delta\beta}. \tag{5.5.7b}$$

Equations 5.5.4b and 5.5.7a are known as the *Gauss–Weingarten* (or structure) equations, and constitute a first-order partial differential system for the unknown basis vectors $\mathbf{e}_{\alpha}$ and $\mathbf{n}$, provided $g_{\alpha\beta}(q)$ and $b_{\alpha\beta}(q)$ are known. (If, further, the latter satisfy the so-called *Gauss* and *Mainardi–Codazzi* conditions, see Problem 5.5.1, below, then these six functions determine the surface to within a general rigid-body motion; just like the curvatures $\kappa_{(\nu)}$ and $\kappa_{(\beta)}$ do for a curve.)* The above show that

- The two fundamental quadratic surface forms, Equations 5.5.3e and 5.5.7c:

$$i \equiv d\mathbf{r}\cdot d\mathbf{r} = g_{\alpha\beta}\,dq^{\alpha}\,dq^{\beta} \quad \text{and} \quad ii \equiv d\mathbf{r}\cdot d\mathbf{n} = -b_{\alpha\beta}\,dq^{\alpha}\,dq^{\beta} \tag{5.5.7c}$$

  possess *invariant* geometrical significance, independent of the q's.
- The $b_{\alpha\beta}$ *cannot* be determined from the $g_{\alpha\beta}$ — they are tied to both $\mathbf{n}$ and the surface. Or, the relation between the surface $S_2 \equiv S$ and the surrounding Euclidean space E_3 is *not* completely determined by the first fundamental form i (I); it also requires the second such form ii (II); and similarly for a higher dimensional surface S_n inside an E_N ($N > n$).**

In view of the above, we can easily verify that *the derivative of a surface vector is not a surface vector; in general, it also has a component normal to the surface.* Indeed, differentiating the generic *surface* vector $\mathbf{V}$, i.e., $\mathbf{V}(q) = V^{\alpha}\,\mathbf{e}_{\alpha}$, we obtain, successively,

$$d\mathbf{V} = (\partial\mathbf{V}/\partial q^\alpha)dq^\alpha = dq^\alpha(V^\beta\,\mathbf{e}_\beta)_{,\alpha} = dq^\alpha(V^\beta{}_{,\alpha}\,\mathbf{e}_\beta + V^\beta\,\mathbf{e}_{\beta,\alpha})$$

$$= dq^\alpha[V^\beta{}_{,\alpha}\,\mathbf{e}_\beta + V^\beta(A^\gamma{}_{\beta\alpha}\,\mathbf{e}_\gamma + b_{\alpha\beta}\,\mathbf{n})] \quad \text{(by Equation 5.5.4b)}$$

$$= (V^\alpha{}_{,\beta} + A^\alpha{}_{\gamma\beta}\,V^\gamma)\mathbf{e}_\alpha\,dq^\beta + (V^\alpha\,b_{\beta\alpha}\,\mathbf{n})dq^\beta$$

$$\equiv (V^\alpha|_\beta\,\mathbf{e}_\alpha + V^\alpha\,b_{\alpha\beta}\,\mathbf{n})dq^\beta \equiv d\mathbf{V}_{(t)} + d\mathbf{V}_{(n)}, \tag{5.5.8}$$

where

---

* **Additional expressions/interpretations of $b_{\alpha\beta}$:**

i. Let P' be a point adjacent to the S-point P in the surrounding space E_3, a distance z apart from it along the local S-normal $\mathbf{n}$. Then, since $\mathbf{OP'} = \mathbf{OP} + z\,\mathbf{n}$,

$$\mathbf{e}'_\alpha \equiv \partial(\mathbf{OP'})/\partial q^\alpha = \partial(\mathbf{OP})/\partial q^\alpha + z(\partial\mathbf{n}/\partial q^\alpha) = \mathbf{e}_\alpha + z\,\mathbf{n}_{,\alpha}$$

$$\Rightarrow g'_{\alpha\beta} \equiv \mathbf{e}'_\alpha\cdot\mathbf{e}'_\beta = \ldots$$

$$= \mathbf{e}_\alpha\cdot\mathbf{e}_\beta + z(\mathbf{e}_\alpha\cdot\mathbf{n}_{,\beta} + \mathbf{e}_\beta\cdot\mathbf{n}_{,\alpha}) + z^2\,\mathbf{n}_{,\alpha}\cdot\mathbf{n}_{,\beta} = g_{\alpha\beta} - 2zb_{\alpha\beta} + z^2\,b_{\alpha\gamma}\,b^\gamma{}_\beta,$$

i.e., *$b_{\alpha\beta}$ and $b^\gamma{}_\beta$ are the increments of the metric $g_{\alpha\beta}$ when we pass from S to a parallel surface S′ a distance z apart.* Such formulae are useful in shell theory.

ii. Equation 5.5.7a can be rewritten, alternatively, as $(\partial\mathbf{r}/\partial q^\alpha)\cdot(\partial\mathbf{n}/\partial q^\beta) \equiv \mathbf{e}_\alpha\cdot\mathbf{n}_{,\beta} = -b_{\alpha\beta}$ or as $[(\partial\mathbf{r}/\partial q^\alpha)\,dq^\alpha]\cdot[(\partial\mathbf{n}/\partial q^\beta)\,dq^\beta] \equiv (\mathbf{e}_\alpha\cdot\mathbf{n}_{,\beta})\,dq^\alpha\,dq^\beta = -\,b_{\alpha\beta}\,dq^\alpha\,dq^\beta \equiv ii$, or finally (since $\mathbf{N} = \mathbf{n}$):

$$ii \equiv d\mathbf{r}\cdot d\mathbf{n} = -b_{\alpha\beta}dq^\alpha dq^\beta \quad \left[= \left(d\mathbf{R}/\sqrt{m}\right)d\mathbf{N} \equiv II/\sqrt{m}\right]: \tag{5.5.7c}$$

second fundamental (quadratic) *surface form* (an invariant $\Rightarrow b_{\alpha\beta}$ is a tensor)

iii. Last, $(\ldots)_{,\alpha}$-differentiating the $\mathbf{n}$-definition, while recalling the Gauss–Weingarten equations, we obtain [with $g \equiv \det(g_{\alpha\beta}) = |\mathbf{e}_1\times\mathbf{e}_2|^2$]:

$$\mathbf{n}_{,\alpha} = \left[(\mathbf{e}_1\times\mathbf{e}_2)/\sqrt{g}\right]_{,\alpha}$$

$$= \left[1/\sqrt{g}\right]_{,\alpha}(\mathbf{e}_1\times\mathbf{e}_2) + \left[1/\sqrt{g}\right](\mathbf{e}_{1,\alpha}\times\mathbf{e}_2 + \mathbf{e}_1\times\mathbf{e}_{2,\alpha}) = -b_{\alpha\beta}\mathbf{e}^\beta,$$

i.e.,

$$b_{\alpha\beta} = -|\mathbf{e}_1\times\mathbf{e}_2|^{-1}[(\mathbf{e}_{1,\alpha}\times\mathbf{e}_2 + \mathbf{e}_1\times\mathbf{e}_{2,\alpha})\cdot\mathbf{e}_\beta]. \tag{5.5.7d}$$

** There is, still yet, a *third* invariant surface form. From Equation 5.5.7a we find, successively,

$$\mathbf{n}_{,\alpha}\cdot\mathbf{n}_{,\beta} = (-b^\gamma{}_\alpha\,\mathbf{e}_\gamma)\cdot(-b^\delta{}_\beta\,\mathbf{e}_\delta) = b^\gamma{}_\alpha\,b^\delta{}_\beta(\mathbf{e}_\gamma\cdot\mathbf{e}_\delta) = b^\gamma{}_\alpha\,b^\delta{}_\beta(g_{\gamma\delta}) = b_{\alpha\delta}\,b^\delta{}_\beta \equiv c_{\alpha\beta},$$

or

$$iii \equiv d\mathbf{n}\cdot d\mathbf{n} = c_{\alpha\beta}\,dq^\alpha\,dq^\beta \quad [= d\mathbf{N}\cdot d\mathbf{N} \equiv III]: \tag{5.5.7f}$$

third fundamental (quadratic) surface form (an invariant $\Rightarrow c_{\alpha\beta}$ is a surface tensor).

$$d\mathbf{V}_{(t)} \equiv (V^{\alpha}|_{\beta}\, dq^{\beta})\mathbf{e}_{\alpha} \equiv DV^{\alpha}\, \mathbf{e}_{\alpha}\text{: } \textit{tangential} \text{ (to surface) part of } d\mathbf{V}, \tag{5.5.8a}$$

$$d\mathbf{V}_{(n)} \equiv (V^{\alpha}\, b_{\alpha\beta}\, dq^{\beta})\mathbf{n}\text{: } \textit{normal} \text{ (to surface) part of } d\mathbf{V}. \tag{5.5.8b}$$

5.5.3 Curve on a Surface

Next, let us consider a *curve* C on our surface S. Then, the unit tangent vector to C at a generic (C- *and* S-) point P, in the sense of increasing C-arc-lengths s, τ (= $\mathbf{t}$: corresponding unit vector, on tangent plane to S at P), is given, as in Section 5.3, by

$$\tau \equiv d\mathbf{r}/ds = (\partial\mathbf{r}/\partial q^{\alpha})(dq^{\alpha}/ds) = (dq^{\alpha}/ds)\mathbf{e}_{\alpha} \quad [\Rightarrow \mathbf{v} = (ds/dt)\tau]. \tag{5.5.9}$$

For a variation along a *surface curve* C: $q^{\alpha} = q^{\alpha}(s)$ (where s: C-arc length), we have $\mathbf{V} \to \mathbf{V}[q^{\alpha}(s)] = \mathbf{V}(s)$ and $dq^{\alpha} \to [\partial q^{\alpha}(s)/\partial s]ds$, and so Equations 5.5.8a and b reduce to

$$d\mathbf{V}_{(t)} \equiv [dV^{\alpha}/ds + A^{\alpha}{}_{\beta\gamma}\, V^{\beta}(dq^{\gamma}/ds)]\mathbf{e}_{\alpha}\, ds$$

$$= (dV^{\alpha} + A^{\alpha}{}_{\beta\gamma}\, V^{\beta}\, dq^{\gamma})\mathbf{e}_{\alpha} \equiv DV^{\alpha}\, \mathbf{e}_{\alpha}, \tag{5.5.10a}$$

$$d\mathbf{V}_{(n)} \equiv [V^{\alpha}\, b_{\alpha\beta}(dq^{\beta}/ds)]\mathbf{n}\, ds. \tag{5.5.10b}$$

We say that the surface vector $\mathbf{V}$ is *parallel transported* along an S-curve C, if $d\mathbf{V}_{(t)} = \mathbf{0} \to DV^{\alpha} = 0$; and we say that it is *constant* if $V^{\alpha}|_{\beta} = 0$.

Next, let $\mathbf{V} \to \tau$. Then $V^{\alpha} \to \tau^{\alpha} = dq^{\alpha}/ds$ and, therefore, Equations 5.5.8 through 8b specialize to

$$d\mathbf{V}_{(t)}/ds \to d\tau_{(t)}/ds =$$

$$[(d^2q^{\alpha}/ds^2) + A^{\alpha}{}_{\beta\gamma}(dq^{\beta}/ds)(dq^{\gamma}/ds)]\mathbf{e}_{\alpha} \equiv \mathbf{k}_{(g)}, \tag{5.5.11a}$$

$$d\mathbf{V}_{(n)}/ds \to d\tau_{(n)}/ds = [b_{\alpha\beta}(dq^{\alpha}/ds)(dq^{\beta}/ds)]\mathbf{n} \equiv \mathbf{k}_{(n)}; \tag{5.5.11b}$$

[$\Rightarrow d^2\mathbf{r}/dt^2 = d\tau/ds = k_{(n)}\,\mathbf{n} + k_{(g)}(\mathbf{n} \times \tau)$, i.e., $k_{(n)} = \mathbf{n} \cdot (d\tau/ds)$, $k_{(g)} = (\mathbf{n} \times \tau) \cdot (d\tau/ds)$] also, by the first of Frenet's equations:

$$d\tau/ds = \kappa_{(\nu)}\, \nu \equiv \kappa_{(\nu)} \tag{5.5.11c}$$

(In general, ν: normal to C at $P \neq \mathbf{n}$: normal to S at P.) From the above, and since $d\tau/ds = d\tau_{(t)}/ds + d\tau_{(n)}/ds$, we readily obtain

- $\kappa_{(\nu)}(\nu \cdot \mathbf{n}) = (d^2\mathbf{r}/ds^2) \cdot \mathbf{n} = \kappa_{(\nu)} \cos\theta$ (θ: angle between $\mathbf{n}$ and ν)

$$= b_{\alpha\beta}\,(dq^{\alpha}/ds)(dq^{\beta}/ds) = (b_{\alpha\beta}\, dq^{\alpha}\, dq^{\beta}) \,/\, (g_{\alpha\beta}\, dq^{\alpha}\, dq^{\beta})$$

$$\equiv -ii/i = -(d\mathbf{r}/ds) \cdot (d\mathbf{n}/ds) = k_{(n)} \quad [= \mathbf{k}_{(n)} \cdot \mathbf{n}]. \tag{5.5.12}$$

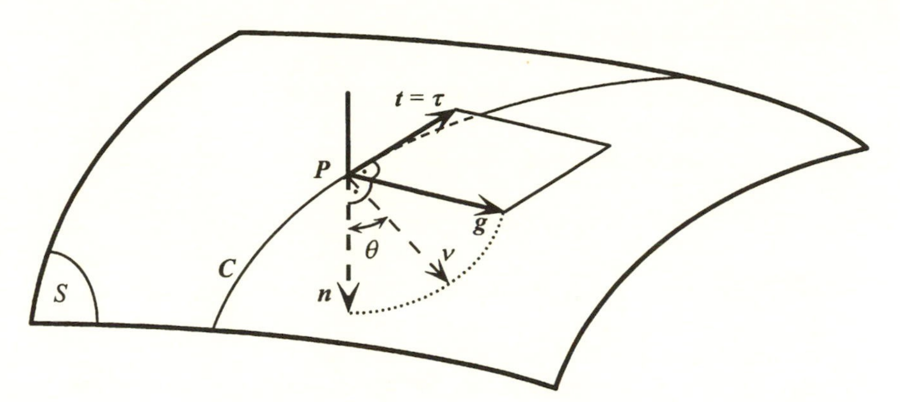

FIGURE 5.3 Geometry of curve C on a surface S. At P: $\mathbf{n} = \mathbf{N}$: unit normal to S, ν: unit (principal) normal to C, $\mathbf{t} = \tau = \mathbf{T}$: unit tangent to C, on tangent plane to S, $\mathbf{g} = \mathbf{G}$: unit normal to projection of C on tangent plane to S: $g = n \times \tau$.

The invariant $k_{(n)}$ is the *normal curvature* of C at P, that is, the first (or principal) curvature of the normal section of S through $\mathbf{n}$, there, C_n (i.e., the plane curve resulting by cutting S with the normal plane of $\tau = \mathbf{t}$ and $\mathbf{n}$, at P); and θ is the angle between the (generally) "oblique" surface curve C and the normal plane (angle between $\mathbf{n}$ and ν in Figure 5.3).*

- The surface vector

$$\mathbf{k}_{(g)} \equiv d\tau_{(t)}/ds = \kappa_{(\nu)}\, \nu - k_{(n)}\, \mathbf{n}$$

$$= k_{(g)}{}^{\alpha}\, \mathbf{e}_{\alpha} = k_{(g)}\, \mathbf{g} \equiv k_{(g)}(\mathbf{n} \times \tau), \tag{5.5.13a}$$

i.e.,

$$k_{(g)}{}^{\alpha} \equiv (d^2q^{\alpha}/ds^2) + A^{\alpha}{}_{\beta\gamma}(dq^{\beta}/ds)(dq^{\gamma}/ds), \tag{5.5.13b}$$

$$[\Rightarrow |\mathbf{k}_{(g)}| \equiv k_{(g)} = (g_{\alpha\beta}\, k_{(g)}{}^{\alpha}\, k_{(g)}{}^{\beta})^{1/2}, \quad \text{since } |\mathbf{g}| = 1] \tag{5.5.13c}$$

is the vector of *geodesic*, or *tangential*, *curvature* of C at P, i.e., *the (first) curvature of the projection of C onto the tangent plane to S there* (Liouville, c. 1850). Since $(\tau = \mathbf{t}, \mathbf{n}, \mathbf{g})$ constitute an orthonormal and dextral basis at P, we can say that Equations 5.5.11a through c represent the *decomposition of the principal curvature vector* ($\kappa_{(\nu)}$), *of the generic oblique S-curve C, into components along the tangent* ($\mathbf{k}_{(g)}$) *and normal planes* ($\mathbf{k}_{(n)}$) *to S:* $\kappa_{(\nu)} = \mathbf{k}_{(g)} + \mathbf{k}_{(n)}$.

* The relation $k_{(n)} = \kappa_{(\nu)} \cos\theta$, or in terms of the corresponding radii of curvature: $\rho_{(n)} \equiv 1/k_{(n)}$ and $\rho_{(\nu)} \equiv 1/\kappa_{(\nu)}$, i.e., $\rho_{(\nu)} = \rho_{(n)} \cos\theta$, constitutes *Meusnier's theorem*: *of all oblique surface curves through* τ, *the normal one has the largest radius of curvature,* $\rho_{(n)} = \max\{\rho_{(\nu)}\}$.

Definition: Surface curves along which $k_{(g)} = 0$ (i.e., no lateral, or transverse, curving of the curve) are called *geodesic lines* of S; they are, simultaneously, its *shortest* and *straightest* curves (recall Equation 3.7.2ff; also, see Example 5.5.1, below). Hence, along such lines, $\boldsymbol{\kappa}_{(v)} = \mathbf{k}_{(n)} \Rightarrow \kappa_{(v)} = \pm\, k_{(n)}$ (since $\boldsymbol{\nu} = \pm\mathbf{n}$); and as Equations 5.5.11a or 5.5.13b show, their equations are

$$k_{(g)} = 0: \quad (d^2q^\alpha/ds^2) + A^\alpha{}_{\beta\gamma}(dq^\beta/ds)(dq^\gamma/ds) = 0 \quad (\alpha, \beta, \gamma = 1, 2) \qquad (5.5.14)$$

These two second-order differential equations determine the geodesic surface coordinates $q^\alpha(s)$ uniquely for given initial: *point* $(q^\alpha)_{\text{initial}}$ and direction $(dq^\alpha/ds)_{\text{initial}}$, or for two neighboring points (i.e., with surface data only).

Finally, let us calculate the variation of $\mathbf{V} \cdot \boldsymbol{\tau}$ along an S-curve C. We have, successively,

$$d(\mathbf{V} \cdot \boldsymbol{\tau}) = d\mathbf{V} \cdot \boldsymbol{\tau} + \mathbf{V} \cdot d\boldsymbol{\tau}$$

$$= (d\mathbf{V}_{(t)} + d\mathbf{V}_{(n)}) \cdot \boldsymbol{\tau} + \mathbf{V} \cdot (d\boldsymbol{\tau}_{(t)} + d\boldsymbol{\tau}_{(n)}) = \ldots = d\mathbf{V}_{(t)} \cdot \boldsymbol{\tau} + \mathbf{V} \cdot d\boldsymbol{\tau}_{(t)}.$$

If, further, $d\mathbf{V}_{(t)} = \mathbf{0}$, then (invoking Equation 5.5.11a) $d(\mathbf{V} \cdot \boldsymbol{\tau}) = \mathbf{V} \cdot d\boldsymbol{\tau}_{(t)} = (\mathbf{V} \cdot \mathbf{k}_{(g)})ds$; hence, along a geodesic line, since $k_{(g)} = 0$, $d(\mathbf{V} \cdot \boldsymbol{\tau}) = 0 \Rightarrow \mathbf{V} \cdot \boldsymbol{\tau} = \text{constant}$. In sum, *if V has a constant length, then under such a transport it makes a constant angle with a geodesic line.* Let the reader verify this when S is the Euclidean plane E_2.

Example 5.5.1. Euler–Lagrange Form of Equations 5.5.11a through c

Due to $\boldsymbol{\tau} = \mathbf{e}_\alpha(dq^\alpha/ds) = \tau^\alpha\, \mathbf{e}_\alpha = \tau_\alpha\, \mathbf{e}^\alpha$, where

$$\tau_\alpha = g_{\alpha\beta}\, \tau^\beta = g_{\alpha\beta}(dq^\beta/ds) = \partial f/\partial(dq^\alpha/ds) = \partial f/\partial\tau^\beta, \qquad \text{(a)}$$

$$2f \equiv g_{\alpha\beta}(dq^\alpha/ds)(dq^\beta/ds) = g_{\alpha\beta}\, \tau^\alpha\, \tau^\beta = |\boldsymbol{\tau}| = 1, \qquad \text{(b)}$$

we can rewrite $d\boldsymbol{\tau}/ds$ as follows:

$$\boldsymbol{\kappa}_{(v)} = d\boldsymbol{\tau}/ds = d/ds(\tau_\alpha\, \mathbf{e}^\alpha) = (d\tau_\alpha/ds)\mathbf{e}^\alpha + \tau_\alpha\, \mathbf{e}^\alpha{}_{,\beta}(dq^\beta/ds)$$

$$= (d\tau_\alpha/ds)\mathbf{e}^\alpha + \tau_\alpha(-A^\alpha{}_{\beta\gamma}\, \mathbf{e}^\gamma + b^\alpha{}_\beta\, \mathbf{n})(dq^\beta/ds)$$

$$= [(d\tau_\alpha/ds) - A^\gamma{}_{\alpha\beta}(dq^\beta/ds)\tau_\gamma]\mathbf{e}^\alpha + [\tau_\alpha\, b^\alpha{}_\beta(dq^\beta/ds)]\mathbf{n}$$

$$= \{d/ds[\partial f/\partial(dq^\alpha/ds)] - (dq^\beta/ds)[g_{\gamma\delta}\,(dq^\delta/ds)]A^\gamma{}_{\beta\alpha}\}\mathbf{e}^\alpha$$

$$+ \{(dq^\beta/ds)[g_{\alpha\gamma}\,(dq^\gamma/ds)b^\alpha{}_\beta]\}\mathbf{n}, \qquad \text{(c)}$$

or, since $g_{\gamma\delta}\, A^\gamma{}_{\beta\alpha} = A_{\delta,\alpha\beta}$ and

$$A_{\delta,\alpha\beta}(dq^{\beta}/ds)(dq^{\delta}/ds) = (1/2)(g_{\beta\delta,\alpha} + g_{\alpha\delta,\beta} - g_{\alpha\beta,\delta})(dq^{\beta}/ds)(dq^{\delta}/ds)$$

$$= (1/2)(g_{\beta\delta,\alpha})(dq^{\beta}/ds)(dq^{\delta}/ds) = \partial f/\partial q^{\alpha},$$

[since $g_{\alpha\delta,\beta}(dq^{\beta}/ds)(dq^{\delta}/ds) = g_{\alpha\beta,\delta}(dq^{\beta}/ds)(dq^{\delta}/ds)$] finally,

$$\kappa_{(v)}\, \mathrm{v} = \{d/ds[\partial f/\partial(dq^{\alpha}/ds)] - (\partial f/\partial q^{\alpha})\}\mathbf{e}^{\alpha} + [b_{\alpha\beta}(dq^{\alpha}/ds)(dq^{\beta}/ds)]\mathbf{n}$$

$$\equiv E_{\alpha}(f)\mathbf{e}^{\alpha} + [b_{\alpha\beta}(dq^{\alpha}/ds)(dq^{\beta}/ds)]\mathbf{n}. \qquad \text{(d)}$$

Thus, the geodesic Equation 5.5.14 have the Euler–Lagrange form $E_{\alpha}(f) = 0$.

Example 5.5.2 Extension of Equations 5.5.8 through 5.5.8b to General Nonsurface Vectors

Let

$$\mathbf{V} = V^{\alpha}\, \mathbf{e}_{\alpha} + V^{n}\, \mathbf{n} = V_{\alpha}\, \mathbf{e}^{\alpha} + V_{n}\, \mathbf{n} \quad \text{(i.e., } V^{n} = V_{n}\text{)}. \qquad \text{(a)}$$

Then,

i. $$\partial \mathbf{V}/\partial q^{\alpha} \equiv \mathbf{V}_{,\alpha} = \ldots = (V_{\beta}|_{\alpha} - b_{\alpha\beta}\, V_{n})\mathbf{e}^{\beta} + (V_{n,\alpha} + b^{\beta}{}_{\alpha}\, V_{\beta})\mathbf{n}$$

$$= (V_{\beta}|_{\alpha} - b_{\alpha\beta}\, V^{n})\mathbf{e}^{\beta} + (V^{n}{}_{,\alpha} + b_{\beta\alpha}\, V^{\beta})\mathbf{n}; \qquad \text{(b)}$$

ii. $$\partial \mathbf{V}/\partial n \equiv \mathbf{V}_{,n} = \ldots = V_{\alpha,n}\, \mathbf{e}^{\alpha} + V_{n,n}\, \mathbf{n} = V^{\alpha}{}_{,n}\, \mathbf{e}_{\alpha} + V^{n}{}_{,n}\, \mathbf{n}, \qquad \text{(c)}$$

where $$V_{\beta}|_{\alpha} \equiv V_{\beta,\alpha} - A^{\gamma}{}_{\beta\alpha}\, V_{\gamma}, \quad V^{\beta}|_{\alpha} \equiv V^{\beta}{}_{,\alpha} + A^{\beta}{}_{\alpha\gamma}\, V^{\gamma}. \qquad \text{(d)}$$

Problem 5.5.1

Consider the equations of Gauss (5.5.4b) and Weingarten (5.5.7a):

$$\mathbf{e}_{\alpha,\beta} = \mathbf{e}_{\beta,\alpha} = A^{\gamma}{}_{\alpha\beta}\, \mathbf{e}_{\gamma} + b_{\alpha\beta}\, \mathbf{n}, \quad \mathbf{n}_{,\alpha} = -b_{\alpha\beta}\, \mathbf{e}^{\beta} = -b^{\beta}{}_{\alpha}\, \mathbf{e}_{\beta}. \qquad \text{(a)}$$

i. Show that

$$(\mathbf{e}_{\alpha,\beta})_{,\gamma} - (\mathbf{e}_{\alpha,\gamma})_{,\beta} = [R^{\delta}{}_{\alpha\gamma\beta} - g^{\lambda\delta}(b_{\alpha\beta}\, b_{\gamma\lambda} - b_{\alpha\gamma}\, b_{\beta\lambda})]\mathbf{e}_{\delta}$$

$$+ [(A^{\delta}{}_{\alpha\gamma}\, b_{\delta\beta} - A^{\delta}{}_{\alpha\beta}\, b_{\gamma\delta}) - (b_{\alpha\beta,\gamma} - b_{\alpha\gamma,\beta})]\mathbf{n}, \qquad \text{(b)}$$

where

$$R^{\delta}{}_{\alpha\gamma\beta} \equiv A^{\delta}{}_{\alpha\beta,\gamma} - A^{\delta}{}_{\alpha\gamma,\beta} + A^{\varepsilon}{}_{\alpha\beta}\, A^{\delta}{}_{\varepsilon\gamma} - A^{\varepsilon}{}_{\alpha\gamma}\, A^{\delta}{}_{\varepsilon\beta}: \qquad \text{(b1)}$$

Riemann-Christoffel tensor of R_n (R–C); here $R_2 \equiv S$ (recall Equation 3.10.1b) and, therefore, the *integrability conditions* of Equation a_1, $(\mathbf{e}_{\alpha,\beta})_{,\gamma} = (\mathbf{e}_{\alpha,\gamma})_{,\beta,}$ lead to

$$R^{\delta}{}_{\alpha\gamma\beta} = g^{\lambda\delta} R_{\lambda\alpha\gamma\beta} = g^{\lambda\delta}(b_{\alpha\beta}\, b_{\gamma\lambda} - b_{\alpha\gamma}\, b_{\beta\lambda}) \quad (\textit{Gauss}\ \text{equations}) \qquad \text{(c1)}$$

$$b_{\alpha\beta,\gamma} - b_{\alpha\gamma,\beta} = A^{\delta}{}_{\alpha\gamma}\, b_{\delta\beta} - A^{\delta}{}_{\alpha\beta}\, b_{\gamma\delta} \quad (\textit{Mainardi–Codazzi}\ \text{equations}) \qquad \text{(c2)}$$

or, compactly (since $A^{\delta}{}_{\alpha\gamma} = A^{\delta}{}_{\gamma\alpha}$) $b_{\alpha\beta}|_{\gamma} = b_{\alpha\gamma}|_{\beta}$; i.e., if $b_{\alpha\beta} = 0$, then $R^{\delta}{}_{\alpha\gamma\beta} \equiv [(\mathbf{e}_{\alpha,\beta})_{,\gamma} - (\mathbf{e}_{\alpha,\gamma})_{,\beta}] \cdot \mathbf{e}^{\delta}$ (definition of *R–C* tensor).

The partial differential equations (c1 and 2) are very important in differential geometry: they relate the coefficients of the two fundamental surface forms, $g_{\alpha\beta}$ and $b_{\alpha\beta}$.

ii. Verify that (in ordinary surfaces): (a) There is only *one* independent Gauss equation, and only *two* independent Mainardi–Codazzi equations ($b_{\alpha\alpha}|_{\beta} = b_{\alpha\beta}|_{\alpha}$, $\alpha \neq \beta$, no sum on α); and (b) the integrability conditions of Equation a_2, $(\mathbf{n}_{,\beta})_{,\alpha} = (\mathbf{n}_{,\alpha})_{,\beta}$, do not produce any additional independent relations. (For further results, see e.g., Kreyszig, 1959, pp. 142ff.)

Remark: As pointed out earlier (following Equations 5.5.5b ff), Equations c1through c3 express extrinsic properties of a Riemannian R_2 embedded in a higher-dimensional space. If the latter is a Euclidean E_3, then *its* Riemann–Christoffel tensor (Sections 3.10 and 3.11) vanishes: $R^{l}{}_{ksr} = 0$ ($l, k, s, r = 1, 2, 3$); and this yields the R_2-equations (c1 and 2) *and* some additional conditions for E_3. (Conversely, it can be shown that if the latter and Equations c1 and c2 hold, then $R^{l}{}_{ksr} = 0$.)

5.5.4 Acceleration

Now, resuming the discussion of kinematics from Equations 5.5.2a and b, the acceleration of our particle P on S is

$$\mathbf{a} \equiv d\mathbf{v}/dt = d(v^{\alpha}\, \mathbf{e}_{\alpha})/dt = d/dt[(dq^{\alpha}/dt)\mathbf{e}_{\alpha}]$$

$$= (d^2q^{\alpha}/dt^2)\mathbf{e}_{\alpha} + (dq^{\alpha}/dt)[\mathbf{e}_{\alpha,\beta}(dq^{\beta}/dt)]$$

$$= (d^2q^{\alpha}/dt^2)\mathbf{e}_{\alpha} + (A^{\gamma}{}_{\alpha\beta}\, \mathbf{e}_{\gamma} + b_{\alpha\beta}\, \mathbf{n})(dq^{\alpha}/dt)\,(dq^{\beta}/dt)$$

(invoking Equations 5.5.4b ff)

$$= [d^2q^{\gamma}/dt^2 + A^{\gamma}{}_{\alpha\beta}(dq^{\alpha}/dt)(dq^{\beta}/dt)]\mathbf{e}_{\gamma} + [b_{\alpha\beta}(dq^{\alpha}/dt)(dq^{\beta}/dt)]\mathbf{n}$$

$$= a^{\gamma}\, \mathbf{e}_{\gamma} + a^{n}\, \mathbf{n} = a_{\gamma}\, \mathbf{e}^{\gamma} + a_{n}\, \mathbf{n} \equiv \mathbf{a}_{S\ (\text{surface lying})} + \mathbf{a}_{n\ (\text{normal to } S)}; \qquad (5.5.15)$$

i.e.,

$$a^{\gamma} = d^2q^{\gamma}/dt^2 + A^{\gamma}{}_{\alpha\beta}(dq^{\alpha}/dt)(dq^{\beta}/dt)$$

$$= dv^{\gamma}/dt + A^{\gamma}{}_{\alpha\beta}\, v^{\alpha}\, v^{\beta} \quad (\neq dv^{\gamma}/dt). \tag{5.5.15a}$$

Similarly, the *normalized* acceleration of P (on correspondingly normalized S and C), $\mathbf{A}$, is

$$\mathbf{A} \equiv d\mathbf{V}/dt = \sqrt{m}\,\mathbf{a} = \ldots = \mathbf{A}_S + \mathbf{A}_N, \tag{5.5.16}$$

where

$$\mathbf{A}_S \equiv [d^2q^{\gamma}/dt^2 + \Gamma^{\gamma}{}_{\alpha\beta}(dq^{\alpha}/dt)(dq^{\beta}/dt)\mathbf{E}_{\gamma}$$

$$\equiv (dV^{\gamma}/dt + \Gamma^{\gamma}{}_{\alpha\beta}\, V^{\alpha}\, V^{\beta})\mathbf{E}_{\gamma} \equiv A^{\gamma}\,\mathbf{E}_{\gamma}, \tag{5.5.16a}$$

$$\mathbf{A}_N \equiv [B_{\alpha\beta}(dq^{\alpha}/dt)(dq^{\beta}/dt)]\mathbf{N} \equiv \mathrm{A}_N\,\mathbf{N} \equiv A^N\,\mathbf{N} \quad (\mathbf{N} = \mathbf{n}); \tag{5.5.16b}$$

$\Gamma^{\gamma}{}_{\alpha\beta}$: surface Christoffels, like the $A^{\gamma}{}_{\alpha\beta}$,
but based on $M_{\alpha\beta}$, $M^{\alpha\beta}$ (recall Equation 5.5.3c) (5.5.16c)

$$B_{\alpha\beta} = B_{\beta\alpha} \equiv \mathbf{E}_{\alpha,\beta} \cdot \mathbf{N} = \left(\sqrt{m}\,\mathbf{e}_{\alpha,\beta}\right) \cdot \mathbf{n} = \sqrt{m}\, b_{\beta\alpha}. \tag{5.5.16d}$$

It is not hard to verify *Lagrange's* famous *kinematico–inertial identity* (recall Equation 5.2.3e):

$$A_{\delta} = M_{\delta\gamma}\, A^{\gamma} = M_{\delta\gamma}[d^2q^{\gamma}/dt^2 + \Gamma^{\gamma}{}_{\alpha\beta}(dq^{\alpha}/dt)(dq^{\beta}/dt)]$$

$$= \ldots = d/dt(\partial T/\partial V^{\delta}) - \partial T/\partial q^{\delta} \equiv E_{\delta}(T): \tag{5.5.17a}$$

covariant surface components of $\mathbf{A}$; also

$$\mathbf{A}_N \equiv \{B_{\alpha\beta}[(dq^{\alpha}/dS)(dS/dt)]\,[(dq^{\beta}/dS)(dS/dt)]\}\,\mathbf{N} \equiv [(dS/dt)^2 K_{(N)}]\,\mathbf{N}$$

$$K_{(N)} \equiv B_{\alpha\beta}(dq^{\alpha}/dS)(dq^{\beta}/dS) = (B_{\alpha\beta}\, dq^{\alpha} dq^{\beta}) \,/\, (M_{\gamma\delta}\, dq^{\gamma}\, dq^{\delta}) \equiv II/\, I:$$

Normalized normal curvature of S at P

$$\left(= \sqrt{m}\, b_{\alpha\beta}\, dq^{\alpha}\, dq^{\beta} / \sqrt{m}\sqrt{m}(ds)^2 = k_{(n)}/\sqrt{m}\right). \tag{5.5.17b}$$

Next, along an S-curve C, through P, $\mathbf{A}_T$ transforms further as follows [recall that $(dS/dt)^2 = 2T$]:

$$\mathbf{A}_S = A^{\gamma}\,\mathbf{E}_{\gamma} = [(d^2S/dt^2)(dq^{\gamma}/dS) + (dS/dt)^2 K_{(G)}{}^{\gamma}]\mathbf{E}_{\gamma}$$

(recalling Equations 5.3.22f ff)

$$= (d^2S/dt^2)\mathbf{T} + [(dS/dt)^2 K_{(G)}]\mathbf{G} \equiv \mathbf{A}_{S,T} + \mathbf{A}_{S,G} \tag{5.5.18a}$$

[$\mathbf{T} = \mathbf{t}$, $\mathbf{G} = \mathbf{g}$: normalized "counterparts" of $\mathbf{t}$, $\mathbf{g} = \mathbf{n} \times \boldsymbol{\tau}$ (Figure 5.3)]
where

$$K_{(G)}{}^{\gamma} = d^2q^{\gamma}/dS^2 + \Gamma^{\gamma}{}_{\alpha\beta}(dq^{\alpha}/dS)(dq^{\beta}/dS): \tag{5.5.18b}$$

Contravariant components of normalized geodesic curvature, of C at P,

$$\mathbf{K}_G = K_{(G)}{}^{\gamma}\,\mathbf{E}_{\gamma} = K_{(G)}\,\mathbf{G}$$

$$[\Rightarrow K_{(G)} = (M_{\alpha\beta}\,K_{(G)}{}^{\alpha}\,K_{(G)}{}^{\beta})^{1/2}\text{: normalized geodesic curvature}] \tag{5.5.18c}$$

In sum, here we can write:

$$\mathbf{A} = \mathbf{A}_{N\ (\text{normal to } S)} + \mathbf{A}_{S,T\ (\text{tangent to } S \text{ and } C)}$$

$$+ \mathbf{A}_{S,G\ (\text{tangent to } S \text{ and normal to projection of } C \text{ on tangent plane to } S)}$$

$$= \mathbf{A}_{\nu\ (\text{principal normal to } C)} + \mathbf{A}_{S,T\ (\text{tangent to } S \text{ and } C)}. \tag{5.5.19}$$

5.5.5 Forces, Equations of Motion

Recalling Section 5.4, we can write the Newton–Euler law for P as

$$m\,\mathbf{a} = \mathbf{F}_{(\text{total } impressed \text{ force on } P)} + \mathbf{Z}_{(\text{total } constraint\ reaction \text{ on } P)} \quad [\equiv \mathbf{f}_{(\text{total force on } P)}] \tag{5.5.20}$$

or, in normalized form:

$$\mathbf{A} = \mathbf{Q} + \boldsymbol{\Lambda}, \tag{5.5.21}$$

where

$$\mathbf{A} \equiv \sqrt{m}\,\mathbf{a}, \quad \mathbf{Q} \equiv \mathbf{F}/\sqrt{m},$$

$$\boldsymbol{\Lambda} \equiv \mathbf{Z}/\sqrt{m} \tag{5.5.21a}$$

(to avoid confusion with the normalized position vector $\mathbf{R}$). In general, both $\mathbf{Q}$ ($\mathbf{F}$) and Λ ($\mathbf{Z}$) have both surface and normal components. If Λ ($\mathbf{Z}$) has no surface component (a *physical,* or *constitutive,* postulate) then it is called *ideal.* Now, projecting Equation 5.5.21 onto the *tangent* plane to S at P, we get

$$\mathbf{A} \cdot \mathbf{E}_\alpha = \mathbf{Q} \cdot \mathbf{E}_\alpha + \Lambda \cdot \mathbf{E}_\alpha. \tag{5.5.22}$$

For ideal surfaces: $\Lambda = 0\, \mathbf{E}_\alpha + \Lambda^N \mathbf{N} \equiv \Lambda\, \mathbf{N} \Rightarrow \Lambda \cdot \mathbf{E}_\alpha = 0$. Then, Equation 5.5.22 reduces to

$$\mathbf{A} \cdot \mathbf{E}_\alpha = \mathbf{Q} \cdot \mathbf{E}_\alpha. \tag{5.5.23}$$

The reactionless Equations 5.5.23 we shall term *kinetic Maggi equations.** Let us bring them to Lagrangean form. We have, successively,

$$2T = \mathbf{V} \cdot \mathbf{V} = M_{\alpha\beta}(dq^\alpha/dt)(dq^\beta/dt) \equiv M_{\alpha\beta}\, V^\alpha\, V^\beta \quad [= (dS/dt)^2]$$

$$\Rightarrow \partial T/\partial V^\alpha = \mathbf{V} \cdot (\partial \mathbf{V}/\partial V^\alpha) = \mathbf{V} \cdot (\partial \mathbf{R}/\partial q^\alpha) = \mathbf{V} \cdot \mathbf{E}_\alpha, \tag{5.5.24a}$$

$$\Rightarrow d/dt(\partial T/\partial V^\alpha) = \mathbf{A} \cdot \mathbf{E}_\alpha + \mathbf{V} \cdot [d/dt(\partial \mathbf{R}/\partial q^\alpha)]$$

$$= \mathbf{A} \cdot \mathbf{E}_\alpha + \mathbf{V} \cdot \partial/\partial q^\alpha(d\mathbf{R}/dt) = \mathbf{A} \cdot \mathbf{E}_\alpha + \mathbf{V} \cdot (\partial \mathbf{V}/\partial q^\alpha), \tag{5.5.24b}$$

i.e.,

$$d/dt(\partial T/\partial V^\alpha) = \partial T/\partial q^\alpha + \mathbf{A} \cdot \mathbf{E}_\alpha \Rightarrow E_\alpha(T) = \mathbf{A} \cdot \mathbf{E}_\alpha \equiv A_\alpha; \tag{5.5.24c)**$$

and with

$$\mathbf{Q} \cdot \mathbf{E}_\alpha + \Lambda \cdot \mathbf{E}_\alpha = Q_\alpha + \Lambda_\alpha = Q_\alpha, \tag{5.5.25}$$

Equation 5.5.23 finally becomes

$$A_\alpha = E_\alpha(T) = Q_\alpha, \tag{5.5.26}$$

and constitutes a system of two (coupled and generally nonlinear) ordinary differential equations for $q^\alpha(t)$, to be solved in conjunction with four *initial* conditions: $(q^\alpha)_{\text{initial}}$, $(dq^\alpha/dt)_{\text{initial}}$: given. The contravariant form of Equation 5.5.26 is found simply by dot multiplication of it with $M^{\beta\alpha}$: $A^\alpha = Q^\alpha$. Next, projecting Equation 5.5.21 along the *normal* to S at P we get the *kinetostatic Maggi equations:*

$$\mathbf{A} \cdot \mathbf{N} = \mathbf{Q} \cdot \mathbf{N} + \Lambda \cdot \mathbf{N} \quad \text{or} \quad A^N = Q^N + \Lambda^N; \tag{5.5.27}$$

---

* Throughout this book, by kinetic we shall mean *equations not containing any constraint reactions*. The solution of the system consisting of such equations, and the equations of constraint yields the *motion* of the problem. The reactions can then be found easily (more in Section 7.3 ff).

** In sum, for holonomic coordinates

$$q^\alpha\!: \ \partial \mathbf{R}/\partial q^\alpha = \partial \mathbf{V}/\partial V^\alpha = \ldots = \mathbf{E}_\alpha,$$

$$d/dt(\partial \mathbf{R}/\partial q^\alpha) = \partial/\partial q^\alpha(d\mathbf{R}/dt) \Rightarrow E_\alpha(\mathbf{V}) \equiv d/dt(\partial \mathbf{V}/\partial V^\alpha) - \partial \mathbf{V}/\partial q^\alpha = \mathbf{0}. \tag{5.5.24d}$$

or, since by Equations 5.5.16b and 5.5.17b: $A^N = B_{\alpha\beta}(dq^\alpha/dt)(dq^\beta/dt) = (dS/dt)^2 K_{(N)}$, finally,

$$(dS/dt)^2 = (Q^N + \Lambda^N) / K_{(N)} = [B_{\alpha\beta}(dq^\alpha/dt)(dq^\beta/dt)]/K_{(N)}. \quad (5.5.28)$$

Once the motion has been found, from Equation 5.5.26, the above yields Λ^N:

$$\Lambda^N = A^N - Q^N = B_{\alpha\beta}(dq^\alpha/dt)(dq^\beta/dt) - Q^N(t, q, dq/dt) \quad (5.5.29)$$

(i.e., function of time and initial conditions)

The above show that if $Q_\alpha = 0$, then $A^\alpha, A_\alpha = 0$; i.e., recalling Equations 5.5.18a,

$$(d^2S/dt^2)(dq^\alpha/dS) + (dS/dt)^2 K_{(G)}{}^\alpha = 0, \quad (5.5.30a)$$

or

$$(dT/dS)(dq^\alpha/dS) + (2T)K_{(G)}{}^\alpha = 0; \quad (5.5.30b)$$

from which it follows that if, further, $dT/dS = 0$, then $|\mathbf{V}|$ = constant and $K_{(G)} = 0$ (geodesic path).

Example 5.5.3. Particle P on an Ideal Surface

Let the constraint be, in rectangular Cartesian coordinates y^k ($k = 1, 2, 3$), $\phi(y^k) = 0$. It is not hard to see that the equations of motion are

$$m(d^2y^k/dt^2) = F^k + \Lambda^k, \quad F^k = F^k(t, y, dy/dt) \quad (a)$$

where, assuming ideal constraints, the constraint reaction Λ^k equals

$$\Lambda^k = \lambda(\partial\phi/\partial y^k) \equiv \lambda\ \phi^k, \quad \text{or, vectorially, } \Lambda = \lambda(\partial\phi/\partial\mathbf{r}) \equiv \lambda \text{ grad } \phi. \quad (b)$$

and λ is a Lagrangean multiplier (more in Chapter 7). To eliminate and/or calculate λ, we proceed as follows:

i. We differentiate $\phi = 0$ totally in time twice (to create accelerations d^2y^k/dt^2):

$$d\phi/dt = 0: (\partial\phi/\partial y^k)(dy^k/dt) \equiv \phi_k(dy^k/dt) = 0 \quad (\text{i.e., } \partial\phi/\partial y^k \equiv \phi_{,k} \equiv \phi_k) \quad (c1)$$

$$d^2\phi/dt^2 = 0: \phi_k(d^2y^k/dt^2) + \phi_{kl}(dy^k/dt)(dy^l/dt) = 0,$$

$$\Rightarrow \phi_k(d^2y^k/dt^2) = -\phi_{kl}(dy^k/dt)(dy^l/dt); \quad (c2)$$

then

ii. Dot Equation a, under Equation b, with ϕ_k:

$$m(d^2y^k/dt^2)\phi_k = F^k\,\phi_k + \lambda\,\phi^k\,\phi_k; \tag{d}$$

and then,

iii. Combining Equation c2 with Equation d and solving for λ, we, finally, obtain

$$\lambda = -[m\,\phi_{kl}(dy^k/dt)(dy^l/dt) + F^k\,\phi_k] \,/\, \phi_k\,\phi^k. \tag{e}$$

Lastly,

iv. Substituting this result into Equation a, with Equation b, we get the purely kinetic *Jacobi–Synge* equations (see also Example 7.4.3, for the general case):

$$m(d^2y^h/dt^2) = F^h - \{[m\,\phi_{kl}\,(dy^k/dt)(dy^l/dt) + F^k\,\phi_k] \,/\phi_k\,\phi^k\}\,\phi^h \tag{f}$$

(known function of t, y, dy/dt) which are to be solved in conjunction with the constraint-satisfying initial conditions:

$$\phi(y^k{}_{\text{initial}}) = 0, \quad \phi_k(y^k{}_{\text{initial}})(dy^k/dt)_{\text{initial}} = 0. \tag{g}$$

Particle on a Rough Surface. In this case the tangential component of the surface reaction $\mathbf{T}$ (commonly known as sliding solid vs. solid, or *dry*, friction, of coefficient μ) is given by

$$\mathbf{T} = T^k\,\mathbf{e}_k = -\mu|\mathbf{N}|[\mathbf{v}/|\mathbf{v}|] \equiv -\mu|\mathbf{N}|\mathbf{u}_v = -\mu|\mathbf{N}|[(v^k\,\mathbf{e}_k)/(g_{hs}\,v^h\,v^s)^{1/2}], \tag{h}$$

$\mathbf{N}$: total *normal* force on P, $\mathbf{v} = (v^k \equiv dq^k/dt)$: velocity of P relative to S, $\mathbf{u}_v$: unit vector along $\mathbf{v}$.

5.6 GENERAL n-DIMENSIONAL (RIEMANNIAN) SURFACES

Here, continuing from Sections 2.12 and 3.8, we generalize some of the ideas and equations of the preceding sections to n-dimensional surfaces ($n = 3, 4, \ldots < \infty$.). Also, we examine the problem of the possibility of *embedding* such general curved (Riemannian) spaces into a surrounding flat (Euclidean) space of higher dimension, and the conditions for a general space to be flat. Here, too, only the dynamically relevant parts of Riemannian geometry are treated.

5.6.1 Riemannian Space (R_n) Inside a Euclidean Space (E_N: $n < N < \infty$.)

Let us consider a Euclidean space E_N covered by the curvilinear coordinate system $x \equiv (x^k$: $n = 1, 2, \ldots, N)$. In there we can define all usual tensor operations, e.g., calculate Christoffels, covariant derivatives, etc. The square of the *element of distance (arc-length)* between two generic adjacent E_N-points, $P(x^k)$ and $P'(x^k + dx^k)$,

$$ds^2 = g_{kl}(x)\, dx^k\, dx^l \quad \text{(positive definite)}, \tag{5.6.1}$$

can always be expressed as the *sum of squares* of coordinate differentials:

$$ds^2 = (dy^1)^2 + (dy^2)^2 + \ldots + (dy^N)^2$$

[i.e., $g_{kl} \to$ constant (rectilinear affine coordinates, say, z^k)

$\to$ diagonal (rectangular Cartesian coordinates y^k): δ_{kl}]; (5.6.1a)

and the position vector of a typical E_N-point P, $\mathbf{R}$, relative to some fixed origin in that space O, can be expressed as: $\mathbf{R} = \mathbf{R}(y) = \mathbf{R}[y(x)] = \mathbf{R}(x)$. Now, the Riemannian space R_n, inside our E_N, is defined by the N scalar "parametric equations," in terms of the n curvilinear coordinates, or *positional parameters*, $q \equiv (q^\alpha$: $\alpha = 1, 2, \ldots, n)$:

$$y^k = y^k(q^1, \ldots, q^n) \equiv y^k(q) \quad \text{or} \quad x^k = x^k(q^1, \ldots, q^n) \equiv x^k(q). \qquad (5.6.2)^*$$

Thus, an ordinary surface S_2, or S, is an R_2 inside our ordinary space E_3. Since the q are *not* the components of a vector, there is no finite position vector lying wholly in R_n; i.e., $\mathbf{R}$ is an E_N-vector. However, the *infinitesimal displacement vector* $\mathbf{PP'}$

$$\mathbf{PP'} \equiv d\mathbf{P} \to d\mathbf{r} = \mathbf{e}_\alpha\, dq^\alpha, \tag{5.6.3}$$

where $\mathbf{R} = \mathbf{R}[x(q)] \equiv \mathbf{r}(q) \Rightarrow d\mathbf{R} = d\mathbf{r}$ and $\mathbf{e}_\alpha \equiv \partial\mathbf{P}/\partial q^\alpha \equiv \partial\mathbf{r}/\partial q^\alpha$: natural basis at $P(q)$, is an R_n-vector.** Then the square of the elementary arc–length, between the generic neighboring R_n-points $P(q^\alpha)$ and $P'(q^\alpha + dq^\alpha)$, equals

$$ds^2 = (dy^1)^2 + \ldots + (dy^N)^2 = [(\partial y^1/\partial q^\alpha)dq^\alpha]^2 + \ldots + [(\partial y^N/\partial q^\alpha)dq^\alpha]^2$$

$$= [(\partial\mathbf{r}/\partial q^\alpha)dq^\alpha] \cdot [(\partial\mathbf{r}/\partial q^\beta)dq^\beta] \equiv g_{\alpha\beta}(q)\, dq^\alpha\, dq^\beta \quad \text{(positive definite)}, \tag{5.6.4}$$

where

$$g_{\alpha\beta}(q) = g_{\alpha\beta} \equiv \mathbf{e}_\alpha \cdot \mathbf{e}_\beta\text{: covariant surface metric tensor, in the } q^\alpha.$$

Or,

---

* In this section: a. Lower case *Latin* indices run from 1 to N (i.e., E_N-values); b. Lower case *Greek* indices run from 1 to n (i.e., R_n-values); c. Upper case *Latin* indices run from $n + 1$ to N (i.e., $E_N - R_n \equiv E_{N-n}$-values).
** Or, we may start with $\mathbf{R} = \mathbf{R}(q^k) = \mathbf{R}(q^\alpha, q^A)$, where $q^A \equiv (q^{n+1}, \ldots, q^N)$, so that (q^α, q^A) is another set of curvilinear coordinates for E_N (just like the x^k), and then view $\mathbf{r}$ as the value of $\mathbf{R}$ for q^A = constant $\equiv q^A{}_o$, say, $q^A{}_o = 0$; i.e., $\mathbf{R}(q^\alpha, 0) = \mathbf{r}(q^\alpha)$.

$I_n \equiv ds^2$: *first* fundamental R_n-(surface) form,

$$\Rightarrow g_{\alpha\beta}: \textit{second-rank} \text{ covariant tensor (by Quotient rule).} \qquad (5.6.4a)$$

Before we examine the differences between the Euclidean ds^2 (Equations 5.6.1 and 5.6.1a), and the Riemannian ds^2 (Equation 5.6.4), let us summarize some of the algebra and analysis of R_n–vectors:

- The *contravariant* components of the R_n–metric tensor, $g^{\alpha\beta}$, are defined by

$$g^{\beta\gamma}\, g_{\alpha\beta} = \delta^{\gamma}{}_{\alpha} \quad [\text{since Det}(g_{\alpha\beta}) \neq 0], \qquad (5.6.5)$$

- The *reciprocal*, or *dual*, basis to $\{P, \mathbf{e}_\alpha\}$, $\{P, \mathbf{e}^\alpha\}$, is defined by

$$\mathbf{e}^\alpha = g^{\alpha\beta}\, \mathbf{e}_\beta \Leftrightarrow \mathbf{e}_\beta = g_{\beta\alpha}\, \mathbf{e}^\alpha; \quad \text{clearly: } \mathbf{e}_\alpha \cdot \mathbf{e}^b = g^{\beta}{}_{\alpha} = \delta^{\beta}{}_{\alpha}. \qquad (5.6.6)$$

- A vector $\mathbf{V}(P)$ is called a *surface* vector to R_n if it lies on its local tangent plane $\mathbf{T}_n(\mathbf{P})$, i.e., if it can be expressed as

$$\mathbf{V} = V^\alpha\, \mathbf{e}_\alpha = V_\alpha\, \mathbf{e}^\alpha, \quad \text{with } V^\alpha = g^{\alpha\beta}\, V_\beta \Leftrightarrow V_\beta = g_{\beta\alpha}\, V^\alpha; \qquad (5.6.7)$$

  in which case (recalling Equations 2.12.3ff), its (non-negative) *length*, or *norm*, equals

$$|\mathbf{V}| \equiv V = (g_{\alpha\beta}\, V^\alpha\, V^\beta)^{1/2} = (g^{\alpha\beta}\, V_\alpha\, V_\beta)^{1/2} = (V_\alpha\, V^\alpha)^{1/2}. \qquad (5.6.7a)$$

- Thc properties of a general R_n divide into *internal*, or *intrinsic* (i.e., completely determinable by $g_{\alpha\beta}$) and *external*, or *extrinsic*, or *extended* (i.e., those that also depend on the properties of the surrounding space; be that Euclidean or even another Riemannian). The need for such a distinction appears below.

Covariant Derivative of a Surface ($\mathbf{R}_n$–) Vector. Between the neighboring surface points $P(q)$ and $P'(q + dq)$, the general surface vector field $\mathbf{V} = \mathbf{V}(q^\alpha)$ undergoes the following first-order change $d\mathbf{V}$:

$$\begin{aligned} d\mathbf{V} &= (\partial \mathbf{V}/\partial q^\alpha) dq^\alpha \equiv \mathbf{V}_{,\alpha}\, dq^\alpha = (V^\beta\, \mathbf{e}_\beta)_{,\alpha}\, dq^\alpha \\ &= (V^\beta{}_{,\alpha}\, \mathbf{e}_\beta + V^\beta\, \mathbf{e}_{\beta,\alpha}) dq^\alpha = (V^\beta{}_{,\alpha}\, \mathbf{e}_\beta) dq^\alpha + V^\beta (A^\gamma{}_{\beta\alpha}\, \mathbf{e}_\gamma + b_{\beta\alpha}\, \mathbf{n}) dq^\alpha \\ &= (V^\beta{}_{,\alpha} + A^\beta{}_{\gamma\alpha}\, V^\gamma) \mathbf{e}_\beta\, dq^\alpha + (V^\beta\, b_{\beta\alpha}) \mathbf{n}\, dq^\alpha \equiv d\mathbf{V}_{(t)} + d\mathbf{V}_{(n)}, \end{aligned} \qquad (5.6.8)$$

where

$$d\mathbf{V}_{(t)} \equiv (V^\beta|_\alpha\, dq^\alpha)\mathbf{e}_\beta = (V^\beta|_\alpha\, \mathbf{e}_\beta)dq^\alpha = (V_\beta|_\alpha\, \mathbf{e}^\beta)dq^\alpha \equiv \mathbf{V}|_\alpha\, dq^\alpha:$$

$d\mathbf{V}$-increment *along* tangent to surface, (5.6.8a)

$$d\mathbf{V}_{(n)} \equiv (V^\beta\, b_{\beta\alpha}\, dq^\alpha)\mathbf{n} = (V^\beta\, b_{\beta\alpha}\, \mathbf{n})dq^\alpha:$$

$d\mathbf{V}$-increment *normal* to surface; (5.6.8b)

and the *surface covariant derivatives* are

$$V^\beta|_\alpha \equiv V^\beta{}_{,\alpha} + A^\beta{}_{\gamma\alpha}\, V^\gamma \quad \text{and} \quad V_\beta|_\alpha \equiv V_{\beta,\alpha} - A^\gamma{}_{\beta\alpha}\, V_\gamma, \tag{5.6.9a}$$

where

$$A^\alpha{}_{\beta\gamma} \equiv g^{\alpha\delta}\, A_{\delta,\beta\gamma} \equiv g^{\alpha\delta}\big[(g_{\delta\beta,\gamma} + g_{\delta\gamma,\beta} - g_{\beta\gamma,\delta})/\, 2\big]: \tag{5.6.9b}$$

first-kind ($A_{\delta,\beta\gamma}$) and *second*-kind ($A^\alpha{}_{\beta\gamma}$) kind surface Christoffels (built entirely from the surface metric coefficients $g_{\alpha\beta}$).

In the above, we have made use of the following basic results (extensions of corresponding equations of Section 5.5 to a general R_n):

i. $(\ldots)_{,\beta}$-differentiating $0 = \mathbf{e}_\alpha \cdot \mathbf{n} = g_{\alpha\gamma}\, \mathbf{e}^\gamma \cdot \mathbf{n}$, we find:

$$b_{\alpha\beta} = b_{\beta\alpha} \equiv \mathbf{e}_{\alpha,\beta} \cdot \mathbf{n} = -\mathbf{e}_\alpha \cdot \mathbf{n}_{,\beta}$$

$$(= \mathbf{e}_{\beta,\alpha} \cdot \mathbf{n} = -\mathbf{e}_\beta \cdot \mathbf{n}_{,\alpha} \Rightarrow 2b_{\alpha\beta} = -(\mathbf{e}_\alpha \cdot \mathbf{n}_{,\beta} + \mathbf{e}_\beta \cdot \mathbf{n}_{,\alpha}), \tag{5.6.10a}$$

$$b^\alpha{}_\beta \equiv g^{\alpha\gamma}\, b_{\gamma\beta} = -g^{\alpha\gamma}\, \mathbf{e}_\gamma \cdot \mathbf{n}_{,\beta} = -\mathbf{e}^\alpha \cdot \mathbf{n}_{,\beta}, \tag{5.6.10b}$$

$$b^{\alpha\beta} \equiv g^{\alpha\gamma}\, g^{\beta\delta}\, b_{\gamma\delta} \quad \text{(equations of \textit{Weingarten} in } R_n); \tag{5.6.10c}$$

and, therefore,

$$\mathbf{e}_{\alpha,\beta} = \mathbf{e}_{\beta,\alpha} = A^\gamma{}_{\alpha\beta}\, \mathbf{e}_\gamma + b_{\alpha\beta}\, \mathbf{n} \quad \text{(equations of \textit{Gauss} in } R_n). \tag{5.6.10d}$$

ii. From Equation 5.6.10d, we get

$$\mathbf{e}_{\alpha,\beta} \cdot \mathbf{e}_\delta = (A^\gamma{}_{\alpha\beta}\, \mathbf{e}_\gamma) \cdot \mathbf{e}_\delta = A^\gamma{}_{\alpha\beta}\, g_{\gamma\delta} = A_{\delta,\alpha\beta}, \tag{5.6.11a}$$

and so, successively,

$$(\mathbf{e}^\alpha \cdot \mathbf{e}_\gamma)_{,\beta} = \delta^\alpha{}_{\gamma,\beta} = 0 \Rightarrow \mathbf{e}^\alpha{}_{,\beta} \cdot \mathbf{e}_\gamma = -\mathbf{e}^\alpha \cdot \mathbf{e}_{\gamma,\beta} = \ldots = -A^\alpha{}_{\gamma\beta}, \tag{5.6.11b}$$

and, similarly,

$$(\mathbf{e}^\alpha \cdot \mathbf{n})_{,\beta} = 0 \;\Rightarrow\; \mathbf{e}^\alpha{}_{,\beta} \cdot \mathbf{n} = -\,\mathbf{e}^\alpha \cdot \mathbf{n}_{,\beta} = b^\alpha{}_\beta, \tag{5.6.11c}$$

i.e., finally,

$$\mathbf{e}^{\alpha}{}_{,\beta} = -A^{\alpha}{}_{\beta\gamma}\,\mathbf{e}^{\gamma} + b^{\alpha}{}_{\beta}\,\mathbf{n}. \qquad (5.6.11\text{d})$$

On the other hand, applying Equations 5.6.8a and b for $\mathbf{V} \to \mathbf{e}_{\alpha}$, we readily find

$$d_{(t)}\mathbf{e}_{\alpha} = A^{\gamma}{}_{\alpha\beta}\,\mathbf{e}_{\gamma}\,dq^{\beta} = A_{\gamma,\alpha\beta}\,\mathbf{e}^{\gamma}\,dq^{\beta} \quad \text{and} \quad d_{(n)}\mathbf{e}_{\alpha} = b_{\alpha\beta}\,\mathbf{n}\,dq^{\beta}; \qquad (5.6.12)$$

and similarly for $d_{(t)}\mathbf{e}^{\alpha}$ and $d_{(n)}\mathbf{e}^{\alpha}$, i.e., Equations 5.6.10d and 5.6.11d.

5.6.2 Differences Between Euclidean and Riemannian Arc-Lengths

We recall (Section 2.12) that in E_N there exists a transformation from the N curvilinear coordinates (q^{α}, q^{A}) to the N retangular Cartesian ones (y^k) so that ds^2 = sum of squares of the dy^k [More generally, ds^2 = quadratic form with *constant* coefficients. Every such (real) form can be further reduced by a (real) linear transformation to $ds^2 = \varepsilon_k(dy^k)^2$, where $\varepsilon_k = \pm 1$. If ds^2 is *positive definite*, then $\varepsilon_k = +1$, and E_N is called *properly* (or *purely*) Euclidean; if ds^2 is *indefinite*, then E_N is called *pseudo*-Euclidean.] In R_n, however, $ds^2 = g_{\alpha\beta}\,dq^{\alpha}dq^{\beta}$ *cannot*, in general, be expressed as a sum of squares of the (differences of) n new variables throughout their entire domain of definition, i.e., *globally*. The pythagorization of ds^2 can be achieved only *locally*, that is, in the neighborhood of an R_n-point P; or, geometrically, a curved surface can be approximated locally by its tangent plane. There we can do this (nonuniquely) by means of the following *linear* and *homogeneous* differential transformation:

$$dq^{\alpha'} = A^{\alpha'}{}_{\alpha}(q)dq^{\varepsilon} \Rightarrow ds^2 = \text{sum of squares of the } dq^{\alpha'}. \qquad (5.6.13)$$

But there is a catch: the Pfaffian Expressions (5.6.13) are, in general, *nonexact*, or *nonperfect differentials*; that is, the $dq^{\varepsilon'}$ are differentials of pseudo-, or quasi-, or nonholonomic, "coordinates" $q^{k'}$ (like the q^{K} of Section 3.12ff). Hence, a Phythagorean representation of the arc–length is only locally possible; unless the $A^{k'}{}_{k}$ are either constant, or satisfy certain integrability conditions, which amount to R_n being flat, i.e., $R_n \to E_N$.*

5.6.3 Problem of Embedding, or Immersing

Given an E_N we can always find it in an R_n; for example, by introducing $N - n$ (independent and generally nonlinear) *constraints* among its N coordinates y^k (rect-

* *Other Definitions of Curved Space*: Consider an ordinary surface, and three noncollinear points A, B, C on it. *If the sum of the three angles of the triangle formed by joining these points with geodesics of that surface does not equal 180°, then that surface is curved.* For example, consider the triangle ABC formed on the Earth's surface with C the North Pole and AC, BC arcs of great circles (meridians). Clearly, the sum of its three angles exceeds 180°. This is in essence, although on a considerably smaller scale, the actual geodetic experiments first carried out by C.F. Gauss (early 19th century) to detect the possible non–Euclideanness of the Earth's surface.

angular Cartesian) or x^k (general curvilinear). Conversely, however, if an R_n is defined by its first fundamental form $I_n \equiv ds^2 = g_{\alpha\beta}dq^\alpha dq^\beta$, where the given $n(n + 1)/2$ (independent and well-behaved) functions $g_{\alpha\beta} = g_{\alpha\beta}(q)$, what is the dimension N of the "larger" Euclidean space E_N containing it? Analytically, we seek N functions $y^k = y^k(q^\alpha)$, such that $I_n = \Sigma(dy^k)^2 = dy^k\, dy^k = (y^k{}_{,\alpha}dq^\alpha)(y^k{}_{,\beta}\, dq^\beta) = g_{\alpha\beta}\, dq^\alpha\, dq^\beta$ for arbitrary dq^α, dq^β. This leads immediately to the following $n(n + 1)/2$ first-order partial differential equations for $y^k(q^\alpha)$:

$$(\partial y^k/\partial q^\alpha)(\partial y^k/\partial q^\beta) = g_{\alpha\beta}(q) \quad (k = 1, \ldots, N;\ \alpha = 1, \ldots, n). \tag{5.6.14}$$

Therefore, unless $N \geq N_{\text{minimum}} \equiv N_{\min} \equiv n(n + 1)/2$ (for > we will have an *overdetermined* system), no real solutions are possible — *E_N must have a sufficiently high dimension.* For example, in an ordinary surface $n = 2 \Rightarrow N_{\min} = 2(2 + 1)/2 = 3$; while for $n = 3 \Rightarrow N_{\min} = 6$; i.e., we cannot embed an R_3 in an E_4! Geometrically, the embedding problem means finding a vector $\mathbf{r} = \mathbf{r}(q^\alpha)$ in E_N in terms of the $N_{\min}$ functions $g_{\alpha\beta}(q)$; i.e., finding the N projections (e.g., rectangular Cartesian components) of $\mathbf{r}$ along the N coordinate axes of E_N. If $N = N_{\min}$, then we say that R_n is *applicable* inside E_n; for example, R_2 is applicable inside E_3; and R_3 is applicable, not inside E_4, but inside E_6.*

5.6.4 Problem of Equivalence, or Integrability

Here we find the restrictions on the R_n-metric $g_{\alpha\beta}(q)$ so that there exist a (global) coordinate system q' in which the metric tensor has constant components throughout that space, i.e., conditions for a curved space to be flat. Clearly (Sections 3.3 and 3.8), such restrictions lead to the identical (i.e., global) vanishing of the Christoffels $\Gamma' = \Gamma'(q')$, which in turn results in the following:

> **Theorem I:** A necessary and sufficient condition for a symmetric tensor $g_{\alpha\beta}(q)$ [with $\text{Det}(g_{\alpha\beta}) \neq 0$] to reduce, under proper coordinate transformation, to a tensor with constant components throughout the associated manifold R_n is that the corresponding, i.e., $g_{\alpha\beta}$-built, Riemann–Christoffel tensor vanish identically.**

Analytically [and with the simplifying notational change $q' \rightarrow z \equiv (z^1, \ldots, z^n)$], we are seeking restrictions on the $n(n + 1)/2$ independent components $g_{\alpha\beta}(q)$ so that there exists another (rectilinear/affine) coordinate system $z = z(q)$ (of class C^2 in R_n) in which: $g_{\alpha\beta}(q) \rightarrow g_{\alpha\beta}(z) = \text{constant} \equiv h_{\alpha\beta}$. From this it follows that $\Gamma^\alpha{}_{\beta\gamma}(q) \equiv A^\alpha{}_{\beta\gamma}(z) \rightarrow \Gamma^\alpha{}_{\beta\gamma}(z) \equiv H^\alpha{}_{\beta\gamma} = 0$, throughout R_n. Conversely, if $H^\alpha{}_{\beta\gamma} = 0$, then (by Ricci's lemma) $0 = h_{\alpha\beta}|_\gamma = h_{\alpha\beta,\gamma} \Rightarrow h_{\alpha\beta} = \text{constant}$. Hence,

---

* (a) In special cases, however, an R_n is applicable inside an E_N even though $N < N_{\min}$; e.g., if $R_n = E_n$, then $N = n$ will suffice (see Levi-Civita, 1926, pp. 122ff; Weatherburn, 1963, pp. 50–52). (b) From the above we also conclude that an R_n may have, at least, $N_{\min} - n = n(n - 1)/2$ *normals* in the surrounding E_N; e.g., for $N = 3$, $n = 2 \Rightarrow 1$ normal.

** These conditions also guarantee: (a) the integrability of $d\mathbf{e}_\alpha = \Gamma^\beta{}_{\alpha\varepsilon}\, dq^\varepsilon \mathbf{e}_\beta$ or $\mathbf{e}_{\alpha,\varepsilon} = \Gamma^\beta{}_{\alpha\varepsilon}\mathbf{e}_\beta$ i.e., $(\mathbf{e}_{\alpha,\beta})_{,\varepsilon} = (\mathbf{e}_{\alpha,\varepsilon})_{,\beta}$ (recall Problem 3.11.7) and (b) that the q^αs in Equation 5.6.13 are genuine, or holonomic, coordinates. (They are special cases of the *Frobenius* integrability conditions (Sections 6.7 through 6.9).

Theorem II (Nontensorial): A necessary and sufficient condition for $g_{\alpha\beta}(z) \equiv h_{\alpha\beta}$ = constant, in some coordinate system z, is that the corresponding Christoffels $\Gamma^{\alpha}{}_{\beta\gamma}(z) \equiv H^{\alpha}{}_{\beta\gamma}$ vanish identically, i.e., everywhere in R_n.*

From this we can, further, deduce the following (ultimately tensorial, i.e., absolute) conditions for $z^{\alpha} = z^{\alpha}(q^{\beta})$: as seen in Section 3.3, the corresponding Christoffels $\Gamma^{\alpha}{}_{\beta\gamma}(q) \equiv A^{\alpha}{}_{\beta\gamma}$ and $\Gamma^{\alpha}{}_{\beta\gamma}(q' \to z) \equiv H^{\alpha}{}_{\beta\gamma}$ are related by

$$\partial^2 z^{\alpha}/\partial q^{\beta}\partial q^{\gamma} = A^{\delta}{}_{\beta\gamma}(\partial z^{\alpha}/\partial q^{\delta}) - H^{\alpha}{}_{\varepsilon\xi}(\partial z^{\varepsilon}/\partial q^{\beta})(\partial z^{\xi}/\partial q^{\gamma}), \qquad (5.6.15a)$$

which, since $H^{\alpha}{}_{\varepsilon\xi} = 0$, yields the following *second*-order (Partial Differential Equation) system:

$$\partial^2 z^{\alpha}/\partial q^{\beta}\partial q^{\gamma} = A^{\delta}{}_{\beta\gamma}(\partial z^{\alpha}/\partial q^{\delta}), \qquad (5.6.15b)$$

or, temporarily suppressing (for extra clarity) the *free* index α, the equivalent *first*-order system:

$$\partial u_{\beta}/\partial q^{\gamma} = A^{\delta}{}_{\beta\gamma}(q)u_{\delta},$$

$$\partial z/\partial q^{\beta} = u_{\beta} = u_{\beta}(q) \quad \text{(instead of } \partial z^{\alpha}/\partial q^{\beta} = u^{\alpha}{}_{\beta}\text{, etc.)} \qquad (5.6.15c)$$

or, with the compact notations

$$z = f^1, \quad u_1 = f^2, \ \ldots, \quad u_n = f^{n+1} \equiv f^M, \quad \text{or } (z, u) \to f$$

and

$$(u_{\beta}, A^{\delta}{}_{\beta\gamma}\, u_{\delta}) \to F^D{}_{\alpha}(f, q) \quad (D = 1, \ldots, M \equiv n + 1;\ \alpha = 1, \ldots, n), \qquad (5.6.15d)$$

finally,

$$\partial f^D/\partial q^{\alpha} = F^D{}_{\alpha}(f^1, \ldots, f^D; q^1, \ldots, q^n) \equiv F^D{}_{\alpha}(f, q) \qquad (5.6.15e)$$

(known functions of their arguments; defined over R_n and well behaved).

The above can also be written in the equivalent Pfaffian form:

$$df^D = (\partial f^D/\partial q^{\alpha})dq^{\alpha} = F^D{}_{\alpha}(f, q)dq^{\alpha} \equiv F^D{}_{\alpha}\, dq^{\alpha}. \qquad (5.6.15f)$$

Now, the system (5.6.15f) is said to be *completely integrable*, or *holonomic*, if for any set of initial values q_o and f_o [for which the $F^D{}_{\alpha}(\ldots)$ are analytic] there exists a unique set of M functions, the $f(q)$, that satisfy its equations (5.6.15f), and take on

---

* In a general R_n, the Γs can be made to vanish at any given of its points P_o and in special "geodesic coordinates" s; i.e., locally $\Gamma(s_o) = 0$. See, e.g., Sokolnikoff (1951, pp. 163–165).

these initial values q_o, f_o. The necessary and sufficient conditions for such integrability are $\partial^2 f^D/\partial q^\beta \partial q^\alpha = \partial^2 f^D/\partial q^\alpha \partial q^\beta$; or, due to Equation 5.6.15e and since $f = f(q)$ (and with $D, D' = 1, \ldots, M$; $\alpha, \beta = 1, \ldots, n$):

$$\partial F^D{}_\alpha/\partial q^\beta + (\partial F^D{}_\alpha/\partial f^{D'})(\partial f^{D'}/\partial q^\beta)$$

$$= \partial F^D{}_\beta/\partial q^\alpha + (\partial F^D{}_\beta/\partial f^{D'})(\partial f^{D'}/\partial q^\alpha), \qquad (5.6.16a)$$

or, by Equation 5.6.15e,

$$\partial F^D{}_\alpha/\partial q^\beta + (\partial F^D{}_\alpha/\partial f^{D'})\ F^{D'} = \partial F^D{}_\beta/\partial q^\alpha + (\partial F^D{}_\beta/\partial f^{D'})\ F^{D'}. \qquad (5.6.16b)$$

Applying the above theorem, Equations 5.6.16a and b, to our problem, i.e., with the identifications (where now we reintroduce all indices, as in Equation 5.6.15b)

$$\partial f^1/\partial q^\alpha = F^1{}_\alpha \to u_\alpha \quad (\alpha = 1, \ldots, n) \qquad (5.6.17a)$$

$$\partial f^k/\partial q^\alpha = F^k{}_\alpha \to A^\beta{}_{k-1,\alpha}\ u_\beta \quad (\alpha, \beta = 1, \ldots, n;\ k = 2, 3, \ldots, n+1) \qquad (5.6.17b)$$

we obtain the two sets of equations:

$$A^\alpha{}_{\beta\gamma}\ u_\alpha = A^\alpha{}_{\gamma\beta}\ u_\alpha \quad \text{and} \quad R^\alpha{}_{\beta\gamma\delta}\ u_\alpha = 0 \quad (\alpha, \beta, \gamma, \delta = 1, \ldots, n). \qquad (5.6.17c)$$

The *first* set (Equation 5.6.17c$_1$), is identically satisfied because in our R_n: $A^\alpha{}_{\beta\gamma} = A^\alpha{}_{\gamma\beta}$ (no torsion); while the *second* (Equation 5.6.17c$_2$), states that (since, in general, $u_\alpha \neq 0$) the system (5.6.15b or c) will be holonomic if $R^\alpha{}_{\beta\gamma\delta}$ vanishes identically (no curvature).* Then, as stated earlier,

$$I_n \equiv (ds)^2 = g_{\alpha\beta}(q)\ dq^\alpha\ dq^\beta$$

$$\to h_{\alpha\beta}\ dz^\alpha\ dz_\beta\text{: general rectilinear form } (h_{\alpha\beta}\text{: constant in } R_n). \qquad (5.6.18)$$

If, in addition, I_n is *positive definite,* then it can be reduced by a linear (nonsingular) transformation $z \to y$ (rectangular Cartesian coordinates) to the *Pythagorean* form $I_n = (dy^1)^2 + \ldots + (dy^n)^2$.**

5.7 PERTURBATION OF TRAJECTORIES IN CONFIGURATION SPACE, AND THEIR STABILITY***

It is in the analytical investigations in connection with stability of motion that the use of the tensorial notation becomes of greatest importance. The appearance of the Rie-

* Recall closing remarks in Section 3.11.
** For further details and insights, see, e.g., Eisenhart (1947, pp. 114–122), Sokolnikoff (1951, pp. 96–100), Veblen (1927, pp. 69–71).
*** This section may be omitted in a first reading; or may be read after Chapter 7.

mannian curvature tensor in the course of the analysis makes it difficult to believe that similar results could be obtained without the use of this method.

Synge (1926, p. 78)

We begin with a few basic definitions:

Trajectory: A curve (here in configuration space) along which the coordinates are given as functions of time: $q^k = q^k(t)$ $(k = 1, \ldots, n)$.

Natural Trajectory: One that corresponds to a motion under *given* forces (i.e., no unknown constraint reactions), according to the equations of dynamics.

5.7.1 The Perturbation Equation

Now, generalizing formally from Section 5.5, we consider a *fundamental*, or undisturbed, natural trajectory, C: $q^k = q^k(t)$, of a figurative particle P (which may represent a scleronomic and holonomic mechanical system — both terms to be detailed in Chapter 6), in a Riemannian space R_n, under given forces Q^k, and with equations of motion:

$$A^k = DV^k/Dt \equiv V^l\, V^k|_l \equiv (dq^l/dt)\,[(dq^k/dt)|_l]$$

$$= d^2q^k/dt^2 + \Gamma^k{}_{rs}(dq^r/dt)(dq^s/dt) = Q^k \quad \text{(Latin indices: } 1, \ldots, n). \qquad (5.7.1)$$

Let q^k be the coordinates of a generic C-point $P(t)$ and $q^k + \delta q^k \equiv q^k + \eta^k$ those of the corresponding *simultaneous*, or *isochronous*, point $P + \delta P(t)$ on a perturbed, or *disturbed*, also natural trajectory, $C + \delta C \equiv C'$, for a generic time t: $t_1 \le t \le t_2$ (Figure 5.4).*

Below we derive the *linearized* equation for $\eta^k = \eta^k(t)$, i.e., for small and isochronous perturbation vectors. Further, we shall call C *stable*, in the *kinematical sense* of Synge et al., *if the magnitude of* η^k, η, *remains small for all time and for arbitrary initial conditions.* From the (closed) curvilinear quadrilateral $ABCD$ (Figure 5.4), we have $\mathbf{AB} + \mathbf{BE} = \mathbf{AD} + \mathbf{DE}$, or, to the first order (and with all derivatives evaluated at A),

$$(dq^k/dt)dt + (\eta^k + d\eta^k) = \eta^k + \{(dq^k/dt) + [\partial/\partial q^l(dq^k/dt)]\eta^l\}dt,$$

or, finally,

---

* Analytically, we are dealing with a single infinity of trajectories that forms a two-dimensional surface in R_n, and (as in the calculus of variations) has the representation $q^k = q^k(\sigma, \varepsilon)$, where σ: parameter that varies along each trajectory (i.e., tells *where we are on a given C*) and ε: parameter that remains constant along each trajectory but varies from trajectory to trajectory (i.e., tells *which trajectory we are on*). Then, if A is a generic C-point,

$$dq^k = (\partial q^k/\partial\sigma)_A\, d\sigma, \qquad \delta q^k = (\partial q^k/\partial\varepsilon)_A\, d\varepsilon$$

and under an isochronous mapping: $A = A(\sigma, \varepsilon) \rightarrow A'(\sigma, \varepsilon + \delta\varepsilon) \equiv A'$, on C'. Usually, $\sigma \rightarrow s$: arc-length along C, or $\sigma \rightarrow t$: time.

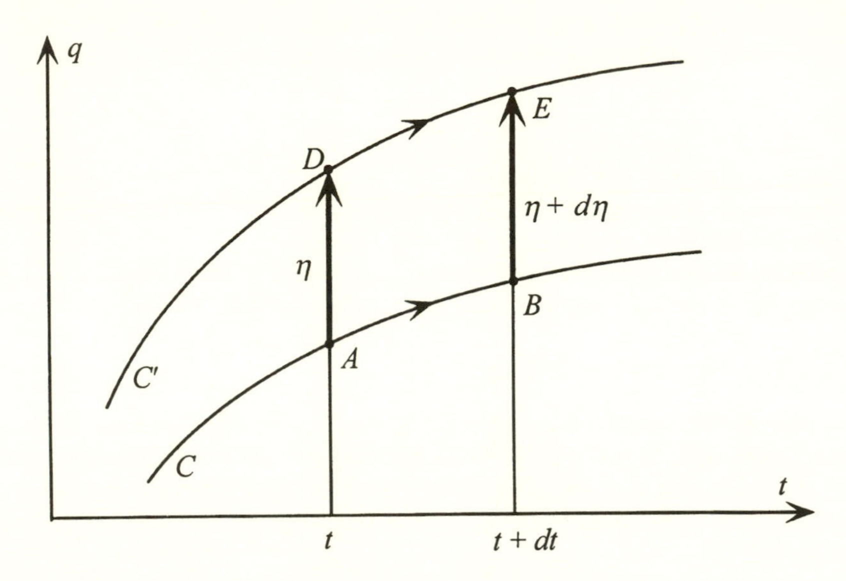

FIGURE 5.4 Isochronous, or kinematical, perturbation $\delta q \equiv \eta$ from a fundamental natural trajectory C to a neighboring natural trajectory $C' = C + \delta C$:

$$\mathbf{A} \equiv P\text{: } q^k(t) \equiv q^k,\ \mathbf{B} \equiv P + dP\text{: } q^k(t + dt) \equiv q^k + (dq^k/dt)_A dt;\ \mathbf{D} \equiv P + \delta P\text{: } q^k + \eta^k(t),$$

$$\mathbf{E} \equiv P + dP + \delta(P + dP) = P + \delta P + d(P + \delta P)\text{: } q^k(t + dt) + \eta^k(t + dt).$$

$$d\eta^k/dt = [\partial/\partial q^l(dq^k/dt)]\eta^l \equiv (\partial V^k/\partial q^l)\eta^l \equiv V^k{}_{,l}\,\eta^l. \tag{5.7.2}$$

Now, although Equation 5.7.2 is nontensorial, yet, as the above derivation clearly shows, it holds in any coordinate system and, therefore, also in a rectangular Cartesian system. But then, as tensor theory teaches, *in passing from the latter to a general curvilinear system, we must replace Equation 5.7.2 by*

$$D\eta^k/Dt = V^k|_l\,\eta^l. \tag{5.7.3}$$

This is the fully tensorial *equation of propagation*, or *evolution, of the small isochronous perturbation from the fundamental path C.*

A second, far more insightful form of it is obtained by eliminating $V^k|_l$ from it using Equation 5.7.1: (D/Dt)-differentiating both sides of Equation 5.7.3, while invoking basic tensor analytical results from Chapter 3, we get

$$\begin{aligned} D^2\eta^k/Dt^2 &= [D/Dt(V^k|_l)]\eta^l + V^k|_l(D\eta^l/Dt) \\ &= (V^k|_l|_r\, V^r)\eta^l + V^k|_l(V^l|_s\,\eta^s) \\ &= [(V^k|_r|_l + R^k{}_{brl}\, V^b)V^r]\eta^l + V^k|_l(V^l|_s\,\eta^s) \end{aligned}$$

(since $R^k{}_{brl} = -R^k{}_{blr}$, by Equation 3.11.4c)

$$= (R^k{}_{rbl}\, V^b\, V^r)\eta^l + (V^k|_r|_l\, V^r + V^k|_r\, V^r|_l)\eta^l$$

$$(\text{since } R^k{}_{brl}\, V^b\, V^r = R^k{}_{rbl}\, V^b\, V^r)$$

$$= (R^k{}_{rbl}\, V^b\, V^r)\eta^l + (V^k|_r\, V^r)|_l\, \eta^l$$

$$(\text{but by Equation 5.7.1 } V^k|_r\, V^r = A^k = Q^k)$$

or, rearranging terms (and since $-R^k{}_{rbl} = +R^k{}_{rlb}$), we finally obtain

$$D^2\eta^k/Dt^2 + (R^k{}_{rlb}\, V^r\, V^b)\eta^l = Q^k|_l\, \eta^l \quad \text{(Synge, 1926)}, \tag{5.7.4}$$

which is the definitive linear(ized) second-order tensorial differential equation for the isochronous perturbation vector $\eta^k(t)$. Its coefficients depend on the fundamental path, through the V^k and $R^k{}_{rlb}$; and also, through $Q^k|_l$, on the *force perturbation* between C and C'. The remarkable thing about Equation 5.7.4 is the appearance of the curvature term $+ (R^k{}_{rlb}\, V^r\, V^b)\eta^l$.

As for *stability*: *if all solutions of Equation 5.7.4 remain small for all time, then the fundamental trajectory C is called stable.*

Finally, it is not hard to see that the *covariant* form of the perturbation equation is:

$$g_{sk}(D^2\eta^k/Dt^2) + (R_{srlb}\, V^r\, V^b)\eta^l = Q_s|_l\, \eta^l. \tag{5.7.5)*$$

5.7.2 The Energy Integral

For wholly potential and explicitly time-independent forces, excess of total energy, $E \equiv T + \Pi$ (T: kinetic energy, Π: potential energy) between the undisturbed and undisturbed motion, i.e., the energy variation, ΔE, is

$$\Delta E \equiv E(C') - E(C) = (T + \Pi)' - (T + \Pi). \tag{5.7.6}$$

But

$$2T' = g'_{kl}[(V^k + d\eta^k/dt)(V^l + d\eta^l/dt)]$$

$$\approx 2T + (g_{kl,r}\, V^k\, V^l)\eta^r + 2(g_{kl}\, V^k)(d\eta^l/dt) \quad (\text{to first order in } \eta,\ d\eta/dt)$$

$$= 2T + [(\Gamma_{k,rl} + \Gamma_{l,rk})V^k\, V^l]\eta^r + 2(g_{kl}\, V^k)(d\eta^l/dt) \quad (\text{invoking Equation 3.3.2b})$$

---

* Equation 5.7.4 with $Q^k = 0 \Rightarrow Q^k|_l\, \eta^l = 0$, and with arc–length s instead of t, is called the equation of *geodesic deviation* (Levi–Civita, 1926). The reason is that then (with $V^k \to dq^k/ds \equiv \tau^k$)

$$D^2\eta^k/Ds^2 = -(R^k{}_{rlb}\, \tau^r\, \tau^b)\eta^l = (R^k{}_{rbl}\, \tau^r\, \tau^b)\eta^l: \tag{5.7.4a}$$

measure of change in the separation of the neighboring geodesics C and C'; or, in kinematic terms, of the relative acceleration between the two figurative particles P and P' as they traverse C and C', respectively. In flat space, clearly, $R_{srlb} = 0$ and, therefore, $D\eta^k/Ds = \text{constant} \equiv c_1$, from which, choosing *rectilinear coordinates*, it follows: $\eta^k = c_1 s + c_{2\ (\text{second integration constant})}$; i.e., *in flat space geodesics starting from a common point ($c_2 = 0$) always diverge from each other.*

and, therefore, to the same accuracy,

$$\Delta T \equiv T' - T = (\Gamma_{k,rl}\, V^k\, V^l)\eta^r + (g_{kl}\, V^k)(D\eta^l/Dt - \Gamma^l{}_{rs}\, V^s\eta^r) = g_{kl}\, V^k(D\eta^l/Dt)$$

(with some dummy index changes, and since $g_{kl}\,\Gamma^l{}_{rs} = \Gamma_{k,rs}$) and since $\Pi' \approx \Pi + \Pi_{,k}\,\eta^k$, finally, the *linearized energy perturbation equation is:*

$$(g_{kl}\, V^k)D\eta^l/Dt + \Pi_{,k}\,\eta^k = V_l(D\eta^l/Dt) + \Pi_{,l}\,\eta^l = \Delta E. \tag{5.7.7}$$

5.7.3 Alternative Forms

i. With $dq^k/dt \equiv V^k = V\,\tau^k \Rightarrow g_{kl}\,\tau^k\;\tau^l = 1$ (where, here, $S = s$, and $V \equiv dS/dt = ds/dt$) the perturbation equation (5.7.4) assumes the "normalized" form:

$$D^2\eta^k/Dt^2 + V^2(R^k{}_{rlb}\,\tau^r\,\tau^b)\eta^l = Q^k|_l\,\eta^l; \tag{5.7.8}$$

and similarly for Equation 5.7.5.

ii. Let ρ^k be a *unit* vector in the direction of η^k, i.e., $\eta^k = \eta\,\rho^k$, $g_{kl}\,\rho^k\,\rho^l = 1$. Then,

$$D\eta^k/Dt = (d\eta/dt)\rho^k + \eta(D\rho^k/Dt), \tag{5.7.9a}$$

$$D^2\eta^k/Dt^2 = (d^2\eta/dt^2)\rho^k + 2(d\eta/dt)(D\rho^k/Dt) + \eta(D^2\rho^k/Dt^2); \tag{5.7.9b}$$

from which

$$g_{ks}(D^2\eta^k/Dt^2)\rho^s = d^2\eta/dt^2 + 2(d\eta/dt)g_{ks}\,(D\rho^k/Dt)\rho^s + \eta\; g_{ks}(D^2\rho^k/Dt^2)\rho^s$$

$$[\text{and due to } g_{ks}(D\rho^k/Dt)\rho^s = 0 \quad (\text{since } Dg_{ks} = 0)$$

$$\Rightarrow g_{ks}(D^2\rho^k/Dt^2)\rho^s + g_{ks}(D\rho^k/Dt)(D\rho^s/Dt) = 0]$$

$$= d^2\eta/dt^2 - \eta\; g_{ks}(D\rho^k/Dt)(D\rho^s/Dt); \tag{5.7.9c}$$

and, therefore, dotting Equation 5.7.5 with ρ^s and utilizing in the result the right side of the above, finally yields the following invariant equation for the isochronous *perturbation magnitude* $\eta = \eta(t)$:

$$d^2\eta/dt^2 + [R_{kmsl}\,\rho^k\,V^m\,\rho^s\,V^l - g_{ks}(D\rho^k/Dt)(D\rho^s/Dt)]\eta$$

$$= (Q_k|_s\,\rho^k\,\rho^s)\eta. \tag{5.7.10}$$

Similarly, with the help of Equation 5.7.9a, the energy equation (5.7.7) can be rewritten as

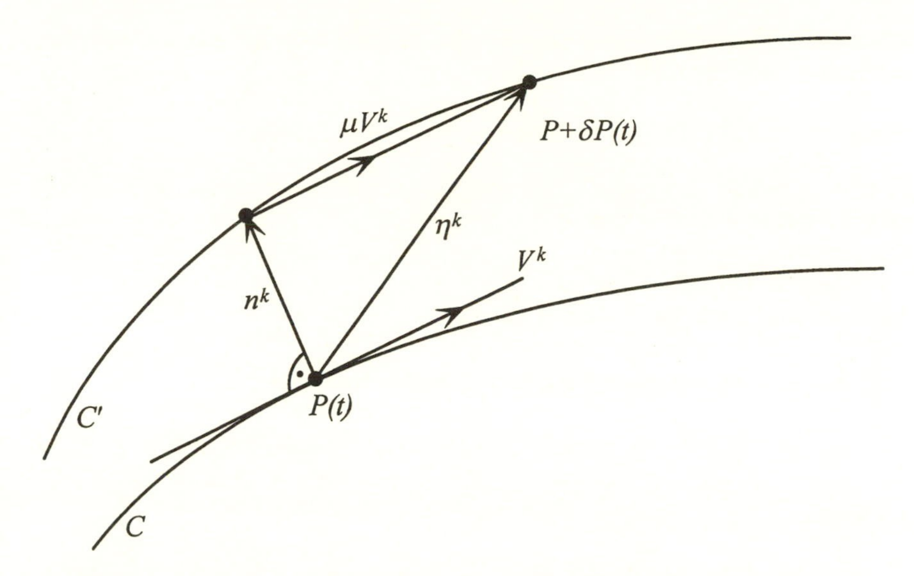

FIGURE 5.5 Isochronous perturbation $\delta q \equiv \eta$, vs. perturbation normal to the velocity n, from a fundamental natural trajectory C to a neighboring natural trajectory C'.

$$(d\eta/dt)g_{kl}\, V^k\, \rho^l + [g_{kl}\, V^k(D\rho^l/Dt) + \Pi_{,l}\, \rho^l]\eta = \Delta E. \qquad (5.7.11)$$

5.7.4 Normal (Nonisochronous) Perturbations

Other correspondences between the points of C and C' are possible. Let us consider, instead of the isochronous perturbation η^k, the *normal* one $n^k = n^k(t)$, defined by $n^k V_k = n_k V^k = 0$. Then (Figure 5.5),

$$\eta^k = n^k + \mu\, V^k, \qquad \mu \equiv \eta^k\, V_k/V^2. \qquad (5.7.12)^*$$

To find the equation satisfied by $n^k(t)$, we substitute η^k from Equation 5.7.12 into Equation 5.7.4, while recalling that $A^k = Q_k$, $A^k = Q^k$. The final result is

$$D^2n^k/Dt^2 + (d^2\mu/dt^2)V^k + 2(d\mu/dt)Q^k + R^k{}_{rlb}\, V^r\, n^l\, V^b = Q^k|_l\, n^l \quad (5.7.13)$$

(the *second* and *third* terms are new terms, vis-à-vis Equation 5.7.4) which along with $n^kV_k = 0$ constitute a system of $n + 1$ equations for the $n + 1$ unknowns $n^k(t)$ and $\mu(t)$. Another invariant form in terms of the normal perturbation magnitude $n = n(t)$ can be obtained easily from Equation 5.7.13; just like Equation 5.7.10 from Equations 5.7.4 and 5.7.5 (see, e.g., Synge, 1936, pp. 15–16). Still more general nonisochronous, or *noncontemporaneous*, mappings between the points of C and C' can be envisioned, leading to corresponding special definitions of stability of C. For

---

* If the magnitude of $n^k(t)$ remains small, throughout a time interval of interest, C and C' stay close to each other there, and C is called *orbitally stable*. In such a case, the magnitude of $\eta^k(t)$ may or may *not* remain small.

these topics, the reader should consult texts on stability, nonlinear and/or celestial mechanics, ordinary differential equations.

Example 5.7.1. Perturbed Motion and Trajectory Stability

Force-free motion of a particle P, of unit mass, on the surface of a fixed sphere, of origin O and radius r. Let us find the equations of its fundamental path and examine their stability.

Fundamental Path. Here, $2T = V^2 = (ds/dt)^2, \quad \Pi = 0 \Rightarrow Q_k = 0,$ (a)

with Gaussian coordinates the ordinary spherical coordinates:

$$\text{longitude: } 0 \leq \theta \equiv q^1 = \text{angle}(Oz, OP) \leq \pi, \quad \text{latitude: } 0 \leq \phi \equiv q^2 \leq 2\pi, \tag{b}$$

we have

$$ds^2 = r^2(d\theta^2 + \sin^2\theta\, d\phi^2) \Rightarrow g_{11} = r^2,\ g_{12} = g_{21} = 0,\ g_{22} = r^2 \sin^2\theta, \tag{c}$$

and, therefore, the corresponding *nonvanishing* first- and second-kind Christoffels are

$$\Gamma_{1,22} = -r^2 \sin\theta \cos\theta, \quad \Gamma_{2,12} = \Gamma_{2,21} = r^2 \sin\theta \cos\theta, \tag{d1}$$

$$\Rightarrow \Gamma^1{}_{22} = -\sin\theta \cos\theta, \quad \Gamma^2{}_{12} = \Gamma^2{}_{21} = \text{cotan}\,\theta. \tag{d2}$$

Hence, the equations of the fundamental path (here geodesics) are

$$q^1: \quad d^2\theta/dt^2 - \sin\theta\cos\theta\,(d\phi/dt)^2 = 0, \tag{e1}$$

$$q^2: \quad d^2\phi/dt^2 + 2\,\text{cotan}\,\theta\,(d\phi/dt)(d\theta/dt) = 0. \tag{e2}$$

The last one can be rewritten as $d/dt[(d\phi/dt)\sin^2\theta] = 0$, and integrates readily to

$$\sin^2\theta\,(d\phi/dt) = \text{constant of integration} \equiv c. \tag{f}$$

Choosing, with no loss of generality, $\phi_{\text{initially}}, (d\phi/dt)_{\text{initially}} = 0$, we find $c = 0 \Rightarrow d\phi(t)/dt = 0$, and so Equation e1 reduces to $d^2\theta/dt^2 = 0 \Rightarrow \theta(t) = c_1 t + c_2$ ($c_{1,2}$: integration constants); i.e., *a great circle described at a constant rate.*
Perturbations. Here $n = 2$, and therefore the sole nonvanishing curvature tensor component is

$$R^1{}_{212} = (\Gamma^1{}_{22})_{,1} - \Gamma^2{}_{12}\,\Gamma^1{}_{22} = \sin^2\theta \Rightarrow R_{1212} = g_{1k}\,R^k{}_{212} = r^2\sin^2\theta, \tag{g1}$$

$$R^2{}_{112} = g^{2k}\,R_{k112} = (1/g_{22})\,R_{2112} = -(R_{1212}/g_{22}) = -(r^2\sin^2\theta/\,r^2\sin^2\theta) = -1. \tag{g2}$$

Thus, with $\delta q^1 \equiv \delta\theta \equiv \eta^1 \equiv x$, $\delta q^2 \equiv \delta\phi \equiv \eta^2 \equiv y$, and dq^k/dt: $d\theta/dt = c_1$, $d\phi/dt = 0$, the perturbation equations become (with c_3, c_4, c_5, c_6: additional integration constants):

$$D^2x/Dt^2 = 0 \Rightarrow c_1{}^2(d^2x/d\theta^2) = 0 \Rightarrow x = c_3\theta + c_4, \tag{h1}$$

$$D^2y/Dt^2 + [R^2{}_{121}\,(d\theta/dt)^2]y = D^2y/Dt^2 + c_1{}^2y = 0 \quad (\text{since } R^2{}_{121} = -R^2{}_{112} = 1)$$

[here, $Dy/Dt = dy/dt + \Gamma^2{}_{21}\, y\, c_1$, $D^2y/Dt^2 = d/dt(Dy/Dt) + \Gamma^2{}_{21}\,(Dy/Dt)\, c_1$, etc.]

$$\Rightarrow c_1{}^2[(d^2y/d\theta^2) + 2\cot\text{an}\,\theta\,(dy/d\theta) - y] + c_1{}^2\, y = 0$$

$$\Rightarrow y = c_5 - c_6 \cot\text{an}\,\theta; \tag{h2}$$

or, further,

$$x(\theta = 0) = 0, \quad x(\theta = \pi) = 0 \Rightarrow c_3, c_4 = 0, \quad \text{i.e., } x = 0, \tag{i1}$$

$$y(\theta)\text{: finite for all } \theta \Rightarrow c_6 = 0, \quad \text{i.e., } y = c_5 = \delta\theta(0); \tag{i2}$$

and so the magnitude of the isochronous perturbation vector is

$$\eta = (g_{kl}\,\eta^k\,\eta^l)^{1/2} = (r^2 \sin^2\theta\; c_5{}^2)^{1/2} = r\, c_5 \sin\theta, \text{ i.e., stability.} \tag{j}$$

Example 5.7.2. Perturbation Equations in Intrinsic Variables

The discussion of stability of an n DOF system trajectory necessitates a deeper study of Riemannian geometry; in particular the n-dimensional generalization of the Frenet–Serret (F–S) formulae and corresponding intrinsic, or path, variables. Due to space limitations, we shall restrict ourselves to the simpler but useful such cases of *two* and *three* DOF.

Two DOF. Let C and C' be, respectively, the fundamental and perturbed paths. Assuming for convenience wholly *potential* forces, i.e., $Q_k = -\partial\Pi/\partial q^k$ (Π: potential function), the equations of a motion of a particle P of unit mass *along* C are (Sections 5.3 and 5.4):

i. Along the (unperturbed) *tangent* (τ) (with $S = s \Rightarrow ds/dt \equiv V$, to avoid confusion with the normal ν)

$$d^2s/dt^2 = dV/dt \equiv A_\tau = -\partial\Pi/\partial s$$

$$= -(\partial\Pi/\partial q^k)(dq^k/ds) = -(\partial\Pi/\partial q^k)(\tau^k) \equiv -\Pi_{,k}\,\tau^k, \tag{a1}$$

ii. Along the (unperturbed) normal (ν), with $\kappa_{(\nu)} \equiv \kappa$ and $\rho_{(\nu)} \equiv 1/\kappa \equiv \rho$,

$$(ds/dt)[\kappa(ds/dt)] = \kappa\ V^2 \equiv A_\nu = -\partial\Pi/\partial\nu$$

$$= -(\partial\Pi/\partial q^k)(dq^k/d\nu) = -(\partial\Pi/\partial q^k)(\nu^k) \equiv -\Pi_{,k}\ \nu^k; \quad \text{(a2)}$$

while the first two F–S formulae specialize to (here all Latin indices run from 1 to 2):

- $d\tau/dt = (d\tau/ds)(ds/dt) = (\kappa\ \nu)(ds/dt) \equiv \omega\ \nu$, [$\omega \equiv \kappa\ V$: component of "angular velocity" of τ (i.e., $\omega\ dt = ds/\rho$ (see Figure 5.2)] or

$$D\tau^k/Dt = \omega\ \nu^k:\ d\tau^k/dt + \Gamma^k{}_{dh}[(ds/dt)\tau^d]\tau^h$$

$$= \omega\ \nu^k \quad [dq^h/dt = (ds/dt)\tau^h]; \quad \text{(b1)}$$

- $d\nu/dt = (d\nu/ds)(ds/dt) = (-k\ \tau)(ds/dt) \equiv -\omega\ \tau$, or

$$D\nu^k/Dt = -\omega\ \tau^k: \quad d\nu^k/dt + \Gamma^k{}_{dh}[(ds/dt)\tau^d]\nu^h = -\omega\ \tau^k. \quad \text{(b2)}$$

Then, the most general two DOF perturbation vector $\delta\mathbf{r} \equiv \delta\mathbf{q} \equiv \eta(t)$ will have the representation:

$$\eta = \eta^k\ \mathbf{e}_k = a\ \tau + b\ \nu \Rightarrow \eta^k = a\ \tau^k + b\ \nu^k \quad (\mathbf{e}_k \equiv \partial\mathbf{r}/\partial q^k); \quad \text{(c1)}$$

and so, $d/dt(\ldots)$-differentiating Equation c1 and invoking the F–S equations, we find,

$$d\eta/dt = (D\eta^k/Dt)\mathbf{e}_k = (da/dt)\tau + a(d\tau/dt) + (db/dt)\nu + b(d\nu/dt)$$

$$= \ldots = [(da/dt) - \omega\ b]\tau + [(db/dt) + \omega\ a]\nu, \quad \text{(c2)}$$

i.e.,

$$(da/dt) - \omega\ b = (d\eta/dt)\cdot\tau = [(D\eta^k/Dt)\mathbf{e}_k]\cdot(\tau^h\ \mathbf{e}_h)$$

$$= g_{kh}(D\eta^k/Dt)\tau^h \quad (\textit{tangential perturbational}\ \text{velocity}), \quad \text{(d1)}$$

$$(db/dt) + \omega\ a = (d\eta/dt)\cdot\nu = [(D\eta^k/Dt)\mathbf{e}_k]\cdot(\nu^h\ \mathbf{e}_h)$$

$$= g_{kh}(D\eta^k/Dt)\nu^h \quad (\textit{normal perturbational}\ \text{velocity}). \quad \text{(d2)}$$

$d/dt(\ldots)$-differentiating Equation c2 once more and proceeding as in the above, we find

$$d^2\eta/dt^2 = (D^2\eta^k/Dt^2)\mathbf{e}^k$$

$$= \ldots = [(d^2a/dt^2) - 2(db/dt)\omega - b(d\omega/dt) - a\,\omega^2]\tau$$

$$+ [(d^2b/dt^2) + 2(da/dt)\omega + a(d\omega/dt) - b\,\omega^2]\nu, \qquad \text{(e1)}$$

i.e.,

$$g_{kh}(D^2\eta^k/Dt^2)\tau^h = (d^2a/dt^2) - 2(db/dt)\omega - b(d\omega/dt) - a\,\omega^2$$

$$(\textit{tangential perturbational}\ \text{acceleration}), \qquad \text{(e2)}$$

$$g_{kh}(D^2\eta^k/Dt^2)\nu^h = (d^2b/dt^2) + 2(da/dt)\omega + a(d\omega/dt) - b\,\omega^2$$

$$(\textit{normal perturbational}\ \text{acceleration}). \qquad \text{(e3)}$$

Next, with the help of Equations e2 and e3, we are going to project the earlier (covariant) perturbation equation in the q-variables, Equation 5.7.5,

$$g_{kl}(D^2\eta^l/Dt^2) + (R_{khsl}\,V^h\,V^l)\eta^s = Q_k|_l\,\eta^l = -(\Pi_{,k})|_l\,\eta^l \equiv -\Pi_{kl}\,\eta^l$$

$$[\Pi_{,k}|_l = \Pi_{k,l} - \Gamma^h{}_{kl}\,\Pi_{,h} = \Pi_{l,k} - \Gamma^h{}_{lk}\,\Pi_{,h} = \Pi_l|_k \equiv \Pi_{kl} = \Pi_{lk}], \qquad \text{(f1)}$$

along the (unperturbed) tangential and normal directions. But, first, we notice that since $dq^h/dt \equiv V^k = (ds/dt)\,\tau^h \equiv V\tau^h$ and thanks to Equation c1, the key *second* left-side term of Equation f1 reduces to

$$R_{khsl}\,V^h\,\eta^s\,V^l = V^2\,R_{khsl}\,\tau^h(a\,\tau^s + b\,\nu^s)\tau^l$$

$$= V^2(a\,R_{khsl}\,\tau^h\,\tau^s\,\tau^l + b\,R_{khsl}\,\tau^h\,\nu^s\,\tau^l)$$

(by Equation 3.11.4b, the *first* sum vanishes)

$$= V^2\,b\,R_{khsl}\,\tau^h\,\nu^s\,\tau^l; \qquad \text{(f2)}$$

and so the *two DOF (covariant) perturbation equation* is

$$g_{kl}(D^2\eta^l/Dt^2) + V^2\,b\,R_{khsl}\,\tau^h\,\nu^s\,\tau^l \equiv -\Pi_{kl}\,\eta^l. \qquad \text{(f3)}$$

Now, dotting the above with τ^k while invoking Equation e2 and recalling that (by Equation 3.11.4a) $R_{khsl} = -R_{hksl} \Rightarrow R_{khsl}\ \tau^k\,\tau^h\,\nu^s\,\tau^l = 0$, we obtain the *tangential* perturbation equation (no R-effects!):

$$(d^2a/dt^2) - 2(db/dt)\omega - b(d\omega/dt) - a\,\omega^2 + (a\,\tau^k + b\,\nu^k)\tau^l\,\Pi_{kl} = 0; \qquad \text{(g1)}$$

and dotting it with ν^k, while invoking Equation e3 and the definition of Riemannian/sectional curvature K (Equation 3.11.7a ff) with

$$\mathbf{A} \to \nu\ (A^k \to \nu^k) \text{ and } \mathbf{B} \to \tau\ (B^k \to \tau^k)$$

$$\Rightarrow \mathbf{A} \cdot \mathbf{B} = \nu \cdot \tau = 0, \quad \mathbf{A} \cdot \mathbf{A} = \nu \cdot \nu = 1, \quad \mathbf{B} \cdot \mathbf{B} = \tau \cdot \tau = 1,$$

i.e.,

$$K \equiv R(\nu, \tau) / G(\nu, \tau) = R(\nu, \tau) / 1 \equiv R = R_{khsl}\, \nu^k\, \tau^h\, \nu^s\, \tau^l$$

$$= R_{1212} / (g_{11}\, g_{22} - g_{12}{}^2) \qquad (q^k\text{-variables}), \tag{g2}$$

we find the *normal* perturbation equation (K-term: curvature effects):

$$(d^2b/dt^2) + 2(da/dt)\omega + a(d\omega/dt) - b\,\omega^2$$

$$+ V^2\, b\, K + (a\, \tau^k + b\, \nu^k)\nu^l\, \Pi_{kl} = 0. \tag{g3}$$

Three DOF. Now the equations of motion along the fundamental path C are (Latin indices: 1, 2, 3):

i. Along the (unperturbed) *tangent* (τ):

$$d^2s/dt^2 = -(\partial\Pi/\partial q^k)\tau^k \equiv \Pi_{,k}\, \tau^k, \tag{h1}$$

ii. Along the (unperturbed) *normal* (ν):

$$V\, \omega = -(\partial\Pi/\partial q^k)\nu^k \equiv \Pi_{,k}\, \nu^k, \tag{h2}$$

iii. Along the (unperturbed) *binormal* (β):

$$0 = -(\partial\Pi/\partial q^k)\beta^k \equiv \Pi_{,k}\, \beta^k; \tag{h3}$$

while the (full) F–S equations there are (with $\kappa_{(n)} \equiv \kappa$, $\kappa_{(\beta)} \equiv \kappa'$; $\omega \equiv \kappa\, V$, $\omega' \equiv \kappa'\, V$):

$$d\tau/dt = \omega\, \nu \tag{i1}$$

$$\text{or} \quad D\tau^k/Dt \equiv d\tau^k/dt + V\, \Gamma^k{}_{dh}\, \tau^d\, \tau^h = \omega\, \nu^k;$$

$$d\nu/dt = -\omega\, \tau + \omega'\, \beta, \tag{i2}$$

$$\text{or} \quad D\nu^k/Dt \equiv d\nu^k/dt + V\, \Gamma^k{}_{dh}\, \tau^d\, \nu^h = -\omega\, \tau^k + \omega'\, \beta^k;$$

$$d\beta/dt = -\omega' \nu, \tag{i3}$$

$$\text{or} \quad D\beta^k/Dt \equiv d\beta^k/dt + V\, \Gamma^k{}_{dh}\, \tau^d\, \beta^h = -\omega' \nu^k.$$

Then, *the most general three DOF perturbation vector* $\delta\mathbf{r} \equiv \delta\mathbf{q} \equiv \eta(t)$ will have the representation:

$$\eta = \eta^k\, \mathbf{e}_k = a\, \tau + b\, \nu + c\, \beta \Rightarrow \eta^k = a\, \tau^k + b\, \nu^k + c\, \beta^k; \tag{j1}$$

and so, $(\ldots)^{\cdot}$-differentiating Equation j1 twice, and reasoning as in the two DOF case, we find

$$g_{kh}(D\eta^k/Dt)\tau^h = (da/dt) - \omega\, b \quad (\textit{tangential perturbational}\ \text{velocity}) \tag{j2}$$

$$g_{kh}(D\eta^k/Dt)\nu^h = (db/dt) + \omega\, a - \omega'\, c \quad (\textit{normal perturbational}\ \text{velocity}) \tag{j3}$$

$$g_{kh}(D\eta^k/Dt)\beta^h = (dc/dt) + \omega'\, b \quad (\textit{binormal perturbational}\ \text{velocity}); \tag{j4}$$

$$g_{kh}(D^2\eta^k/Dt^2)\tau^h = (d^2a/dt^2) - 2(db/dt)\omega - b(d\omega/dt) - a\, \omega^2 - c\, \omega\, \omega'$$

$$(\textit{tangential perturbational}\ \text{acceleration}) \tag{k1}$$

$$g_{kh}(D^2\eta^k/Dt^2)\nu^h = (d^2b/dt^2) + 2(da/dt)\omega - 2(dc/dt)\omega' + a(d\omega/dt) - c(d\omega'/dt)$$

$$- b[\omega^2 + (\omega')^2] \quad (\textit{normal perturbational}\ \text{acceleration}) \tag{k2}$$

$$g_{kh}(D^2\eta^k/Dt^2)\beta^h = (d^2c/dt^2) + 2(db/dt)\omega' + b(d\omega'/dt) + a\, \omega\, \omega' - c(\omega')^2$$

$$(\textit{binormal perturbational}\ \text{acceleration}). \tag{k3}$$

The perturbation equation in the q-variables is now

$$g_{kl}(D^2\eta^l/Dt^2) + V^2\, R_{khsl}\, \tau^h(a\, \tau^s + b\, \nu^s + c\, \beta^s)\tau^l = -\Pi_{kl}\, \eta^l. \tag{l1}$$

Again, the *second* left-side term of the above reduces to

$$V^2(a\, R_{khsl}\, \tau^h\, \tau^s\, \tau^l + b\, R_{khsl}\, \tau^h\, \nu^s\, \tau^l + c\, R_{khsl}\, \tau^h\, \beta^s\, \tau^l)$$

[since $R_{khsl} = -R_{khls}$, the *first* sum vanishes]

$$= V^2(b\, R_{khsl}\, \tau^h\, \nu^s\, \tau^l + c\, R_{khsl}\, \tau^h\, \beta^s\, \tau^l). \tag{l2}$$

Next, let us project Equation l1, with Equation l2, along τ, ν, and β:

i. Dotting it with τ^k, while noting that since $R_{khsl} = -R_{hksl} \Rightarrow R_{hksl}\,\tau^k\,\tau^h\,\nu^s\,\tau^l = 0$, etc., and invoking Equation k1, we obtain the *tangential* perturbation equation (no R–effects!):

$$(d^2a/dt^2) - 2(db/dt)\omega - b(d\omega/dt) - a\,\omega^2 - c\,\omega\,\omega'$$

$$+ (a\,\tau^k + b\,\nu^k + c\,\beta^k)\tau^l\,\Pi_{kl} = 0. \tag{m1}$$

ii. Dotting Equations l1 and l2 with ν^k, while invoking Equation k2, we obtain the *normal* perturbation equation:

$$(d^2b/dt^2) + 2(da/dt)\omega - 2(dc/dt)\omega' + a(d\omega/dt) - c(d\omega'/dt)$$

$$- b[\omega^2 + (\omega')^2] + V^2(b\,R_{khsl}\,\nu^k\,\tau^h\,\nu^s\,\tau^l + c\,R_{khsl}\,\nu^k\,\tau^h\,\beta^s\,\tau^l)$$

$$+ (a\,\tau^k + b\,\nu^k + c\,\beta^k)\nu^l\,\Pi_{kl} = 0. \tag{m2}$$

The $V^2(\ldots)$ terms can be rewritten more concisely as follows:

$$R_{khsl}\,\nu^k\,\tau^h\,\nu^s\,\tau^l \equiv R(\nu, \tau) = K\,G(\nu, \tau)$$

$$= K[(\nu \cdot \nu)(\tau \cdot \tau) - (\nu \cdot \tau)(\nu \cdot \tau)] = K,$$

$R_{khsl}\,\nu^k\,\tau^h\,\beta^s\,\tau^l \equiv A(\nu, \beta)$: *Bi-Gaussian curvature*

$$[\Rightarrow K = R(\nu, \tau) = A(\nu, \nu)],$$

i.e., finally,

$$V^2(\ldots) = V^2[b\,A(\nu, \nu) + c\,A(\nu, \beta)] = V^2[b\,K + c\,A(\nu, \beta)]. \tag{m3}$$

iii. Dotting Equations l1 and l2 with β^k, while invoking Equation k3, yields the *binormal* perturbation equation:

$$(d^2c/dt^2) + 2(db/dt)\omega' + b(d\omega'/dt) + a\,\omega\,\omega' - c(\omega')^2$$

$$+ V^2(b\,R_{khsl}\,\beta^k\,\tau^h\,\nu^s\,\tau^l + c\,R_{khsl}\,\beta^k\,\tau^h\,\beta^s\,\tau^l)$$

$$+ (a\,\tau^k + b\,\nu^k + c\,\beta^k)\beta^l\,\Pi_{kl} = 0. \tag{m4}$$

Again, with the help of the earlier $A(\ldots)$-notation, the $V^2(\ldots)$ terms can be rewritten as follows:

$$V^2(\ldots) = V^2[b\,A(\beta, \nu) + c\,A(\beta, \beta)] = V^2\,[c\,A(\nu, \beta) + cK]. \tag{m5}$$

Example 5.7.3. Energy Perturbation Equations in Intrinsic Variables

Two DOF. Due to $dq^k/dt = V\,\tau^k$ and Equations c1 and d1 of Example 5.7.2, the earlier energy integral (Equation 5.7.7):

$$(g_{kl}\,V^k)D\eta^l/Dt + \Pi_{,k}\,\eta^k = \Delta E, \tag{a}$$

becomes

$$V[(da/dt) - \omega\,b] + (a\,\tau^k + b\,\nu^k)\Pi_{,k} = \Delta E. \tag{b}$$

Also, with the help of (the right sides of) Equations a1 and a2 of Example 5.7.2, the above transforms to

$$V(da/dt) - (dV/dt)a - 2V\,\omega\,b = \Delta E. \tag{c}$$

Three DOF. Due to Equation j1 ff of Example 5.7.2, the energy integral (Equation a) becomes

$$V[(da/dt) - \omega\,b] + (a\,\tau^k + b\,\nu^k + c\,\beta^k)\,\Pi_{,k} = \Delta E; \tag{d}$$

and, again, with the help of the undisturbed equations of motion (Equations h1 through h3 of Example 5.7.2), Equation d transforms to

$$V(da/dt) - (dV/dt)a - 2V\,\omega\,b = \Delta E. \tag{e}$$

[$d/dt(\ldots)$–differentiating Equation d and invoking Equations h1 through h3 of Example 5.7.2, we recover Equation m1 of Example 5.7.2.]

Finally, $d/dt(\ldots)$–differentiating Equation h2 of Example 5.7.2 and invoking Equations h1, h3, and i1 through i3 of Example 5.7.2 produces

$$V(d\omega/dt) - 2\Pi_{,k}\,\tau^k\,\omega + \nu^k\,\tau^l\,V\,\Pi_{kl} = 0; \tag{e1}$$

while $d/dt(\ldots)$–differentiating Equations h1 through h3 of Example 5.7.2, etc., results in

$$V\omega\omega' + \beta^k\,\tau^k\,V\,\Pi_{kl} = 0. \tag{e2}$$

Problem 5.7.1

Show that by $d/dt(\ldots)$–differentiating Equation b of Example 5.7.3 and then utilizing into it: dV/dt, V, τ^k, ν^k from Equations a1, a2, b1, and b2 of Example 5.7.2, and simplifying, one recovers Equation g1 of Example 5.7.3.

Problem 5.7.2

Show that by $d/dt(\ldots)$–differentiating Equation a2 of Example 5.7.2 while invoking Equation a1 of Example 5.7.2, one obtains

$$(dV/dt)\omega + V(d\omega/dt) = \Pi_{,k}\, \tau^k\, \omega - \Pi_{kl}\, \tau^k\, \nu^l\, V, \tag{a1}$$

or

$$2(dV/dt)\omega + V(d\omega/dt) = -\Pi_{kl}\, \tau^k\, \nu^l\, V. \tag{a2}$$

Then show with the help of Equation c of Example 5.7.3, Equation a2, and Equation b1 of Example 5.7.2, that the b-equation (Equation g3 of Example 5.7.2) transforms to

$$(d^2b/dt^2) + [V^2\, K + 3\, V^2\, \kappa^2 + \nu^k\, \nu^l\, \Pi_{kl}]b + 2\, \kappa\, \Delta E = 0; \tag{b}$$

an equation depending only on b and on C–quantities. After $b(t)$ has been found from it, $a(t)$ can be determined by integrating the first-order equation (Equation c of Example 5.7.3)

$$(da/dt) - \left[(dV/dt)/V\right]a = 2\, \omega\, b + \Delta E/V. \tag{c}$$

Problem 5.7.3

With the help of Equations d, e1, and e2 of Example 5.7.3, show that Equations m2 through m5 of Example 5.7.2 can be rewritten as the following equivalent set:

$$d^2b/dt^2 + [V^2\, A(\nu, \nu) + 3\omega^2 - (\omega')^2 + \Pi_{kl}\, \nu^k\, \nu^l]b$$

$$+ [V^2\, A(\nu,\beta) - d\omega'/dt + \Pi_{kl}\, \nu^k\, \beta^l]c - 2\omega'(dc/dt) + 2\kappa\, \Delta E = 0, \tag{a}$$

$$d^2c/dt^2 + [V^2\, A(\beta, \beta) - (\omega')^2 + \Pi_{kl}\, \beta^k\, \beta^l]c$$

$$+ [V^2 A(\beta, \nu) + d\omega'/dt + \Pi_{kl}\, \nu^k\, \beta^l\,]b + 2\omega'(db/dt) = 0, \tag{b}$$

which does not contain a.

6 Lagrangean Mechanics: Kinematics

6.1 INTRODUCTION

This chapter begins our systematic treatment of Lagrangean or analytical mechanics (AM) by presenting the fundamental geometrical and kinematical concepts and equations of constrained and discrete (or discretizable) mechanical systems, such as *position*, *velocity*, *kinematically admissible* and *virtual displacements*; *geometrical* or *positional* and *kinematical* or (here) *velocity constraints* and their integrability (*holonomicity*) or absence thereof (*nonholonomicity*), (including the *transitivity* equations and associated Hamel–Volterra coefficients (Section 3.13) and the fundamental integrability theorem of Frobenius) and stationarity (*scleronomicity*) or *nonstationarity* (*rheonomicity*); also their geometrical interpretation in physical and generalized (configuration/event) spaces, in both particle and system variables. The above constitute the indispensable underpinnings for the understanding of the next chapter on Lagrangean kinetics (i.e., equations of motion of the so-constrained systems). As with the rest of mechanics, the importance of kinematics can never be overestimated.*

6.2 HOLONOMIC CONSTRAINTS

6.2.1 Basic Definitions, System (or Generalized) Coordinates

Let us consider an initially free, or unconstrained, mechanical system S of N particles with inertial positions (recall Section 4.2 ff, with P replaced by p) $\mathbf{r}_p = \mathbf{r}_p(t)$

$$\mathbf{r}_p = x_p\,\mathbf{i} + y_p\,\mathbf{j} + z_p\,\mathbf{k} \equiv r_{p,c}\,\mathbf{u}_c \equiv r_{pc}\,\mathbf{u}_c \quad (p = 1, \ldots, N;\ c = 1, 2, 3). \qquad (6.2.1)^{**}$$

Now, we shall say that S is subject to H ($< 3N$) *positional*, or *geometrical*, or *finite*, or *holonomic* constraints, applied to it either externally or internally, if, throughout the motion, its $\mathbf{r}_p$, in addition to obeying the Newton–Euler equations of motion (plus initial/boundary temporal conditions), are also restricted by the H functionally known but independent equations (of class C^2 in an x, y, z, t domain)

* For complementary reading, we recommend (alphabetically): Hamel (1949), Lur'e (1968), Maißer (1981, 1982, 1983–4), Neimark and Fufaev (1972), Papastavridis (1998), Pars (1965), Poliahov et al. (1985), Prange (1935), Synge (1936, 1960), Vujicic (1981, 1990).

** At this point, the convention according to which uppercase Latin and/or Greek indices signify nonholonomic variables/components (Sections 3.12ff) is suspended. But it will be resumed later in this chapter.

$$f^h = f^h(\mathbf{r}_p, t) = 0 \quad [h = 1, \ldots, H\,(< 3N)] \tag{6.2.2a}$$

or in *component* form [recall Equations 4.2.2 ff and $y^S \equiv \sqrt{M_S}\,\xi^S$ (no sum on S)]:

$$f^h(x_p, y_p, z_p; t) \equiv f^h(r_{pc}, t)$$

$$\equiv f^h(\xi^S, t) \equiv f^h(y^S, t) = 0 \quad (S = 1, \ldots, 3N). \tag{6.2.2b}*$$

By independent we mean that the functions f^h are not related by one or more equations of the form $\phi(f^h) = 0$, or $\phi(f^h, t) = 0$. In this case,

$$\mathrm{rank}(\partial f^h/\partial \xi^S) = \mathrm{rank}(\partial f^h/\partial y^S) = H \quad \text{(constraint independence)} \tag{6.2.2c}$$

in the domain of definition of the ξ or y and t. We say that our so-constrained system has $n \equiv 3N - H\ (> 0)$ *degrees of freedom* (DOF – i.e., H freedoms down from the original $3N$),** which means, analytically, that we can use the H independent constraints (6.2.2b) to express the $3N$ dependent position coordinates, say, the ξ^S (or y^S) in terms of n *independent* (or *unconstrained*, or *minimal*) positional parameters, henceforth called *system*, or *Lagrangean*, coordinates, and denoted by

$$q = \{q^1 = q^1(t), \ldots, q^n = q^n(t)\} \equiv \{q^k = q^k(t);\ k = 1, \ldots, n\}. \tag{6.2.3}***$$

For example, we may express the first H (dependent) ξ or y in terms of the last $3N - H \equiv n$ (independent) of them and time:

$$\xi^D = Z^D(\xi^I, t) \quad \text{or} \quad y^D = \Psi^D(y^I, t) \tag{6.2.3a}$$

($D = 1, \ldots, H$: **D**ependent; $I = H + 1, \ldots, 3N$, or $1, \ldots, n$: **I**ndependent).

More generally, we can choose, in E_{3N}, new curvilinear coordinates:

$$\phi = (\phi^D, \phi^I), \quad \text{where } \phi^D \equiv (\phi^1, \ldots, \phi^H) \quad \text{and} \quad \phi^I \equiv (\phi^{H+1}, \ldots, \phi^{3N}), \tag{6.2.3b}$$

such that in them the H constraints read simply:

---

* Other, perhaps more suggestive terms, for constraints are *conditions* (victorian English: *equations of condition*, German: *bedingungen*) and *connections* or *couplings* (French: *liaisons*, German: *bindungen*, Greek: σύνδεσμοι, Russian: *svyaz'*). The Greek term *holonomic* (Hertz, 1894) means whole, or *integral*, law; i.e., *finite*, nondifferential, equations of constraint; as opposed to *nonholonomic*, i.e., *nonintegrable differential*, such equations (see below). Also, here only equality, or *bilateral*, constraints are examined; i.e., inequality, or *unilateral*, constraints are excluded.

** If $H = 3N$, the $r_{pc} = r_{pc}(t)$ would, in general, be incompatible with the equations of motion (and their initial conditions), while, on the other extreme, the case $H = 0$ *signifies* the (trivial) case of the complete absence of holonomic constraints; i.e., we shall always assume: $0 < h, H < 3N$.

*** The common term *generalized* coordinates, for the qs (due to Kelvin and Tait, 1860s, and intended by them, in those pretensorial days, to signify general, i.e., possibly *curvilinear*, coordinates) is correct but does not capture the essence of the concept, which is *system* coordinates.

$$\phi^1 \equiv f^1(y, t) = 0, \ldots, \phi^H \equiv f^H(y, t) = 0; \quad \text{i.e., } \phi^D \equiv f^D(y, t) = 0; \tag{6.2.3c}$$

and

$$\phi^{H+1} \equiv f^{H+1}(y, t) \neq 0, \ldots, \phi^{3N} \equiv f^{3N}(y, t) \neq 0;$$

i.e.,

$$\phi^I \equiv f^I(y, t) \neq 0, \tag{6.2.3d}$$

$$\phi^{3N+1} \equiv f^{3N+1}(y, t) \equiv t \neq 0; \tag{6.2.3e}$$

where *the f^I are n new arbitrary functions (of class C^2 in some domain of interest, of y and t) but such that when the (assumed admissible) transformations Equations 6.2.3c and d are solved for the y, in terms of the $\phi(\rightarrow \phi^I)$ and t, and the results are inserted back into the constraints (6.2.2b) they satisfy them identically:*

$$y^S = y^S(\phi^D, \phi^I, t) = y^S(\phi^I, t) \equiv y^S(q^1, \ldots, q^n, t) \equiv y^S(q^k, t) \equiv y^S(q^\alpha), \tag{6.2.4a}$$

and

$$\sum (\partial f^h / \partial y^S)(\partial y^S / \partial q^k) = \partial f^h / \partial q^k = 0 \quad [h = 1, \ldots, H; k = 1, \ldots, n] \tag{6.2.4b}$$

$$\text{rank}(\partial y^S / \partial q^k) = n \quad \text{(independence of constraints);} \tag{6.2.4c}$$

where we have renamed, for convenience, $\phi^I \equiv q^k$, $\phi^{3N+1} \equiv q^{n+1} \equiv t$; and, from now on, *all Greek indices run from 1 to n+1.**

Thus, with the introduction of the $n \equiv 3N - H$ *unconstrained qs* we are able to *absorb*, or *build in*, the H constraints into our system and, thus, can write

$$\mathbf{r}_p = \mathbf{r}_p[y(q, t)] \Rightarrow \mathbf{r}_p = \mathbf{r}_p(q^1, \ldots, q^n, t) \equiv \mathbf{r}_p(q^k, t) \equiv \mathbf{r}_p(q^\alpha). \tag{6.2.5)**$$

The ability to represent all possible configurations of every system particle by a *finite number of (now) independent parameters and time,* à la Equation 6.2.5, is absolutely essential (nonnegotiable) to AM. As long as it holds, the original assump-

---

* (a) In view of Equations 6.2.3c, the ϕ^D are sometimes called "equilibrium" coordinates; and this idea proves extremely useful to kinetics, even for nonholonomic constraints (see below and next chapter). (b) It is not hard to see that Equations 6.2.3a correspond to the following ϕ-choice:

$$\phi^D \equiv y^D - \Psi^D(y^I, t) = 0 \quad \text{and} \quad \phi^I \equiv y^I \neq 0.$$

** Similarly, as shown later in this chapter, by introducing appropriate *nonholonomic coordinates* we shall be able to absorb or build in *nonholonomic constraints* into our system. This process of building in *positional* constraints via *holonomic* system coordinates and *velocity* constraints via *nonholonomic* system velocities is the most essential feature of Lagrangean kinematics!

tion of discrete mass points, or particles, is not really necessary — we could have started with a rigid *continuum* as well and assumed Equation 6.2.5; e.g., a rigid body moving about a fixed point, whether assumed discrete or continuous, needs *three* Lagrangean coordinates for its positional (here angular) description (recall Example 1.3.3.1, ii). In view of this, we may write $\mathbf{r}(p, t)$ or $\mathbf{r}(p, q^k, t)$, instead of $\mathbf{r}_p(t)$ or $\mathbf{r}_p(q^k, t)$, where p denotes the name, or tag, of the particle, like a special discrete and/or continuous variable; or, further, regarding p as the position of that particle at some initial time t_o, we can write: $\mathbf{r}(p, t_o) \equiv \mathbf{r}_o \Rightarrow \mathbf{r} = \mathbf{r}(\mathbf{r}_o, t)$ or $\mathbf{r} = \mathbf{r}(\mathbf{r}_o, q, t)$. Most often, the explicit dependence on p, $\mathbf{r}_o$, shall be omitted: $\mathbf{r}_p(q^k, t) \Rightarrow \mathbf{r}(q^k, t) \equiv \mathbf{r}(q^\alpha)$; a practice that will allow us to concentrate on system indices and summations.

6.2.2 Scleronomic vs. Rheonomic Constraints

If the holonomic constraints (6.2.2a ff) *do not contain the time explicitly,* i.e., if

$$\partial f^h/\partial t = 0 \Rightarrow f^h(y^S) = 0 \Rightarrow df^h/dt = \sum (\partial f^h/\partial \mathbf{r}) \cdot (d\mathbf{r}/dt) = 0, \quad (6.2.6a)$$

$$\text{and} \quad \mathbf{r} = \mathbf{r}(q^k) \Rightarrow d\mathbf{r}/dt = (\partial \mathbf{r}/\partial q^k)(dq^k/dt) \equiv v^k\, \mathbf{e}_k, \quad (6.2.6b)$$

these constraints are called *stationary* and the system *scleronomic*; if they do, they are called *nonstationary* and the system *rheonomic.** Typically, rheonomic representations like $\mathbf{r}(q^k, t)$ appear when we express the inertial positions $\mathbf{r}$ in terms of Lagrangean coordinates relative to a noninertial frame of *known* motion. (That *the Lagrangean formalism treats inertial and noninertial system coordinates alike* is one of the great advantages of that method.)

6.2.3 Additional Holonomic Constraints

If the n qs are further constrained by the H' ($< n$) independent holonomic conditions:

$$g^{h'}(q, t) = 0 \quad (h' = 1, \ldots, H') \quad (6.2.7)$$

then, repeating the above procedure, we express the qs in terms of $n - H' \equiv n'$ new Lagrangean coordinates $q' \equiv (q^{k'}; k' = 1, \ldots, n')$, i.e.,

$$q^k = q^k(q^{k'}, t), \quad \text{with } \operatorname{rank}(\partial q/\partial q') = n' \quad [\mathbf{r}(q^k, t) \Rightarrow \mathbf{r}(q^{k'}, t)]. \quad (6.2.7a)$$

The only possible complication is that if the system is scleronomic in the qs, it may become rheonomic in the q's.

---

* After Boltzmann (early 1900s). Respectively, from the Greek *rhéo* (ρέω): to flow, i.e., changing conditions; and *sklerós* (σκληρός): hard, unyielding, i.e., invariable conditions. This is a secondary classification of constraints, compared to that into holonomic and nonholonomic (see Section 6.6).

6.2.4 Geometrical Interpretation of Holonomic Constraints

6.2.4.1 Configuration Space

As discussed in Sections 4.2 and 4.3, *before* the imposition of the constraints (6.2.2a and b) the figurative system particle $P(S)$, or simply P, unrelated to any particular system point and having position vector

$$\mathbf{R} = \sum y^S\, \mathbf{u}_S = \mathbf{R}(y) \quad (\text{where } \mathbf{u}_S \equiv \partial\mathbf{R}/\partial y^S\text{: unit vectors}) \tag{6.2.8a}$$

is free to roam all over the *unconstrained configuration space* of the system, i.e., the Euclidean space E_{3N}. *After* these constraints are imposed on S, the above becomes

$$\mathbf{R} = \mathbf{R}[y^S(q^k, t)] = \mathbf{R}(q^k, t) = \mathbf{R}(q^\alpha) \equiv \mathbf{q}(q^k, t) \equiv \mathbf{q}(q^\alpha). \tag{6.2.8b}$$

Geometrically this means that now P (the tip of $\mathbf{R}$) is forced to remain on an E_{3N}-immersed (generally curved) n-dimensional surface R_n called *constrained configuration space* of S, and described by the curvilinear coordinates q^k. In sum, *the imposition of general (nonlinear) holonomic constraints on an initially unconstrained system turns its configuration space from Euclidean (flat) to non-Euclidean (curved).**

Hence, the problem of holonomically constrained dynamics consists in finding the motion of the figurative particle P, of unit mass, on R_n under various impressed and constraint-induced reactive forces, and this explains the importance of the motion of a particle on a general n-surface (Chapter 5).

6.2.4.2 Extended Configuration Space

Instead of the preceding "dynamical" manifolds E_{3N}, R_n, we may use the *configuration and time,* or extended, or kinematical, space-time manifolds, i.e., unions of E_{3N}, R_n with time $q^{n+1} \equiv t$: E_{3N+1} (unconstrained) and R_{n+1} (constrained), respectively. R_n is suitable for scleronomic systems, while R_{n+1} is suitable for rheonomic systems. Here is why: The stationary H equations $f^h(\mathbf{r}) = f^h(y) = 0$ define, in E_{3N}, a *fixed* (nonmoving) and *rigid* (nondeforming) n-surface; while the nonstationary H equations $f^h(\mathbf{r}, t) = f^h(y, t) = 0$ define, in E_{3N}, a *nonstationary* (moving) and *nonrigid* (deforming) n-surface. But, in E_{3N+1}, these same nonstationary constraints define a stationary and rigid $(n+1)$-surface; and hence the relativity of these terms! Further, through each R_n-point, there is an $(n-1)$-ple infinity of kinematically possible system

* (a) If these constraints are *linear* in the y, then R_n is an n-dimensional *hyperplane*. (b) That curved space turns out to be a *Riemannian* (or, sometimes, *pseudo-Riemannian*) manifold, with metric based on the kinetic energy of the system, which is why we denoted it as R_n. Also, it should be clear by now that this "curveness" of the constrained configuration space has nothing to do with the possible use of curvilinear coordinates in E_{3N}.

paths, each with an arbitrary rate of traverse (i.e., dq/dt), while through each R_{n+1}-point there is an n-ple infinity of such paths, but these latter are not traversed in time, since there is no motion in R_{n+1}.*

6.3 VELOCITY, ADMISSIBLE AND VIRTUAL DISPLACEMENTS, AND ACCELERATION IN PARTICLE AND HOLONOMIC SYSTEM VARIABLES

- The absolute, or inertial, *velocity* of a typical system particle, with Lagrangean positional representation $\mathbf{r} = \mathbf{r}(q^k, t) \equiv \mathbf{r}(q^\alpha)$ is

$$\mathbf{v} \equiv d\mathbf{r}/dt = (\partial\mathbf{r}/\partial q^k)(dq^k/dt) + \partial\mathbf{r}/\partial t = (\partial\mathbf{r}/\partial q^\alpha)(dq^\alpha/dt)$$

$$\equiv v^k\,\mathbf{e}_k + v^{n+1}\,\mathbf{e}_{n+1} \equiv v^k\,\mathbf{e}_k + \mathbf{e}_0 \equiv v^\alpha\,\mathbf{e}_\alpha, \tag{6.3.1}$$

where

$$\mathbf{e}_k \equiv \partial\mathbf{r}/\partial q^k, \quad \mathbf{e}_{n+1} \equiv \partial\mathbf{r}/\partial q_{n+1} \equiv \partial\mathbf{r}/\partial t \equiv \mathbf{e}_0;$$

$$v^k \equiv dq^k/dt, \quad v^{n+1} \equiv v^0 \equiv dq^{n+1}/dt \equiv dt/dt = 1. \tag{6.3.1a}$$

- The (absolute, or inertial) first-order, or elementary, *kinematically admissible,* or *possible,* and *virtual* displacements of that particle, $d\mathbf{r}$ and $\delta\mathbf{r}$, respectively, are

$$d\mathbf{r} = (\partial\mathbf{r}/\partial q^k)dq^k + (\partial\mathbf{r}/\partial t)dt = \mathbf{e}_k\,dq^k + \mathbf{e}_0\,dt \equiv \mathbf{e}_\alpha\,dq^\alpha, \tag{6.3.2a}$$

$$\delta\mathbf{r} \equiv (d\mathbf{r})\,|_{dt\to\delta t=0} = (\partial\mathbf{r}/\partial q^k)\delta q^k = \mathbf{e}_k\,\delta q^k \tag{6.3.2b}$$

[= linear and *homogeneous* part of $\mathbf{r}(q + \delta q, t) - \mathbf{r}(q, t)$; i.e., for a *fixed* time].

- The (absolute, or inertial) *acceleration* of that particle, $\mathbf{a}$, is

$$\mathbf{a} \equiv d\mathbf{v}/dt$$

$$= \ldots = (d^2q^k/dt^2)\mathbf{e}_k + (dq^k/dt)(dq^l/dt)\mathbf{e}_{k,l} + 2(dq^k/dt)\mathbf{e}_{k,n+1} + \mathbf{e}_{n+1,n+1}$$

$$\equiv (dv^k/dt)\mathbf{e}_k + v^k\,v^l\,\mathbf{e}_{k,l} + 2v^k\,\mathbf{e}_{k,0} + \mathbf{e}_{0,0}. \tag{6.3.3}$$

The n+1 holonomic (= gradient), covariant and, generally, neither unit nor orthogonal, vectors $\{\mathbf{e}_\alpha\} \equiv \{\mathbf{e}_k, \mathbf{e}_{n+1} \equiv \mathbf{e}_0\}$, functions of the particle *and*

---

* Additional differences between R_n and R_{n+1} will be given later, in connection with nonholonomic constraints and kinematically admissible/virtual displacements. Also, other generalized spaces are used in mechanics (e.g., phase space), but they will not be discussed here.

the system configuration and time, i.e., *particle and system* vectors, are fundamental to all subsequent considerations.*

The definitions (6.3.1 through 6.3.3) are perfectly general and hold whether the n qs are inertial or not, and even if the qs and/or their $d/dt(\ldots)$-derivatives become constrained later. (However, if the latter does not happen, then $\delta\mathbf{r} = \mathbf{0}$ leads immediately to $\delta q^k = 0$.)

Next, let us examine the *system* counterparts of the above, that is, the inertial velocity, kinematically admissible and virtual displacements, and acceleration of the figurative system particle P, i.e., of the tip of Equation 6.2.8b. These are defined, respectively, as follows:

$$\mathbf{V} \equiv d\mathbf{R}/dt = (\partial\mathbf{R}/\partial q^k)(dq^k/dt) + \partial\mathbf{R}/\partial t = (\partial\mathbf{R}/\partial q^\alpha)(dq^\alpha/dt)$$

$$\equiv V^k\,\mathbf{E}_k + V^{n+1}\,\mathbf{E}_{n+1} \equiv V^k\,\mathbf{E}_k + \mathbf{E}_0 \equiv V^\alpha\,\mathbf{E}_\alpha, \tag{6.3.4a}$$

where

$$\mathbf{E}_k \equiv \partial\mathbf{R}/\partial q^k, \quad \mathbf{E}_{n+1} \equiv \partial\mathbf{R}/\partial q_{n+1} \equiv \partial\mathbf{R}/\partial t \equiv \mathbf{E}_0;$$

$$V^k \equiv dq^k/dt (= v^k), \quad V^{n+1} \equiv V^0 \equiv dq^{n+1}/dt \equiv dt/dt = 1 \quad (= v^{n+1} = v^0):$$

Holonomic (contravariant) components of system velocity. (6.3.4b)

$$d\mathbf{R} \equiv d\mathbf{q} = (\partial\mathbf{R}/\partial q^k)dq^k + (\partial\mathbf{R}/\partial t)dt = \mathbf{E}_k\,dq^k + \mathbf{E}_0\,dt \equiv \mathbf{E}_\alpha\,dq^\alpha; \tag{6.3.4c}$$

$$\delta\mathbf{R} \equiv \delta\mathbf{q} = (d\mathbf{R})\,|_{dt\to\delta t=0} = (\partial\mathbf{R}/\partial q^k)\delta q^k = \mathbf{E}_k\,\delta q^k; \tag{6.3.4d}$$

$$\mathbf{A} \equiv d\mathbf{V}/dt$$

$$= \ldots = (d^2q^k/dt^2)\mathbf{E}_k + (dq^k/dt)(dq^l/dt)\mathbf{E}_{k,l} + 2(dq^k/dt)\mathbf{E}_{k,n+1} + \mathbf{E}_{n+1,n+1}$$

$$= (dV^k/dt)\mathbf{E}_k + V^k\,V^l\,\mathbf{E}_{k,l} + 2V^k\,\mathbf{E}_{k,0} + \mathbf{E}_{0,0}; \tag{6.3.4e}$$

i.e., the $\{\mathbf{E}_\alpha\} \equiv \{\mathbf{E}_k, \mathbf{E}_{n+1} \equiv \mathbf{E}_0\}$ are the system counterparts of the $\{\mathbf{e}_\alpha\} \equiv \{\mathbf{e}_k, \mathbf{e}_{n+1} \equiv \mathbf{e}_0\}$ and, like the latter, they are basic to the Lagrangean method.

From Equations 6.3.1 ff and 6.2.4a ff, we readily obtain, respectively,

$$\partial\mathbf{r}/\partial q^k = \partial\mathbf{v}/\partial v^k = \partial\mathbf{a}/\partial(dv^k/dt) = \ldots = \mathbf{e}_k, \tag{6.3.5a}$$

$$\partial\mathbf{R}/\partial q^k = \partial\mathbf{V}/\partial V^k = \partial\mathbf{A}/\partial(dV^k/dt) = \ldots = \mathbf{E}_k, \tag{6.3.5b}$$

---

* These vectors were, most likely, introduced by Somoff (early 1870s, publ. 1879); but were popularized as *begleitvektoren* (i.e., accompanying, or attendant, vectors) by K. Heun (early 1900s).

and

$$\partial \mathbf{v}/\partial q^k = \partial/\partial q^k(d\mathbf{r}/dt) = d/dt(\partial \mathbf{r}/\partial q^k) = d/dt(\partial \mathbf{v}/\partial v^k) = d\mathbf{e}_k/dt, \quad (6.3.5c)$$

$$\partial \mathbf{V}/\partial q^k = \partial/\partial q^k(d\mathbf{R}/dt) = d/dt(\partial \mathbf{R}/\partial q^k) = d/dt(\partial \mathbf{V}/\partial V^k) = d\mathbf{E}_k/dt; \quad (6.3.5d)$$

and from these we obtain the following basic *integrability* conditions:

$$\mathbf{e}_{k,l} = \mathbf{e}_{l,k}, \quad \mathbf{e}_{k,0} = \mathbf{e}_{0,k}, \quad \text{and} \quad E_k(\mathbf{v}) \equiv d/dt(\partial \mathbf{v}/\partial v^k) - \partial \mathbf{v}/\partial q^k = \mathbf{0}, \quad (6.3.5e)$$

$$\mathbf{E}_{k,l} = \mathbf{E}_{l,k}, \quad \mathbf{E}_{k,0} = \mathbf{E}_{0,k}, \quad \text{and} \quad E_k(\mathbf{V}) \equiv d/dt(\partial \mathbf{V}/\partial V^k) - \partial \mathbf{V}/\partial q^k = \mathbf{0}; \quad (6.3.5f)$$

where

$$E_k(\ldots) \equiv d/dt\left[\partial(\ldots)/\partial(dq^k/dt)\right] - \partial(\ldots)/\partial q^k \equiv d/dt\left[\partial(\ldots)/\partial v^k\right] - \partial(\ldots)/\partial q^k:$$

Euler–Lagrange operator in *holonomic* variables. (6.3.5g)

The above equations (6.3.5a through f) are kinematical *identities* that hold always, as long as the q^k are holonomic coordinates; i.e., as with the earlier definitions (6.3.1 ff), they are independent of any past or future additional holonomic and/or nonholonomic constraints imposed on our system.

Clearly, the $n+1$ vectors $\{\mathbf{E}_\alpha\}$ are tangent to the q^α coordinates through (q, t), i.e., P, and (recalling the discussion in Sections 2.9, 2.12, and 5.5) span the *tangent space* to R_{n+1} there: $\mathbf{T}(\mathbf{R}_{n+1}, \mathbf{P}) \equiv \mathbf{T}_{n+1}(\mathbf{P})$; while the n vectors $\{\mathbf{E}_k\}$ span the corresponding R_n-space: $\mathbf{T}(\mathbf{R}_n, \mathbf{P}) \equiv \mathbf{T}_n(\mathbf{P})$. Let us find the relation between the $\{\mathbf{e}_\alpha\}$ and $\{\mathbf{E}_\alpha\}$. Reverting temporarily to the discrete notation, we have (with $S = 1, \ldots, 3N$ and $p = 1, \ldots, N$):

$$\begin{aligned}\mathbf{R} &= \sum y^S \mathbf{u}_S = \sum \sqrt{M_S}\, \xi^S \mathbf{u}_S = \sum \sqrt{m_p}\,(x_p \mathbf{i} + y_p \mathbf{j} + z_p \mathbf{k}) \\ &= \sum \sqrt{m_p}\, \mathbf{r}_p,\end{aligned} \quad (6.3.6a)$$

or, in *continuum* notation:

$$\mathbf{R} = \mathcal{S} \sqrt{dm}\, \mathbf{r}. \quad (6.3.6b)*$$

And since all these particle components are functions of the qs and t, we get

* The Lagrangean symbol $\mathcal{S}(\ldots)$ signifies *material summation*, over the system particles, for a fixed time; like a *Stieltje's* integral, so it can handle uniformly both continuous and discrete situations. Those uncomfortable with it may replace it with the more familiar Leibnizian $\int(\ldots)$.

$$\partial \mathbf{R}/\partial q^\alpha = \sum (\partial y^S/\partial q^\alpha)\mathbf{u}_S = \sum \sqrt{M_S}(\partial \xi^S/\partial q^\alpha)\mathbf{u}_S$$
$$= \sum \sqrt{m_p}[(\partial x_p/\partial q^\alpha)\mathbf{i} + \ldots] = \sum \sqrt{m_p}(\partial \mathbf{r}_p/\partial q^\alpha); \tag{6.3.7a}$$

or, in *continuum* notation:

$$\partial \mathbf{R}/\partial q^\alpha = \mathbf{S}\sqrt{dm}(\partial \mathbf{r}/\partial q^\alpha), \quad \text{i.e., } \mathbf{E}_\alpha = \mathbf{S}\sqrt{dm}\,\mathbf{e}_\alpha. \tag{6.3.7b}$$

6.4 NONHOLONOMIC COORDINATES, VELOCITIES, ETC.

6.4.1 Basic Definitions, Quasi-coordinates

Next (and with an eye toward the *absorbing* or *building in* of future Pfaffian, possibly nonholonomic, constraints, and thus express the above kinematical vectors $\mathbf{v}$, $\mathbf{V}$, $\delta\mathbf{r}$, $\delta\mathbf{R}$, etc., in terms of independent velocity, virtual displacement, etc., "parameters," for subsequent use in kinetics), and recalling Section 3.12ff, we introduce at $T_{n+1}(P)$ the $n + 1$, generally nonholonomic, contravariant system *velocities* $\{\omega^\Lambda\} = \{\omega^L, \omega^{N+1}\}$ by the following linear (affine) and invertible relations:

- $\omega^L \equiv d\theta^L/dt \equiv A^L{}_\lambda(dq^\lambda/dt) \equiv A^L{}_\lambda V^\lambda$

$$= A^L{}_l V^l + A^L{}_{n+1} V^{n+1} = A^L{}_l V^l + A^l{}_{n+1} \quad (\text{since } V^{n+1} = 1) \tag{6.4.1a)*}$$

$$\omega^{N+1} \equiv d\theta^{N+1}/dt \equiv A^{N+1}{}_\lambda V^\lambda$$
$$= A^{N+1}{}_l V^l + A^{N+1}{}_{n+1} V^{n+1} = (0)V^l + A^{N+1}{}_{n+1} V^{n+1}$$
$$= \delta^{N+1}{}_{n+1} V^{n+1} = (1)(1) = dt/dt = 1 \quad (\textit{isochrony}); \tag{6.4.1b}$$

$$\Rightarrow V^l = A^l{}_\Lambda \omega^\Lambda = A^l{}_L \omega^L + A^l{}_{N+1} \omega^{N+1}$$
$$= A^l{}_L \omega^L + A^l{}_{N+1} \quad (\text{by Equation 6.4.1b}), \tag{6.4.1c}$$

$$V^{n+1} = A^{n+1}{}_\Lambda \omega^\Lambda = A^{n+1}{}_L \omega^L + A^{n+1}{}_{N+1} \omega^{N+1}$$
$$= (0)\omega^L + \delta^{n+1}{}_{N+1} \omega^{N+1} = (1)(1) = 1; \tag{6.4.1d}$$

---

* Due to Hamel (1903–4; and 1938, for the nonlinear case). Also, at this point we reintroduce and extend the notation of Sections 3.12ff as follows:

- Uppercase (nonholonomic) **Latin** indices range from 1 to $N = n$, while
- Uppercase (nonholonomic) **Greek** indices range from 1 to $N + 1 = n + 1$.

Here, N signifies the nonholonomic counterpart of n, and has nothing to do with the earlier number of system particles. We hope that the appropriate meaning of it should be clear from the context. Equations 6.4.1b and d essentially state that $\omega^{N+1} = V^{n+1} = 1$ (always); and similarly with Equations 6.4.2b and d: $d\theta^{N+1} = dq^{n+1} = dt$ (always); and Equations 6.4.3b and d: $\delta\theta^{N+1} = \delta q^{n+1} = \delta t = 0$ (always).

or, compactly,

$$\omega^{\Lambda} \equiv A^{\Lambda}{}_{\lambda}\, V^{\lambda} \Leftrightarrow V^{\lambda} = A^{\lambda}{}_{\Lambda}\, \omega^{\Lambda}; \tag{6.4.1e}$$

Now, in general,

$$\partial A^{L}{}_{l}/\partial q^{\lambda} - \partial A^{L}{}_{\lambda}/\partial q^{l} \equiv A^{L}{}_{l,\lambda} - A^{L}{}_{\lambda,l} \neq 0,$$

for at least one value of l, λ. (6.4.1f)

Then, as discussed in Equations 3.12.3 ff, $A^{L}{}_{\lambda}\, dq^{\lambda}$ is an *inexact*, or *nontotal*, Pfaffian expression, and θ^{L} is a nonholonomic coordinate, or *quasicoordinate*. If, on the other hand, $A^{L}{}_{l,\lambda} = A^{L}{}_{\lambda,l}$, identically for all l, λ, then θ^{L} is just another holonomic or Lagrangean coordinate; and if that holds for all L, then the $\{\theta^{L}\}$ are just another set of Lagrangean coordinates, like the $\{q^{l}\}$; i.e., there exists an admissible global transformation: $\theta^{L} = \theta^{L}(q^{l}, q^{n+1} \equiv t) \Leftrightarrow q^{l} = q^{l}(\theta^{L}, \theta^{N+1} \equiv t)$. Below we develop a kinematics that is uniformly valid for both holonomic and nonholonomic coordinates.

In analogy to Equations 6.4.1a through e and Section 6.3, we define the corresponding *kinematically admissible* system *displacements* by

- $d\theta^{L} = A^{L}{}_{\lambda}\, dq^{\lambda} = A^{L}{}_{l}\, dq^{l} + A^{L}{}_{n+1}\, dq^{n+1} = A^{L}{}_{l}\, dq^{l} + A^{L}{}_{n+1}\, dt,$ (6.4.2a)

$$d\theta^{N+1} = A^{N+1}{}_{\lambda}\, dq^{\lambda} = A^{N+1}{}_{l}\, dq^{l} + A^{N+1}{}_{n+1}\, dq^{n+1}$$

$$= (0)dq^{l} + (\delta^{N+1}{}_{n+1})dt = (1)dt = dt \text{ (isochrony);} \tag{6.4.2b}$$

$$\Rightarrow dq^{l} = A^{l}{}_{\Lambda}\, d\theta^{\Lambda} = A^{l}{}_{L}\, d\theta^{L} + A^{l}{}_{N+1}\, d\theta^{N+1}$$

$$= A^{l}{}_{L}\, d\theta^{L} + A^{l}{}_{N+1}\, dt, \tag{6.4.2c}$$

$$dq^{n+1} = A^{n+1}{}_{\Lambda}\, d\theta^{\Lambda} = A^{n+1}{}_{L}\, d\theta^{L} + A^{n+1}{}_{N+1}\, d\theta^{N+1}$$

$$= (0)\, d\theta^{L} + (\delta^{n+1}{}_{N+1})\, d\theta^{N+1} = (1)\, dt = dt; \tag{6.4.2d}$$

or, compactly,

$$d\theta^{\Lambda} \equiv A^{\Lambda}{}_{\lambda}\, dq^{\lambda} \Leftrightarrow dq^{\lambda} = A^{\lambda}{}_{\Lambda}\, d\theta^{\Lambda}; \tag{6.4.2e}$$

and the *virtual* system displacements by

- $\delta\theta^{L} = A^{L}{}_{\lambda}\, \delta q^{\lambda} = A^{L}{}_{l}\, \delta q^{l} + A^{L}{}_{n+1}\, \delta q^{n+1}$

$$= A^{L}{}_{l}\, \delta q^{l} + A^{L}{}_{n+1}\, \delta t = A^{L}{}_{l}\, \delta q^{l} \quad (\delta t = 0\text{: } \textit{time constraint!}) \tag{6.4.3a}$$

$$\delta\theta^{N+1} = A^{N+1}{}_{\lambda}\,\delta q^{\lambda} = A^{N+1}{}_{l}\,\delta q^{l} + A^{N+1}{}_{n+1}\,\delta q^{n+1}$$

$$= (0)\delta q^{l} + (\delta^{N+1}{}_{n+1})\delta q^{n+1} = (1)\delta t = 0; \qquad (6.4.3b)$$

$$\Rightarrow \delta q^{l} = A^{l}{}_{\Lambda}\,\delta\theta^{\Lambda} = A^{l}{}_{L}\,\delta\theta^{L} + A^{l}{}_{N+1}\,\delta\theta^{N+1}$$

$$= A^{l}{}_{L}\,\delta\theta^{L} + A^{l}{}_{N+1}\,\delta t = A^{l}{}_{L}\,\delta\theta^{L} + A^{l}{}_{N+1}(0) = A^{l}{}_{L}\,\delta\theta^{L}, \qquad (6.4.3c)$$

$$\delta q^{n+1} = A^{n+1}{}_{\Lambda}\,\delta\theta^{\Lambda} = A^{n+1}{}_{L}\,\delta\theta^{L} + A^{n+1}{}_{N+1}\,\delta\theta^{N+1}$$

$$= (0)\delta\theta^{L} + (\delta^{n+1}{}_{N+1})\delta\theta^{N+1} = (1)\delta t = 0; \qquad (6.4.3d)$$

or, compactly,

$$\delta\theta^{L} \equiv A^{L}{}_{l}\,\delta q^{l} \Leftrightarrow \delta q^{l} = A^{l}{}_{L}\,\delta\theta^{L}. \qquad (6.4.3e)$$

6.4.2 Properties of Pfaffian Transformations

i. The $(n+1) \times (n+1)$ coefficient matrices $(A^{\Lambda}{}_{\lambda})$ and $(A^{\lambda}{}_{\Lambda})$, whose elements are assumed (at least) once piecewise continuously differentiable functions of the $q^{\lambda} = \{q^{l}, t\}$, in some R_{n+1}-region of interest, are mutually inverse ($\Rightarrow$ nonsingular) and, therefore, satisfy the following *compatibility* conditions:

$$A^{\Lambda}{}_{\lambda}\,A^{\lambda}{}_{\Gamma} = \delta^{\Lambda}{}_{\Gamma} \quad \text{and} \quad A^{\Lambda}{}_{\lambda}\,A^{\beta}{}_{\Lambda} = \delta^{\beta}{}_{\lambda}. \qquad (6.4.4)$$

Also, due to the special forms of these coefficients, dictated by the general requirements of analytical mechanics:

$$A^{N+1}{}_{l} = \delta^{N+1}{}_{l} = 0,$$

$$A^{N+1}{}_{n+1} = \delta^{N+1}{}_{n+1} = 1 \quad (\text{compactly } A^{N+1}{}_{\lambda} = \delta^{N+1}{}_{\lambda}), \qquad (6.4.4a)$$

$$A^{n+1}{}_{L} = \delta^{n+1}{}_{L} = 0,$$

$$A^{n+1}{}_{N+1} = \delta^{n+1}{}_{N+1} = 1 \quad (\text{compactly } A^{\lambda}{}_{N+1} = \delta^{\lambda}{}_{N+1}). \qquad (6.4.4b)$$

- The *first* of Equations 6.4.4 yields the two groups of such conditions (recall footnote in Example 2.3.1):

$$A^{L}{}_{l}\,A^{l}{}_{K} + A^{L}{}_{n+1}\,A^{n+1}{}_{K} = A^{L}{}_{l}\,A^{l}{}_{K} + A^{L}{}_{n+1}(0)$$

$$= \delta^{L}{}_{K}, \quad \text{i.e., } A^{L}{}_{l}\,A^{l}{}_{K} = \delta^{L}{}_{K}, \qquad (6.4.4c)$$

and

$$A^L{}_l\, A^l{}_{N+1} + A^L{}_{n+1}\, A^{n+1}{}_{N+1} = A^L{}_l\, A^l{}_{N+1} + A^L{}_{n+1}(\delta^{n+1}{}_{N+1}) = \delta^L{}_{N+1}(= 0),$$

i.e.,

$$A^L{}_l\, A^l{}_{N+1} = -\delta^{n+1}{}_{N+1}\, A^L{}_{n+1} \equiv -A^L{}_{N+1}; \tag{6.4.4d}$$

- And, similarly, the *second* of them yields:

$$A^L{}_l\, A^k{}_L = \delta^k{}_l \quad \text{and} \quad A^l{}_L\, A^L{}_{n+1} = -\delta^{N+1}{}_{n+1}\, A^l{}_{N+1} \equiv -A^l{}_{n+1}. \tag{6.4.4e, f}$$

Equations 6.4.4c and e state that the "spatial" parts of the coefficient matrices, $(A^L{}_l)$ and $(A^l{}_L)$, *are independently mutually inverse* — a fact of importance in kinetics (Chapter 7).

ii. From the above, we readily conclude that (also, revoking the results of Section 3.12ff):

$$\partial\theta^L/\partial q^l \equiv \partial\omega^L/\partial V^l = \partial(d\theta^L)/\partial(dq^l) = \partial(\delta\theta^L)/\partial(\delta q^l) \equiv A^L{}_l, \tag{6.4.5a}$$

$$\partial q^l/\partial\theta^L \equiv \partial V^l/\partial\omega^L = \partial(dq^l)/\partial(d\theta^L) = \partial(\delta q^l)/\partial(\delta\theta^L) \equiv A^l{}_L; \tag{6.4.5b}$$

$$\partial\theta^{N+1}/\partial q^l \equiv A^{N+1}{}_l = \delta^{N+1}{}_l = 0, \quad \partial\theta^{N+1}/\partial q^{n+1} \equiv A^{N+1}{}_{n+1} = \delta^{N+1}{}_{n+1} = 1, \tag{6.4.5c}$$

$$\partial q^{n+1}/\partial\theta^L \equiv A^{n+1}{}_L = \delta^{n+1}{}_L = 0, \quad \partial q^{n+1}/\partial\theta^{N+1} \equiv A^{n+1}{}_{N+1} = \delta^{n+1}{}_{N+1} = 1; \tag{6.4.5d}$$

and also recall the (symbolic) *quasi-chain rule*:

$$\partial(\ldots)/\partial\theta^L \equiv [\partial(\ldots)/\partial q^\lambda](\partial q^\lambda/\partial\theta^L) = A^l{}_L[\partial(\ldots)/\partial q^l] \tag{6.4.6a}$$

(invoking Equations 6.4.5b and d_1)

$$\partial(\ldots)/\partial q^l \equiv [\partial(\ldots)/\partial\theta^\Lambda](\partial\theta^\Lambda/\partial q^l) = A^L{}_l[\partial(\ldots)/\partial\theta^L] \tag{6.4.6b}$$

(invoking Equations 6.4.5a and c_1); and (in analogy with Equations 6.4.6a and b, respectively):

$$\begin{aligned}\partial(\ldots)/\partial\theta^{N+1} &\equiv [\partial(\ldots)/\partial q^\lambda]\,(\partial q^\lambda/\partial\theta^{N+1}) \\ &= [\partial(\ldots)/\partial q^l]\,(\partial q^l/\partial\theta^{N+1}) + [\partial(\ldots)/\partial q^{n+1}]\,(\partial q^{n+1}/\partial\theta^{N+1}) \\ &= A^l{}_{N+1}[\partial(\ldots)/\partial q^l] + A^{n+1}{}_{N+1}[\partial(\ldots)/\partial q^{n+1}] \\ &= A^l{}_{N+1}[\partial(\ldots)/\partial q^l] + \partial(\ldots)/\partial t \quad [\neq \partial(\ldots)/\partial t],\end{aligned} \tag{6.4.6c}$$

$$\partial(\ldots)/\partial q^{n+1} \equiv \partial(\ldots)/\partial t \equiv [\partial(\ldots)/\partial\theta^\Lambda]\,(\partial\theta^\Lambda/\partial q^{n+1})$$

$$= [\partial(\ldots)/\partial\theta^L]\,(\partial\theta^L/\partial q^{n+1}) + [\partial(\ldots)/\partial\theta^{N+1}]\,(\partial\theta^{N+1}/\partial q^{n+1})$$

$$= A^L{}_{n+1}[\partial(\ldots)/\partial\theta^L] + A^{N+1}{}_{n+1}[\partial(\ldots)/\partial\theta^{N+1}]$$

$$= A^L{}_{n+1}[\partial(\ldots)/\partial\theta^L] + \partial(\ldots)/\partial\theta^{N+1}; \qquad (6.4.6d)$$

which are compatible with the differential (kinematically admissible) operator invariance:

$$d(\ldots) = [\partial(\ldots)/\partial q^\lambda]dq^\lambda = [\partial(\ldots)/\partial q^l]dq^l + [\partial(\ldots)/\partial t]dt$$

$$= [\partial(\ldots)/\partial q^l]\,(A^l{}_L\, d\theta^L + A^l{}_{N+1}\, dt) + [\partial(\ldots)/\partial t]dt$$

$$= \{[\partial(\ldots)/\partial q^l]A^l{}_L\}d\theta^L + \{[\partial(\ldots)/\partial q^l]A^l{}_{N+1} + [\partial(\ldots)/\partial t]\}dt$$

$$= \{[\partial(\ldots)/\partial q^l]A^l{}_L\}d\theta^L + [\partial(\ldots)/\partial\theta^{N+1}]d\theta^{N+1}$$

$$= [\partial(\ldots)/\partial\theta^L]d\theta^L + [\partial(\ldots)/\partial\theta^{N+1}]d\theta^{N+1}$$

$$= [\partial(\ldots)/\partial\theta^\Lambda]d\theta^\Lambda \quad (\text{with } d\theta^{N+1} = dt). \qquad (6.4.6e)^*$$

6.4.3 Particle and System Kinematics in Quasi-variables

With the help of the preceding results, and the useful (and henceforth frequently employed) notation:

$$f = f(t,\, q,\, dq/dt) = f[t,\, q,\, dq(t,\, q,\, \omega)\,/\,dt] = f^*(t,\, q,\, \omega) = f^*, \quad (6.4.7)$$

we readily obtain the following basic representations for the (inertial) *particle* velocity, acceleration, kinematically admissible and virtual displacements, respectively:

- $\mathbf{v}^* = v^\lambda\, \mathbf{e}_\lambda = (A^\lambda{}_\Lambda\, \omega^\Lambda)\mathbf{e}_\lambda = (A^\lambda{}_\Lambda\, \mathbf{e}_\lambda)\omega^\Lambda \equiv \omega^\Lambda\, \mathbf{e}_\Lambda$ (we recall that $v^\lambda = V^\lambda$)

$$= \omega^L\, \mathbf{e}_L + \omega^{N+1}\, \mathbf{e}_{N+1} \equiv \omega^L\, \mathbf{e}_L + \mathbf{e}_{N+1} \quad (\text{definition of } \mathbf{e}_{N+1}), \qquad (6.4.8a)$$

where

$$\mathbf{e}_\Lambda = A^\lambda{}_\Lambda\, \mathbf{e}_\lambda \Leftrightarrow \mathbf{e}_\lambda = A^\Lambda{}_\lambda\, \mathbf{e}_\Lambda, \qquad (6.4.8b)$$

---

* However, some authors set: $\partial(\ldots)/\partial\theta^{N+1} \equiv [\partial(\ldots)/\partial q^l](\partial q^l/\partial\theta^{N+1}) = A^l{}_{N+1}[\partial(\ldots)/\partial q^l]$.

from which it follows:

$$\mathbf{e}_L = A^l{}_L\ \mathbf{e}_l + A^{n+1}{}_L\ \mathbf{e}_{n+1} = A^l{}_L\ \mathbf{e}_l \Leftrightarrow \mathbf{e}_l = \ldots = A^L{}_l\ \mathbf{e}_L, \tag{6.4.8c}$$

$$\mathbf{e}_{N+1} = A^\lambda{}_{N+1}\ \mathbf{e}_\lambda = A^l{}_{N+1}\ \mathbf{e}_l + A^{n+1}{}_{N+1}\ \mathbf{e}_{n+1} = A^l{}_{N+1}\ \mathbf{e}_l + \delta^{n+1}{}_{N+1}\ \mathbf{e}_{n+1}$$

$$= \ldots = -A^L{}_{N+1}\ \mathbf{e}_L + \delta^{n+1}{}_{N+1}\ \mathbf{e}_{n+1} \quad \text{(by Equation 6.4.4d)} \tag{6.4.8d}$$

$$\mathbf{e}_{n+1} = A^\Lambda{}_{n+1}\ \mathbf{e}_\Lambda = A^L{}_{n+1}\ \mathbf{e}_L + A^{N+1}{}_{n+1}\ \mathbf{e}_{N+1} = A^L{}_{n+1}\ \mathbf{e}_L + \delta^{N+1}{}_{n+1}\ \mathbf{e}_{N+1}$$

$$= \ldots = -A^l{}_{n+1}\ \mathbf{e}_l + \delta^{N+1}{}_{n+1}\ \mathbf{e}_{N+1} \quad \text{(by Equation 6.4.4f)} \tag{6.4.8e}$$

- $\mathbf{a}^* = d/dt(\omega^L\ \mathbf{e}_L + \mathbf{e}_{N+1})$

$$= (d\omega^L/dt)\mathbf{e}_L + \ldots \quad [\text{where} \ldots \equiv \text{no other } (d\omega/dt) \text{ terms}] \tag{6.4.9}$$

- $d\mathbf{r}^* = \ldots = d\theta^L\ \mathbf{e}_L + d\theta^{N+1}\ \mathbf{e}_{N+1} \equiv d\theta^L\ \mathbf{e}_L + dt\ \mathbf{e}_{N+1}$ (6.4.10)

- $\delta\mathbf{r}^* = \ldots = \delta\theta^L\ \mathbf{e}_L;$ (6.4.11)

and completely analogously for their system counterparts:

- $\mathbf{V}^* = \omega^\Lambda\ \mathbf{E}_\Lambda = \ldots = \omega^L\ \mathbf{E}_L + \mathbf{E}_{N+1} \quad [\text{i.e., } v^\Lambda \equiv V^\Lambda \equiv \omega^\Lambda]$ (6.4.12a)

where

$$\mathbf{E}_\Lambda = A^\lambda{}_\Lambda\ \mathbf{E}_\lambda \Leftrightarrow \mathbf{E}_\lambda = A^\Lambda{}_\lambda\ \mathbf{E}_\Lambda, \tag{6.4.12b}$$

$$\mathbf{E}_L = A^l{}_L\ \mathbf{E}_l + A^{n+1}{}_L\ \mathbf{E}_{n+1} = A^l{}_L\ \mathbf{E}_l \Leftrightarrow \mathbf{E}_l = \ldots = A^L{}_l\ \mathbf{E}_L, \tag{6.4.12c}$$

$$\mathbf{E}_{N+1} = A^\lambda{}_{N+1}\ \mathbf{E}_\lambda = \ldots = -A^L{}_{N+1}\ \mathbf{E}_L + \delta^{n+1}{}_{N+1}\ \mathbf{E}_{n+1} \tag{6.4.12d}$$

$$\mathbf{E}_{n+1} = A^\Lambda{}_{n+1}\ \mathbf{E}_\Lambda = \ldots = -A^l{}_{n+1}\ \mathbf{E}_l + \delta^{N+1}{}_{n+1}\ \mathbf{E}_{N+1} \tag{6.4.12e}$$

- $\mathbf{A}^* = d/dt(\omega^L\ \mathbf{E}_L + \mathbf{E}_{N+1})$

$$= (d\omega^L/dt)\mathbf{E}_L + \ldots \quad (\text{where} \ldots \equiv \text{no other } (d\omega/dt) \text{ terms}) \tag{6.4.13}$$

- $d\mathbf{R}^* = \ldots = d\theta^L\ \mathbf{E}_L + d\theta^{N+1}\ \mathbf{E}_{N+1} \equiv d\theta^L\ \mathbf{E}_L + dt\ \mathbf{E}_{N+1}$ (6.4.14)

- $\delta\mathbf{R}^* = \ldots = \delta\theta^L\ \mathbf{E}_L.$ (6.4.15)

The $\{\omega^\Lambda\} = \{\omega^L, \omega^{N+1}\}$ are the *nonholonomic contravariant components of the system velocity,* or *system quasi-velocities,* in the local nonholonomic R_{n+1}-basis $\{\mathbf{P}, \mathbf{E}_\Lambda\} \equiv$

$\{\mathbf{P}; \mathbf{E}_L, \mathbf{E}_{N+1}\}$ in its local tangent space $\mathbf{T}_{n+1}(\mathbf{P})$, while the $\{\delta\theta^L\}$ are the corresponding components of the system virtual displacement along the $\mathbf{T}_n(\mathbf{P}) \equiv \mathbf{T}_{n+1}(\mathbf{P})\,|_{t=\text{constant}}$ basis $\{\mathbf{P}, \mathbf{E}_L\}$.

6.5 THE TRANSITIVITY EQUATIONS

Recalling the theory of Section 3.14 ff, with $d_2(\ldots) \to d(\ldots)$ and $d_1(\ldots) \to \delta(\ldots)$, or by direct differentiation of Equations 6.4.2e and 6.4.3e and recalling that $\delta\theta^{N+1} = \delta t = 0$, we obtain the following fundamental *transitivity equations*, corresponding to the transformation $A^\Lambda{}_\lambda$, $A^\lambda{}_\Lambda$,

$$d(\delta\theta^K) - \delta(d\theta^K) = \gamma^K{}_{L\Lambda}\, d\theta^\Lambda\, \delta\theta^L + A^K{}_k[(d(\delta q^k) - \delta(dq^k)]$$

$$= \gamma^K{}_{LR}\, d\theta^R\, \delta\theta^L + \gamma^K{}_{L,N+1}\, d\theta^{N+1}\, \delta\theta^L + A^K{}_k[(d(\delta q^k) - \delta(dq^k)]$$

$$\equiv (\gamma^K{}_{LR}\, d\theta^R + \gamma^K{}_L\, dt)\delta\theta^L + A^K{}_k[(d(\delta q^k) - \delta(dq^k)], \quad (6.5.1a)$$

or, dividing with dt [which does not interact with $\delta(\ldots)$]

$$d/dt(\delta\theta^K) - \delta\omega^K = \gamma^K{}_{L\Lambda}\, \omega^\Lambda\, \delta\theta^L + A^K{}_k[d/dt(\delta q^k) - \delta(dq^k/dt)]$$

$$= \gamma^K{}_{LR}\, \omega^R\, \delta\theta^L + \gamma^K{}_{L,N+1}\, \omega^{N+1}\, \delta\theta^L + A^K{}_k[d/dt(\delta q^k) - \delta(dq^k/dt)]$$

$$\equiv h^K{}_L\, \delta\theta^L + A^K{}_k[d/dt(\delta q^k) - \delta V^k], \quad (6.5.1b)$$

where

$$h^K{}_L \equiv \gamma^K{}_{LR}\, \omega^R + \gamma^K{}_{L,N+1}\, \omega^{N+1}$$

$$\equiv \gamma^K{}_{LR}\, \omega^R + \gamma^K{}_L: \textit{two-}\text{index } \textit{Hamel coefficients}. \quad (6.5.1c)$$

For the $(N+1)$th "coordinate," we similarly find

$$d(\delta t) - \delta(dt) = \gamma^{N+1}{}_{\Lambda\Psi}\, d\theta^\Psi\, \delta\theta^\Lambda = \gamma^{N+1}{}_{L\Psi}\, d\theta^\Psi\, \delta\theta^L = 0, \quad (6.5.2a)$$

$$\Rightarrow \gamma^{N+1}{}_{\Lambda\Psi} = 0 \Rightarrow \gamma^{N+1}{}_{L\Psi} = 0 \quad (6.5.2b)$$

or

$$d(\delta t) - \delta(dt) = 0 - \delta(dt) = 0 \Rightarrow \delta t = 0 \quad \textit{and} \quad \delta(dt) = 0. \quad (6.5.2c)$$

In concrete problems, the (antisymmetric) *three-index* Hamel coefficients:

$$\{\gamma^K{}_{L\Lambda} = -\gamma^K{}_{\Lambda L}\} = \{\gamma^K{}_{LS} = -\gamma^K{}_{SL},\ \gamma^K{}_{L,N+1} = -\gamma^K{}_{N+1,L} \equiv \gamma^K{}_L\}$$

do not have to be calculated from Equations 3.13.5 and 3.13.7c through e, i.e.,

$$\gamma^K{}_{LS} = (A^K{}_{l,s} - A^K{}_{s,l})A^l{}_L A^s{}_S, \tag{6.5.3a}$$

$$\begin{aligned}\gamma^K{}_{L,N+1} \equiv \gamma^K{}_L &= (A^K{}_{\lambda,\sigma} - A^K{}_{\sigma,\lambda})A^\lambda{}_L\, A^\sigma{}_{N+1} \\ &= (A^K{}_{l,s} - A^K{}_{s,l})A^l{}_L\, A^s{}_{N+1} + (A^K{}_{l,n+1} - A^K{}_{n+1,l})A^l{}_L\, A^{n+1}{}_{N+1} \\ &\equiv (\partial A^K{}_l/\partial q^s - \partial A^K{}_s/\partial q^l)A^l{}_L\, A^s{}_{N+1} + (\partial A^K{}_l/\partial t - \partial A^K{}_{n+1}/\partial q^l)A^l{}_L, \end{aligned} \tag{6.5.3b}$$

but can be read off most conveniently as coefficients of $(\ldots)\, d\theta^R\, \delta\theta^L$ and $(\ldots)\, dt\, \delta\theta^L$ in the "Frobenius bilinear covariants" Equations 6.5.1a or b.

Inverting Equations 6.5.1a and b with the help of Equation 6.4.4, as in Equation 3.14.1e, while invoking Equation 6.5.2b, and with the standard dynamics notation $d/dt(\ldots) \equiv (\ldots)^{\cdot}$, we easily find

$$d(\delta q^k) - \delta(dq^k) = A^k{}_K\{[d(\delta\theta^K) - \delta(d\theta^K)] - \gamma^K{}_{L\Lambda}\, d\theta^\Lambda\, \delta\theta^L\} \tag{6.5.4a}$$

$$\begin{aligned}\Rightarrow (\delta q^k)^{\cdot} - \delta V^k &= A^k{}_K\{[(\delta\theta^K)^{\cdot} - \delta\omega^K] - \gamma^K{}_{L\Lambda}\, \omega^\Lambda\, \delta\theta^L\} \\ &= A^k{}_K\{[(\delta\theta^K)^{\cdot} - \delta\omega^K] - (\gamma^K{}_{LR}\, \omega^R + \gamma^K{}_L)\delta\theta^L\} \\ &= A^k{}_K\{[(\delta\theta^K)^{\cdot} - \delta\omega^K] - h^K{}_L\, \delta\theta^L\}. \end{aligned} \tag{6.5.4b}$$

Remarks

i. In view of the bilinear (covariant-tensor-like) transformation (Equation 3.13.5)

$$\gamma^R{}_{KL} \equiv (A^R{}_{k,l} - A^R{}_{l,k})A^k{}_K\, A^l{}_L \equiv 2A^k{}_K\, A^l{}_L\, A^R{}_{[k,l]} \equiv A^k{}_K\, A^l{}_L\, \gamma^R{}_{kl}$$

$$\Rightarrow \gamma^R{}_{kl} = A^K{}_k\, A^L{}_l\, \gamma^R{}_{KL}, \tag{6.5.5}$$

if the $\gamma^R{}_{kl}$ vanish, so do the $\gamma^R{}_{KL}$, and vice versa. Therefore, the necessary and sufficient exactness conditions (Equation 6.4.1f): $\gamma^L{}_{l\lambda} = 0$ can be replaced by $\gamma^L{}_{K\Lambda} = 0$.

ii. In the transitivity equations we usually take, in accordance with the Calculus of Variations, $d(\delta q^k) = \delta(dq^k)$, $(\delta q^k)^{\cdot} = \delta V^k$. However, as shown in kinetics (Chapter 7), *all fundamental principles and equations of motion of dynamics are independent of any such assumptions;* i.e., $d(\delta q^k) = \delta(dq^k)$ are *possible* but *non necessary.* If we accept them, then Equations 6.5.1a and b reduce, respectively, to their standard forms:

$$d(\delta\theta^K) - \delta(d\theta^K) = \gamma^K{}_{LR}\, d\theta^R\, \delta\theta^L + \gamma^K{}_L\, dt\, \delta\theta^L, \tag{6.5.6a}$$

$$(\delta\theta^K)^{\cdot} - \delta\omega^K = \gamma^K{}_{L\Lambda}\, \omega^\Lambda\, \delta\theta^L \equiv h^K{}_L\, \delta\theta^L. \tag{6.5.6b}$$

iii. Finally, the transformation properties of the γ's, under $d\theta \to d\theta'$, etc. are given by Equation 3.14.2d; or they can be obtained by use of Equations 6.5.4a and b between $dq \leftrightarrow d\theta$ and $dq \leftrightarrow d\theta'$.

6.5.1 Nonintegrability Conditions

Let us examine the nonholonomic variable counterparts of Equations 6.3.5a through g. From Equations 6.4.8a through 6.4.11 and 6.4.12a through 6.4.15 we readily obtain, respectively,

- $\partial \mathbf{r}/\partial\theta^K = \partial \mathbf{v}^*/\partial\omega^K = \partial \mathbf{a}^*/\partial(d\omega^K/dt) = \ldots = \mathbf{e}_K,$ (6.5.7a)

- $\partial \mathbf{R}^*/\partial\theta^K = \partial \mathbf{V}^*/\partial\omega^K = \partial \mathbf{A}^*/\partial(d\omega^K/dt) = \ldots = \mathbf{E}_K.$ (6.5.7b)*

Here, however, contrary to Equations 6.3.5c and d, by applying chain rule to $\mathbf{v}^*(t, q, \omega) = \mathbf{v}(t, q, dq/dt)$ while invoking the quasi-chain rule and Equation 6.5.7c, we find

- $$\partial/\partial\theta^K(d\mathbf{r}/dt) \equiv \partial \mathbf{v}^*/\partial\theta^K \equiv (\partial \mathbf{v}^*/\partial q^k)(\partial v^k/\partial\omega^K)$$
$$= [\partial \mathbf{v}/\partial q^k + (\partial \mathbf{v}/\partial v^l)(\partial v^l/\partial q^k)]\,(\partial v^k/\partial\omega^K)$$
$$= (\partial \mathbf{v}/\partial q^k)(\partial v^k/\partial\omega^K) + (\partial \mathbf{v}/\partial v^l)(\partial v^l/\partial\theta^K) \qquad (6.5.8a)$$

(applying again quasi-chain rule to the second term)

- $$d/dt(\partial \mathbf{r}/\partial\theta^K) = d/dt(\partial \mathbf{v}^*/\partial\omega^K) = d/dt[(\partial \mathbf{v}/\partial v^k)(\partial v^k/\partial\omega^K)]$$
$$= [d/dt(\partial \mathbf{v}/\partial v^k)]\,(\partial v^k/\partial\omega^K)$$
$$+ (\partial \mathbf{v}/\partial v^k)\,[d/dt(\partial v^k/\partial\omega^K)], \qquad (6.5.8b)$$

and therefore, subtracting Equation 6.5.8a from Equation 6.5.8b side by side we get

$$E_K(\mathbf{v}^*) \equiv d/dt(\partial \mathbf{v}^*/\partial\omega^K) - \partial \mathbf{v}^*/\partial\theta^K$$
$$= [d/dt(\partial \mathbf{v}/\partial v^k) - \partial \mathbf{v}/\partial q^k]\,(\partial v^k/\partial\omega^K) +$$
$$[d/dt(\partial v^k/\partial\omega^K) - \partial v^k/\partial\theta^K]\,(\partial \mathbf{v}/\partial v^k)$$
$$= E_k(\mathbf{v})(\partial v^k/\partial\omega^K) + E_K(v^k)(\partial \mathbf{v}/\partial v^k)$$
$$= (\mathbf{0})(\partial v^k/\partial\omega^K) + E_K(v^k)(\partial \mathbf{v}/\partial v^k) \neq \mathbf{0}, \qquad (6.5.8c)^{**}$$

---

* We note the difference between the *symbolic definition* Equations 6.4.6a and b, etc. and the covariant vector *transformation* (Equation 6.4.8b); i.e., $\mathbf{v}^* = \mathbf{v}^*(t, q, \omega) = \mathbf{v}(t, q, dq/dt \equiv v) = \mathbf{v}$

$$\Rightarrow \partial \mathbf{v}^*/\partial\omega^K = (\partial \mathbf{v}/\partial v^k)\,[\partial(dq^k)/\partial(d\theta^K)] \equiv (\partial \mathbf{v}/\partial v^k)(\partial v^k/\partial\omega^K) \equiv A^k{}_K(\partial \mathbf{v}/\partial v^k). \qquad (6.5.7c)$$

** (See following page for footnote.)

(by Equations 6.3.5e and f) in general; or, finally, in terms of the (particle and system) vector of *nonholonomic deviation* $\mathbf{g}_K$:

$$\mathbf{g}_K \equiv E_K(\mathbf{v}^*) \equiv d\mathbf{e}_K/dt - \partial\mathbf{v}^*/\partial\theta^K \quad [= d/dt(\partial\mathbf{r}/\partial\theta^K) - \partial/\partial\theta^K(d\mathbf{r}/dt)]$$

$$= E_K(v^k)\mathbf{e}_k \equiv W^k{}_K\, \mathbf{e}_k$$

$$= E_K(v^k)[(\partial\omega^L/\partial v^k)\mathbf{e}_L] = E_K(v^k)(A^L{}_k\, \mathbf{e}_L) \equiv -H^L{}_K\, \mathbf{e}_L, \tag{6.5.9a}$$

where

$$W^k{}_K \equiv E_K(v^k) = -(\partial v^k/\partial\omega^L)H^L{}_K \equiv -A^k{}_L\, H^L{}_K, \tag{6.5.9b}$$

$$H^L{}_K \equiv -(\partial\omega^L/\partial v^k)W^k{}_K \equiv -A^L{}_k\, W^k{}_K. \qquad (6.5.9c)^*$$

Since $\mathbf{e}_K = \mathbf{e}_K(t, q)$, the above yield, explicitly,

$$\mathbf{g}_K = [(\partial\mathbf{e}_K/\partial q^k)v^k + \partial\mathbf{e}_K/\partial t] - \partial/\partial\theta^K(\omega^L\, \mathbf{e}_L + \mathbf{e}_{N+1})$$

$$= [(\partial\mathbf{e}_K/\partial\theta^L)(\partial\omega^L/\partial v^k)v^k + \partial\mathbf{e}_K/\partial t] - [\omega^L(\partial\mathbf{e}_L/\partial\theta^K) + \partial\mathbf{e}_{N+1}/\partial\theta^K]$$

[noticing that: $(\partial\omega^L/\partial v^k)v^k = A^L{}_k\, v^k = \omega^L - A^L{}_{n+1}$]

$$= [(\partial\mathbf{e}_K/\partial\theta^L)(\omega^L - A^L{}_{n+1}) - \omega^L(\partial\mathbf{e}_L/\partial\theta^K)] + \partial\mathbf{e}_K/\partial t - \partial\mathbf{e}_{N+1}/\partial\theta^K$$

** Where $E_K(\ldots) \equiv d/dt[\partial(\ldots)/\partial\omega^K] - \partial(\ldots)/\partial\theta^K$; or, sometimes, $E_K{}^*(\ldots)$: Euler–Lagrange operator in *nonholonomic* variables; *not to be confused with the magnitude of a vector* $\mathbf{E}_K$.

* **Alternative Derivation of Equations 6.5.9a–c.** We have, successively,

$$\mathbf{g}_K \equiv d\mathbf{e}_K/dt - \partial\mathbf{v}^*/\partial\theta^K = d/dt[(\partial v^k/\partial\omega^K)\mathbf{e}_k] - (\partial\mathbf{v}^*/\partial q^k)(\partial v^k/\partial\omega^K)$$

$$= d/dt(\partial v^k/\partial\omega^K)\mathbf{e}_k + (\partial v^k/\partial\omega^K)(d\mathbf{e}^k/dt) - [\partial/\partial q^k(v^l\, \mathbf{e}_l + \mathbf{e}_{n+1})](\partial v^k/\partial\omega^K)$$

$$= \ldots + \ldots -[(\partial v^l/\partial q^k)\mathbf{e}_l + v^l(\partial\mathbf{e}_l\,/\partial q^k) + \partial\mathbf{e}_{n+1}/\partial q^k](\partial v^k/\partial\omega^K)$$

$$= \ldots + \ldots -(\partial v^l/\partial q^k)(\partial v^k/\partial\omega^K)\mathbf{e}_l - v^l\,(\partial v^k/\partial\omega^K)(\partial\mathbf{e}_l/\partial q^k) - (\partial v^k/\partial\omega^K)(\partial\mathbf{e}_{n+1}/\partial q^k)$$

[invoking the integrability conditions (Equation 6.3.5e) in *fourth* and *fifth* terms]

$$= \ldots + \ldots -[(\partial\mathbf{e}_k/\partial q^l)v^l + \partial\mathbf{e}_k/\partial t](\partial v^k/\partial\omega^K) - (\partial v^l/\partial\theta^K)\mathbf{e}_l$$

$$= \ldots + \ldots -(\partial v^k/\partial\omega^K)(d\mathbf{e}_k/dt) - (\partial v^k/\partial\theta^K)\mathbf{e}_k$$

[the *second* and third *terms* add up to zero]

$$= [d/dt(\partial v^k/\partial\omega^K) - (\partial v^k/\partial\theta^K)]\mathbf{e}_k \equiv E_K(v^k)\mathbf{e}_k \equiv W^k{}_K\, \mathbf{e}_k.$$

$$= (\partial \mathbf{e}_K/\partial\theta^L - \partial \mathbf{e}_L/\partial\theta^K)\omega^L + \{[-(\partial \mathbf{e}_K/\partial\theta^L)A^L{}_{n+1} + \partial \mathbf{e}_K/\partial t] - \partial \mathbf{e}_{N+1}/\partial\theta^K\}$$

[by Equations 6.4.4f and 6.4.6c: $-(\partial \mathbf{e}_K/\partial\theta^L)A^L{}_{n+1} = -(\partial \mathbf{e}_K/\partial q^l)(\partial v^l/\partial\omega^L)A^L{}_{n+1}$

$$\equiv -(\partial \mathbf{e}_K/\partial q^l)(A^l{}_L\, A^L{}_{n+1}) = -(\partial \mathbf{e}_K/\partial q^l)(-\delta^{N+1}{}_{n+1}\, A^l{}_{n+1})$$

$$= A^l{}_{N+1}(\partial \mathbf{e}_K/\partial q^l)\delta^{N+1}{}_{n+1}, \quad \text{and since } \omega^{N+1} = 1]$$

$$= (\partial \mathbf{e}_K/\partial\theta^L - \partial \mathbf{e}_L/\partial\theta^K)\,\omega^L + (\partial \mathbf{e}_K/\partial\theta^{N+1} - \partial \mathbf{e}_{N+1}/\partial\theta^K)\omega^{N+1}, \qquad (6.5.10a)$$

or, finally, and recalling the noncommutativity relations (Equation 3.15.3e),

$$\mathbf{g}_K = (\partial \mathbf{e}_K/\partial\theta^\Delta - \partial \mathbf{e}_\Delta/\partial\theta^K)\omega^\Delta$$

$$= (\gamma^\Psi{}_{\Delta K}\,\mathbf{e}_\Psi)\omega^\Delta = (\gamma^L{}_{\Delta K}\,\mathbf{e}_L)\omega^\Delta \quad \text{(recalling Equation 6.5.2b)}$$

$$= -(\gamma^L{}_{K\Delta}\,\omega^\Delta)\mathbf{e}_L = -h^L{}_K\,\mathbf{e}_L \quad \text{(recalling Equation 6.5.1c)} \qquad (6.5.10b)$$

i.e., $H^L{}_K = h^L{}_K$.*

Problem 6.5.1

Verify the following *additional expresssions* for Hamel's coefficients:

- $\gamma^K{}_{LS} = (A^k{}_L\, A^s{}_S - A^s{}_L\, A^k{}_S)(\partial A^K{}_k/\partial q^s) = (\partial A^k{}_S/\partial\theta^L - \partial A^k{}_L/\partial\theta^S)A^K{}_k$; (a)
- $\gamma^K{}_{L,N+1} \equiv \gamma^K{}_L = (A^k{}_L\, A^\beta{}_{N+1} - A^\beta{}_L\, A^k{}_{N+1})(\partial A^K{}_k/\partial q^\beta)$

$$= (\partial A^k{}_{N+1}/\partial\theta^L - \partial A^k{}_L/\partial\theta_{N+1})A^K{}_k \qquad \text{(b)}$$

(recalling Equations 3.13.7c through e)

Now we are ready to handle additional Pfaffian (possibly nonholonomic) constraints. But before doing that, and as a concrete illustration of the use of quasi-variables

---

* (a) It is not hard to see that

$$\partial \mathbf{e}_K/\partial\theta^{N+1} - \partial \mathbf{e}_{N+1}/\partial\theta^K = \gamma^L{}_{N+1,K}\,\mathbf{e}_L = -\gamma^L{}_K\,\mathbf{e}_L. \qquad (6.5.10c)$$

(b) Equations 3.15.3e also show that

$$\partial \mathbf{E}_K/\partial\theta^\Delta - \partial \mathbf{E}_\Delta/\partial\theta^K = \gamma^\Psi{}_{\Delta K}\,\mathbf{E}_\Psi. \qquad (6.5.10d)$$

(c) Also, it can be shown that

$$H^L{}_K = h^L{}_K = (\partial v^k/\partial\omega^K)[d/dt(\partial\omega^L/\partial v^k) - \partial\omega^L/\partial q^k] \equiv A^k{}_K\, E_k(\omega^L). \qquad (6.5.10e)$$

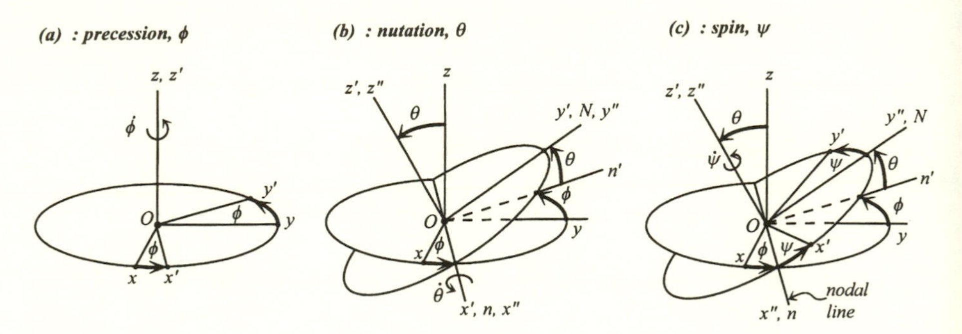

FIGURE 6.1 Eulerian angles: $\phi \to \theta \to \psi$ (or $3 \to 1 \to 3$): We assume that, originally, O–XYZ and O–$X'Y'Z'$ coincide. Then we subject O–$X'Y'Z'$ to the following rotation sequence: ϕ (about OZ) $\to$ θ (about new position of OX') $\to$ ψ [about new (second) position of OZ']. *Restrictions:* $0 \leq \phi, \psi \leq 2\pi$ and $0 < \theta < \pi$ (see Equations b2 and b4).

and their transitivity equations to kinematics, let us examine the rotations of a rigid body via Euler's angles.

6.5.2 Comprehensive Examples and Problems on Rigid-Body Kinematics

Example 6.5.1. Rigid-Body Rotation about a Fixed Point

Eulerian Angles and Associated Quasi-variables, and Transitivity Equations

Let us consider a rigid body B moving about a (body and space) fixed point O. To describe that motion (which, as shown in rigid body kinematics, is a *rotation about* O) we shall use the following two rectangular Cartesian sets of axes: (a) a moving *body-fixed* one (i.e., noninertial) M: O–$X'Y'Z'$, with associated orthonormal and dextral basis O–$\mathbf{I'J'K'} \equiv O$–$\mathbf{U}_{K'}$ ($K' = 1, 2, 3$; with $\mathbf{U}$ for unit) and (b) a *space-fixed* one (i.e., inertial) F: O–XYZ, with associated orthonormal and dextral basis O–$\mathbf{IJK}$ $\equiv O$–$\mathbf{U}_K$ ($K = 1, 2, 3$).* Now, the position (angular orientation) and motion of B relative to F is described in several, theoretically equivalent, ways in terms of *three* independent parameters. As such we shall use here the relatively common and useful *Eulerian angles,* to be denoted by: ϕ (*precession*) $\to$ θ (*nutation*) $\to$ ψ (*proper,* or *eigen-spin*); in that order, which is known to kinematicians as the "$3 \to 1 \to 3$" type, one out of 12 possible such rotation sequences (Figure 6.1, standard U.S. notation).

It is shown in mechanics ** that the various components of the inertial *angular velocity vector* of B (i.e., the angular velocity of M relative to F):

---

* (a) For the purposes of kinematics, O–XYZ does not have to be inertial, and so the following results are, really, free of such restrictions. (b) In some cases, other "intermediate," or "accessory," axes O–$X''Y''Z''$, are used (more on this later).

** See any good text on mechanics (intermediate level, at least); e.g., Rosenberg (1977, pp. 77–82), Spiegel (1967, pp. 267–268).

$$\boldsymbol{\omega} = \omega_X \mathbf{U}_X + \omega_Y \mathbf{U}_Y + \omega_Z \mathbf{U}_Z \quad (\textit{fixed} \text{ axes components}) \tag{a1}$$

$$= \omega_{X'} \mathbf{U}_{X'} + \omega_{Y'} \mathbf{U}_{Y'} + \omega_{Z'} \mathbf{U}_{Z'} \quad (\textit{moving/body} \text{ axes components}) \tag{a2}$$

are related to the (contravariant) Eulerian rates $d\phi/dt \equiv v^\phi$, $d\theta/dt \equiv v^\theta$, $d\psi/dt \equiv v^\psi$ (along the corresponding *nonorthogonal* axes through O) by the following linear and homogeneous transformations:

- *Fixed/Inertial (Space-Fixed) Axes Components* (only ϕ and θ are involved):

$$\omega_X = (0)v^\phi + (\cos\phi)v^\theta + (\sin\phi\ \sin\theta)v^\psi,$$

$$\omega_Y = (0)v^\phi + (\sin\phi)v^\theta + (-\cos\phi\ \sin\theta)v^\psi,$$

$$\omega_Z = (1)v^\phi + (0)v^\theta + (\cos\theta)v^\psi; \tag{b1}$$

$$v^\phi = (-\text{cotan}\,\theta\ \sin\phi)\omega_X + (-\text{cotan}\,\theta\ \cos\phi)\omega_Y + (1)\omega_Z,$$

$$v^\theta = (\cos\phi)\omega_X + (\sin\phi)\omega_Y + (0)\omega_Z,$$

$$v^\psi = (\sin\phi/\sin\theta)\omega_X + (-\cos\phi/\sin\theta)\omega_Y + (0)\omega_Z. \tag{b2}$$

- *Moving/Noninertial (Body–Fixed) Axes Components* (only θ and ψ are involved):

$$\omega_{X'} = (\sin\psi\ \sin\theta)v^\phi + (\cos\psi)v^\theta + (0)v^\psi,$$

$$\omega_{Y'} = (\cos\psi\ \sin\theta)v^\phi + (-\sin\psi)v^\theta + (0)v^\psi,$$

$$\omega_{Z'} = (\cos\theta)v^\phi + (0)v^\theta + (1)v^\psi; \tag{b3}$$

$$v^\phi = (\sin\psi/\sin\theta)\omega_{X'} + (\cos\psi/\sin\theta)\omega_{Y'} + (0)\omega_{Z'},$$

$$v^\theta = (\cos\psi)\omega_{X'} + (-\sin\psi)\omega_{Y'} + (0)\omega_{Z'},$$

$$v^\psi = (-\text{cotan}\,\theta\ \sin\psi)\omega_{X'} + (-\text{cotan}\,\theta\ \cos\psi)\omega_{Y'} + (1)\omega_{Z'}. \tag{b4}$$

The above show that the $\omega_{X,Y,Z}$ and $\omega_{X',Y',Z'}$ are *quasi-velocities*; for example, from the first of Equations b3, we readily verify that

$$\partial(\sin\psi\ \sin\theta)/\partial\theta = \sin\psi\ \cos\theta \neq \partial(\cos\psi)/\partial\phi = 0, \text{ etc.}; \tag{b5}$$

i.e., *no ordinary (finite) angle $\chi_{X'}$ exists such that $\omega_{X'} = d\chi_{X'}/dt$*, or $\chi_{X'}$ is a *quasi-coordinate* (although the *system* is holonomic — see Section 6.6); and similarly for the rest. Therefore, these components can be taken as ω^K (here, due to the orthonormality of both body- and space-fixed axes, covariant and contravariant components are equal to each other); while the Eulerian angles ϕ, θ, ψ can be taken as q^k. This remarkable fact is a direct result of the well-known *noncommutativity* of finite rotations (that is why the *order* of the latter is important). Let us calculate the corresponding transitivity equations and Hamel coefficients:

- *Moving/Body Axes*: from the first of Equation b3, and since this system is scleronomic (see also Section 6.6 below), replacing $\omega_{X'}$ with $d\chi_{X'}$ or $\delta\chi_{X'}$, etc. and v_ϕ with $d\phi$ or $\delta\phi$, etc., and invoking the commutation rule "$d\delta(\ldots) = \delta d(\ldots)$" for the holonomic coordinates ϕ, θ, ψ, we find

$$d/dt(\delta\chi_{X'}) - \delta\omega_{X'} = d/dt[(\sin\theta\ \sin\psi)\delta\phi + (\cos\psi)\delta\theta]$$

$$- \delta[(\sin\theta\ \sin\psi)v^\phi + (\cos\psi)v^\theta]$$

$$= (v^\theta\ \delta\phi - v^\phi\ \delta\theta)\cos\theta\ \sin\psi$$

$$+ (v^\psi\ \delta\phi - v^\phi\ \delta\psi)\sin\theta\ \cos\psi + (v^\theta\ \delta\psi - v^\psi\ \delta\theta)\sin\psi$$

and substituting into it $v^{\phi,\theta,\psi}$ in terms of $\omega_{X',Y',Z'}$ and $\delta\phi$, $\delta\theta$, $\delta\psi$ in terms of $\delta\chi_{X',Y',Z'}$, from Equation b4, we get after some straightforward algebra:

$$d/dt(\delta\chi_{X'}) - \delta\omega_{X'} = (0)\delta\chi_{X'} + (\omega_{Z'})\delta\chi_{Y'} + (-\omega_{Y'})\delta\chi_{Z'}, \qquad \text{(c1)}$$

and, similarly,

$$d/dt(\delta\chi_{Y'}) - \delta\omega_{Y'} = (-\omega_{Z'})\delta\chi_{X'} + (0)\delta\chi_{Y'} + (\omega_{X'})\delta\chi_{Z'}, \qquad \text{(c2)}$$

$$d/dt(\delta\chi_{Z'}) - \delta\omega_{Z'} = (\omega_{Y'})\delta\chi_{X'} + (-\omega_{X'})\delta\chi_{Y'} + (0)\delta\chi_{Z'}; \qquad \text{(c3)}$$

from which we readily read off the following nonvanishing Hamel coefficients:

$$\gamma^{X'}{}_{Y'Z'} = -\gamma^{X'}{}_{Z'Y'} = 1, \quad \gamma^{Y'}{}_{Z'X'} = -\gamma^{Y'}{}_{X'Z'} = 1, \quad \gamma^{Z'}{}_{X'Y'} = -\gamma^{Z'}{}_{Y'X'} = 1; \qquad \text{(c4)}$$

i.e., compactly (with X', Y', Z' $\rightarrow$ K', L', R' $\rightarrow$ 1, 2, 3):

$$\gamma^{K'}{}_{L'R'}|_{body\ \text{axes}} = \varepsilon_{K'L'R'}\text{: Levi-Civita permutation symbol (Equation 1.2.5)} \qquad \text{(c5)}$$

- *Fixed/Inertial Axes:* Working similarly with Equations b1 and b2, we find

$$d/dt(\delta\chi_X) - \delta\omega_X = (0)\delta\chi_X + (-\omega_Z)\delta\chi_Y + (\omega_Y)\delta\chi_Z, \qquad \text{(d1)}$$

$$d/dt(\delta\chi_Y) - \delta\omega_Y = (\omega_Z)\delta\chi_X + (0)\delta\chi_Y + (-\omega_X)\delta\chi_Z, \qquad \text{(d2)}$$

$$d/dt(\delta\chi_Z) - \delta\omega_Z = (-\omega_Y)\delta\chi_X + (-\omega_X)\delta\chi_Y + (0)\delta\chi_Z; \tag{d3}$$

$$\gamma^X{}_{YZ} = -\gamma^X{}_{ZY} = -1, \quad \gamma^Y{}_{ZX} = -\gamma^Y{}_{XZ} = -1, \quad \gamma^Z{}_{XY} = -\gamma^Z{}_{YX} = -1; \tag{d4}$$

i.e., compactly (with $X, Y, Z \to K, L, R \to 1, 2, 3$):

$$\gamma^K{}_{LR}|_{fixed\ \mathrm{axes}} = -\varepsilon_{KLR}\text{: Levi-Civita permutation symbol.} \tag{d5}$$

Also, as pointed out in Section 3.13, the above γ-values are independent of the particular $\omega_{...} \leftrightarrow v_{...}$ relations used in their derivation, i.e., Equations b1 through b4.

It is not hard to show that Equations c1 through c3 and d1 through d3 can be put in the following *vector forms*, respectively,

- *Body axes:* $d'(\delta\chi)/dt - \delta'\omega = \delta\chi \times \omega,$ (e1)
- *Fixed axes:* $d(\delta\chi)/dt - \delta\omega = \omega \times \delta\chi;$ (e2)

where

$d\chi/\delta\chi$: vector of kinematically admissible/virtual *infinitesimal rotation* of B relative to F (in general, a *quasi-vector*, i.e., χ may not exist) (e3)

$\omega \equiv d\chi/dt$: vector of *angular velocity* of B relative to F, (e4)

$$d(\text{vector}) = d'(\text{vector}) + d\chi \times (\text{vector}), \tag{e5}$$

$$\delta(\text{vector}) = \delta'(\text{vector}) + \delta\chi \times (\text{vector}), \tag{e6}$$

$d(\ldots)/\delta(\ldots)$: differential/virtual variation relative to *fixed* axes, (e7)

$d'(\ldots)/\delta'(\ldots)$: differential/virtual variation relative to *body* axes. (e8)*

Euler–Lagrange Form of the Transitivity Equations

For a general scleronomic system:

$$\omega = \omega(q^k, v^k \equiv dq^k/dt) \quad (k = 1, 2, 3); \tag{f1}$$

e.g., $q^k = \phi, \theta, \psi$; and, of course, other angular coordinate choices are possible. Therefore,

$$\delta\omega = (\partial\omega/\partial q^k)\delta q^k + (\partial\omega/\partial v^k)\delta v^k. \tag{f2}$$

* For a direct, *ad hoc,* derivation of Equations e1 and 2 see Hamel (1949, pp. 488–489). The earliest component derivation of Equations e1 is due to Lagrange himself (early 1810s; published in Vol. 2 of 2nd edition of his *Mécanique Analytique*, 1815).

But also, since ω is *linear and homogeneous* in the v^k (by the homogeneous function theorem):

$$\omega = (\partial\omega/\partial v^k)v^k \equiv d\chi/dt = (\partial\chi/\partial q^k)v^k, \tag{f3}$$

i.e.,

$$\partial\chi/\partial q^k \equiv \partial\omega/\partial v^k \equiv \mathbf{c}_k \tag{f4}$$

(symbolic derivative; since, in general, no χ exists), from which

$$\delta\chi = (\partial\chi/\partial q^k)\delta q^k \equiv \mathbf{c}_k\, \delta q^k. \tag{f5}$$

Substituting now $\delta\omega$ from Equation f2 and $\delta\chi$ from Equation f5 into Equation e2, while invoking Equation f4 and the $d(\delta q^k) = \delta(dq^k)$ rule, yields:

Left side: $d/dt(\mathbf{c}_k\, \delta q^k) - \delta\omega = \ldots = (d\mathbf{c}_k/dt - \partial\omega/\partial q^k)\delta q^k$

Right side: $\omega \times (\mathbf{c}_k\, \delta q^k)$,

and so finally, since the δq are arbitrary, we obtain the Euler–Lagrange transitivity equation:

$$E_k(\omega) \equiv d/dt(\partial\omega/\partial v^k) - \partial\omega/\partial q^k = \omega \times (\partial\omega/\partial v^k); \tag{f6}$$

which shows that ω *is a vectorial quasi-velocity* (unless $\omega \times \mathbf{c}_k = \mathbf{0}$).

An Additional Form of the Rotational Transitivity Equations

This additional form occurs if we substitute Equation f5 into the *differential form* of Equation e2 (i.e., multiplied with dt):

$$d(\delta\chi) - \delta(d\chi) = d\chi \times \delta\chi. \tag{g1}$$

Left side (with all Latin indices ranging from 1 to 3): we find, successively,

$$d(\delta\chi) = d(\mathbf{c}_k\, \delta q^k) = d\mathbf{c}_k\, \delta q^k + \mathbf{c}_k\, d(\delta q^k) = [(\partial\mathbf{c}_k/\partial q^l)dq^l]\delta q^k + \mathbf{c}_k\, d(\delta q^k),$$

$$\delta(d\chi) = \delta(\mathbf{c}_k\, dq^k) = [(\partial\mathbf{c}_k/\partial q^l)\delta q^l]dq^k + \mathbf{c}_k\, \delta(dq^k),$$

and so subtracting side by side (and changing dummy indices appropriately) we get:

$$d(\delta\chi) - \delta(d\chi) = (\partial\mathbf{c}_k/\partial q^l - \partial\mathbf{c}_l/\partial q^k)dq^l\, \delta q^k. \tag{g2}$$

Right side:

$$d\chi \times \delta\chi = (\mathbf{c}_l\, dq^l) \times (\mathbf{c}_k\, \delta q^k) = (\mathbf{c}_l \times \mathbf{c}_k)\, dq^l\, \delta q^k. \tag{g3}$$

Therefore, equating the right sides of Equations g2 and g3 we finally obtain the *nonintegrability* conditions:

$$\partial \mathbf{c}_k/\partial q^l - \partial \mathbf{c}_l/\partial q^k \equiv 2\mathbf{c}_{[k,l]} \equiv 2\partial_{[l}\, \mathbf{c}_{k]} = \mathbf{c}_l \times \mathbf{c}_k; \tag{g4}$$

which show that if $\mathbf{c}_l \times \mathbf{c}_k = \mathbf{0}$ the $\{\mathbf{c}_k\}$ form a natural holonomic basis (see Integrability below).

An Alternative Derivation of Equation g4

We consider a *rigid* holonomic basis $\{\mathbf{e}_k(q)\}$. The rigidity constraint (*not* to be confused with *Ricci's* lemma: $Dg_{kl} = 0$) means that $\mathbf{e}_k \cdot \mathbf{e}_l = g_{kl}$ *remains constant.* Hence, successively (and using the comma notation for partial derivatives),

$$0 = dg_{kl} = d\mathbf{e}_k \cdot \mathbf{e}_l + \mathbf{e}_k \cdot d\mathbf{e}_l = (\mathbf{e}_{k,s}\, dq^s) \cdot \mathbf{e}_l + \mathbf{e}_k \cdot (\mathbf{e}_{l,s}\, dq^s),$$

from which, since the dqs are independent, we get

$$\mathbf{e}_{k,s} \cdot \mathbf{e}_l + \mathbf{e}_k \cdot \mathbf{e}_{l,s} = 0 \quad \text{(recall Example 3.8.3).} \tag{g5}$$

But from the well-known kinematical formula of Poisson: $d(\ldots)/dt = \omega \times (\ldots) = (\mathbf{c}_s v^s) \times (\ldots)$, where $(\ldots) \equiv$ *any vector,* applied to $\mathbf{e}_k$ we find:

$$d\mathbf{e}_k/dt = (\partial \mathbf{e}_k/\partial q^s)v^s = (\mathbf{c}_s \times \mathbf{e}_k)\, v^s \Rightarrow \partial \mathbf{e}_k/\partial q^s \equiv \mathbf{e}_{k,s} = \mathbf{c}_s \times \mathbf{e}_k. \tag{g6}$$

Then, crossing Equation g6 with $\mathbf{e}^k$ and summing over k, we obtain (with some easily understood notation):

$$\mathbf{e}^k \times (\partial \mathbf{e}_k/\partial q^s) = \mathbf{e}^k \times (\mathbf{c}_s \times \mathbf{e}_k) = (\mathbf{e}^k \cdot \mathbf{e}_k)\mathbf{c}_s - (\mathbf{e}^k \cdot \mathbf{c}_s)\mathbf{e}_k$$

$$= (\delta^k{}_k)\mathbf{c}_s - (\mathbf{c}_s)^k \mathbf{e}_k = 3\mathbf{c}_s - \mathbf{c}_s = 2\mathbf{c}_s,$$

$$\Rightarrow \mathbf{c}_s = (1/2)[\mathbf{e}^k \times (\partial \mathbf{e}_k/\partial q^s)]. \tag{g7}$$

Similarly, from: $0 = dg^l{}_k = d(\delta^l{}_k) = d(\mathbf{e}^l \cdot \mathbf{e}_k) = d\mathbf{e}^l \cdot \mathbf{e}_k + \mathbf{e}^l \cdot d\mathbf{e}_k$, we find

$$\mathbf{e}_{k,s} \cdot \mathbf{e}^l + \mathbf{e}_k \cdot \mathbf{e}^l{}_{,s} = 0, \tag{g8}$$

(invoking Equation g6) $\Rightarrow (\mathbf{c}_s \times \mathbf{e}_k) \cdot \mathbf{e}^l + \mathbf{e}_k \cdot \mathbf{e}^l{}_{,s} = \mathbf{e}_k \cdot (\mathbf{e}^l \times \mathbf{c}_s) + \mathbf{e}_k \cdot \mathbf{e}^l{}_{,s} = 0,$

i.e., finally,

$$\mathbf{e}^l{}_{,s} = \mathbf{c}_s \times \mathbf{e}^l. \tag{g9}$$

From Equation g7 we find, successively,

$$\mathbf{c}_{s,l} = (1/2)(\mathbf{e}^k{}_{,l} \times \mathbf{e}_{k,s} + \mathbf{e}^k \times \mathbf{e}_{k,s,l}) \quad \text{and} \quad \mathbf{c}_{l,s} = (1/2)(\mathbf{e}^k{}_{,s} \times \mathbf{e}_{k,l} + \mathbf{e}^k \times \mathbf{e}_{k,l,s})$$

and so, subtracting the above side by side, and then invoking Equations g6 and g9, we get

$$\mathbf{c}_{s,l} - \mathbf{c}_{l,s} = (1/2)(\mathbf{e}^k{}_{,l} \times \mathbf{e}_{k,s} - \mathbf{e}^k{}_{,s} \times \mathbf{e}_{k,l}) = (1/2)(\mathbf{e}^k{}_{,l} \times \mathbf{e}_{k,s} + \mathbf{e}_{k,l} \times \mathbf{e}^k{}_{,s})$$

or

$$2(\mathbf{c}_{s,l} - \mathbf{c}_{l,s}) = (\mathbf{c}_l \times \mathbf{e}^k) \times (\mathbf{c}_s \times \mathbf{e}_k) + (\mathbf{c}_l \times \mathbf{e}_k) \times (\mathbf{c}_s \times \mathbf{e}^k)$$

(then using well-known vector algebra identities, and some simple *ad hoc* notation)

$$= \mathbf{e}^k[\mathbf{c}_l \cdot (\mathbf{c}_s \times \mathbf{e}_k)] - \mathbf{c}_l[\mathbf{e}^k \cdot (\mathbf{c}_s \times \mathbf{e}_k)] + \mathbf{e}_k[\mathbf{c}_l \cdot (\mathbf{c}_s \times \mathbf{e}^k)] - \mathbf{c}_l[\mathbf{e}_k \cdot (\mathbf{c}_s \times \mathbf{e}^k)]$$

$$= \mathbf{e}^k[\mathbf{e}_k \cdot (\mathbf{c}_l \times \mathbf{c}_s)] + \mathbf{e}_k[\mathbf{e}^k \cdot (\mathbf{c}_l \times \mathbf{c}_s)]$$

$$\equiv \mathbf{e}^k(\mathbf{e}_k \cdot \mathbf{b}_{ls}) + \mathbf{e}_k(\mathbf{e}^k \cdot \mathbf{b}_{ls}) \equiv \mathbf{e}^k(\mathbf{b}_{ls})_k + \mathbf{e}_k(\mathbf{b}_{ls})^k$$

$$= 2\mathbf{b}_{ls} \equiv 2(\mathbf{c}_l \times \mathbf{c}_s), \quad \text{q.e.d.} \tag{g10}$$

This can be transformed further as in Example 3.8.3: With the metric $c_{kl} \equiv \mathbf{c}_k \cdot \mathbf{c}_l$ and affinities $C^l{}_{ks}$ defined by

$$d\mathbf{c}_k = \mathbf{c}_{k,s}\, dq^s \equiv (C^l{}_{ks}\, dq^s)\mathbf{c}_l \Rightarrow \mathbf{c}_{k,s} = C^l{}_{ks}\, \mathbf{c}_l, \tag{g11}$$

we get,

$$\mathbf{c}_{k,s} - \mathbf{c}_{s,k} = C^l{}_{ks}\, \mathbf{c}_l - C^l{}_{sk}\, \mathbf{c}_l = (C^l{}_{ks} - C^l{}_{sk})\mathbf{c}_l \equiv (C_{l,ks} - C_{l,sk})\mathbf{c}^l$$

$$\equiv 2C^l{}_{[ks]}\, \mathbf{c}_l \equiv 2C_{l,[ks]}\, \mathbf{c}^l \quad (\equiv 2S^l{}_{ks}\, \mathbf{c}_l \equiv 2S_{l,ks}\, \mathbf{c}^l). \tag{g12}$$

Hence, Equations g4 and g10 become:

$$2C^l{}_{[ks]}\, \mathbf{c}_l = \mathbf{c}_s \times \mathbf{c}_k \Rightarrow 2C^l{}_{[ks]} = \mathbf{c}^l \cdot (\mathbf{c}_s \times \mathbf{c}_k), \tag{g13}$$

or

$$2C_{l,[ks]}\, \mathbf{c}^l = \mathbf{c}_s \times \mathbf{c}_k$$

$$\Rightarrow 2C_{l,[ks]} = \mathbf{c}_l \cdot (\mathbf{c}_s \times \mathbf{c}_k) \equiv (\mathbf{c}_l, \mathbf{c}_s, \mathbf{c}_k)$$

$$[= 1, \text{ if } \mathbf{c}_{l,s,k} \text{ orthonormal and dextral, etc.}]. \tag{g14}$$

Integrability

The above show that if $\mathbf{c}_s \times \mathbf{c}_k = \mathbf{0}$ ($k, s = 1, 2, 3$) then $d\chi$ is integrable; i.e., if all the $\mathbf{c}_k$ are mutually *parallel*:

$$\mathbf{c}_k = c_k\, \mathbf{u}_{\text{(constant unit vector)}} \Rightarrow \mathbf{c}_{k,l} = c_{k,l}\, \mathbf{u} = c_{l,k}\, \mathbf{u} = \mathbf{c}_{l,k}, \tag{h1}$$

in which case,

$$d\chi = \mathbf{c}_k\, dq^k = (c_k\, \mathbf{u})dq^k = (c_k\, dq^k)\mathbf{u}$$

$$\equiv d\chi(q)\mathbf{u}: \text{ integrable angle differential.} \tag{h2}$$

The physical meaning of the above is made clearer by the following coordinate transformation: $q^{1'} = q^{1'}(\chi)$, $q^{k'} = q^{k'}(q^k)$. Then, $\chi = \chi(q^{1'})$ and

$$d\chi = \mathbf{c}_{k'}\, dq^{k'} = [(\partial q^k/\partial q^{k'})\mathbf{c}_k]dq^{k'} = \mathbf{u}\; d\chi = \mathbf{u}[(d\chi/dq^{1'})dq^{1'}]$$

$$\Rightarrow \mathbf{c}_{1'} = (d\chi/dq^{1'})\mathbf{u}, \quad \mathbf{c}_{k'} = (\partial q^k/\partial q^{k'})\mathbf{c}_k = \mathbf{0} \quad (k' = 2', 3'). \tag{h3}$$

In sum, *only rotations with one DOF, with one angular vector, are integrable.*

Semimobile Axes

Frequently, to simplify the equations of motion, we choose components of the various dynamical quantities along orthogonal axes that are *neither space nor body fixed.* One such intermediate, or semimobile, triad associated naturally with the earlier $\phi \to \theta \to \psi$ transformation is defined by the following orthonormal and dextral basis $O\text{–}\mathbf{U}_{X''Y''Z''}$ (Figure 6.1):

$$\mathbf{U}_{X''} \equiv \mathbf{U}_n = \mathbf{U}_X \cos\phi + \mathbf{U}_Y \sin\phi \qquad \text{(nodal line)} \tag{i1}$$

$$\mathbf{U}_{Y''} \equiv \mathbf{U}_N = \mathbf{U}_{n'} \cos\theta + \mathbf{U}_Z \sin\theta$$

$$\equiv (-\mathbf{U}_X \sin\phi + \mathbf{U}_Y \cos\psi)\cos\theta + \mathbf{U}_Z \sin\theta \tag{i2}$$

$$\mathbf{U}_{Z''} \equiv \mathbf{U}_{Z'} = -\mathbf{U}_{n'} \sin\theta + \mathbf{U}_Z \cos\theta$$

$$\equiv (-\mathbf{U}_X \sin\phi + \mathbf{U}_Y \cos\psi)\sin\theta + \mathbf{U}_Z \cos\theta. \tag{i3}$$

From Figure 6.1, we easily deduce that the angular velocity of the associated semimobile axes $O\text{–}X''Y''Z''$ relative to the fixed ones $O\text{–}XYZ$ (or F), $\boldsymbol{\Omega}$, equals

$$\boldsymbol{\Omega} = (d\theta/dt)\mathbf{U}_n + (d\phi/dt)\mathbf{U}_Z \equiv \nu^\theta\, \mathbf{U}_n + \nu^\phi\, \mathbf{U}_Z$$

$$= \nu^\theta\, \mathbf{U}_n + (\nu^\phi \sin\theta)\mathbf{U}_N + (\nu^\phi \cos\theta)\mathbf{U}_{Z'}$$

$$\equiv \Omega_{X''}\, \mathbf{U}_{X''} + \Omega_{Y''}\, \mathbf{U}_{Y''} + \Omega_{Z''}\, \mathbf{U}_{Z''}, \tag{j1}$$

while that of the body-fixed axes $O\text{–}X'Y'Z'$, ω, is

$$\omega = \Omega + (d\psi/dt)\mathbf{U}_{Z'} = v^{\theta}\,\mathbf{U}_n + (v^{\phi}\sin\theta)\mathbf{U}_N + (v^{\phi}\cos\theta + v^{\psi})\mathbf{U}_{Z'}$$

$$\equiv \omega_{X''}\,\mathbf{U}_{X''} + \omega_{Y''}\,\mathbf{U}_{Y''} + \omega_{Z''}\,\mathbf{U}_{Z''}; \tag{j2}$$

and since the system is scleronomic

$$\delta\chi_{X''} = \delta\theta, \quad \delta\chi_{Y''} = \sin\theta\delta\phi, \quad \delta\chi_{Z''} = \cos\theta\delta\phi + \delta\psi, \tag{j3}$$

where $\omega_{X''} \equiv \delta\chi_{X''}/dt$, etc.

We leave it to the reader to verify (by carrying out similar calculations as for Equations c1 through d5) that:

$$d/dt(\delta\chi_{X''}) - \delta\omega_{X''} = 0 \quad (\chi_{X''} = \theta \text{ is holonomic coordinate; i.e., } \gamma^{X''}{}_{\dots} = 0) \tag{k1}$$

$$d/dt(\delta\chi_{Y''}) - \delta\omega_{Y''} = (\text{cotan}\,\theta)(\omega_{X''}\,\delta\chi_{Y''} - \omega_{Y''}\,\delta\chi_{X''}), \tag{k2}$$

$$d/dt(\delta\chi_{Z''}) - \delta\omega_{Z''} = \omega_{Y''}\,\delta\chi_{X''} - \omega_{X''}\,\delta\chi_{Y''}; \tag{k3}$$

and so the *nonvanishing* γs are

$$\gamma^{Y''}{}_{Y''X''} = -\gamma^{Y''}{}_{X''Y''} = \text{cotan}\,\theta, \quad \gamma^{Z''}{}_{X''Y''} = -\gamma^{Z''}{}_{Y''X''} = 1. \tag{k4}$$

Problem 6.5.2

Show that Equation f6 also holds if $\omega = \omega(t, q^k, v^k)$ (*rheonomic* case).

Example 6.5.2. General Rigid-Body Motion

Next, let us assume that the body B is in *general motion*, i.e., no point of it is fixed in space. It is shown in kinematics, that then *the most general displacement of its points consists of a translation of an arbitrarily chosen point of it, called base point or pole, say P, plus a rotation of the body about an axis through P*; a rotation whose characteristics are independent of P. Let us, therefore, find the additional kinematical features due to the translation of P. With $\mathbf{OP} \equiv \mathbf{R}_P \equiv \mathbf{R} = R_K\mathbf{U}_K$, the (inertial) velocity of P is $\mathbf{V}_P \equiv \mathbf{V} \equiv d\mathbf{R}/dt$, and, therefore,

$$\mathbf{V} \cdot \mathbf{U}_K = dR_K/dt \equiv dq_K/dt \equiv V_K = dq^K/dt \equiv V^K: \tag{a1}$$

holonomic components of $\mathbf{V}$ along the *space*-fixed (inertial) axes $O\text{–}XYZ$ (in spite of the uppercase indices!),

$$\mathbf{V} \cdot \mathbf{U}_{K'} \equiv V_{K'} = V^{K'}: \tag{a2}$$

nonholonomic components of $\mathbf{V}$ along the *body*-fixed (moving) axes $P\text{–}X'Y'Z'$; where

$$V_{K'} = A_{K'K}\, V_K \Leftrightarrow V_K = A_{KK'}\, V_{K'}, \tag{a3}$$

$$A_{K'K} = A_{KK'} \equiv \cos(PK', PK) = \cos(PK, PK') \quad [P\text{–}XYZ\text{: parallel to } O\text{–}XYZ] \tag{a4}$$

from which, since the system is scleronomic, we conclude that the virtual displacement of P is

$$\delta\mathbf{R} = \delta\Xi_{K'}\, \mathbf{U}_{K'} \quad (\text{since } \mathbf{U}_{K'} = \mathbf{U}^{K'}) \Rightarrow \delta\mathbf{R} \cdot \mathbf{U}_{K'} \equiv \delta\Xi_{K'} \quad (= A_{K'K}\, \delta R_K) \tag{a5}$$

where $d\Xi_{K'}/dt \equiv V_{K'}$, i.e., in general, *the* $\Xi_{K'}$ *are the quasi-coordinates of* P *relative to* $P\text{–}X'Y'Z'$.

Therefore, and since (recalling Poisson's kinematical formula Equation g6 of the preceding Example),

$$\delta\mathbf{U}_{K'} = \delta\chi \times \mathbf{U}_K \quad \text{and} \quad d\mathbf{U}_{K'} = d\chi \times \mathbf{U}_{K'}, \tag{b1}$$

where

$$\delta\chi = \delta\phi\, \mathbf{U}_Z + \delta\theta\, \mathbf{U}_n + \delta\psi\, \mathbf{U}_{Z'} \quad (= \delta\phi\, \mathbf{U}_Z + \delta\theta\, \mathbf{U}_{X''} + \delta\psi\, \mathbf{U}_{Z''})$$

$$= \delta\chi_{X'}\, \mathbf{U}_{X'} + \delta\chi_{Y'}\, \mathbf{U}_{Y'} + \delta\chi_{Z'}\, \mathbf{U}_{Z'} \equiv \delta\chi_{K'}\, \mathbf{U}_{K'}, \tag{b2}$$

we find

$$d/dt(\delta\Xi_{K'}) = d/dt(\delta\mathbf{R} \cdot \mathbf{U}_{K'}) = [d(\delta\mathbf{R})/dt] \cdot \mathbf{U}_{K'} + \delta\mathbf{R} \cdot (d\mathbf{U}_{K'}/dt)$$

$$= [d(\delta\mathbf{R})/dt] \cdot \mathbf{U}_{K'} + \delta\mathbf{R} \cdot (\omega \times \mathbf{U}_{K'}), \tag{b3}$$

and, similarly,

$$\delta(d\Xi_{K'}/dt) \equiv \delta V_{K'} = \delta(\mathbf{V} \cdot \mathbf{U}_{K'}) = \delta\mathbf{V} \cdot \mathbf{U}_{K'} + \mathbf{V} \cdot \delta\mathbf{U}_{K'}$$

$$= \delta\mathbf{V} \cdot \mathbf{U}_{K'} + \mathbf{V} \cdot (\delta\chi \times \mathbf{U}_{K'}); \tag{b4}$$

and, therefore, subtracting Equation b4 from Equation b3 side by side, while noting that $\delta\mathbf{V} = \delta(d\mathbf{R}/dt) = d/dt(\delta\mathbf{R})$, we get successively:

$$d/dt(\delta\Xi_{K'}) - \delta(d\Xi_{K'}/dt) = \delta\mathbf{R} \cdot (\omega \times \mathbf{U}_{K'}) - \mathbf{V} \cdot (\delta\chi \times \mathbf{U}_{K'})$$

$$= (\delta\mathbf{R} \times \omega) \cdot \mathbf{U}_{K'} - (\mathbf{V} \times \delta\chi) \cdot \mathbf{U}_{K'}$$

$$= (\delta\mathbf{R} \times \omega - \mathbf{V} \times \delta\chi) \cdot \mathbf{U}_{K'}, \tag{b5}$$

or in vector form (with d'/δ': differentials/variations relative to *body-fixed* axes):

$$d'/dt(\delta\mathbf{R}) - \delta'(d\mathbf{R}/dt) = \delta\mathbf{R} \times \omega - \mathbf{V} \times \delta\chi. \tag{b6}$$

In $P\text{–}X'Y'Z'$ components, the above reads

$$d/dt(\delta\Xi_{X'}) - \delta(d\Xi_{X'}/dt) = (\omega_{Z'}\, \delta\Xi_{Y'} - \omega_{Y'}\, \delta\Xi_{Z'}) - [(d\Xi_{Y'}/dt)\delta\chi_{Z'} - (d\Xi_{Z'}/dt)\delta\chi_{Y'}] \tag{b7}$$

$$d/dt(\delta\Xi_{Y'}) - \delta(d\Xi_{Y'}/dt) = (\omega_{X'}\, \delta\Xi_{Z'} - \omega_{Z'}\, \delta\Xi_{X'}) - [(d\Xi_{Z'}/dt)\delta\chi_{X'} - (d\Xi_{X'}/dt)\delta\chi_{Z'}] \tag{b8}$$

$$d/dt(\delta\Xi_{Z'}) - \delta(d\Xi_{Z'}/dt) = (\omega_{Y'}\, \delta\Xi_{X'} - \omega_{X'}\, \delta\Xi_{Y'}) - [(d\Xi_{X'}/dt)\delta\chi_{Y'} - (d\Xi_{Y'}/dt)\delta\chi_{X'}]; \tag{b9}$$

from which the corresponding nonvanishing Hamel coefficients can be easily read off.

Problem 6.5.3. Alternative Derivation of the Transitivity Equations

Applying the earlier equations (Equations e5 and e6 of Example 6.5.1),

$$d(\textbf{vector})/dt = d'(\textbf{vector})/dt + \omega \times (\textbf{vector}) \quad \text{for} \quad \delta\mathbf{R}, \tag{a}$$

$$\delta(\textbf{vector}) = \delta'(\textbf{vector}) + \delta\chi \times (\textbf{vector}) \quad \text{for} \quad d\mathbf{R}/dt, \tag{b}$$

and then subtracting side by side, while recalling that $d/dt(\delta\mathbf{R}) = \delta(d\mathbf{R}/dt) \equiv \delta\mathbf{V}$, obtain the above transitivity equation (Equation b6 of Example 6.5.2).

Problem 6.5.4

Show that relative to the "semi-fixed" basis $P\text{–}\mathbf{U}_n\mathbf{U}_{n'}\mathbf{U}_Z$, we have

$$V_n \equiv d\Xi_n/dt = \mathbf{V} \cdot \mathbf{U}_n = (dR_X/dt) \cos\phi + (dR_Y/dt) \sin\phi, \tag{a1}$$

$$V_{n'} \equiv d\Xi_{n'}/dt = \mathbf{V} \cdot \mathbf{U}_{n'} = -(dR_X/dt) \sin\phi + (dR_Y/dt) \cos\phi, \tag{a2}$$

$$V_Z \equiv d\Xi_Z/dt = \mathbf{V} \cdot \mathbf{U}_Z \quad \text{(i.e., a holonomic velocity);} \tag{a3}$$

and, accordingly, the transitivity equation for V_n is

$$d/dt(\delta\Xi_n) - \delta(d\Xi_n/dt) = -[(d\phi/dt)\delta\Xi_X - (d\Xi_X/dt)\delta\phi] \sin\phi$$

$$+ [(d\phi/dt)\delta\Xi_Y - (d\Xi_Y/dt)\delta\phi] \cos\phi$$

$$= (d\phi/dt)\delta\Xi_{n'} - (d\Xi_{n'}/dt)\delta\phi$$

$$= (1/\sin\theta)\,[\omega_{Y'}\,\delta\Xi_{n'} - (d\Xi_{n'}/dt)\delta\chi_{Y'}], \qquad \text{(b)}$$

from which the corresponding nonvanishing γs can be read off; and similarly for those of $V_{n'}$, V_Z.

6.6 ADDITIONAL PFAFFIAN CONSTRAINTS

Now, let us assume that our system S, in addition to the hitherto applied to it H [$< 3N_{\text{(-umber of particles)}}$] holonomic constraints (built in, or absorbed, into our Lagrangean description by the $n \equiv 3N - H$ independent holonomic system coordinates q^k), is subjected to the *additional* and *independent* m ($< n$) *linear first-order kinematical,* or *linear velocity,* or *Pfaffian* constraints:

- $$\Omega^D \equiv C^D{}_k(dq^k/dt) + C^D \equiv C^D{}_k\, V^k + C^D{}_{n+1}$$

$$= C^D{}_\beta\, V^\beta = 0 \quad (D = 1, \ldots, m), \qquad \text{(6.6.1)*}$$

where

* (a) The reasons for making D *uppercase* (i.e., nonholonomic) and placing it *up* will become clear later.
(b) In *particle* form, linear velocity constraints appear in the *discrete* form:

$$\sum \mathbf{B}^D{}_p \cdot \mathbf{v}_p + B^D = 0 \quad (p = 1, \ldots, N) \qquad \text{(6.6.1b)}$$

($B^D{}_p$, B^D: known/given functions of the $\mathbf{r}_p$ and t) or, in *continuum* form,

$$\Omega^D \equiv S\,\mathbf{B}^D \cdot \mathbf{v} + B^D = 0 \quad [\mathbf{B}^D = \mathbf{B}^D(t, \mathbf{r}),\ B^D = B^D(t, \mathbf{r})]. \qquad \text{(6.6.1c)}$$

Substituting into Equations 1b, 1c, the velocity representation Equation 6.2.1: $\mathbf{v} = v^k\, \mathbf{e}_k + \mathbf{e}_{n+1}$, where: $v^k = V^k \equiv dq^k/dt$, we get the *system* form (6.6.1):

$$\Omega^D = S\,\mathbf{B}^D \cdot (V^k \mathbf{e}_k + \mathbf{e}_{n+1}) + B^D = \ldots = C^D{}_k V^k + C^D{}_{n+1}, \qquad \text{(6.6.1d)}$$

where

$$C^D{}_k = S\,\mathbf{B}^D \cdot \mathbf{e}_k \equiv S\,\mathbf{B}^D \cdot (\partial\mathbf{r}/\partial q^k), \qquad \text{(6.6.1e)}$$

$$C^D = S\,\mathbf{B}^D \cdot \mathbf{e}_{n+1} + B^D \equiv S\,\mathbf{B}^D \cdot (\partial\mathbf{r}/\partial t) + B^D. \qquad \text{(6.6.1f)}$$

$$C^D{}_k = C^D{}_k(t, q), \quad C^D \equiv C^D{}_{n+1} = C^D(t, q) \quad \text{[of class } C^1 \text{ in some } (t, q)\text{-region]};$$

and

$$\text{rank}(C^D{}_k) = m \quad \text{[independence of the constraint equations].} \qquad (6.6.1a)$$

6.6.1 Holonomicity vs. Nonholonomicity

Clearly, *every holonomic constraint can be brought to the velocity form* (6.6.1). Indeed, $d/dt(\ldots)$-differentiating Equation 6.2.2a: $f^h = f^h(t, \mathbf{r}_P) = 0$ or $f^h = f^h(t, \mathbf{r}) = 0$ we obtain:

$$df^h/dt = S(\partial f^h/\partial \mathbf{r}) \cdot \mathbf{v} + \partial f^h/\partial t = 0, \qquad (6.6.2)$$

i.e., Equation 6.6.1c with $B^h \rightarrow \partial f^h/\partial \mathbf{r}$ and $B^h \rightarrow \partial f^h/\partial t$. The converse, however, is not true: *every Pfaffian constraint system (6.6.1) cannot be brought to the finite form* Equation 6.2.2a ff.

- If it can, either as it stands or after multiplication of its equations with *integrating factors* and combination among them, i.e., if, and *independently of any subsequent kinetic considerations,* it can be replaced by the finite or geometrical system of equations $h^D(t, q) = 0$, then that system of constraints (and associated mechanical system S) is called *completely* or *unconditionally integrable*, or simply *holonomic* [H, disguised in the kinematical form (6.6.1)]. Then, we may replace its so-constrained n qs by another set of $3N - (h + m) \equiv n - m$ new unconstrained or independent holonomic coordinates (à la Section 6.2ff); and say that now S has $n - m$ *global*, or *positional*, *DOF.*
- If, on the other hand, 6.6.1 cannot be brought to finite form, *that system of constraints (and associated mechanical system) is called nonholonomic* (NH). Then we say that S still has n global DOF, but $n - m$ *local*, or *velocity*, DOF; and, as discussed below, *can go from anywhere to anywhere else in configuration space, but not along any path we want;* i.e., for such constraints, all configurations/events are still possible but not all velocities — the 6.6.1 impose local restrictions, not global restrictions as the holonomic conditions.*

  This is a profound distinction among constraints and corresponding mechanical systems; *both H and NH systems satisfy the same mechanical principles (Sections 4.3, 7.4, and 7.5), but their equations of motion and their solutions are markedly different* (Section 7.4).**

---

* For reasons that will become clear in Kinetics, *the term DOF is usually reserved for the number of independent* δqs, i.e., for $n - m \equiv f$, whether Equations 6.6.1 are holonomic or not.

** Later we present necessary and sufficient conditions for the holonomicity of a Pfaffian system of equations (Frobenius' theorem). This is a different, and far more complicated, matter than the earlier exactness conditions of linear and homogeneous differential transformations (Equation 6.4.1f).

- It can be shown that the linear independence and integrability (or absence thereof) of the Pfaffian constraints (6.6.1) is preserved under admissible transformations: $q \to q' = q'(t, q)$.

6.6.2 Scleronomicity vs. Rheonomicity

If the Pfaffian constraints have the form: $C^D{}_k(q)\, V^k = 0$, i.e., if $\partial C^D{}_k/\partial t = 0$ *and* $C^D = 0$, then {in analogy with the velocity form of the stationary holonomic constraints (6.2.6a ff): $f^D(q) = 0 \Rightarrow [\partial f^D(q)/\partial q^k]\, V^k = 0$} they are called *stationary*, and the system *scleronomic*; if not, they are called *nonstationary*, and the system *rheonomic*.

6.6.3 Catastaticity vs. Acatastaticity

A final, related, classification of Pfaffian constraints, made necessary by the analytical structure of AM (and due to Pars, 1965, p. 16) is the following: if $C^D = 0$, these constraints are called *catastatic* (i.e., homogeneous); if not, they are called *acatastatic* (i.e., nonhomogeneous). Thus, stationary Pfaffian constraints are always catastatic; but catastatic Pfaffian constraints may be nonstationary, e.g., $C^D{}_k(t, q)\, V^k = 0$.

Finally, as with the $q \leftrightarrow \theta$ transformations (Section 6.4), the *kinematically admissible* and *virtual* form of Equation 6.6.1 are, respectively (with $d\Theta_K \equiv \Omega^K\, dt$, etc.):

- $d\Theta^D \equiv C^D{}_k\, dq^k + C^D\, dt \equiv C^D{}_k\, dq^k + C^D{}_{n+1}\, dt = C^D{}_\beta\, dq^\beta = 0,$ (6.6.3)

- $\delta\Theta^D \equiv C^D{}_k\, \delta q_k = C^D{}_k\, \delta q^k + C^D{}_{n+1}\,(0) = C^D{}_\beta\, \delta q^\beta = 0$ (6.6.4)

  (i.e., $\delta t = 0$, always).

6.6.4 Nonholonomic Constraints (Equations 6.6.1)

These are most naturally handled with the following choice of quasi-variable transformation coefficients, at the R_{n+1}-point (q, t):

$$A^D{}_k \equiv C^D{}_k \quad \text{and} \quad A^D{}_{n+1} \equiv C^D{}_{n+1} \equiv C^D. \tag{6.6.5}$$

$$(\Rightarrow \Theta^k \equiv \theta^k,\ \Omega^k \equiv \omega^k)$$

Then the local $V^\beta \leftrightarrow \omega^\beta$ transformation (Equations 6.4.1a through e) take the forms:

- *Velocities:*

$$\omega^D \equiv A^D{}_k\, V^k + A^D{}_{n+1} = A^D{}_\beta\, V^\beta = 0 \quad (\text{since } V^{n+1} = v^{n+1} = 1) \tag{6.6.6a}$$

$$\omega^I \equiv A^I{}_k\, V^k + A^I{}_{n+1} = A^I{}_\beta\, V^\beta \neq 0 \quad (I = m+1, \ldots, n) \tag{6.6.6b}$$

where the $(n - m) \times (n + 1)$ coefficients $\{A^I{}_k, A^I{}_{n+1} \equiv A^I\}$ are (as well behaved as needed but otherwise) arbitrary, except that *when the system* of Equations 6.6.6a and b *is inverted,* i.e., solved for the V^β:

$$V^\beta = A^\beta{}_\Psi\,\omega^\Psi = A^\beta{}_K\,\omega^K + A^\beta{}_{N+1}\,\omega^{N+1} \equiv A^\beta{}_K\,\omega^K + A^\beta \equiv A^\beta{}_I\,\omega^I + A^\beta:$$

$$V^k = A^k{}_K\,\omega^K + A^{n+1} = A^k{}_I\,\omega^I + A^{n+1},$$

$$V^{n+1} = A^{n+1}{}_K\,\omega^K + A^{n+1} = (0)\omega^K + \delta^{n+1}{}_{N+1} = 1 \quad (= \omega^{N+1}), \tag{6.6.6c}$$

and these expressions are inserted back into the constraints (6.6.1) they satisfy them identically; and this inversion of Equations 6.6.6a and b and subsequent *representation of V^k as a linear combination of n – m independent quasi-velocities* is the reason for introducing the ω^I. But these particular "equilibrium" quasi-velocities (due to Hamel, 1903–4), like their holonomic counterparts (Equations 6.2.3b ff), have the advantage that in them *the constraints decouple* to the simple form:

$$\omega^D = 0; \tag{6.6.7}$$

whereas Equation 6.6.1 involve all the V^ks. Choices of local transformations other than Equation 6.6.5, in general, do not result in such simple constraint descriptions.

Similarly, and recalling Equations 6.4.2a through 6.4.3e, we find that the *kinematically admissible* and *virtual* forms of Equations 6.6.6a and 6b are, respectively (with $d\theta^K \equiv \omega^K\,dt$, etc.):

- $$d\theta^D \equiv A^D{}_k\,dq^k + A^D{}_{n+1}\,dt = A^D{}_\beta\,dq^\beta = 0 \quad (\text{since } dq^{n+1} = dt) \tag{6.6.8a}$$

$$d\theta^I \equiv A^I{}_k\,dq^k + A^I{}_{n+1}\,dt = A^I{}_\beta\,dq^\beta \neq 0 \quad (I = m+1, \ldots, n) \tag{6.6.8b}$$

with inverse

$$dq^\beta = A^\beta{}_\Psi\,d\theta^\Psi = A^\beta{}_K\,d\theta^K + A^\beta{}_{N+1}\,d\theta^{N+1} \equiv A^\beta{}_K\,d\theta^K + A^\beta\,dt:$$

$$dq^k = A^k{}_K\,d\theta^K + A^{n+1}\,dt = A^k{}_I\,d\theta^I + A^{n+1}\,dt,$$

$$dq^{n+1} = A^{n+1}{}_K\,d\theta^K + A^{n+1}{}_{N+1}\,dt$$

$$= (0)d\theta^K + (\delta^{n+1}{}_{N+1})dt = dt \quad (= d\theta^{N+1}); \tag{6.6.8c}$$

- $$\delta\theta^D \equiv A^D{}_k\,\delta q^k + A^D{}_{n+1}\,\delta t = A^D{}_k\,\delta q^k = 0$$

$$[= A^D{}_\beta\,\delta q^\beta,\ \delta q^{n+1} = \delta t = 0] \tag{6.6.9a}$$

$$\delta\theta^I \equiv A^I{}_k\,\delta q^k + A^I{}_{n+1}\,\delta t = A^I{}_k\,\delta q^k \neq 0$$

$$[= A^I{}_\beta\,\delta q^\beta;\ I = m+1, \ldots, n] \tag{6.6.9b}$$

with inverse

$$\delta q^\beta = A^\beta{}_K\,\delta\theta^K + A^\beta{}_{N+1}\,\delta\theta^{N+1} = A^\beta{}_I\,\delta\theta^I + A^\beta\,\delta t \quad (= A^\beta{}_\Psi\,\delta\theta^\Psi):$$

$$\delta q^k = A^k{}_K\,\delta\theta^K + A^{n+1}\,\delta t = A^k{}_I\,\delta\theta^I,$$

$$\delta q^{n+1} = A^{n+1}{}_K\,\delta\theta^K + A^{n+1}{}_{N+1}\,\delta t$$

$$= (0)\delta\theta^K + (\delta^{n+1}{}_{N+1})(0) = \delta\theta^{N+1} = 0. \qquad (6.6.9c)^*$$

Example 6.6.1. Special Choices of Quasi-coordinates

The above show that we have an "$(n-m)$–freedom" in the selection of the ω's. Here is what happens for the following choices:

i. *Hamel* (1903–4):

$$\omega^D \equiv A^D{}_\beta\,V^\beta = A^D{}_k\,V^k + A^D{}_{n+1}\,V^{n+1} = 0, \qquad (a1)$$

$$\omega^I = \delta^I{}_i\,V^i \neq 0 \quad (I, i{:}\ m+1, \ldots, n;\ \text{i.e., } \omega^I = V^i = dq^i/dt). \qquad (a2)$$

Then, as a direct calculation shows

$$\gamma^{N+1}{}_{KL} = 0, \quad \gamma^{N+1}{}_{K,N+1} = -\gamma^{N+1}{}_{N+1,K} \equiv \gamma^{N+1}{}_K = 0; \quad \text{i.e., } \gamma^{N+1}{}_{K\Psi} = 0, \quad (b1)$$

$$\Rightarrow d(\delta\theta^{N+1}) - \delta(d\theta^{N+1}) = d(\delta t) - \delta(dt) = \gamma^{N+1}{}_{\Psi K}\,d\theta^K\,\delta\theta^\Psi$$

$$= \gamma^{N+1}{}_{LK}\,d\theta^K\,\delta\theta^L + \gamma^{N+1}{}_{N+1,K}\,d\theta^K\,\delta\theta^{N+1} = 0; \qquad (b2)$$

$$\gamma^I{}_{KL} = 0, \quad \gamma^I{}_{K,N+1} = -\gamma^I{}_{N+1,K} \equiv \gamma^I{}_K = 0; \quad \text{i.e., } \gamma^I{}_{K\Psi} = 0; \qquad (b3)$$

$$\Rightarrow d(\delta\theta^I) - \delta(d\theta^I) = \gamma^I{}_{\Psi K}\,d\theta^K\,\delta\theta^Y$$

$$= \gamma^I{}_{LK}\,d\theta^K\,\delta\theta^L + \gamma^I{}_{N+1,K}\,d\theta^K\,d\theta^{N+1} = 0. \qquad (b4)$$

Specialization: Clearly, if $\partial A^K{}_k/\partial t$, $\partial A^k{}_K/\partial t = 0$ and $A^K{}_{n+1}$, $A^K{}_{N+1} = 0$ (i.e., stationary constraints $\Rightarrow$ scleronomic system), then $\gamma^K{}_{L,N+1} = -\gamma^K{}_{N+1,L} = 0$.

ii. Frequently, the Pfaffian constraints (6.6.1) have the following form:

---

* From now on, and unless specified otherwise, we shall assume the following index ranges:

$D, D', D'', \ldots; d, d', d'', \ldots = 1, \ldots, m$ **(Dependent)**
$I, I', I'', \ldots; i, i', i'', \ldots = m+1, \ldots, n$ **(Independent)**

$$V^d = B^d{}_i(t, q)V^i + B^d{}_{n+1}(t, q) \equiv B^d{}_i(t, q)V^i + B^d(t, q) \quad [\equiv B^d{}_{i+1}(t, q)V^{i+1}]$$

$$(i + 1 = m + 1, \ldots, n + 1); \tag{c1)*}$$

i.e., the constraints (6.6.1) are solved for the *first m* V^ds (dependent) in terms of the *last n – m* V^is (independent); and similarly for the kinematically admissible and virtual forms. These can be viewed as the following special case of the Hamel choice (Equations 6.6.6a and 6b):

$$\begin{aligned}
\omega^D &\equiv \delta^D{}_d\,(V^d - B^d{}_i\, V^i - B^d) \\
&\equiv \delta^D{}_d\, V^d - \delta^D{}_d\, B^d{}_i\, V^i - \delta^D{}_d\, B^d \\
&\equiv V^D - B^D{}_i\, V^i - B^D \quad [\equiv V^D - B^D{}_{i+1}\, V^{i+1}] \\
&= 0,
\end{aligned} \tag{c2}$$

$$\omega^I \equiv \delta^I{}_i\, V^i \neq 0 \quad (\text{i.e., } \omega^I = V^i) \quad \text{and} \quad \omega^{N+1} = \delta^{N+1}{}_{n+1}\, V^{n+1} = 1; \tag{c3}$$

$$[\text{or, compactly, } \omega^{I+1} = \delta^{I+1}{}_{i+1}\, V^{i+1} \quad (\text{where: } I + 1 = m + 1, \ldots, n + 1)]$$

with inverse:

$$\begin{aligned}
V^d &= \delta^d{}_D(\omega^D + B^D{}_I\, \omega^I + B^D) \\
&\equiv \delta^d{}_D\, \omega^D + \delta^d{}_D\, B^D{}_I\, \omega^I + \delta^d{}_D\, B^D \\
&\equiv \omega^d + B^d{}_I\, \omega^I + B^d \quad [\equiv \omega^d + B^d{}_{I+1}\, \omega^{I+1}]
\end{aligned} \tag{c4}$$

$$V^i = \delta^i{}_I\, \omega^I \quad \text{and} \quad V^{n+1} = \delta^{n+1}{}_{N+1}\, \omega^{N+1}. \tag{c5}$$

[This elaborate notation is chosen here to stress the fact that *these V^ks have now become quasi-velocities,* even though their notation might suggest otherwise. That is why some authors denote them as $V^{(k)}$.]
Equations c2 through c5 state that, in this case, the transformation coefficients are

$$A^\Lambda{}_\lambda: \quad A^D{}_d = \delta^D{}_d, \quad A^D{}_i = -B^D{}_i, \quad A^D{}_{n+1} = -B^D{}_{n+1} \equiv -B^D$$

$$A^I{}_d = \delta^I{}_d = 0, \quad A^I{}_i = \delta^I{}_i, \quad A^I{}_{n+1} = \delta^I{}_{n+1} = 0$$

$$A^{N+1}{}_d = \delta^{N+1}{}_d = 0, \quad A^{N+1}{}_i = \delta^{N+1}{}_i = 0, \quad A^{N+1}{}_{n+1} = \delta^{N+1}{}_{n+1} = 1; \tag{c6}$$

* Originally employed by Chaplygin (in 1895), Hadamard (in 1895), and Voronets (in 1901); see e.g., Neimark and Fufaev (1972, Chapter 3).

$$A^{\lambda}{}_{\Lambda}: \quad A^d{}_D = \delta^d{}_D, \quad A^d{}_I = B^d{}_I, \quad A^d{}_{N+1} = B^d{}_{N+1} \equiv B^d$$

$$A^i{}_D = \delta^i{}_D = 0, \quad A^i{}_I = \delta^i{}_I, \quad A^i{}_{N+1} = \delta^i{}_{N+1} = 0$$

$$A^{n+1}{}_D = \delta^{n+1}{}_D = 0, \quad A^{n+1}{}_I = \delta^{n+1}{}_I = 0, \quad A^{n+1}{}_{N+1} = \delta^{n+1}{}_{N+1} = 1; \tag{c7}$$

[where $B^d{}_I \equiv \delta^d{}_D\, \delta^i{}_I\, B^D{}_i = \delta^i{}_I\, B^d{}_i = \delta^d{}_D\, B^D{}_I$] and, as a result, the Hamel coefficients become

$$\gamma^I{}_{\Lambda\Psi} = 0, \quad \gamma^D{}_{D'Y} = 0, \tag{c8}$$

and

$$\gamma^D{}_{II'} = 2A^D{}_{[k,\lambda]}\, A^k{}_I\, A^{\lambda}{}_{I'}$$

$$= 2A^D{}_{[k,\lambda]}\, B^k{}_I\, B^{\lambda}{}_{I'} = B^D{}_{I',k}\, B^k{}_I - B^D{}_{I,\lambda}\, B^{\lambda}{}_{I'} \equiv B^D{}_{I',(I)} - B^D{}_{I,(I')}, \tag{c9}$$

$$\gamma^D{}_{I,N+1} = \dots = B^D{}_{N+1,(I)} - B^D{}_{I,(N+1)}, \tag{c10}$$

where (symbolic quasi-chain rule)

$$(\dots)_{,(I)} \equiv \partial(\dots)/\partial q^{(I)} \equiv B^{\lambda}{}_I[\partial(\dots)/\partial q^{\lambda}]$$

$$= B^d{}_I[\partial(\dots)/\partial q^d] + B^i{}_I\,[\partial(\dots)/\partial q^i] + B^{n+1}{}_I[\partial(\dots)/\partial q^{n+1}]$$

$$= B^d{}_I[\partial(\dots)/\partial q^d] + \delta^i{}_I\,[\partial(\dots)/\partial q^i] + \delta^{n+1}{}_I[\partial(\dots)/\partial t]$$

$$= B^d{}_I[\partial(\dots)/\partial q^d] + \delta^i{}_I\,[\partial(\dots)/\partial q^i] + (0)[\partial(\dots)/\partial t]$$

$$\equiv \partial(\dots)/\partial q^I + B^d{}_I[\partial(\dots)/\partial q^d], \tag{c11}$$

$$(\dots)_{,(N+1)} \equiv \partial(\dots)/\partial q^{(N+1)} \equiv \partial(\dots)/\partial(t) \equiv B^{\lambda}{}_{N+1}\,[\partial(\dots)/\partial q^{\lambda}]$$

$$= B^d{}_{N+1}[\partial(\dots)/\partial q^d] + B^i{}_{N+1}[\partial(\dots)/\partial q^i] + B^{n+1}{}_{N+1}\,[\partial(\dots)/\partial q^{n+1}]$$

$$= B^d[\partial(\dots)/\partial q^d] + \delta^i{}_{N+1}[\partial(\dots)/\partial q^i] + \delta^{n+1}{}_{N+1}[\partial(\dots)/\partial t]$$

$$= B^d[\partial(\dots)/\partial q^d] + (0)[\partial(\dots)/\partial q^i] + (1)[\partial(\dots)/\partial t]$$

$$= \partial(\dots)/\partial t + B^d[\partial(\dots)/\partial q^d]. \tag{c12}$$

In "holonomic" coordinates, the above read (with W for *Woronetz*, or *Voronets*):

$$\gamma^D{}_{II'} \Rightarrow -W^d{}_{ii'} = W^d{}_{i'i} \equiv \partial B^d{}_{i'}/\partial q^{(i)} - \partial B^d{}_i/\partial q^{(i')}$$

$$= [\partial B^d{}_{i'}/\partial q^i + B^{d'}{}_i(\partial B^d{}_{i'}/\partial q^{d'})] - [\partial B^d{}_i/\partial q^{i'} + B^{d'}{}_{i'}(\partial B^d{}_i/\partial q^{d'})]$$

$$= (\partial B^d{}_{i'}/\partial q^i - \partial B^d{}_i/\partial q^{i'}) + [B^{d'}{}_i(\partial B^d{}_{i'}/\partial q^{d'}) - B^{d'}{}_{i'}(\partial B^d{}_i/\partial q^{d'})], \qquad \text{(c13)}$$

$$\gamma^D{}_{I,N+1} \equiv \gamma^D{}_I \Rightarrow -W^d{}_{i,n+1} \equiv -W^d{}_i$$

$$\equiv \partial B^d{}_{n+1}/\partial q^{(i)} - \partial B^d{}_i/\partial q^{(n+1)} \equiv \partial B^d{}_{n+1}/\partial q^{(i)} - \partial B^d{}_i/\partial(t)$$

$$= [\partial B^d{}_{n+1}/\partial q^i + B^{d'}{}_i(\partial B^d{}_{n+1}/\partial q^{d'})] - [\partial B^d{}_i/\partial q^{n+1} + B^{d'}{}_{n+1}(\partial B^d{}_i/\partial q^{d'})]$$

$$= (\partial B^d{}_{n+1}/\partial q^i - \partial B^d{}_i/\partial q^{n+1}) + [B^{d'}{}_i(\partial B^d{}_{n+1}/\partial q^{d'}) - B^{d'}{}_{n+1}(\partial B^d{}_i/\partial q^{d'})]$$

$$\equiv [\partial B^d/\partial q^i + B^{d'}{}_i(\partial B^d/\partial q^{d'})] - [\partial B^d{}_i/\partial t + B^{d'}(\partial B^d{}_i/\partial q^{d'})]$$

$$= (\partial B^d/\partial q^i - \partial B^d{}_i/\partial t) + [B^{d'}{}_i(\partial B^d/\partial q^{d'}) - B^{d'}(\partial B^d{}_i/\partial q^{d'})]. \qquad \text{(c14)}$$

[If $B^d{}_i = B^d{}_i(q^{i'})$ and $B^d = 0$, which is the case dealt by Chaplygin, the only surviving transitivity coefficients are (with T for *Tsaplygine, or Chaplygin*)]:

$$-W^d{}_{ii'} \Rightarrow -T^d{}_{ii'} \equiv \partial B^d{}_{i'}/\partial q^i - \partial B^d{}_i/\partial q^{i'}. \qquad \text{(c15)}$$

Finally, enforcing the constraints Equations 6.6.7, 6.6.8a and 6.6.9a in the "standard" transitivity equations (6.5.6a and b), we find the useful *constrained transitivity equations*:

$$\textit{Dependent:} \quad d(\delta\theta^D) - \delta(d\theta^D) = \gamma^D{}_{II'}\, d\theta^{I'}\, \delta\theta^I + \gamma^D{}_I\, dt\, \delta\theta^I, \qquad \text{(6.6.10a)}$$

$$\textit{Independent:} \quad d(\delta\theta^I) - \delta(d\theta^I) = \gamma^I{}_{I'I''}\, d\theta^{I''}\, d\theta^{I'} + \gamma^I{}_{I'}\, dt\, d\theta^I; \qquad \text{(6.6.10b)}$$

or, dividing with dt

$$(\delta\theta^D)^{\cdot} - \delta\omega^D = \gamma^D{}_{II'}\, \omega^{I'}\, \delta\theta^I + \gamma^D{}_I\, \delta\theta^I \equiv h^D{}_I\, \delta\theta^I, \qquad \text{(6.6.11a)}$$

$$(\delta\theta^I)^{\cdot} - \delta\omega^I = \gamma^I{}_{I'I''}\, \omega^{I''}\, \delta\theta^{I'} + \gamma^I{}_{I'}\, \delta\theta^{I'} \equiv h^I{}_{I'}\, \delta\theta^{I'}. \qquad \text{(6.6.11b)}$$

The above show that since, in general, $d(\delta\theta^D) \neq \delta(d\theta^D)$, we *cannot* have both $d(\delta\theta^D) = 0$ *and* $\delta(d\theta^D) = 0$ (even though $d\theta^D = 0$ and $\delta\theta^D = 0$); *it is either one or the other.**

* This has important consequences in the formulation of *time-integrated equations* ("variational principles" à la Hamilton, etc.) for nonholonomic systems. See, e.g., Papastavridis (1997).

6.7 THEOREM OF FROBENIUS

The preceding transitivity equations are intimately connected with the holonomicity, or absence thereof, of the Pfaffian constraints (6.6.8a and 6.6.9a). The issue is answered by the following and fundamenal and nontrivial theorem of Frobenius.

Theorem of Frobenius (1877): The necessary and sufficient condition for the holonomicity of the m $(< N)$ independent Pfaffian constraints:

$$A^D{}_K\, U^K = 0 \quad [D = 1, \ldots, m\ (< N);\ K = 1, \ldots, N] \tag{6.7.1}$$

where $A^D{}_K = A^D{}_K\,(x^1, x^2, \ldots, x^N) \equiv A^D{}_K\,(x)$ and $rank\,(A^D{}_K) = m$, that is, for that system to have m independent integrals $H^D(x) = \text{constant}$, is the vanishing of the m bilinear forms:

$$F^D \equiv [(\partial A^D{}_K/\partial x^L) - (\partial A^D{}_L/\partial x^K)]\,V^L\, W^K \quad (K, L = 1, \ldots, N) \tag{6.7.2}$$

identically (in the xs) and simultaneously (for *all* Ds) for any two solutions $V \equiv (V_1, \ldots, V_N)$ and $W \equiv (W_1, \ldots, W_N)$ of the constraints, i.e., for arbitrary U^K–pairs V^K, W^K satisfying

$$A^D{}_K\, V^K = 0, \quad A^D{}_K W^K = 0. \tag{6.7.3}$$

(Since $m < N$, Equations 6.7.1 will, in general, have more than one set of solutions.)*

Adapting the above to our problem, i.e., with the identifications:

$$x \to t,\, q, \quad N \to n + 1,$$

$$V^K \to dq^\lambda, \quad W^K \to \delta q^k \quad [k = 1, \ldots, n;\ \lambda = 1, \ldots, n + 1] \tag{6.7.4a}$$

results readily in the following *basic kinematical theorem:* if

$$0 = d(\delta\theta^D) - \delta(d\theta^D) = d(A^D{}_k\, \delta q^k) - \delta(A^D{}_\lambda\, dq^\lambda) \quad [= dA^D{}_k\, \delta q^k - \delta A^D{}_\lambda\, dq^\lambda]$$

$$= (\partial A^D{}_k/\partial q^\lambda - \partial A^D{}_\lambda/\partial q^k)dq^\lambda\, \delta q^k$$

$$= [(\partial A^D{}_k/\partial q^l - \partial A^D{}_l/\partial q^k)dq^l + (\partial A^D{}_k/\partial t) - (\partial A^D{}_{n+1}/\partial q^k)dt]\delta q^k, \tag{6.7.4b}$$

for $dq^\lambda \equiv (dq^k,\, dq^{n+1} \equiv dt)$ and $\delta q^\lambda \equiv (\delta q^k,\, \delta q^{n+1} \equiv \delta t = 0)$ arbitrary solutions of

$$d\theta^D \equiv A^D{}_\lambda\, dq^\lambda = A^D{}_k\, dq^k + A^D{}_{n+1}\, dq^{n+1} \equiv A^D{}_k\, dq^k + A^D\, dt = 0,$$

$$\delta\theta^D \equiv A^D{}_\lambda\, \delta q^\lambda = A^D{}_k\, \delta q^k + A^D{}_{n+1}\, \delta q^{n+1} = A^D{}_k\, \delta q^k = 0, \tag{6.7.4c}$$

---

* For readable proofs, see, e.g., (alphabetically): Cartan (1922, ch. 10), Chetaev (1989, pp. 319–326), Favard (1957, Ch. 3), Forsyth [1890, Chs. 1, 2 (pp. 1–79), 30 (pp. 299–332)], Guldberg (1927), Lovelock and Rund (1975, pp. 145–156), Pascal (1927); also Maißer (1981, 1983–4).

then the constraint system (6.6.8a and 6.6.9a) or Equations 6.7.4c is holonomic.*

In this form, however, the theorem is not always easy to apply. To alleviate that we proceed to reformulate it as follows:

i. First, we notice that the *general solutions* of Equation 6.7.4c are, respectively,

$$dq^k = P^k{}_I\, E^I + P^k{}_{N+1}\, E^{N+1} \equiv P^k{}_I\, E^I + P^k\, E^{N+1} \quad \text{and} \quad \delta q^k = P^k{}_I\, \varepsilon^I, \tag{6.7.5a}$$

where the $\{P^k{}_I, P^k{}_{N+1} \equiv P^k\}$ are locally constant coefficients, while the $\{E^I, E^{N+1}; \varepsilon^I\}$ are *independent parameters*;** and this, with the judicious choice $P^k{}_I \to A^k{}_I$, $P^k \to A^k{}_{N+1} \equiv A^k$ and $E^I \to d\theta^I$, $E^{N+1} = dt$ and $\varepsilon^I \to \delta\theta^I$ assumes the useful form:

$$dq^k = A^k{}_I\, d\theta^I + A^k\, dt \quad \text{and} \quad \delta q^k = A^k{}_I\, \delta\theta^I; \tag{6.7.5b}$$

ii. Then, substituting Equations 6.7.5b into Equation 6.7.4b we easily conclude that the necessary and sufficient condition for the holonomicity of the system of (6.7.4c) is the identical vanishing of

$$d(\delta\theta^D) - \delta(d\theta^D) = \gamma^D{}_{II'}\, d\theta^{I'}\, \delta\theta^I + \gamma^D{}_I\, dt\, \delta\theta^I, \tag{6.7.5c}$$

where, recalling Equations 6.5.3a and b,

$$\gamma^D{}_{II'} = (\partial A^D{}_k/\partial q^l - \partial A^D{}_l/\partial q^k) A^k{}_I\, A^l{}_{I'} = (A^k{}_I\, A^l{}_{I'} - A^l{}_I\, A^k{}_{I'})(\partial A^D{}_k/\partial q^l)$$

$$= A^D{}_k[A^l{}_I(\partial A^k{}_{I'}/\partial q^l) - A^l{}_{I'}(\partial A^k{}_I/\partial q^l)] \equiv A^D{}_k(\partial A^k{}_{I'}/\partial\theta^I - \partial A^k{}_I/\partial\theta^{I'}),$$

---

* Lack of space prevents us from a comprehensive discussion of holonomicity/nonholonomicity. The best we can do here is provide the following summary of basic definitions and concepts:

- Equations $d\theta^D \equiv A^D{}_k dq^k = 0$ (i.e., Equation 6.7.4c but with $A^D = 0$, for convenience but no loss in generality) have an integral if m *integrating factors* $\mu_D = \mu_D(q)$ (not all zero) exist such that $\mu_D\, d\theta^D = \mu_D\, A^D{}_k\, dq^k = df(q)$ $(= 0) \Rightarrow f = f(q) =$ constant $\equiv C$; and they are called *completely integrable*, or *holonomic*, if m independent integrals $f^D(q) = C^D$ exist; or, equivalently, if their "solution manifold" (locally at least) is an m–parameter family of $(n - m)$–dimensional hypersurfaces, in an n-dimensional space: $q^k = q^k(u^l, C^D)$ [u^l: $(n - m)$ surface coordinates, C^D: m surface parameters] Then, $\mu^D{}_{D'}\, d\theta^{D'} = df^{D'}$ $(= 0) \Rightarrow f^D = f^D(q) = C^D$.
- The determinant $\mu \equiv |\, \mu^D{}_{D'}\,|$ is the *multiplicator* of $d\theta^D = 0$.
- *Theorem*: If μ is a multiplicator of $d\theta^D = 0$, so is $\mu\phi$, where ϕ is any function of $f^1, \ldots, f^m$; i.e., there is an *infinity* of multiplicators.
- *Theorem*: For the holonomicity of $d\theta^D = 0$ it is *necessary* that its Pfaffians $d\theta^D$ be representable as $d\theta^D = A^D{}_{D'}\, df^{D'}$ $(= 0)$ [definition of coefficients $A^D{}_{D'}$, with $\mathrm{Det}(A^D{}_{D'}) \neq 0$]. Inverting the above yields $df^{D'} = A^{D'}{}_D\, d\theta^D$ $(= 0)$ (definition of "inverse coefficients" $A^D{}_{D'}$). The m^2 coefficients $A^D{}_{D'}$ (or $A^{D'}{}_D$) are now the integrating factors.

** This fundamental step is due to the Italian mechanician G.A. Maggi (1896, 1901, 1903).

$$\gamma^D{}_{I,N+1} \equiv \gamma^D{}_I = (\partial A^D{}_k/\partial q^l - \partial A^D{}_l/\partial q^k)A^k{}_I\, A^l + (\partial A^D{}_k/\partial t - \partial A^D/\partial q^k)A^k{}_I$$

$$\equiv (\partial A^D{}_k/\partial q^\lambda - \partial A^D{}_\lambda/\partial q^k)A^k{}_I\, A^\lambda$$

$$= (A^k{}_I\, A^\lambda - A^\lambda{}_I\, A^k)(\partial A^D{}_k/\partial q^\lambda)$$

$$= A^D{}_k[A^\lambda{}_I(\partial A^k/\partial q^\lambda) - A^\lambda(\partial A^k{}_I/\partial q^\lambda)]$$

$$\equiv A^D{}_k(\partial A^k/\partial \theta^I - \partial A^k{}_I/\partial \theta^{N+1}); \tag{6.7.5d}$$

which, *since the $d\theta^I$, dt, $\delta\theta^I$ are independent* (and this is the crux of the argument) leads to the following *alternative form of Frobenius' theorem*: The necessary and sufficient conditions for the holonomicity of the system of Pfaffian constraints (6.7.4c) are:

$$\gamma^D{}_{II'} = 0, \quad \gamma^D{}_I = 0 \quad [D = 1, \ldots, m;\ I, I' = m + 1, \ldots, n] \tag{6.7.5e)*$$

Since for each D, (6.7.5e$_1$) yields $(n - m)\ (n - m - 1)/2 \equiv f(f - 1)/2$ independent equations (where $f \equiv n - m$: number of DOF), while (6.7.5e$_2$) yields f independent equations, Equations 6.7.5e stand for a total of *$mf(f - 1)/2 + mf = mf(f + 1)/2$ independent holonomicity conditions.***

The above show clearly that holonomicity, or absence thereof, is a *system* property, i.e., *each Pfaffian equation is tested against the entire system* (6.7.4c); and if $\gamma^D{}_{\ldots} \neq 0$ for some D, then by attaching the corresponding Pfaffian equation(s) to another Pfaffian system the new $\gamma^D{}_{\ldots}$ may be made to vanish; i.e., *additional Pfaffian constraints may turn an originally (individually) nonholonomic constraint into a holonomic one, as part of the new system.* [This contrasts sharply with the exactness conditions of Pfaffian *forms* (not equations), which, as we have seen (e.g., Equations 6.4.1f and 6.5.5 ff) is an *individual* form property.]

Example 6.7.1

Let us apply Frobenius' theorem to check the holonomicity of

$$d\theta^1 \equiv A\, dx + B\, dy + C\, dz = 0, \quad \text{where } A, B, C\text{: functions of } x, y, z. \tag{a}$$

By Equations 6.7.4b, for this to happen we must have (using commas for partial differentiations):

---

* Most likely due to the French mathematician E. Cartan (late 1890s) and the German mechaniker G. Hamel (1903–4, 1924). For additional, readable, forms of Frobenius' theorem, see v. Weber (1899–1916, pp. 315ff).

** A *geometrical interpretation* of the constraints and the form 6.7.5e is given in the next section.

$$0 = d(A\ \delta x + B\ \delta y + C\ \delta z) - \delta(A\ dx + B\ dy + C\ dz)$$

$$= (dA\ \delta x + dB\ \delta y + dC\ \delta z) - (\delta A\ dx + \delta B\ dy + \delta C\ dz)$$

$$= [(A_{,x}\ dx + A_{,y}\ dy + A_{,z}\ dz)\delta x + \ldots]$$

$$- [(A_{,x}\ \delta x + A_{,y}\ \delta y + A_{,z}\ \delta z)dx + \ldots]$$

$$= (A_{,y} - B_{,x})(dy\ \delta x - dx\ \delta y) + (A_{,z} - C_{,x})(dz\ \delta x - dx\ \delta z)$$

$$+ (B_{,z} - C_{,y})(dz\ \delta y - dy\ \delta z), \tag{b}$$

for all $d(x, y, z)$ and $\delta(x, y, z)$ satisfying Equation a. In terms of the following vectors:

$$\mathbf{V} \equiv (A, B, C) \quad \text{and} \quad d\mathbf{r} \equiv (dx, dy, dz), \quad \delta\mathbf{r} \equiv (\delta x, \delta y, \delta z), \tag{c}$$

where (x, y, z) and (A, B, C) are assumed rectangular Cartesian components, Equation b reads, simply,

$$\text{curl}\ \mathbf{V} \cdot (d\mathbf{r} \times \delta\mathbf{r}) = 0, \quad \text{under} \quad \mathbf{V} \cdot d\mathbf{r} = 0 \quad \text{and} \quad \mathbf{V} \cdot \delta\mathbf{r} = 0. \tag{d}$$

But Equations $d_{2,3}$ state that $\mathbf{V}$ is perpendicular to both $d\mathbf{r}$ and $\delta\mathbf{r}$; or, equivalently, that $\mathbf{V}$ is *parallel* to $d\mathbf{r} \times \delta\mathbf{r}$ ($\neq \mathbf{0}$, assumed), i.e., $d\mathbf{r} \times \delta\mathbf{r} = \lambda\mathbf{V}$ (λ: real scalar), and so Equation d_1 leads to

$$\mathbf{V} \cdot \text{curl}\ \mathbf{V} = A(C_{,y} - B_{,z}) + B(A_{,z} - C_{,x}) + C(B_{,x} - A_{,y}) = 0 \tag{h}$$

[Euler (1770); also, recall Equations 3.13.8a through 3.13.8q].

- If Equation h holds *identically*, then Equation a can be brought to the form $df(x, y, z) = 0$ or $\mu(x, y, z)\ df(x, y, z) = 0$, in which case its solution is a one-parameter (or ∞^1) family of *integral surfaces* $F(x, y, z) = \text{constant} \equiv c_{(\text{family parameter})}$ orthogonal to $\mathbf{V}$, at each point (x, y, z).
- If Equation h is satisfied, *but not identically*, then the solution of Equation a does *not* contain an arbitrary constant; i.e, there exist only a *finite* number of them (possibly zero). And since *such solutions cannot be made to pass through any desired space point* (x, y, z), *they cannot be classified as holonomic* ($\equiv$ completely integrable). For example, applying Equation h to $(yz)dx + (z)dy + (-1)dz = 0$ or $(z)dx + (zx)dy + (-1)dz = 0$ yields the sole integral $z = 0$, while applying it to $(y)\ dx + (2x)\ dy + (-1)\ dz = 0$ yields no solution at all.
- If Equation h *does not hold*, then the general solution of Equation a consists of a one-parameter family of curves $\psi(x, y) = c_{(\text{arbitrary constant})}$ on an arbitrarily selected surface $\phi(x, y, z) = 0$; these curves being, at each point (x, y, z), orthogonal to the vector $\mathbf{V}$.

- Clearly, condition Equation h remains unchanged under cyclic permutation of x, y, z and, *simultaneously*, of A, B, C; or, whether we consider z, or x, or y as the unknown function. Also, it is not hard to see that (like its generalizations presented in the next Problem) it remains invariant under admissible variable changes: $(x, y, z) \to [x' = x'(x, y, z), y' = y'(x, y, z), z' = z'(x, y, z)]$. Then, $A\, dx + B\, dy + C\, dz \to A'\, dx' + B'\, dy' + C'\, dz'$, and Equation h holds but with $x, y, z; A, B, C$ replaced with $x', y', z'; A', B', C'$.
- From Equation h we readily conclude that the *Pfaffian equation* $A(x, y)\, dx + B(x, y)\, dy = 0$ is always holonomic [it has an *infinity* of integrating factors $\mu(x, y)$]. Mechanically, this means that a scleronomic system with *two* Lagrangean coordinates is always holonomic; or, the simplest nonholonomic system has *three* such coordinates and one Pfaffian constraint.

Problem 6.7.1

i. Show that the necessary and sufficient condition for the holonomicity of the single Pfaffian equation in n variables:

$$d\theta \equiv A_k\, dq^k = 0, \quad \text{where } A_k = A_k(q^1, \ldots, q^n) \equiv A_k(q), \tag{a}$$

is the identical satisfaction of [with $(\ldots)_{,k} \equiv \partial(\ldots)/\partial q^k$]:

$$A_k(A_{r,s} - A_{s,r}) + A_r(A_{s,k} - A_{k,s}) + A_s(A_{k,r} - A_{r,k}) = 0, \tag{b}$$

or, compactly,

$$A_k\, A_{[r,s]} + A_r\, A_{[s,k]} + A_s\, A_{[k,r]} = 0, \tag{c}$$

simultaneously (for *all* combinations of $k, r, s = 1, \ldots, n$).

ii. Verify that out of a total of $n(n-1)(n-2)/3!$ such conditions, only $(n-1)(n-2)/2$ are independent (recall Equation 6.7.5e ff); and, further, if $A_k \neq 0$, we only need to satisfy Equation b for all $r, s \neq k$.

iii. Verify that for $n = 3$ Equation b leads to the *single* condition [since then: $(n-1)(n-2)/2 = (3-1)(3-2)/2 = 1$]:

$$e^{krs}\, A_k\, A_{s,r} \equiv e^{krs}\, A_k\, \partial_r\, A_s = 0 \Rightarrow A_1\, A_{[2,3]} + A_2\, A_{[3,1]} + A_3\, A_{[1,2]} = 0, \tag{d}$$

which constitutes a compact formulation of Equation h of Example 6.7.1 with proper variable renaming.

Problem 6.7.2

Verify that if the Pfaffian equation has the form:

$$dz = B_I\, dq^I, \quad \text{where } B_I = B_I\,(q, z) \text{ (given functions)},\ q \equiv (q^1, \ldots, q^i), \tag{a}$$

(instead of the "symmetric" form, Equation a, of Problem 6.7.1) then the necessary and sufficient condition for its holonomicity is the identical satisfaction of the $i(i-1)/2$ [$=(n-1)(n-2)/2$, since here $i = n-1$] independent equations:

$$\text{“}dB_I/dq^{I'} = dB_{I'}/dq^I\text{:”}$$

$$\partial B_I/\partial q^{I'} + (\partial B_I/\partial z)(\partial z/\partial q^{I'}) = \partial B_{I'}/\partial q^I + (\partial B_{I'}/\partial z)(\partial z/\partial q^I), \tag{b}$$

or, finally,

$$\partial B_I/\partial q^{I'} + B_{I'}(\partial B_I/\partial z) = \partial B_{I'}/\partial q^I + B_I(\partial B_{I'}/\partial z), \tag{c}$$

for all combinations of $I, I' = 1, \ldots, i$ $(= n-1)$.

Specialization: Applied to the Pfaffian equation $dz = P(x, y, z)\, dx + Q(x, y, z)\, dy$, i.e., with $I = 1, 2$, Equations a reduce to the single holonomicity condition:

$$dP/dy = dQ/dx: \quad \partial P/\partial y + (\partial P/\partial z)Q = \partial Q/\partial x + (\partial Q/\partial z)\, P; \tag{d}$$

an equation which remains invariant under arbitrary changes $z \to z'(x, y, z)$, i.e., $z = z(x, y, z')$.

Remark: Rewriting Equation a of Example 6.7.1 as $dz = (-A/C)\, dx + (-B/C)\, dy$ (assuming $C \neq 0$) and then applying to it Equation d, with $P = -A/C$ and $Q = -B/C$, we recover Equation h of Example 6.7.1, as we should. [Inversely, with $\mathbf{V} \to \mathbf{V}' \equiv (P, Q, -1) \Rightarrow \text{curl}\, \mathbf{V}' = (-Q_z, P_z, Q_x - P_y)$, Equation h of Example 6.7.1 yields

$$\mathbf{V}' \cdot \text{curl}\, \mathbf{V}' = P_y + Q\, P_z - (Q_x + P\, Q_z) = \ldots = (1/C^2)(\mathbf{V} \cdot \text{curl}\, \mathbf{V}) = 0] \tag{h}$$

Problem 6.7.3

Generalizing the results of Problem 6.7.2, show that the necessary and sufficient condition for the holonomicity of the Pfaffian system:

$$dq^D = B^D{}_I\, dq^I \quad [D, D', \ldots = 1, 2, \ldots, m;\ I, I', \ldots = m+1, \ldots, n] \tag{a}$$

where $B^D{}_I = B^D{}_I(q)$ (given functions), $q \equiv (q^1, \ldots, q^n)$, is the identical satisfaction of the system: $dB^D{}_I/dq^{I'} = dB^D{}_{I'}/dq^I$, or *in extenso* (F. Deahna, 1840; J.C. Bouquet, 1872):

$$\partial B^D{}_I/\partial q^{I'} + (\partial B^D{}_I/\partial q^{D'})(\partial q^{D'}/\partial q^{I'})$$

$$= \partial B^D{}_{I'}/\partial q^I + (\partial B^D{}_{I'}/\partial q^{D'})(\partial q^{D'}/\partial q^I), \tag{b}$$

or

$$\partial B^D{}_I/\partial q^{I'} + B^{D'}{}_{I'}(\partial B^D{}_I/\partial q^{D'}) = \partial B^D{}_{I'}/\partial q^I + B^{D'}{}_I(\partial B^D{}_{I'}/\partial q^{D'}); \tag{c}$$

for each $D = 1, \ldots, m$, and each of the $(n - m)(n - m - 1)/2$ possible combinations of $I, I' = m + 1, \ldots, n$.

Remarks:

i. Equation c coincides, in form, with the earlier *necessary and sufficient conditions for a curved space to be flat,* Equations 5.6.16a; with appropriate variable and index renaming.
ii. Since Equation a can be viewed as the following special Hamel choice (recall Equation c1 and following of Example 6.6.1):

$$d\theta^D \equiv dq^D - B^D{}_I \, dq^I = 0, \tag{d}$$

conditions Equations b and c are the corresponding special case of $\gamma^D{}_{II'} = 0$ (recall Equation c13 of Example 6.6.1).
iii. It can be shown (e.g., by induction) that of these conditions, i.e., b or c, only $m(n - m)(n - m - 1)/2 \equiv mf(f - 1)/2$ are independent; in agreement with the general case (6.7.5e$_1$).

Problem 6.7.4

Using Frobenius' theorem, show that the Pfaffian system:

$$d\theta^1 \equiv a \, dx + b \, dy + c \, dz = 0, \quad d\theta^2 \equiv A \, dx + B \, dy + C \, dz = 0, \tag{a}$$

where $a, b, c; A, B, C$ are functions of x, y, z (i.e., $n = 3$, $m = 2$) and Equations $a_{1,2}$ are assumed independent, is *always* holonomic, even if Equations a_1 and a_2 separately (i.e., with $n = 3$, $m = 1$) are nonholonomic; and so, its general solution is the two-parameter family of curves: $f_1(x, y, z) = c_1, f_2(x, y, z) = c_2$ (c_1, c_2: arbitrary constants).

[Generally, it is not hard to show that *a system of $m = n - 1$ (or n) independent Pfaffian equations in n variables is always holonomic.*]

Example 6.7.2

Let us apply Frobenius' theorem to check the holonomicity of the following *two* (independent) scleronomic Pfaffians in *four* variables x, y, ϕ, ψ:

$$d\theta^1 \equiv dx - r \cos\phi \, d\psi = 0, \quad d\theta^2 \equiv dy - r \sin\phi \, d\psi = 0, \tag{a}$$

which express the rolling of a vertical thin circular disk D of radius r on a fixed rough plane [x, y: coordinates of center of D, ϕ: precessional angle of plane of D with, say, $+ xz$-plane, ψ: spin angle (and, of course, nutational angle $\theta = \pi/2$) — see, e.g., Rosenberg (1977, pp. 108–111)]. From Equation a, and invoking Frobenius' expressions (6.7.4b ff), we readily find

$$d(\delta\theta^1) - \delta(d\theta^1) = r \sin\phi \, \delta\psi \, d\phi - r \sin\phi \, \delta\phi \, d\psi, \tag{b}$$

$$d(\delta\theta^2) - \delta(d\theta^2) = -r \cos\phi \, \delta\psi \, d\phi + r \cos\phi \, \delta\phi \, d\psi. \tag{c}$$

Since, in general, $d\phi$, $d\psi$, $\delta\phi$, $\delta\psi \neq 0$, the above bilinear expressions vanish only if $\sin\phi = 0$, $\cos\phi = 0$. But then the constraints reduce to $dx = 0$, $dy = 0$, which is, in general, impossible, i.e., *Equations b and c cannot be satisfied identically.* Hence, by Frobenius' theorem, the system (Equation a) is nonholonomic.

6.8 GEOMETRICAL INTERPRETATION OF PFAFFIAN CONSTRAINTS

As already seen in Sections 6.2, 6.3, and 6.4, at the generic admissible R_{n+1}–point $P(q^k, q^{n+1} \equiv t) \equiv (q^\alpha)$ (with which the figurative system particle $P(S) \equiv P$, instantaneously, coincides), an arbitrary system vector can be resolved either along the *holonomic* basis $\mathbf{E}_\lambda \equiv \partial\mathbf{R}/\partial q^\lambda \equiv \partial\mathbf{q}/\partial q^\lambda$, or along the *nonholonomic* basis $\mathbf{E}_\Lambda \equiv \partial\mathbf{R}/\partial\theta^\Lambda \equiv (\partial\mathbf{R}/\partial q^\lambda)\,(\partial V^\lambda/\partial\omega^\Lambda) = A^\lambda{}_\Lambda\mathbf{E}_\lambda$, where these bases span the *tangent space* $\mathbf{T}(\mathbf{R}_{n+1}, \mathbf{P}) \equiv \mathbf{T}_{n+1}(\mathbf{P})$. Further, the totality of these local spaces, i.e., for all such R_{n+1}–points, constitutes the *tangential bundle* of R_{n+1}, $\mathbf{TB}_{n+1}(\mathbf{R}_{n+1})$. Now,

- At each such point P, the m ($< n$) linearly independent (assumed) nonholonomic Pfaffian constraints (Equations 6.6.6a and 6.6.8a): $\omega^D \equiv A^D{}_k\, V^k + A^D{}_{n+1} = A^D{}_\beta\, V^\beta = 0$, or

$$d\theta^D \equiv A^D{}_k\, dq^k + A^D{}_{n+1}\, dt = A^D{}_\beta\, dq^\beta = 0, \tag{6.8.1}$$

  define, or map, or order, an $[(n-m)+1]$–dimensional local space ("plane element") $T_{n+1,(n-m)+1}(P) \equiv T_{f+1}(P)$ or $T_{I+1}(P)$, where $f \equiv n - m$ (number of local DOF — see below), *the tangent space of kinematically admissible, or possible, system displacements on which the dq^λ must lie $\Rightarrow$ only motions for which the $dq^\lambda/dt \equiv V^\lambda$ lie on that plane, at every point of the system trajectory in R_{n+1}, are possible.* The corresponding vector space $\mathbf{T}_{f+1}(\mathbf{P})$, or $\mathbf{T}_{I+1}(\mathbf{P})$, is spanned by the $f + 1$ independent vectors $\{\mathbf{E}_I, \mathbf{E}_{N+1}\} \equiv \{\mathbf{E}_{I+1}\}$; or, after introduction of an appropriate metric (Section 7.2), by their dual basis $\{\mathbf{E}^I, \mathbf{E}^{N+1}\} \equiv \{\mathbf{E}^{I+1}\}$. In the case of *stationary/scleronomic* constraints, i.e., if $\partial A^D{}_k/\partial t = 0$ and $A^D = 0$, the m equations $A^D{}_k(q)\, V^k = 0$ or $A^D{}_k(q)\, dq^k = 0$ represent a local f–dimensional plane element.
- Similarly, the virtual form of the constraints (Equation 6.6.9a): $\delta\theta^D \equiv A^D{}_k\, \delta q^k = 0$ (and $\delta q^{n+1} \equiv \delta t = 0$) define, at each R_{n+1}–point the f $[= (n + 1) - (m + 1)_{\#\ \text{constraints}}]$–dimensional plane $T_f(P)$ or $T_I(P)$, the *tangent space of virtual system displacements on which the δq^k must lie.* Clearly, $T_f(P)$ is the intersection of $T_{f+1}(P)$ with the hyperplane $dt \to \delta t = 0$, there. Symbolically, $T_f(P) = T_{f+1}(P)|_{dt=0}$. The *totality* of these virtual planes constitutes the *manifold of virtual displacements* embedded in R_{n+1}, $VD_f(R_{n+1})$. [A similar manifold exists for the totality of the kinematically admissible local planes $T_{f+1}(P)$.] And a manifold R_n or R_{n+1} whose tangential bundle is restricted by the m Pfaffian constraints $A^D{}_\beta\, dq^\beta = 0$ is called *nonholonomic manifold* $R_n(NH_f)$ or $R_{n+1}(NH_f)$. (Some authors call the so-restricted bundle *nonholonomic space embedded* in R_n *or* R_{n+1}.)

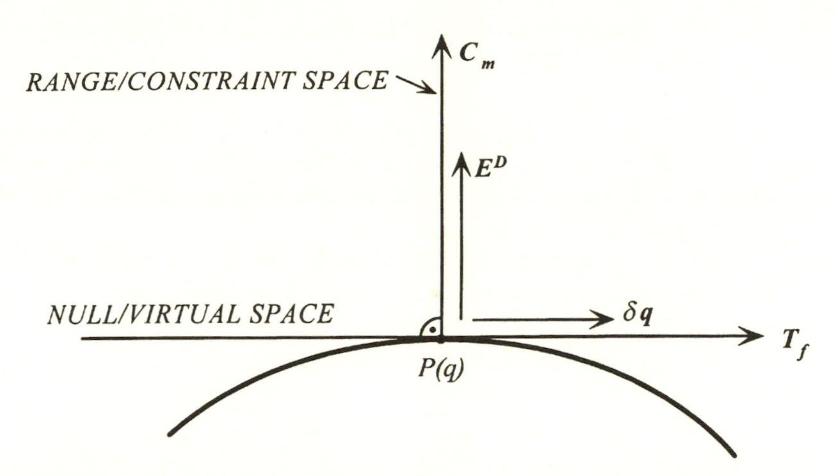

FIGURE 6.2 On the geometrical interpretation of Pfaffian constraints $\delta\theta^D \equiv A^D{}_k\ \delta q^k = 0$: $\mathbf{C}_m$: *range* (or *constraint*) space of *constraint matrix* $(A^D{}_k)$; spanned by $\{\mathbf{E}^D\}$ or $\{\mathbf{E}_D\}$ $\mathbf{T}_{n-m} \equiv \mathbf{T}_f$: *null* (or *virtual*) space of $(A^D{}_k)$; spanned by $\{\mathbf{E}^I\}$ or $\{\mathbf{E}_I\}$.

- The given constraint coefficients $(A^D{}_\beta) \equiv (A^D{}_k, A^D)$ define, at each P, an $(m + 1)$–dimensional *kinematically admissible constraint space*, C_{m+1}, perpendicular to T_{f+1}; and an m–dimensional *virtual constraint space*, C_m, perpendicular to T_f (or *orthogonal complement* of T_f relative to T_n — Figure 6.2). In particular, since (by Equation 6.6.9a)

$$\delta\theta^D = A^D{}_k\ \delta q^k \equiv \mathbf{E}^D \cdot \delta\mathbf{q} = 0 \quad \text{(in } vector \text{ form)} \tag{6.8.2a}$$

or

$$\mathbf{A}^D \cdot \delta\mathbf{q} = \mathbf{0} \quad \text{(in } matrix \text{ form)} \tag{6.8.2b}$$

[where $\mathbf{A}^{\mathrm{D}} = (A^D{}_k)$ and $\delta\mathbf{q}^{\mathrm{T}} = (\delta q^k)$ (row vector)] the $A^D{}_k$ can be viewed as the holonomic covariant components of the *m virtual constraint vectors* $\mathbf{E}^D = (A^D{}_k)$ which are perpendicular to the virtual displacement vector $\delta\mathbf{q} = (\delta q^k)$ and, being independent, constitute a basis for the earlier space $\mathbf{C}_\mathbf{m}$. Similarly, in view of $\delta\theta^I \equiv A^I{}_k\ \delta q^k = \mathbf{E}^I \cdot \delta\mathbf{q} \neq 0$, the $n - m$ vectors $\mathbf{E}^I = (A^I{}_k)$ constitute a basis for the local *virtual vector space* $\mathbf{T}_\mathbf{f}$.*

* In linear algebra terms, $\mathbf{C}_\mathbf{m}$ is the *range* space of the *virtual constraint matrix* $\mathbf{A}^{\mathrm{D}} = (A^D{}_k)$, $\mathbf{T}_\mathbf{f}$ is its *null* space, and the $n \times (n - m) \equiv n \times f$ *virtual displacement matrix* $\mathbf{A}_{\mathrm{I}} = (A^k{}_I)$ is its *orthogonal complement.* The matrix $(A^D{}_k)$ together with its $(n - m) \times n$ *enlargement* $\mathbf{A}^{\mathrm{I}} = (A^I{}_k)$ make up the $\delta q \to \delta\theta$ transformation matrix $\mathbf{A} = (A^K{}_k)$; and may be viewed as consisting of the following *row* vectors [with $(\ldots)^{\mathrm{T}}$ for *transpose* of $(\ldots)$, to save space]: $(\mathbf{A}^{\mathrm{D}})^{\mathrm{T}} = (A^D{}_k)^{\mathrm{T}} = (\mathbf{E}^1 \ldots \mathbf{E}^m)$ (*row* vectors), $(\mathbf{A}^{\mathrm{I}})^{\mathrm{T}} = (A^I{}_k)^{\mathrm{T}} = (\mathbf{E}^{m+1} \ldots \mathbf{E}^n)$ (*row* vectors). Similarly, its *inverse* $\mathbf{A}^{-1} \equiv (A^k{}_K)$ can be partitioned as follows:

$$\mathbf{A}^{-1} = (A^k{}_D | A^k{}_I) \equiv (\mathbf{A}_D | \mathbf{A}_I),$$

$\mathbf{A}_{\mathrm{D}} = (A^k{}_D) = (\mathbf{E}_1 \ldots \mathbf{E}_m)$ (*column* vectors), $\mathbf{A}_{\mathrm{I}} = (A^I{}_k) = (\mathbf{E}_{m+1} \ldots \mathbf{E}_n)$ (*column* vectors).

In sum, and due to the compatibility conditions (Equation 6.4.4ff) $A^K{}_k\, A^k{}_L = \delta^K{}_L$, $A^K{}_k\, A^I{}_K = \delta^I{}_k$, we have the following results:

- $A^I{}_k A^k{}_{I'} = \mathbf{E}^I \cdot \mathbf{E}_{I'} = \delta^I{}_{I'}$; i.e., the $\mathbf{E}^I = (A^I{}_k)$ and $\mathbf{E}_{I'} = (A^k{}_{I'})$ are dual bases in $\mathbf{T}_f$.
- $A^D{}_k A^k{}_{D'} = \mathbf{E}^D \cdot \mathbf{E}_{D'} = \delta^D{}_{D'}$; i.e., the $\mathbf{E}^D = (A^D{}_k)$ and $\mathbf{E}_{D'} = (A^k{}_{D'})$ are dual bases in $\mathbf{C}_m$. Clearly, if the $\mathbf{E}^D$ are *orthonormal*, so are the $\mathbf{E}_D$ and $\mathbf{E}^D = \mathbf{E}_D$; and similarly for the $\mathbf{E}^I$, $\mathbf{E}_I$.
- $A^D{}_k\, A^k{}_I = \mathbf{E}^D \cdot \mathbf{E}_I = \delta^D{}_I = 0$ and $A^I{}_k\, A^k{}_D = \mathbf{E}^I \cdot \mathbf{E}_D = \delta^I{}_D = 0$; i.e., $\mathbf{E}^D$ and $\mathbf{E}_I$, and $\mathbf{E}^I$ and $\mathbf{E}_D$, are mutually orthogonal (Figure 6.2).

Hence, any virtual displacement can be expressed as

$$\delta\mathbf{q} = \delta q^k\, \mathbf{E}_k = \delta\theta^L\, \mathbf{E}_L = \ldots = \delta\theta^I\, \mathbf{E}_I \quad (= \delta\mathbf{R})$$

$$\Rightarrow \delta q^k = A^k{}_L\, \delta\theta^L = A^k{}_D\, \delta\theta^D + A^k{}_I\, \delta\theta^I = 0 + A^k{}_I\, \delta\theta^I, \quad \delta\theta^I = A^I{}_k\, \delta q^k; \qquad (6.8.3)$$

and similarly an arbitrary $\mathbf{T_n}$ – vector $\mathbf{V}$ can be decomposed uniquely into its mutually orthogonal *range* and *null* space components:

$$\mathbf{V} = \mathbf{V}_{\text{range}} + \mathbf{V}_{\text{null}} \equiv \mathbf{V}_{(r)} + \mathbf{V}_{(n)}, \qquad (6.8.4)$$

where

$$\mathbf{V}_{(r)} = V^D\, \mathbf{E}_D = V_D\, \mathbf{E}^D, \quad \mathbf{V}_{(n)} = V^I\, \mathbf{E}_I = V_I\, \mathbf{E}^I,$$

and

$$\mathbf{V}_{(r)} \cdot \mathbf{V}_{(n)} = 0, \qquad (6.8.4a)$$

{ $\Rightarrow \mathbf{E}_D \cdot \mathbf{E}^I = 0$, $\mathbf{E}^D \cdot \mathbf{E}_I = 0$ [from $\mathbf{E}^K \cdot \mathbf{E}_L = \delta^K{}_\Lambda$, for $K \to I(D)$ $L \to D(I)$] and $\mathbf{E}_D \cdot \mathbf{E}_I = 0$, $\mathbf{E}^D \cdot \mathbf{E}^I = 0$ (from the *additional* orthogonality of the range and null spaces), so that

$$V^k = A^k{}_D\, V^D + A^k{}_I\, V^I, \quad V_k = A^D{}_k\, V_D + A^I{}_k\, V_I. \qquad (6.8.4b)$$

6.8.1 Degrees of Freedom Revisited, Accessibility

Let the original system S be subjected to H holonomic constraints and m Pfaffian (possibly nonholonomic) constraints. If the latter are holonomic, then S can be described by $3N - (H + m) = (3N - H) - m \equiv n - m \equiv f$ independent holonomic coordinates; we say that S still has f *global (or positional) degrees of freedom, or* f *DOF in the large;* i.e., its configuration space is f-dimensional. If, however, these velocity constraints are nonholonomic, then S still has n independent holonomic coordinates, i.e., its configuration space is still n-dimensional, but its local freedom has been reduced to f; as we say, now it has *f local (or motional) degrees of freedom, or f DOF in the small* (and still n DOF in the large). Hence, in both cases the number of independent components of its virtual (system) displacement vector $\delta \mathbf{q}$ is f; and it is this latter that from now on will be understood as the number of degrees of freedom.

The above have the following geometrical consequences:

- If the m Pfaffian constraints are nonholonomic, the representative system point $P(q^k)$ can go from any R_n-point P_1 to any other R_n–point P_2, under proper forces and initial conditions, a property known as *accessibility* in configuration space. However, P *cannot* go from P_1 to P_2 via any (R_n–lying) route it wishes. Due to the cumulative effects of the local nonholonomic constraints, the set of kinematically possible paths between P_1 and P_2 is "narrower," or "denser," like a set of railroad tracks that steer the system. In short, the imposition of nonholonomic constraints does *not* affect either the number of positional coordinates of the system (freedom in the large) or its accessibility; for such constraints, all events are kinematically achievable, but not all velocities are possible, and, therefore, *not* all paths between two events in space–time.
- If, on the other hand, these constraints are holonomic, disguised in kinematical (velocity) form, that restricts further the configuration space, from R_n to a "smaller," or "narrower" R_{n-m}, on which P must now remain; i.e., it may not be able to go from any previous P_1 to any other P_2

Example 6.8.1

Let us consider an initially free (i.e., unconstrained) particle P moving in ordinary space E_3. Then number of global DOF = number of local DOF = $n - m = 3 - 0 = 3$, and P can go from any initial E_3-point $P(x_1, y_1, z_1)$ to any other final E_3–point $P(x_2, y_2, z_2)$, where x, y, z: rectangular Cartesian coordinates of P. Next, let us impose on P the Pfaffian constraint:

$$d\theta \equiv A(x, y, z)dx + B(x, y, z)dy + C(x, y, z)dz = 0. \qquad \text{(a)}$$

- If $d\theta = 0$ is holonomic, then it can be reduced to a surface, say $f(x, y, z)$ = constant, and P has to lie on it. In that case, number of global DOF = number of local DOF = $n - m = 3 - 1 = 2$.
- If $d\theta = 0$ is nonholonomic, then P can still go from any point of E_3 to any other point of it, but only along certain paths that, locally, satisfy $d\theta = 0$. In that case, number of global DOF = $n = 3$, but number of local DOF = $n - m = 3 - 1 = 2$ (number of independent componets of virtual displacement).

6.9 GEOMETRICAL INTERPRETATION OF THE FROBENIUS' CONDITIONS (EQUATION 6.7.5e)

6.9.1 FIRST INTERPRETATION

Since (recalling the results of Problem 3.16.3), the Curl(...) tensor of the constraint vectors $\mathbf{E}^D = (A^D{}_k) = A^D{}_k\,\mathbf{E}^k$ is

$$(\text{Curl } \mathbf{E}^D)_{hk} = (\partial A^D{}_k/\partial q^h - \partial A^D{}_h/\partial q^k) \equiv (A^D{}_{k,h} - A^D{}_{h,k}) \equiv (\gamma^D{}_{kh}), \quad (6.9.1)$$

(note the difference in subscript order, between far left and right expressions) Frobenius' conditions (Equation 6.7.5e$_1$) assume the following form, in direct notation (Section 2.11),

$$\gamma^D{}_{II'} = A^h{}_{I'}\,(\partial A^D{}_k/\partial q^h - \partial A^D{}_h/\partial q^k)\,A^k{}_I = A^h{}_{I'}\,\gamma^D{}_{kh}\,A^k{}_I$$

$$= (\mathbf{E}_{I'})^h\,(\text{Curl } \mathbf{E}^D)_{hk}\,(\mathbf{E}_I)^k = \mathbf{E}_{I'} \cdot \text{Curl } \mathbf{E}^D \cdot \mathbf{E}_I = 0. \quad (6.9.2)$$

In words, *for the m Pfaffians* $\mathbf{E}^D \cdot d\mathbf{q} = 0$ *to be holonomic, it is necessary and sufficient that the components of the "curl" tensor of the m constraint (dependent) vectors* $\{\mathbf{E}^D\}$ *along the* $n - m$ *independent (virtual) directions* $\{\mathbf{E}_I\}$ *vanish.**

6.9.2 SECOND INTERPRETATION

In view of the additional γ-expression (Equations 6.7.5d$_1$):

$$\gamma^D{}_{II'} = A^D{}_k[A^h{}_I(\partial A^k{}_{I'}/\partial q^h) - A^h{}_{I'}(\partial A^k{}_I/\partial q^h)] \equiv A^D{}_k(\partial A^k{}_{I'}/\partial\theta^I - \partial A^k{}_I/\partial\theta^{I'}),$$

and the noncommutativity relations (Equation 6.5.10d):

$$\partial\mathbf{E}_K/\partial\theta^\Delta - \partial\mathbf{E}_\Delta/\partial\theta^K = \gamma^\Psi{}_{\Delta K}\,\mathbf{E}_\Psi \Rightarrow \gamma^\Psi{}_{DK} = (\partial\mathbf{E}_K/\partial\theta^\Delta - \partial\mathbf{E}_\Delta/\partial\theta^K) \cdot \mathbf{E}^\Psi, \quad (6.9.3)$$

Frobenius' conditions become

$$\gamma^D{}_{II'} = \mathbf{E}^D \cdot \mathbf{E}_{II'} = 0, \quad (6.9.4)$$

where

$$\mathbf{E}_{II'} = -\mathbf{E}_{I'I} \equiv (A^h{}_I\,A^k{}_{I',h} - A^h{}_{I'}\,A^k{}_{I,h})\mathbf{E}_k \equiv (\partial A^k{}_{I'}/\partial\theta^I - \partial A^k{}_I/\partial\theta^{I'})\mathbf{E}_k$$

$$= [(A^k{}_{I'}\,\mathbf{E}_k)_{,I} - A^k{}_{I'}\,\mathbf{E}_{k,I}] - [(A^k{}_I\,\mathbf{E}_k)_{,I'} - A^k{}_I\,\mathbf{E}_{k,I'}]$$

---

* (a) The directions $\{\mathbf{E}_I\}$ and $\{\mathbf{E}_D\}$ are mutually *complementary*, in the sense of Section 2.1. (b) On the other hand, the *sufficient* conditions $\gamma^D{}_{kh} = (\text{Curl } \mathbf{E}^D)_{hk} = 0$ involve *all the holonomic components* of these vectors. For an interpretation in ordinary three-dimensional space, see also Lodge [1974, p. 201: Problem (6) and p. 303: Problem 8.7(6); notation slightly different than ours] and Problem 6.9.1 below.

$$= [(A^k{}_{I'}\ \mathbf{E}_k)_{,I} - (A^k{}_I\ \mathbf{E}_k)_{,I'}] - [A^k{}_{I'}\ (A^l{}_I\ \mathbf{E}_{k,l}) - A^k{}_I(A^l{}_{I'}\ \mathbf{E}_{k,l})]$$

$$= \ldots = \partial\mathbf{E}_{I'}/\partial\theta^I - \partial\mathbf{E}_I/\partial\theta^{I'} \quad (\text{since } \mathbf{E}_{k,l} = \mathbf{E}_{l,k}); \tag{6.9.5}$$

i.e., *the constraint vectors* $\{\mathbf{E}^D\}$ *must be perpendicular to the independent direction vectors* $\{\mathbf{E}_{II'}\}$*; or, the "curling" of the virtual (null) space basis* $(\mathbf{E}_I\}$ *along the constraint (range) space basis* $\{\mathbf{E}^D\}$ *directions should vanish.*

Similarly, for the "nonstationary/rheonomic" Frobenius' conditions (Equations 6.7.5e$_2$): $\gamma^D{}_I = 0$.

Problem 6.9.1

Using Equations 3.13.8a ff, show that Frobenius' theorem in an orthonormal basis in three dimensions, i.e., $n = 3$ and $m = 1$ [$\Rightarrow f \equiv n - m = 3 - 1 = 2$ (number of local DOF)] results in the single condition (recall Example 6.7.1):

$$\gamma^1{}_{23} = -\gamma^1{}_{32} = -\mathbf{E}^1 \cdot \text{curl } \mathbf{E}^1 = -\mathbf{E}_1 \cdot \text{curl } \mathbf{E}_1 = 0. \tag{a}$$

Additional mechanical examples and problems are detailed in Chapter 7; especially Section 7.7, where their kinematical and kinetical aspects are treated together.

7 Lagrangean Mechanics: Kinetics

7.1 INTRODUCTION

This chapter continues and concludes our compact but comprehensive treatment of Lagrangean or analytical mechanics (AM). It begins with its essential *kinematico–inertial* ingredients, such as *kinetic energy* and (system) *acceleration* or *inertial "force"* and their relation with the *Euler–Lagrange* operator (6.3.5g), and the *forces* involved. Then, the discussion moves to the synthesis of these concepts into the *two basic principles* of AM (Section 4.3): of d'Alembert–Lagrange (LP) and of Lagrange–Hamel of the relaxation of the constraints (Lagrangean multipliers). These principles, in turn, produce all possible *equations of motion*: with reactions (*kinetostatic*) and without reactions (*kinetic*). The equivalent route based on the *central equation of dynamics* of Lagrange–Heun is also presented. The chapter ends with a discussion of the structure of the Lagrangean equations in holonomic variables and their geometrical interpretation in configuration space and a general treatment of *energy rate*, or *power*, equations. The above are presented in both particle vector ("raw" or "brute") forms and system forms, in both holonomic and nonholonomic variables.*

7.2 THE FUNDAMENTAL KINEMATICO–INERTIAL QUANTITIES

7.2.1 KINETIC ENERGY

The kinetic energy T of a system S, relative to an inertial frame of reference F, is defined as (recalling the notations of Sections 4.2 ff, with $P \to p$):

$$2T = \sum m_p \, \mathbf{v}_p \cdot \mathbf{v}_p = \sum m_p \, v_p^{\,2}$$

$$= \sum m_p[(dx_p/dt)^2 + (dy_p/dt)^2 + (dz_p/dt)^2]$$

[Particle model: $p = 1, \ldots, N$ (# particles of S)]

$$= S\, dm \, \mathbf{v} \cdot \mathbf{v} \quad \text{[Continuum model]:} \qquad (7.2.1)$$

* For complementary reading, we recommend (alphabetically): Dobronravov (1948, 1970, 1976), Gantmacher (1970), Golomb (1961), Hamel (1949), Johnsen (1941), Kil'chevskii (1972, 1977), Kilmister (1964, 1966), Lur'e (1968), Maißer (1981, 1982, 1983–4), Neimark and Fufaev (1972), Papastavridis (1998), Pars (1965), Poliahov et al. (1985), Prange (1935), Synge (1936, 1960), Vujanovic and Jones (1989), Vujicic (1981, 1990), Whittaker (1937), Zander (1970).

Positive definite function in the **v**s (i.e., always nonnegative and zero only if all velocities vanish). Substituting into the above the representation **v** in terms of system variables (6.3.1 ff) yields, successively, the following.

7.2.1.1 Holonomic Variables

With $v^\alpha = V^\alpha \equiv dq^\alpha/dt$, and $\alpha = 1, \ldots, n+1$ we obtain successively:

$$2T = S\,[dm(\mathbf{e}_\alpha\, v^\alpha)\cdot(\mathbf{e}_\beta\, v^\beta)] = S\,[dm(\mathbf{e}_k\, v^k + \mathbf{e}_{n+1})\cdot(\mathbf{e}_l\, v^l + \mathbf{e}_{n+1})]$$

$$= \ldots = 2(T_{(2)} + T_{(1)} + T_{(0)}), \tag{7.2.2}$$

where

$$2T_{(2)} = \left(S\, dm\,\mathbf{e}_k\cdot\mathbf{e}_l\right) v^k\, v^l \equiv M_{kl}\, V^k\, V^l, \tag{7.2.2a}$$

$$T_{(1)} = \left(S\, dm\,\mathbf{e}_k\cdot\mathbf{e}_{n+1}\right) v^k\, v^{n+1} \equiv M_{k,n+1}\, V^k\, V^{n+1} \equiv M_k\, V^k, \tag{7.2.2b}$$

$$2T_{(0)} = \left(S\, dm\,\mathbf{e}_{n+1}\cdot\mathbf{e}_{n+1}\right) v^{n+1}\, v^{n+1} \equiv M_{n+1,n+1} V^{n+1}\, V^{n+1} \equiv M_0; \tag{7.2.2c}$$

i.e.,

$$2T = M_{\alpha\beta}\, V^\alpha\, V^\beta = M_{kl}\, V^k\, V^l + 2M_k\, V^k + M_0, \tag{7.2.3}$$

where

$$M_{\alpha\beta} = \mathrm{M}_{\beta\alpha} = M_{\alpha\beta}(q^\gamma) = M_{\alpha\beta}(q^k, t) \equiv M_{\alpha\beta}(q, t) \equiv S\, dm\;\mathbf{e}_\alpha\cdot\mathbf{e}_\beta = \mathbf{E}_\alpha\cdot\mathbf{E}_\beta:$$

Inertia coefficients (variable, although the masses are not). (7.2.3a)*

- Clearly, T is a *positive semidefinite function* in the V^k terms, i.e., it is never negative but may vanish for a nontrivial (nonzero) set of values of the V^k terms. It can also be rewritten as

  $$2T = M_{kl}\, X^k\, X^l + f(t, q), \tag{7.2.4}$$

  where

  $$X^k \equiv V^k - g^k(t, q) \quad \text{and} \quad f(\ldots),\; g^k(\ldots)\text{: functions of } t, q^k; \tag{7.2.4a}$$

i.e., the homogeneous quadratic form $M_{kl}\, X^k\, X^l$ is positive definite in the X^k terms.

- If $\mathbf{r} = \mathbf{r}(q) \Rightarrow \mathbf{e}_{n+1} = \partial\mathbf{r}/\partial t = \mathbf{0}$ (stationary holonomic constraints), then, clearly,

$$2T = 2T_{(2)} = M_{kl}(q)V^k\, V^l\text{: positive definite function in the } V^k \text{ terms.} \quad (7.2.5)$$

7.2.1.2 Nonholonomic Variables

Substituting into T, Equation 7.2.2ff, the transformation relations $V^\lambda = A^\lambda{}_\Lambda\, \omega^\Lambda$ (or, using Equation 6.4.8a in Equation 7.2.1) and with the helpful notation:

$$2T = 2T(t, q, V) \rightarrow 2T[t, q, V(t, q, \omega)] = 2T^*(t, q, \omega) = 2T^*, \quad (7.2.6)$$

we find

$$2T^* = S\, dm(\mathbf{e}_\Gamma\, \omega^\Gamma) \cdot (\mathbf{e}_\Delta\, \omega^\Delta) \equiv M_{\Gamma\Delta}\, \omega^\Gamma\, \omega^\Delta$$

$$= 2(T^*_{(2)} + T^*_{(1)} + T^*_{(0)}), \quad (7.2.7)$$

where

$$2T^*_{(2)} = \left(S dm\, \mathbf{e}_K \cdot \mathbf{e}_L \right) \omega^K\, \omega^L \equiv M_{KL}\, \omega^K\, \omega^L, \quad (7.2.7a)$$

$$T^*_{(1)} = \left(S dm\, \mathbf{e}_K \cdot \mathbf{e}_{N+1} \right) \omega^K \equiv M_{K,N+1}\, \omega^K\, \omega^{N+1} \equiv M_K\, \omega^K, \quad (7.2.7b)$$

$$2T^*_{(0)} = \left(S dm\, \mathbf{e}_{N+1} \cdot \mathbf{e}_{N+1} \right) \omega^{N+1}\, \omega^{N+1} \equiv M_{N+1,N+1}\omega^{N+1}\, \omega^{N+1} \equiv M^*_0; \quad (7.2.7c)$$

---

* (a) In detail (reverting momentarily to the discrete notation of Section 4.2, for extra clarity):

$$M_{kl} \equiv S dm\, \mathbf{e}_k \cdot \mathbf{e}_l = S(\sqrt{dm}\, \mathbf{e}_k) \cdot (\sqrt{dm}\, \mathbf{e}_l) = S[\partial(\sqrt{dm}\, \mathbf{r})/\partial q^k] \cdot [\partial(\sqrt{dm}\, \mathbf{r})/\partial q^k]$$

$$\sum m_p(\partial \mathbf{r}_p/\partial q^k) \cdot (\partial \mathbf{r}_p/\partial q^l) \quad [p = 1, \ldots, N]$$

$$= \sum [\partial(\sqrt{M_s}\xi^s)/\partial q^k] \cdot [\partial(\sqrt{M_s}\xi^s)/\partial q^l] \quad [S = 1, \ldots, 3N]$$

$$= \sum (\partial y^s/\partial q^k) \cdot (\partial y^s/\partial q^l) = (\partial \mathbf{R}/\partial q^k) \cdot (\partial \mathbf{R}/\partial q^l) = \mathbf{E}_k \cdot \mathbf{E}_l,$$

and similarly for $M_{k,n+1} \equiv M_{k0} \equiv M_k$ and $M_{n+1,n+1} \equiv M_{00} \equiv M_0$. (b) In concrete problems, only the $M_{\alpha\beta}(t, q)$ are given; in agreement with Riemannian geometry (Sections 2.12, 3.8, and 5.6).

i.e.,

$$2T^* = M_{KL}\, \omega^K \omega^L + 2M_K\, \omega^K + M^*_0, \tag{7.2.7d}$$

where

$$M_{KL} = M_{LK} \equiv S\, dm\, \mathbf{e}_K \cdot \mathbf{e}_L = S\left(\sqrt{dm}\, \mathbf{e}_K\right)\cdot\left(\sqrt{dm}\, \mathbf{e}_L\right)$$

$$= S\left[\partial\left(\sqrt{dm}\, \mathbf{r}\right)/\partial\theta^K\right]\cdot\left[\partial\left(\sqrt{dm}\, \mathbf{r}\right)/\partial\theta^L\right]$$

$$= \ldots = (\partial\mathbf{R}/\partial\theta^K)\cdot(\partial\mathbf{R}/\partial\theta^L) = \mathbf{E}_K \cdot \mathbf{E}_L, \quad \text{etc.} \tag{7.2.7e}$$

From the invariance of the kinetic energy, Equation 7.2.6, and the above we readily obtain the second-rank tensor transformation equations among the holonomic and nonholonomic inertia coefficients:

$$M_{\Gamma\Delta} = (\partial V^\gamma/\partial\omega^\Gamma)(\partial V^\delta/\partial\omega^\Delta) M_{\gamma\delta} \equiv A^\gamma{}_\Gamma\, A^\delta{}_\Delta\, M_{\gamma\delta}$$

$$\Leftrightarrow M_{\gamma\delta} = (\partial\omega^\Gamma/\partial V^\gamma)(\partial\omega^\Delta/\partial V^\delta) M_{\Gamma\Delta} \equiv A^\Gamma{}_\gamma\, A^\Delta{}_\delta\, M_{\Gamma\Delta}. \tag{7.2.8}$$

- The above show that *an initially scleronomic system* [i.e., in the V^ks: $T = T_{(2)}$; $T_{(1)}$, $T_{(0)} = 0$] *may very well become rheonomic in the ω^Ks;* and, clearly, in general: $T_{(2)} \neq T^*_{(2)}$, etc.
- Under the constraints $\omega^D = 0$, T^* assumes its *constrained form*:

$$2T^*(t, q, \omega^D, \omega^I) \rightarrow 2T^*(t, q, 0, \omega^I) = 2T^*_o(t, q, \omega_I) = 2T^*_o$$

$$= M_{II'}\, \omega^I \omega^{I'} + 2\, M_I\, \omega^I + M^*_0. \tag{7.2.9}$$

7.2.2 Metric in Configuration/Event Space

With the help of T we can now introduce a definite *metric structure* into the earlier manifolds R_n and R_{n+1} (Section 6.2). We assume that the square of the elementary arc-length between the neighboring events (t, q) and $(t + dt, q + dq)$, in the configuration + time manifold R_{n+1}, is given by the *kinematical line element* (with time a privileged manifold coordinate):

$$(dS)^2 = 2T(dt)^2 = M_{\gamma\delta}\, dq^\gamma\, dq^\delta$$

$$= M_{kl}\, dq^k\, dq^l + 2M_{k,n+1}\, dq\, dt + M_{n+1,n+1}\, dt\, dt$$

$$\equiv M_{kl}\, dq^k\, dq^l + 2M_k\, dq\, dt + M_0 (dt)^2, \tag{7.2.10a}$$

i.e.,

$$2T = (dS/dt)^2 = \mathbf{V} \cdot \mathbf{V} = V^2 \quad (7.2.10b)$$

(kinetic energy of particle of *unit* mass), where $V^\beta \equiv dq^\beta/dt = (dq^\beta/dS)(dS/dt) \equiv V\,T^\beta$ (T^β: *unit* tangent vector to system path).

Clearly, since T is *not* positive definite, $(dS)^2$ is not positive definite either; and, therefore, the event space R_{n+1} is not Riemannian, but rather *pseudo-Riemannian.* The equations t = constant define ∞^1 privileged surfaces $R_n(t)$ in R_{n+1}; the system motion may be viewed either as a *world–line* in R_{n+1}, or as a *motion* of a point in the *deformable,* or *rheonomic,* space $R_n(t)$ with line element:

$$(ds)^2 = 2T_{(2)}(dt)^2 = M_{kl}(t, q)dq^k\, dq^l \quad (7.2.11)$$

[i.e., quadratic and homogeneous part of $(dS)^2$ in the dq; or, *spatial* part of $(dS)^2$].

Remark

As will become clear later in this chapter, *it is the spatial metric tensor M_{kl}, and its special conjugate (or inverse) M^{kl}, defined by:*

$$M^{kb}\, M_{bl} = \delta^k{}_l, \quad (7.2.11a)$$

that are the metrics used to raise/lower the indices of the various components of the equations of motion. Then, the so-built geometrical/tensorial formalism produces the correct equations of motion; i.e., it is the latter that dictates the properties of the former and constitutes the ultimate criterion of its correctness. Equation 7.2.11a, however, leads to

$$\delta^\alpha{}_\varepsilon = M^{\alpha\beta}\, M_{\beta\varepsilon} = M^{\alpha k}\, M_{k\varepsilon} + M^{\alpha,n+1}\, M_{n+1,\varepsilon}$$

$$\Rightarrow \delta^b{}_l = M^{bk}\, M_{kl} + M^{b,n+1}\, M_{n+1,l} = \delta^b{}_l + M^{b,n+1}\, M_{n+1,l}$$

$$\Rightarrow M^{b,n+1} = 0, \quad (7.2.11b)$$

$$\Rightarrow \delta^{n+1}{}_{n+1} = M^{n+1,k}\, M_{k,n+1} + M^{n+1,n+1}\, M_{n+1,n+1} = \delta^{n+1}{}_{n+1} + M^{n+1,n+1}\, M_{n+1,n+1}$$

$$\Rightarrow M^{n+1,n+1} = 0, \quad (7.2.11c)$$

since, as Equation 7.2.10a shows, $M_{b,n+1}, M_{n+1,n+1} \neq 0$, i.e., $M^{\alpha,n+1} = 0$; and from the latter we also conclude that:

$$\Gamma^{n+1}{}_{\beta\lambda} \equiv M^{n+1,\varepsilon}\, \Gamma_{\varepsilon,\beta\lambda} = M^{n+1,k}\, \Gamma_{k,\beta\lambda} + M^{n+1,n+1}\, \Gamma_{n+1,\beta\lambda} = 0; \quad (7.2.11d)$$

and similarly for the nonholonomic metric (Equations 7.2.7, 7.2.7a through e, and 7.2.8) and Christoffels.

If the system is *scleronomic,* then its configuration space is a Riemannian manifold, R_n, with positive-definite kinematical line-element:

$$(ds)^2 = 2T(dt)^2 = M_{kl}(q)\, dq^k\, dq^l; \tag{7.2.12}$$

and if, further, the system is *potential,* then an *action line element* may be used:

$$(d\sigma)^2 = (E - \Pi)\, M_{kl}\, dq^k\, dq^l \tag{7.2.13}$$

[E, Π: total and potential energies of system].

7.2.3 Acceleration

The (inertial) system acceleration vector, **A**, is defined, formally, by

$$\begin{aligned}\mathbf{A} &\equiv d\mathbf{V}/dt = d/dt(d\mathbf{R}/dt) = d/dt[(\partial\mathbf{R}/\partial q^\alpha)V^\alpha]\\ &= (\partial\mathbf{R}/\partial q^\alpha)(dV^\alpha/dt) + (\partial^2\mathbf{R}/\partial q^\alpha \partial q^\beta)V^\alpha\, V^\beta\\ &\equiv (dV^\alpha/dt)\mathbf{E}_\alpha + V^\alpha(d\mathbf{E}_\alpha/dt);\end{aligned} \tag{7.2.14}$$*

or, using the tensor analysis machinery (Section 3.5 ff),

$$\mathbf{A} = (DV^\beta/Dt)\mathbf{E}_\beta \equiv A^\beta\, \mathbf{E}_\beta, \tag{7.2.15}$$

where

$$\begin{aligned}A^\beta &\equiv DV^\beta/Dt \equiv V^\beta|_\delta\, V^\delta \equiv (V^\beta{}_{,\delta} + \Gamma^\beta{}_{\delta\zeta}\, V^\zeta)V^\delta\\ &= dV^\beta/dt + \Gamma^\beta{}_{\delta\zeta}\, V^\delta\, V^\zeta,\end{aligned} \tag{7.2.15a}$$

with the Christoffels $\Gamma^\beta{}_{\delta\zeta}$ built from the earlier kinematic metric tensor $M_{\alpha\beta} = M_{\beta\alpha}$ (see below). Similarly, the *covariant* components of **A** are

$$A_\beta = DV_\beta/Dt = dV_\beta/dt - \Gamma^\delta{}_{\beta\zeta}\, V_\delta\, V^\zeta = dV_\beta/dt - \Gamma_{\delta,\beta\zeta}\, V^\delta\, V^\zeta, \tag{7.2.15b}$$

or

$$\begin{aligned}A_\beta &\equiv \mathbf{A}\cdot\mathbf{E}_\beta = (A^\delta\, \mathbf{E}_\delta)\cdot\mathbf{E}_\beta = A^\delta\, M_{\beta\delta}\\ &= M_{\beta\delta}(dV^\delta/dt) + \Gamma_{\beta,\delta\zeta}\, V^\delta\, V^\zeta = M_{\beta k}(dV^k/dt) + \Gamma_{\beta,\delta\zeta}\, V^\delta\, V^\zeta.\end{aligned} \tag{7.2.15c}$$

---

* Since $V^{n+1} = 1$, $dV^{n+1}/dt = 0$, also: $\mathbf{A} = (\partial\mathbf{R}/\partial q^k)(dV^k/dt) + (\partial^2\mathbf{R}/\partial q^k\partial q^l)V^k\, V^l + 2(\partial^2\mathbf{R}/\partial q^k\partial t)V^k + \partial^2\mathbf{R}/\partial t^2$.

In extenso, we have

- **Spatial** part of A_β:

$$A_k = M_{kb}(dV^b/dt) + \Gamma_{k,bh}\, V^b\, V^h + 2\Gamma_{k,b,n+1}\, V^b\, V^{n+1} + \Gamma_{k;n+1,n+1}\, V^{n+1}\, V^{n+1}$$

$$\equiv M_{kb}(dV^b/dt) + \Gamma_{k,bh}\, V^b\, V^h + 2\Gamma_{k,b0}\, V^b + \Gamma_{k,00}. \qquad (7.2.16a)$$

- **Temporal** part of A_β:

$$A_{n+1} = M_{n+1,b}(dV^b/dt) + \Gamma_{n+1,bh}\, V^b\, V^h + 2\Gamma_{n+1;b,n+1}\, V^b\, V^{n+1} + \Gamma_{n+1;n+1,n+1}\, V^{n+1}\, V^{n+1}$$

$$\equiv M_b(dV^b/dt) + \Gamma_{0,bh}\, V^b\, V^h + 2\Gamma_{0,b0}\, V^b + \Gamma_{0,00}. \qquad (7.2.16b)$$

- **Spatial** part of A^β:

$$A^k = dV^k/dt + \Gamma^k{}_{bh}\, V^b\, V^h + 2\Gamma^k{}_{b,n+1}\, V^b\, V^{n+1} + \Gamma^k{}_{n+1,n+1}\, V^{n+1}\, V^{n+1}$$

$$\equiv dV^k/dt + \Gamma^k{}_{bh}\, V^b\, V^h + 2\Gamma^k{}_{b0}\, V^b + \Gamma^k{}_{00}. \qquad (7.2.16c)$$

- **Temporal** part of A^β:

$$A^{n+1} = dV^{n+1}/dt + \Gamma^{n+1}{}_{bh}\, V^b\, V^h + 2\Gamma^{n+1}{}_{b,n+1}\, V^b\, V^{n+1} + \Gamma^{n+1}{}_{n+1,n+1}\, V^{n+1}\, V^{n+1}$$

$$\equiv \Gamma^0{}_{bh}\, V^b\, V^h + 2\Gamma^0{}_{b0}\, V^b + \Gamma^0{}_{00} \quad (= 0, \text{ by Equation 7.7.11d}). \qquad (7.2.16d)$$

The Christoffel symbols of the first and second kind are, respectively (recalling Sections 3.3 and 3.8):

$$\Gamma_{\alpha,\beta\delta} \equiv (1/2)(\partial M_{\alpha\beta}/\partial q^\delta + \partial M_{\alpha\delta}/\partial q^\beta - \partial M_{\beta\delta}/\partial q^\alpha)$$

$$= S\, dm\; \mathbf{e}_\alpha \cdot (\partial \mathbf{e}_\beta/\partial q^\delta) = S\, dm\; \mathbf{e}_\alpha \cdot (\partial \mathbf{e}_\delta/\partial q^\beta)$$

$$= \mathbf{E}_\alpha \cdot (\partial \mathbf{E}_\beta/\partial q^\delta) = \mathbf{E}_\alpha \cdot (\partial \mathbf{E}_\delta/\partial q^\beta), \qquad (7.2.17a)$$

and

$$\Gamma^\alpha{}_{\beta\delta} \equiv M^{\alpha\zeta}\, \Gamma_{\zeta,\beta\delta}$$

(where $M^{\alpha\beta} M_{\beta\gamma} = \delta^\alpha{}_\gamma$, under Equations 7.2.11a through d)

$$= \mathbf{E}^\alpha \cdot (\partial \mathbf{E}_\beta/\partial q^\delta) = \mathbf{E}^\alpha \cdot (\partial \mathbf{E}_\delta/\partial q^\beta); \qquad (7.2.17b)$$

or, *in extenso* (comparing Equations 7.2.16a and c)

$$\Gamma^k{}_{bh} \equiv M^{kd}\,\Gamma_{d,bh}, \quad \Gamma^k{}_{b,n+1} \equiv M^{kd}\,\Gamma_{d;b,n+1}, \quad \Gamma^k{}_{n+1,n+1} \equiv M^{kd}\,\Gamma_{d;n+1,n+1}, \tag{7.2.17c}$$

where (adding a subsemicolon to the Christoffels, just for extra clarity):

$$2\Gamma_{d;b,n+1} \equiv \partial M_{d,n+1}/\partial q^b + \partial M_{db}/\partial q^{n+1} - \partial M_{b,n+1}/\partial q^d, \tag{7.2.17d}$$

or, simply,

$$2\Gamma_{d,b0} = \partial M_d/\partial q^b + \partial M_{db}/\partial t - \partial M_b/\partial q^d$$

$$= (\partial M_d/\partial q^b - \partial M_b/\partial q^d) + \partial M_{db}/\partial t; \tag{7.2.17e}$$

$$\Gamma_{d;n+1,n+1} \equiv (1/2)(\partial M_{d,n+1}/\partial q^{n+1} + \partial M_{d,n+1}/\partial q^{n+1} - \partial M_{n+1,n+1}/\partial q^d), \tag{7.2.17f}$$

or, simply,

$$\Gamma_{d,00} = \partial M_d/\partial t - (1/2)\partial M_0/\partial q^d; \tag{7.2.17g}$$

and

$$\Gamma_{n+1;n+1,n+1} = (1/2)\partial M_{n+1,n+1}/\partial t, \quad \text{or, simply,} \quad \Gamma_{0,00} = (1/2)\partial M_0/\partial t, \tag{7.2.17h}$$

$$\Gamma_{n+1;b,n+1} = (1/2)\partial M_{n+1,n+1}/\partial q^b, \quad \text{or, simply,} \quad \Gamma_{0,b0} = (1/2)\partial M_0/\partial q^b; \tag{7.2.17i}$$

i.e., *the second-kind Christoffels, appearing in A^k (and as shown later, in the equations of motion) are built with the help of M^{kd}, i.e., with the spatial part of $(dS)^2$.* As a result of the above,

$$A_{n+1} = M_{n+1,k}(dV^k/dt) + \Gamma_{n+1,kb}\,V^k\,V^b + (\partial M_{n+1,n+1}/\partial q^k)V^k + (1/2)(\partial M_{n+1,n+1}/\partial t),$$

or, simply,

$$A_0 = M_k(dV^k/dt) + \Gamma_{0,kb}\,V^k\,V^b + (\partial M_0/\partial q^k)V^k + (1/2)\partial M_0/\partial t, \quad \text{etc.} \tag{7.2.17j}$$

7.2.4 Inertia Force: Holonomic Components and Holonomic Euler–Lagrange Operator

The holonomic covariant components of the system *inertia "force,"* I_k, are defined by

$$I_k = S\,dm\;\mathbf{a}\cdot(\partial\mathbf{r}/\partial q^k) \equiv S\,dm\;\mathbf{a}\cdot\mathbf{e}_k.$$

(since $\mathbf{a} = d\mathbf{v}/dt = (Dv^\beta/Dt)\,\mathbf{e}_\beta \equiv (DV^\beta/Dt)\,\mathbf{e}_\beta \equiv A^\beta\,\mathbf{e}_\beta = A^k\,\mathbf{e}_k$, by Equation 7.2.16d)

$$= S\, dm(A^l\, \mathbf{e}_l) \cdot \mathbf{e}_k = A^l \left(S\, dm\ \mathbf{e}_l \cdot \mathbf{e}_k \right) \equiv M_{kl}\, A^l \quad (= A_k). \quad (7.2.18)$$

But also (and to represent these "forces," explicitly, in terms of *velocities*) applying the chain rule to I_k, since $d(\ldots)/dt$ and $S\,(\ldots)$ commute,

$$I_k = d/dt \left(S\, dm\ \mathbf{v} \cdot \mathbf{e}_k \right) - S\, dm\ \mathbf{v} \cdot (d\mathbf{e}_k/dt)$$

[since $\mathbf{e}_k = \partial \mathbf{v}/\partial V^k$ (Equation 6.3.5a) and $d\mathbf{e}_k/dt = \ldots = \partial \mathbf{v}/\partial q^k$ (Equation 6.3.5c)]

$$= d/dt \left[S\, dm\ \mathbf{v} \cdot (\partial \mathbf{v}/\partial V^k) \right] - S\, dm\ \mathbf{v} \cdot (\partial \mathbf{v}/\partial q^k)$$

$$= d/dt(\partial T/\partial V^k) - \partial T/\partial q^k \equiv E_k(T); \quad (7.2.19)^*$$

to be calculated as if the q terms and V terms (and t) in $2T = S\, dm\ \mathbf{v} \cdot \mathbf{v} = 2T(t, q, V \equiv dq/dt)$ were independent variables.

Similarly, the *contravariant* components of the inertia force, I^k, are given by

$$I^k \equiv M^{k\beta}\, E_\beta(T) = M^{k\beta}\, A_\beta = M^{kl}\, A_l \equiv A^k \quad \text{(by Equations 7.2.11a ff)} \quad (7.2.20)$$

In sum,

$$I_k \equiv S\, dm\ \mathbf{v} \cdot \mathbf{e}_k = E_k(T) = A_k. \quad (7.2.21)$$

7.2.5 Inertia Force: Nonholonomic Components and Nonholonomic Euler–Lagrange Operator

7.2.5.1 First Derivation

The nonholonomic covariant components of the system *inertia force*, I_K, are defined by

$$I_K = S\, dm\ \mathbf{a}^* \cdot \mathbf{e}_K \quad (7.2.22)$$

$$\left[= S\, dm\ \mathbf{a} \cdot (A^k{}_K\, \mathbf{e}_k) = A^k{}_K \left(S\, dm\ \mathbf{a} \cdot \mathbf{e}_k \right) = A^k{}_K\, I_k \right]. \quad (7.2.23)^{**}$$

---

* Where, recalling Equation 6.3.5g, $E_k(\ldots) \equiv d/dt[\partial(\ldots)/\partial V^k] - \partial(\ldots)/\partial q^k$: *Euler–Lagrange operator, in holonomic variables.*

** Where, recalling the earlier notation (Equation 6.4.7): $f = f(t, q, dq/dt \equiv V) = f[t, q, dq(t, q, \omega)/dt] = f^*(t, q, \omega) = f^*$.

Here, too, to represent these "forces," explicitly, in terms of velocities, we transform the above as follows (recalling Equation 6.5.7a)

$$I_K = \mathcal{S} dm(d\mathbf{v}^*/dt) \cdot (\partial \mathbf{v}^*/\partial \omega^K)$$

$$= d/dt\left[\mathcal{S} dm\, \mathbf{v}^* \cdot (\partial \mathbf{v}^*/\partial \omega^K)\right] - \mathcal{S} dm\, \mathbf{v}^* \cdot d/dt\, (\partial \mathbf{v}^*/\partial \omega^K)$$

[adding and subtracting $\mathcal{S} dm\ \mathbf{v}^* \cdot (\partial \mathbf{v}^*/\partial \theta^K)$]

$$= d/dt\left[\mathcal{S} dm\, \mathbf{v}^* \cdot (\partial \mathbf{v}^*/\partial \omega^K)\right] - \mathcal{S} dm\, \mathbf{v}^* \cdot (\partial \mathbf{v}^*/\partial \theta^K)$$

$$- \mathcal{S} dm\ \mathbf{v}^* \cdot [d/dt(\partial \mathbf{v}^*/\partial \omega^K) - \partial \mathbf{v}^*/\partial \theta^K]$$

(recalling Equations 6.5.8c through 6.5.10b and 7.2.7 ff)

$$= d/dt(\partial T^*/\partial \omega^K) - \partial T^*/\partial \theta^K - \Gamma_K \equiv E_K(T^*) - \Gamma_K, \qquad (7.2.24)$$

where, successively,

$$\Gamma_K \equiv \mathcal{S} dm\ \mathbf{v}^* \cdot [d/dt(\partial \mathbf{v}^*/\partial \omega^K) - \partial \mathbf{v}^*/\partial \theta^K]$$

$$\equiv \mathcal{S} dm\ \mathbf{v}^* \cdot E_K(\mathbf{v}^*) \equiv \mathcal{S} dm\ \mathbf{v}^* \cdot \mathbf{g}_K$$

$$= \mathcal{S} dm\ \mathbf{v}^* \cdot \{[d/dt(\partial V^k/\partial \omega^K) - \partial V^k/\partial \theta^K]\mathbf{e}_k\}$$

(recalling Equations 6.5.9a through c)

$$= \mathcal{S} dm\ \mathbf{v}^* \cdot (\partial \mathbf{e}_K/\partial \theta^\Lambda - \partial \mathbf{e}_\Lambda/\partial \theta^K)\ \omega^\Lambda = \mathcal{S} dm\ \mathbf{v}^* \cdot (\gamma^L{}_{\Lambda K}\ \mathbf{e}_L)\omega^\Lambda$$

$$= \left(\mathcal{S} dm\, \mathbf{v}^* \cdot \mathbf{e}_L\right)\gamma^L{}_{\Lambda K}\ \omega^\Lambda \equiv (\partial T^*/\partial \omega^L)\gamma^L{}_{\Lambda K}\ \omega^\Delta$$

$$= -\gamma^L{}_{K\Lambda}(\partial T^*/\partial \omega^L)\ \omega^\Delta,$$

i.e., finally,

$$-\Gamma_K = \gamma^L{}_{K\Lambda}(\partial T^*/\partial\omega^L)\omega^\Lambda \equiv h^L{}_K\, P_L: \tag{7.2.24a}$$

corrective term, due to the nonholonomic coordinates, where

$$P_L \equiv \partial T^*/\partial\omega^L = P_L(t, q, \omega)\text{: nonholonomic covariant system momenta.} \tag{7.2.24b}$$

7.2.5.2 Second Derivation

Applying the chain rule to

$$T = T(t,\ q,\ dq/dt \equiv V) = T[t,\ q,\ dq(t,\ q,\ \omega)/dt]$$

$$= T^*(t,\ q,\ \omega) = T^*, \tag{7.2.25a}$$

we readily find

- $\partial T/\partial V^k = (\partial T^*/\partial\omega^K)(\partial\omega^K/\partial V^k) = A^K{}_k(\partial T^*/\partial\omega^K)$ (7.2.25b)

  [$\Leftrightarrow \partial T^*/\partial\omega^K = A^k{}_K(\partial T/\partial V^k)$, i.e., *tensorial* transformation],

- $\partial T/\partial q^k = \partial T^*/\partial q^k + (\partial T^*/\partial\omega^K)(\partial\omega^K/\partial q^k)$, (7.2.25c)

- $\partial T^*/\partial\theta^K \equiv (\partial T^*/\partial q^k)(\partial V^k/\partial\omega^K) = A^k{}_K(\partial T^*/\partial q^k)$ (7.2.25d)

  [$\Leftrightarrow \partial T^*/\partial q^k = A^K{}_k\,(\partial T^*/\partial\theta^K)$, i.e., *nontensorial* transformation].

Therefore, successively,

$$E_k(T) \equiv d/dt(\partial T/\partial V^k) - \partial T/\partial q^k$$

$$= d/dt[A^K{}_k(\partial T^*/\partial\omega^K)] - [\partial T^*/\partial q^k + (\partial T^*/\partial\omega^K)(\partial\omega^K/\partial q^k)]$$

$$= (dA^K{}_k/dt)(\partial T^*/\partial\omega^K) + A^K{}_k\, d/dt(\partial T^*/\partial\omega^K) - A^K{}_k(\partial T^*/\partial\theta^K)$$

$$- (\partial\omega^K/\partial q^k)\,(\partial T^*/\partial\omega^K)$$

$$= A^K{}_k[d/dt(\partial T^*/\partial\omega^K) - \partial T^*/\partial\theta^K] + [d/dt(\partial\omega^K/\partial V^k) - \partial\omega^K/\partial q^k](\partial T^*/\partial\omega^K)$$

(since $A^K{}_k = \partial\omega^K/\partial V^k$)

i.e., finally,

$$E_k(T) = A^K{}_k\, E_K(T^*) + E_k(\omega^K)(\partial T^*/\partial\omega^K)$$

(tensorlike part + nontensorlike part) (7.2.26)*

---

* Where, recalling footnote to Equation 6.5.8c, $E_K(\ldots) \equiv d/dt[\partial(\ldots)/\partial\omega^K] - \partial(\ldots)/\partial\theta^K$: *Euler–Lagrange operator, in nonholonomic variables.*

In words, *the Euler–Lagrange operator does not transform as a (covariant) vector under $dq \leftrightarrow d\theta$ transformations,* unless the following *integrability* conditions are satisfied:

$$E_k(\omega^K) \equiv d/dt(\partial\omega^K/\partial V^k) - (\partial\omega^K/\partial q^k) = 0. \tag{7.2.27a}$$

In our Pfaffian case, the expression $E_k(\omega^K)$ assumes the following form:

$$E_k(\omega^K) \equiv d/dt(\partial\omega^K/\partial V^k) - (\partial\omega^K/\partial q^k) = d/dt(A^K{}_k) - \partial/\partial q^k(A^K{}_\lambda\ V^\lambda)$$

$$= (\partial A^K{}_k/\partial q^\lambda)V^\lambda - (\partial A^K{}_\lambda/\partial q^k)V^\lambda = (\partial A^K{}_k/\partial q^\lambda - \partial A^K{}_\lambda/\partial q^k)V^\lambda$$

$$= (\partial A^K{}_k/\partial q^\lambda - \partial A^K{}_\lambda/\partial q^k)(A^\lambda{}_\Lambda\omega^\Lambda) = 2A^\lambda{}_\Lambda\ A^K{}_{[k,\lambda]}\ \omega^\Lambda$$

$$= A^L{}_k(2A^K{}_{[\theta,\lambda]}\ A^\lambda{}_\Lambda\ A^\theta{}_L)\omega^\Lambda \quad (\text{since } A^L{}_k\ A^\theta{}_L = \delta^\theta{}_k), \tag{7.2.27b}$$

i.e.,

$$E_k(\omega^K) = A^L{}_k\ \gamma^K{}_{L\Lambda}\ \omega^\Lambda \equiv A^L{}_k\ h^K{}_L \quad (\text{recalling Equation 6.5.1c}). \tag{7.2.27c}^*$$

Hence, with some dummy index changes,

$$E_k(\omega^K)(\partial T^*/\partial\omega^K) = A^K{}_k[\gamma^L{}_{K\Lambda}(\partial T^*/\partial\omega^L)\omega^\Lambda] \equiv -A^K{}_k\ \Gamma_K. \tag{7.2.27e}$$

In view of these results, Equation 7.2.26 assumes the covariant vector transformation form:

$$E_k(T) = A^K{}_k[E_K(T^*) - \Gamma_K] \Leftrightarrow E_K(T^*) - \Gamma_K = A^k{}_K\ E_k(T). \tag{7.2.28}$$

The above show that although neither $E_K(T^*)$ *nor* $-\Gamma_K$, *individually,* transform vectorially under local affine velocity transformations $V \to \omega \to \omega' \to \ldots$, yet when *taken together* as $E_K(T^*) - \Gamma_K = I_K$ they do.

In conclusion, we have the covariant vector transformations:

$$I_k = A^K{}_k\ I_K \Leftrightarrow I_K = A^k{}_K\ I_k, \tag{7.2.29a}$$

---

* Equation 7.2.27c readily yields: $E_k(\omega^K)\ A^k{}_L = \gamma^K{}_{L\Lambda}\ \omega^\Lambda$, and so the transitivity (6.5.1b) becomes:

$$(\delta\theta^K)^{\cdot} - \delta\omega^K = (\gamma^K{}_{L\Lambda}\ \omega^\Lambda)\delta\theta^L + A^K{}_k[(\delta q^k)^{\cdot} - \delta V^k]$$

$$= [E_k(\omega^K)\ A^k{}_L]\ \delta\theta^L + A^K{}_k\ [(\delta q^k)^{\cdot} - \delta V^k]$$

$$= E_k(\omega^K)\ \delta q^k + A^K{}_k[(\delta q^k)^{\cdot} - \delta V^k]; \tag{7.2.27d}$$

from which it follows that the necessary and sufficient conditions for θ^K to be a holonomic *coordinate* are $E_k(\omega^K) = 0$, i.e., Equation 7.2.27a.

where

$$I_k = A_k \equiv \int dm\ \mathbf{a} \cdot \mathbf{e}_k = E_k(T) \equiv d/dt(\partial T/\partial V^k) - \partial T/\partial q^k: \quad (7.2.29b)$$

holonomic covariant components of system *inertia force,*

$$I_K = A_K \equiv \int dm\ \mathbf{a} \cdot \mathbf{e}_K = E_K(T^*) - \Gamma_K$$

$$\equiv d/dt(\partial T^*/\partial \omega^K) - \partial T^*/\partial \theta^K + \gamma^L{}_{K\Lambda}(\partial T^*/\partial \omega^L)\omega^\Lambda: \quad (7.2.29c)$$

nonholonomic covariant components of system *inertia force.*

Example 7.2.1. Explicit Forms of I_K, I^K, etc.

With the help of the above we obtain, successively,

$$I_K = A_K = A^\lambda{}_K\, I_\lambda = A^\lambda{}_K\, A_\lambda \quad \text{(recalling that } A^{n+1}{}_K = 0)$$

$$= A^k{}_K\, I_k = A^k{}_K\, A_k$$

$$= A^k{}_K[M_{kb}(dV^b/dt) + \Gamma_{k,\sigma\delta}\, V^\sigma\, V^\delta]$$

$$= A^k{}_K\{M_{kb}[d/dt(A^b{}_\Xi\, \omega^\Xi)] + \Gamma_{k,\sigma\delta}(A^\sigma{}_\Sigma\, \omega^\Sigma)(A^\delta{}_\Delta\, \omega^\Delta)\}$$

$$= A^k{}_K\{M_{kb}\, A^b{}_\Sigma(d\omega^\Sigma/dt) + [\Gamma_{k,\sigma\delta}\, A^\sigma{}_\Sigma\, A^\delta{}_\Delta + M_{kb}\, A^\delta{}_\Delta(\partial A^b{}_\Sigma/\partial q^\delta)\omega^\Sigma\, \omega^\Delta]\}$$

$$= M_{KL}(d\omega^L/dt) + [A^k{}_K\, A^\sigma{}_\Sigma\, A^\delta{}_\Delta\, \Gamma_{k,\sigma\delta} + (M_{kb}\, A^k{}_K)A^\delta{}_\Delta\, A^b{}_{\Sigma,\delta}]\omega^\Sigma\, \omega^\Delta, \quad \text{(a)}$$

or [recalling Equation 3.15.7a and noting that: $M_{kb}\, A^k{}_K = (A^\Gamma{}_k\, A^\Lambda{}_b\, M_{\Gamma\Lambda})A^k{}_K = \delta^\Gamma{}_K A^\Lambda{}_b\, M_{\Gamma\Lambda} = M_{K\Lambda}\, A^\Lambda{}_b$], finally,

$$I_K = M_{KB}(d\omega^B/dt) + \Gamma_{K,\Sigma\Delta}\, \omega^\Sigma\, \omega^\Delta; \quad \text{(b)}$$

i.e., the nonholonomic components I_K have the same form as the holonomic ones I_k.
From these covariant forms, the contravariant ones follow easily:

$$I^K = A^K = A^K{}_\lambda\, A^\lambda = A^K{}_k\, A^k + A^K{}_{n+1}\, A^{n+1} \quad \text{(recalling Equations 7.2.16c and d)}$$

$$= \ldots = d\omega^K/dt + \Gamma^K{}_{\Sigma\Delta}\, \omega^\Sigma\, \omega^\Delta, \quad \text{(d)*}$$

---

* We also notice that

$$\omega^\Lambda = A^\Lambda{}_\lambda\, V^\lambda \quad \text{and} \quad D\omega^\Lambda/Dt = A^\Lambda{}_\lambda(DV^\lambda/Dt) \Leftrightarrow DV^\lambda/Dt = A^\lambda{}_\Lambda(D\omega^\Lambda/Dt), \quad \text{(c)}$$

but $d\omega^\Lambda/dt \neq A^\Lambda{}_\lambda(dV^\lambda/dt)$.

where (recalling Equation 3.15.4a):

$$\Gamma^K{}_{\Sigma\Delta} = A^K{}_k\, A^\sigma{}_\Sigma\, A^\delta{}_\Delta\, \Gamma^k{}_{\sigma\delta} - A^\sigma{}_\Sigma\, A^\delta{}_\Delta\, A^K{}_{\sigma,\delta}$$

$$= A^K{}_k\, A^\sigma{}_\Sigma\, A^\delta{}_\Delta\, \Gamma^k{}_{\sigma\delta} + A^K{}_\sigma\, A^\sigma{}_{\Sigma,\Delta}$$

$$= A^K{}_k\, A^\sigma{}_\Sigma\, A^\delta{}_\Delta\, \Gamma^k{}_{\sigma\delta} + A^K{}_\sigma\, A^\delta{}_\Delta\, A^\sigma{}_{\Sigma,\delta}. \tag{e}$$

- Further, it is not hard to see that (as in all quadratic forms):

$$\Gamma_{K,\Sigma\Delta}\, \omega^\Sigma\, \omega^\Delta = \Gamma_{K,(\Sigma\Delta)}\, \omega^\Sigma\, \omega^\Delta + \Gamma_{K,[\Sigma\Delta]}\, \omega^\Sigma\, \omega^\Delta = \Gamma_{K,(\Sigma\Delta)}\, \omega^\Sigma\, \omega^\Delta, \tag{f}$$

where (recalling the NH Christoffel decompositions, Equations 3.15.5a through 6b, and 3.15.7b and c):

$$\Gamma^K{}_{(\Sigma\Delta)} = A^K{}_k\, A^\sigma{}_\Sigma\, A^\delta{}_\Delta\, \Gamma^k{}_{(\sigma\delta)} - A^\sigma{}_\Sigma\, A^\delta{}_\Delta\, A^K{}_{(\sigma,\delta)}, \tag{f1}$$

$$\Gamma^K{}_{[\Sigma\Delta]} = A^K{}_k\, A^\sigma{}_\Sigma\, A^\delta{}_\Delta\, \Gamma^k{}_{[\sigma\delta]} + (1/2)\gamma^K{}_{\Delta\Sigma}$$

$$\equiv S^K{}_{\Sigma\Delta} + (1/2)\gamma^K{}_{\Delta\Sigma}$$

$$[\equiv (1/2)\ \gamma^K{}_{\Delta\Sigma}, \text{ in our torsionless manifold}]. \tag{f2}$$

- Alternatively, since (recalling Equations 3.15.8a through c):

$$\Gamma_{K,\Sigma\Delta} = [C_{K,\Sigma\Delta} + (1/2)(\gamma_{\Sigma,K\Delta} + \gamma_{\Delta,K\Sigma})]_{\text{symmetric in } \Sigma,\ \Delta}$$

$$+ [(1/2)\gamma_{K,\Delta\Sigma}]_{\text{antisymmetric in } \Sigma,\ \Delta}, \tag{g1}$$

where

$$2C_{K,\Sigma\Delta} = 2C_{K,\Delta\Sigma} \equiv M_{K\Sigma,\Delta} + M_{K\Delta,\Sigma} - M_{\Sigma\Delta,K}, \quad \gamma_{K,\Delta\Sigma} \equiv M_{K\Xi}\, \gamma^\Xi{}_{\Delta\Sigma}, \tag{g2}$$

Expression b transforms, successively,

$$I_K = M_{KB}(d\omega^B/dt) + \Gamma_{K,(\Sigma\Delta)}\, \omega^\Sigma\, \omega^\Delta$$

$$= M_{KB}(d\omega^B/dt) + [C_{K,\Sigma\Delta} + (1/2)(\gamma_{\Sigma,K\Delta} + \gamma_{\Delta,K\Sigma})]\omega^\Sigma\, \omega^\Delta$$

$$= M_{KB}(d\omega^B/dt) + C_{K,\Sigma\Delta}\, \omega^\Sigma\, \omega^\Delta + (1/2)(\gamma_{\Sigma,K\Delta} + \gamma_{\Delta,K\Sigma})\omega^\Sigma\, \omega^\Delta$$

$$= M_{KB}(d\omega^B/dt) + C_{K,\Sigma\Delta}\, \omega^\Sigma\, \omega^\Delta + \gamma_{\Sigma,K\Delta}\, \omega^\Sigma\, \omega^\Delta$$

$$= M_{KB}(d\omega^B/dt) + C_{K,\Sigma\Delta}\, \omega^\Sigma\, \omega^\Delta + \gamma^\Sigma{}_{K\Delta}\, \omega_\Sigma\, \omega^\Delta \quad (\text{i.e., } \omega_\Sigma = P_\Sigma). \tag{g3}^*$$

---

* (See following page for footnote.)

- Similarly, in terms of second-kind NH Christoffels: since

$$\Gamma^K{}_{\Sigma\Delta} = M^{K\Lambda}\,\Gamma_{\Lambda,\Sigma\Delta} = M^{K\Lambda}[C_{\Lambda,\Sigma\Delta} + (1/2)(\gamma_{\Sigma,\Lambda\Delta} + \gamma_{\Delta,\Lambda\Sigma}) + (1/2)\gamma_{\Lambda,\Delta\Sigma}]$$

$$\equiv [C^K{}_{\Sigma\Delta} + (M^{K\Lambda}/2)(\gamma_{\Sigma,\Lambda\Delta} + \gamma_{\Delta,\Lambda\Sigma})]_{\text{symmetric in } \Delta,\, \Sigma}$$

$$+ [(1/2)\gamma^K{}_{\Delta\Sigma}]_{\text{antisymmetric in } \Delta,\, \Sigma,} \tag{h1}$$

we find

$$\Gamma^K{}_{\Sigma\Delta}\,\omega^\Delta\,\omega^\Sigma = [C^K{}_{\Sigma\Delta} + (M^{K\Lambda}/2)(\gamma_{\Sigma,\Lambda\Delta} + \gamma_{\Delta,\Lambda\Sigma})]\omega^\Delta\,\omega^\Sigma. \tag{h2}$$

Remark: Expressions like 7.2.15a through 7.2.16d and b and g3 of Example 7.2.1, implying that to obtain the inertial forces one has to, first, calculate the Christoffels, are good for theoretical arguments, *not for concrete problems.*

In the latter, it is the other way around:

> The I_k, I_K are found as the gradients of T, T^*; i.e., T, $T^* \to E_k(T)$, $E_K(T^*) \Rightarrow I_k = E_k(T)$, $I_K = E_K{}^*(T^*) - \Gamma_K$; and then the Christoffels are read off from them as appropriate coefficients, if needed; and this is an advantage of the Lagrangean formalism.

For example, for a particle of mass m, in spherical polar coordinates $r = q^1$, $\theta = q^2$ (nutation), $\phi = q^3$ (precession), we have

$$(ds)^2 = (dq^1)^2 + (q^1)^2\,(dq^2)^2 + (q^1\,\sin q^2)^2\,(dq^3)^2 \tag{i}$$

---

* But, the *first* term of Equation g3 equals

$$M_{KB}(d\omega^B/dt) = d/dt(M_{K\Sigma}\,\omega^\Sigma) - (dM_{K\Sigma}/dt)\omega^\Sigma$$

$$= d/dt(M_{K\Sigma}\,\omega^\Sigma) - M_{K\Sigma,\Delta}\,\omega^\Delta\,\omega^\Sigma = d/dt(\partial T^*/\partial\omega^K) - M_{K\Sigma,\Delta}\,\omega^\Delta\,\omega^\Sigma; \tag{g4}$$

the *second* term equals

$$C_{K,\Sigma\Delta}\,\omega^\Sigma\,\omega^\Delta \equiv (1/2)(M_{K\Sigma,\Delta} + M_{K\Delta,\Sigma} - M_{\Sigma\Delta,K})\omega^\Sigma\,\omega^\Delta$$

$$= M_{K\Sigma,\Delta}\,\omega^\Delta\,\omega^\Sigma - (1/2)M_{\Sigma\Delta,K}\,\omega^\Sigma\,\omega^\Delta = M_{K\Sigma,\Delta}\,\omega^\Delta\,\omega^\Sigma - \partial T^*/\partial\theta^K; \tag{g5}$$

while the *third* term equals

$$\gamma^\Sigma{}_{K\Delta}\,\omega_\Sigma\,\omega^\Delta \equiv \gamma^\Sigma{}_{K\Delta}(\partial T^*/\partial\omega^\Sigma)\omega^\Delta \equiv h^\Sigma{}_K(\partial T^*/\partial\omega^\Sigma); \tag{g6}$$

finally,

$$I_K = d/dt(\partial T^*/\partial\omega^K) - \partial T^*/\partial\theta^K + \gamma^\Sigma{}_{K\Delta}(\partial T^*/\partial\omega^\Sigma)\omega^\Delta \quad \text{(recalling that } \gamma^{N+1}{}_{\dots} = 0\text{)}$$

$$= d/dt(\partial T^*/\partial\omega^K) - \partial T^*/\partial\theta^K + \gamma^R{}_{K\Delta}(\partial T^*/\partial\omega^R)\omega^\Delta; \quad \text{i.e., Equation 7.2.29c.} \tag{g7}$$

$$\Rightarrow 2T = m(ds/dt)^2 = m[(dq^1/dt)^2 + (q^1)^2 (dq^2/dt)^2 + (q^1 \sin q^2)^2 (dq^3/dt)^2] \tag{i1}$$

$$\Rightarrow mA_1 = E_1(T) = m[(d^2q^1/dt^2) - q^1(dq^2/dt)^2 - q^1(\sin q^2)^2 (dq^3/dt)^2] = m[d^2r/dt^2 - r(d\theta/dt)^2 - r \sin^2 \theta(d\phi/dt)^2], \tag{i2}$$

$$mA_2 = E_2(T) = m[(q^1)^2 (d^2q^2/dt^2) + 2q^1(dq^1/dt)(dq^2/dt) - (q^1)^2 \sin q^2 \cos q^2(dq^3/dt)^2] = m[r^2(d^2\theta/dt^2) + 2r(dr/dt)(d\theta/dt) - r^2 \sin\theta \cos\theta(d\phi/dt)^2], \tag{i3}$$

$$mA_3 = E_3(T) = m[(q^1 \sin q^2)^2 (d^2q^3/dt^2) + 2\, q^1(\sin q^2)^2 (dq^1/dt)(dq^3/dt) + (q^1)^2 \sin\, (2q^2)(dq^2/dt)(dq^3/dt)] = m[r^2 \sin^2\theta(d^2\phi/dt^2) + 2r \sin^2\theta(dr/dt)(d\phi/dt) + r^2 \sin(2\theta)(d\theta/dt)(d\phi/dt)]; \tag{i4}$$

and comparing the above with Equations 7.2.15c and 7.2.16a while noticing that, due to the symmetry $\Gamma_{k,bh} = \Gamma_{k,hb}$, the *bilinear terms in the velocities occurring in Equations i2 through 4 are really* $2\Gamma_{k,bh}(dq^b/dt)(dq^h/dt)$, we readily conclude that the *nonvanishing* first-kind Christoffels are

$$A_1\colon\ \Gamma_{1,22} = -q^1 = -r, \quad \Gamma_{1,33} = -q^1(\sin q^2)^2 = -r \sin^2\theta, \tag{i5}$$

$$A_2\colon\ \Gamma_{2,12} = q^1 = r, \quad \Gamma_{2,33} = -(q^1)^2 (\sin q^2)(\cos q^2) = -r^2 \sin\theta \cos\theta, \tag{i6}$$

$$A_3\colon\ \Gamma_{3,13} = q^1 (\sin q^2)^2 = r \sin^2\theta, \quad \Gamma_{3,23} = (1/2)(q^1)^2 \sin(2q^2) = r^2 \sin(2\theta)/2; \tag{i7}$$

and similarly for the nonvanishing second-kind Christoffels.

Problem 7.2.1

Show that under a local quasi-velocity transformation: $\omega^K \rightarrow \omega^{K'} = A^{K'}{}_K(q^\beta)\,\omega^K + A^{K'}(q^\beta)$, where $T^* \rightarrow T^{**} = T^*$, the following *nontensorial* transformations hold:

$$\Gamma_K \rightarrow \Gamma_{K'} = A^K{}_{K'}\, \Gamma_K + [d/dt(\partial\omega^K/\partial\omega^{K'}) - \partial\omega^K/\partial\theta^{K'}]\, (\partial T^*/\partial\omega^K) \equiv A^K{}_{K'}\, \Gamma_K + E_{K'}(\omega^K)(\partial T^*/\partial\omega^K), \tag{a}$$

$$E_K(T^*) \to E_{K'}(T^{**}) = A^K{}_{K'}\, E_K(T^*) + E_{K'}(\omega^K)(\partial T^*/\partial \omega^K); \qquad \text{(b)}$$

i.e.,

$$E_{K'}(T^{**}) - \Gamma_{K'} = (\partial \omega^K/\partial \omega^{K'})\, [E_K(T^*) - \Gamma_K] = A^K{}_{K'}[E_K(T^*) - \Gamma_K]. \qquad \text{(c)}$$

7.3 THE FORCES

We recall (Section 4.3) that, according to *d'Alembert's physical assumption*, the *total force* on a typical system particle p, $d\mathbf{f}$, is decomposed into two parts: an *impressed* $d\mathbf{F}$ and a *constraint reaction* $d\mathbf{R}$:

$$d\mathbf{f} = d\mathbf{F} + d\mathbf{R} \qquad (7.3.1)$$

where

$d\mathbf{F}$: Total *impressed, or physically given,* force on p. By impressed* we understand (external and/or internal) forces of physical origin and caused by physical effects, e.g., gravity (weight), elastic springs, viscous damping, slipping (or sliding, or kinetic) friction; i.e., *forces expressed by constitutive equations and, hence, containing experimentally determinable and measurable material functions* (or coefficients, constants, etc.); e.g. gravitational constant, elastic moduli, viscous and dry friction coefficients. Such forces are also variously, but not completely equivalently, referred to as *directly applied, active, acting, assigned, known,* etc.

$d\mathbf{R}$: Total *constraint reaction* force on p. By constraint reaction we mean (external and/or internal) forces originating from geometrical (positional) and/or kinematical (motional) constraints; e.g., inextensible cable tensions, internal forces on a rigid body, rolling friction, i.e., forces due to prescribed (external and/or internal) constraints and, hence, *not containing any material coefficients.* They are also referred to as *passive* or *unknown.***

Now, with the help of the earlier covariant particle/system basis vectors $\mathbf{e}_\gamma$, $\mathbf{e}_\Gamma$ we define the *system forces* as follows:

$$Q_\gamma \equiv S\, d\mathbf{F} \cdot \mathbf{e}_\gamma: \qquad (7.3.2a)$$

covariant holonomic components of impressed system force,

$$\Lambda_\gamma \equiv S\, d\mathbf{R} \cdot \mathbf{e}_\gamma: \qquad (7.3.2b)$$

* What Hamel (1949, p. 65) calls *eingeprägte*.

** (See following page for footnote.)

covariant holonomic components of system reaction force;

$$Q_\Gamma \equiv \mathcal{S}\, d\mathbf{F} \cdot \mathbf{e}_\Gamma: \tag{7.3.2c}$$

covariant nonholonomic components of impressed system force,

$$\Lambda_\Gamma \equiv \mathcal{S}\, d\mathbf{R} \cdot \mathbf{e}_\Gamma: \tag{7.3.2d}$$

covariant nonholonomic components of system reaction force.

Clearly, these components are interrelated by

$$Q_\gamma = \mathcal{S}\, d\mathbf{F} \cdot (A^\Gamma{}_\gamma\, \mathbf{e}_\Gamma) = \left(\mathcal{S}\, d\mathbf{F} \cdot \mathbf{e}_\Gamma\right) A^\Gamma{}_\gamma,$$

** a. More general *noncontact* (or indirect, or active) reactions caused by so-called *servo-*, or *control* constraints are possible (and can be successfully handled by Lagrangean mechanics), but they are not examined here, (See, e.g., Papastavridis, *Analytical Mechanics*, in press.)

b. *Rolling* friction should be classified as a *reaction* because it is not determined by a constitutive equation but by a kinematical constraint (vanishing of tangential component of relative velocity of contacting surfaces, at contact point); while *slipping* friction (no geometrical or kinematical constraint here) should be counted as *impressed* because it is determined by the well-known Coulomb–Morin law, which contains a friction coefficient.

c. It should be stressed that the decomposition (7.3.1) is the distinguishing physical feature of AM, and, as such, is fundamentally different from the corresponding decomposition of the momentum mechanics of Newton–Euler. In the latter, $d\mathbf{f}$ is split into a total *external* force $d\mathbf{f}_{\text{ext'l}}$ (= force originating, even partially, from *outside* of the system S), and a total *internal* (or *mutual*) force $d\mathbf{f}_{\text{int'l}}$ (= force due exclusively to the rest of S on P): $d\mathbf{f} = d\mathbf{f}_{\text{ext'l}} + d\mathbf{f}_{\text{int'l}}$. Then, each of these two forces can, in turn, be resolved into impressed and constraint reaction parts:

$$d\mathbf{f}_{\text{ext'l}} = d\mathbf{F}_{\text{ext'l}} + d\mathbf{R}_{\text{ext'l}} \quad \text{and} \quad d\mathbf{f}_{\text{int'l}} = d\mathbf{F}_{\text{int'l}} + d\mathbf{R}_{\text{int'l}}; \tag{7.3.1a}$$

from which, clearly,

$$d\mathbf{f} = (d\mathbf{F}_{\text{ext'l}} + d\mathbf{R}_{\text{ext'l}}) + (d\mathbf{F}_{\text{int'l}} + d\mathbf{R}_{\text{int'l}}) = (d\mathbf{F}_{\text{ext'l}} + d\mathbf{F}_{int'l}) + (d\mathbf{R}_{\text{ext'l}} + d\mathbf{R}_{\text{int'l}})$$

$$= d\mathbf{F} + d\mathbf{R}, \quad \text{where } d\mathbf{F} \equiv d\mathbf{F}_{\text{ext'l}} + d\mathbf{F}_{\text{int'l}},\ d\mathbf{R} \equiv d\mathbf{R}_{\text{ext'l}} + d\mathbf{R}_{\text{int'l}}. \tag{7.3.1b}$$

Other types of mechanics may opt for different decompositions of $d\mathbf{f}$. This fundamental distinction between the methods of Newton–Euler (momentum) and d'Alembert–Lagrange (energy), rarely emphasized in the English language literature, was made explicit during the late 19th to early 20th centuries by such dynamics greats as Stäckel, Heun, Hamel, and Webster. Failure to appreciate this difference right from the start can seriously stifle the correct understanding of the scope and methods of AM.

Such omissions appear in the special system known as *rigid body*: the word *rigid* (i.e., internal geometrical constraints) means that all its internal forces are reactions and, hence, all its internal impressed forces vanish. Further, if such a body is *externally unconstrained*, or *free*, all its external forces are impressed and, hence, all its external reactions vanish. The coincidence of external impressed with external forces, and of internal reactions with internal forces, in this popular and well-known system is, probably, responsible for the frequent confusion and error accompanying the principle of d'Alembert (Section 7.4), even in contemporary expositions of dynamics!

i.e.,

$$Q_\gamma = A^\Gamma{}_\gamma\, Q_\Gamma \Leftrightarrow Q_\Gamma = A^\gamma{}_\Gamma\, Q_\gamma; \tag{7.3.3a}$$

$$\Lambda_\gamma = A^\Gamma{}_\gamma\, \Lambda_\Gamma \Leftrightarrow \Lambda_\Gamma = A^\gamma{}_\Gamma\, \Lambda_\gamma. \tag{7.3.3b}$$

Their *contravariant* components are defined via the following invariant representations:

$$\mathbf{Q} = Q_\gamma\, \mathbf{E}^\gamma = Q^\gamma\, \mathbf{E}_\gamma = Q_\Gamma\, \mathbf{E}^\Gamma = Q^\Gamma\, \mathbf{E}_\Gamma, \tag{7.3.4a}$$

$$\Lambda = \Lambda_\gamma\, \mathbf{E}^\gamma = \Lambda^\gamma\, \mathbf{E}_\gamma = \Lambda_\Gamma\, \mathbf{E}^\Gamma = \Lambda^\Gamma\, \mathbf{E}_\Gamma. \tag{7.3.4b}$$

Also, we can easily verify that

$$Q^\gamma \equiv S\, d\mathbf{F} \cdot \mathbf{e}^\gamma = M^{\gamma\delta}\, Q_\delta$$

$$\Rightarrow Q^k = M^{kl}\, Q_l, \quad Q^{n+1} = 0 \text{ (by Equations 7.2.11a through d)} \tag{7.3.5a}$$

$$\Lambda^\gamma \equiv S\, d\mathbf{R} \cdot \mathbf{e}^\gamma = M^{\gamma\delta}\, \Lambda_\delta \Rightarrow \Lambda^k = M^{kl}\, \Lambda_l,$$

$$\Lambda^{n+1} = 0 \text{ (by Equations 7.2.11a through d)} \tag{7.3.5b}$$

$$Q^\Gamma \equiv S\, d\mathbf{F} \cdot \mathbf{e}^\Gamma = M^{\Gamma\Delta}\, Q_\Delta \Rightarrow Q^K = M^{KL}\, Q_L,$$

$$Q^{N+1} = 0 \text{ (by Equations 7.2.11a through d)} \tag{7.3.5c}$$

$$\Lambda^\Gamma \equiv S\, d\mathbf{R} \cdot \mathbf{e}^\Gamma = M^{\Gamma\Delta}\, \Lambda_\Delta \Rightarrow \Lambda^K = M^{KL}\, \Lambda_L,$$

$$\Lambda^{N+1} = 0 \text{ (by Equations 7.2.11a through d)} \tag{7.3.5d}$$

Additional ways of relating these components will also be given in the following sections. The stage is now set for the final kinetic synthesis of all these concepts, i.e., forces with kinematics.

7.4 THE PHYSICAL SYNTHESIS: LAGRANGE'S PRINCIPLE(S), EQUATIONS OF MOTION

7.4.1 LAGRANGE'S PRINCIPLE (LP)

The starting point here is, as in most other versions of classical mechanics, the Newton–Euler law of motion, in the d'Alembert form (Equations 4.3.1 and 7.3.1):

$$dm\ \mathbf{a} = d\mathbf{F} + d\mathbf{R}. \tag{7.4.1}$$

Dotting each of Equations 7.4.1 with the corresponding particle virtual displacement $\delta\mathbf{r}$, and then summing over all the system particles, for a fixed time, we obtain

$$S(dm\ \mathbf{a} - d\mathbf{F}) \cdot \delta\mathbf{r} + S(-\ d\mathbf{R}) \cdot \delta\mathbf{r} = 0. \tag{7.4.2}$$

This (differential variational) equation does not contain anything physically new; i.e., it results from (7.4.1) by purely mathematical transformations. To make further progress toward the formulation of *reactionless*, or *kinetostatic,* equations of motion, which, as already stated (Sections 4.3 and 7.1), is one of the key objectives of AM, we, now, postulate that the total first-order virtual work of the "lost" forces $\{-d\mathbf{R} = d\mathbf{F} - dm\ \mathbf{a}\}$ vanishes:

$$-\delta' W_{(r)} \equiv S(-d\mathbf{R}) \cdot \delta\mathbf{r} = 0 \Rightarrow \delta' W_{(r)} \equiv S\, d\mathbf{R} \cdot \delta\mathbf{r} = 0. \tag{7.4.3}$$

Then Equation 7.4.2 reduces to the physically new and nontrivial kinetic *principle of Lagrange:*

$$S(dm\ \mathbf{a} - d\mathbf{F}) \cdot \delta\mathbf{r} = 0, \quad \text{or} \quad \delta I = \delta' W, \tag{7.4.4}$$

where

$$\delta I \equiv S\, dm\ \mathbf{a} \cdot \delta\mathbf{r}: \tag{7.4.4a}$$

total first-order virtual work of the *inertial* "forces" $\{-dm\ \mathbf{a}\}$,

$$\delta' W \equiv S\, d\mathbf{F} \cdot \delta\mathbf{r}: \tag{7.4.4b}$$

total first-order virtual work of the *impressed* forces $\{d\mathbf{F}\}$.*

Finally, inserting into the above the particle virtual diaplacement representations (6.3.2b) and (6.4.11): $\delta\mathbf{r} = \mathbf{e}_k\, \delta q^k = \mathbf{e}_K\, \delta\theta^K$, and recalling the definitions (7.2.18 and 7.2.23 ff) and (7.3.2a ff), we obtain their *system* expressions:

* (a) Lagrange (1764). LP is also referred to as *Principle of d'Alembert in Lagrange's Form* or, simply, *Kinetic Principle of Virtual Work.* (b) Equations 7.4.3 and 7.4.4 hold for *equality* (or *bilateral*, or *reversible*) constraints. Otherwise, i.e., for *inequality* (or *unilateral*, or *irreversible*) constraints, they must be enlarged to

$$\delta' W_{(r)} \equiv \delta I - \delta' W \geq 0 \Rightarrow \delta I \geq \delta' W \quad \text{[i.e., for statics: } \delta' W \leq 0 \text{ (Fourier, 1798)]} \tag{7.4.4c}$$

$$\delta I = I_k\,\delta q^k = I^k\,\delta q_k \qquad (\text{where: } \delta q_k = M_{kl}\,\delta q^l \Rightarrow I^k = M^{kl}\,I_l) \tag{7.4.4d}$$

$$= I_K\,\delta\theta^K = I^K\,\delta\theta_K \qquad (\text{where: } \delta\theta_K = M_{KL}\,\delta\theta^L \Rightarrow I^K = M^{KL}\,I_L) \tag{7.4.4e}$$

$$\delta' W = Q_k\,\delta q^k = Q^k\,\delta q_k \qquad (\text{where: } \delta q_k = M_{kl}\,\delta q^l \Rightarrow Q^k = M^{kl}\,Q_l) \tag{7.4.4f}$$

$$= Q_K\,\delta\theta^K = Q^K\,\delta\theta_K \qquad (\text{where: } \delta\theta_K = M_{KL}\,\delta\theta^L \Rightarrow Q^K = M^{KL}\,Q_L). \tag{7.4.4g}$$

Remarks

i. The above concepts and equations constitute the *contemporary* interpretation of d'Alembert's principle, and are due, primarily, to Heun and Hamel (early 20th century). As such, they bear practically zero resemblance to the original *workless* formulation of d'Alembert (1742–3). The latter postulated what, again in modern terms, amounts to *equilibrium* of the $\{-d\mathbf{R}\}$, not in the virtual sense of Equation 7.4.3, but in the elementary sense of zero resultant force and moment (relative to some fixed origin), i.e.,

$$\mathcal{S}(-d\mathbf{R}) = \mathcal{S}(-d\mathbf{R})_{\text{int'l}} = \mathbf{0}$$

and

$$\mathcal{S}\,\mathbf{r} \times (-d\mathbf{R}) = \mathcal{S}\,\mathbf{r} \times (-d\mathbf{R})_{\text{int'l}} = \mathbf{0}. \tag{7.4.5}*$$

ii. In LP, Equation 7.4.3, it is the *sum* $\delta' W_{(r)}$ that vanishes, and not necessarily each of its terms $d\mathbf{R} \cdot \delta\mathbf{r}$, separately; although the latter may happen in special cases. For example, as explained earlier (Section 7.3, footnotes), in a *free* rigid body Equation 7.4.3 reduces to $(\delta' W)_{\text{internal forces}} = 0$, although individually $d\mathbf{R} \cdot \delta\mathbf{r} \to d\mathbf{f}_{\text{int'l}} \cdot \delta\mathbf{r}$ may not vanish. Also, while the $\{dm\,\mathbf{a}\}$ are present wherever a particle is accelerated, the $\{d\mathbf{F}\}$ may act only on a few system particles.

iii. In general, neither $\delta' W_{(r)}$ nor $\delta' W$ are *exact* (or perfect) differentials of some "work" or "force functions" $W_{(r)}$, W, respectively; i.e., they are *quasi-variables*; hence, the special notation $\delta'(\ldots)$. The same holds for δI and the $\delta\theta$, but for them, for convenience, we will make an exception and leave their δ term*s* unaccented.

iv. It must be stressed that LP, Equations 7.4.3 and 7.4.4, is what is known in continuum mechanics as **constitutive** postulates for the *nonphysical* part of $\{d\mathbf{f}\}$, i.e., the reactions $\{d\mathbf{R}\}$; just like Hooke's law in linear elasticity, or the Navier–Stokes law in viscous fluid mechanics.** As such, LP is not a law of nature, that is, it is not on the same level with the

* Let the reader show that for a rigid body, i.e., $\delta\mathbf{r} = \delta\mathbf{r}_p + \delta\boldsymbol{\chi} \times (\mathbf{r} - \mathbf{r}_p)$ [where, recalling Example 6.5.1, p: arbitrary body-fixed point (pole), $\delta\boldsymbol{\chi}$: virtual rotation of body, spatially constant ($\boldsymbol{\chi}$: generally, a quasi-vector), the single but variational *work* Equation 7.4.3 specializes to the two force/moment Equations 7.4.5.

** (See following page for footnote.)

Newton–Euler law $dm\ \mathbf{a} = d\mathbf{f}$ (or the principles of linear and angular momentum, in their various discrete or continuous forms), but it is subservient to them; and just as in continuum mechanics where not all parts of the stress need be elastic, here in AM, too: not all $d\mathbf{R}$s need satisfy LP. Those that do (this book) we call *ideal*; but more general reactions can be handled with proper modifications. In view of the above, the frequently heard expressions "workless, or nonworking, constraints" must be replaced with the more precise *virtually workless constraints* — in general, the reactions *are* working, even when they are virtually nonworking (LP), and this is why the concept of virtualness was invented in AM! Let us see this: the elementary (first-order) work of the reactions under general *kinematically admissible* displacements is (recalling the representation 6.3.2a):

$$d'W_{(r)} \equiv S\, d\mathbf{R} \cdot d\mathbf{r} = S\, d\mathbf{R} \cdot (\mathbf{e}_k\, dq^k + \mathbf{e}_{n+1}\, dt)$$

$$= (d'W_{(r)})' + (d'W_{(r)})'', \tag{7.4.6}$$

where

$$(d'W_{(r)})' \equiv \left(S\, d\mathbf{R} \cdot \mathbf{e}_k \right) dq^k \equiv \Lambda_k\, dq^k\ (\sim dq), \tag{7.4.6a}$$

$$(d'W_{(r)})'' \equiv \left(S\, d\mathbf{R} \cdot \mathbf{e}_{n+1} \right) dq^{n+1} \equiv \Lambda_{n+1}\, dq^{n+1} \equiv \Lambda_0\, dt\ (\sim dt), \tag{7.4.6b}$$

while under equally general *virtual* displacements, the corresponding work is

$$\delta' W_{(r)} \equiv S\, d\mathbf{R} \cdot \delta\mathbf{r} = S\, d\mathbf{R} \cdot (\mathbf{e}_k\, \delta q^k) \equiv \Lambda_k\, \delta q^k\, (\sim \delta q); \tag{7.4.6c}$$

and so, since $(d'W_{(r)})'$ and $\delta' W_{(r)}$ are *mathematically equivalent* $(dq \sim \delta q)$,

$$\delta' W_{(r)} = 0 \Rightarrow (d'W_{(r)})' = 0$$

$$\Rightarrow d'W_{(r)} = (d'W_{(r)})'' = \left[S\, d\mathbf{R} \cdot (\partial \mathbf{r}/\partial t) \right] dt \neq 0 \quad \text{(in general)}. \tag{7.4.6d}$$

In sum, AM is both mathematically and physically different from the mechanics of Newton–Euler. Schematically,

---

** And like the latter, LP is indispensable to the formulation of *dynamically determinate* methods which generate as many equations as needed for the unambiguous determination of the relevant unknowns as functions of time. For additional reasons for the absolute necessity of LP, see, e.g., Gantmacher (1970, pp. 16–23).

Lagrangean AM = Newton–Euler Law + Lagrange's Physical Postulate. (7.4.7)

7.4.2 Principle of Relaxation of Constraints

LP, (7.4.3 and 7.4.4) gets rid of the constraint forces and thus allows us to obtain *reactionless*, or purely *kinetic*, equations; that is, equations which (along with given initial/boundary conditions) determine the *motion* of the system. If, next, we want to calculate, or retrieve, some or all of these reactions, we have two ways to proceed: either

a. invoke the Newton–Euler laws [e.g., $q(t) \rightarrow \mathbf{r}(t) \rightarrow \mathbf{v}(t)$, $\mathbf{a}(t) \rightarrow d\mathbf{R} = dm\,\mathbf{a}(t) - d\mathbf{F}[t, \mathbf{r}(t), \mathbf{v}(t)] = d\mathbf{R}(t)$]; that is, use Lagrange for the motion and Newton–Euler for the reactions, i.e., apply a conceptually mixed, or nonuniform, procedure; or
b. Use Lagrange, *appropriately modified*, for *both* motion and reactions, i.e., apply a conceptually uniform method for both types of unknowns.

This modification has come to be known as the *principle of relaxation of the constraints* (PRC).* It consists in the following: to find certain reactions, we deliberately violate the corresponding constraints, i.e., *we assume that these formerly rigid (or inextensible) constraints become a little flexible, and thus add to $\delta'W$ the virtual work of the associated reactions as if they were impressed forces;* and similarly for δI. Finally, we enforce the named constraints in both the old (kinetic) equations as well as the new ones, i.e., those obtained by the relaxation, which contain the sought reactions and are called *kinetostatic*. In practice this amounts to utilizing the well-known method of *Lagrangean multipliers.* Hence, by applying both LP and PRC (Equations 7.4.3 and 7.4.4, plus multipliers), we succeed in creating a *dynamically determinate* system of equations, i.e., *kinetic + kinetostatic + constraints*, whose solution yields the motion and the reactions; and this general methodology solves, in principle, the problem of constrained system mechanics.** Let us, next, obtain the *system* forms of these equations by applying these principles in terms of the earlier system expressions for the inertial, impressed, and constraint forces (Sections 7.2 and 7.3).

7.4.3 Lagrangean Forms of the Equations of Motion

7.4.3.1 Holonomic Variables

We begin with the basic variational, Equations 7.4.3 and 7.4.4, expressed in terms of the holonomic *system* variables, Equations 7.4.4d and f and 7.4.6c:

---

* This second, and final, general principle of AM is implicit in Lagrange's work (1788), but its full significance was brought to the fore by Hamel (*Befreiungsprinzip,* 1916). The terminology PRC seems to be due to Lawden (1972, p. 54); also Bahar (unpublished Dynamics notes, 1970s).

** As will become clear soon, in holonomic variables that system is, in general, *coupled.* To uncouple it, we choose appropriate *nonholonomic* variables. Then, solving its kinetic equations we find the motion, and, next, substituting the latter in its kinetostatic equations we get the reactions.

$$\delta' W_{(r)} = 0: \quad \Lambda_\alpha\, \delta q^\alpha = \Lambda_k\, \delta q^k = 0 \quad (\text{since } \delta q^{n+1} \equiv \delta t = 0), \tag{7.4.8a}$$

$$\delta I = \delta' W: \quad I_k\, \delta q^k = E_k(T)\, \delta q^k = Q_k\, \delta q^k, \quad \text{or} \quad [E_k(T) - Q_k]\delta q_k = 0, \tag{7.4.8b}$$

All possible equations of motion, in holonomic variables, flow out of Equations 7.4.8a and b. Indeed,

i. If the n δq are *unconstrained* (or free, or unrestricted), the above yield immediately the famous *equations of Lagrange* (1780):

$$\text{Kinetostatic: } \Lambda_k = 0, \tag{7.4.9a}$$

$$\text{Kinetic: } I_k \equiv E_k(T) \equiv d/dt(\partial T/\partial V^k) - \partial T/\partial q^k = Q_k \quad (V^k \equiv dq^k/dt). \tag{7.4.9b}$$

Equation 7.4.9b, among the most acclaimed one in all of theoretical physics, constitutes a set of n second-order, generally nonlinear and coupled, (ordinary) differential equations in the n functions $q^k(t)$; the $Q_k = Q_k[t, q^h, dq^h/dt \equiv V^h]$ are assumed given.

ii. If, however, the n δq are *constrained* by Equations 6.6.9a: $\delta\theta^D \equiv A^D{}_k\, \delta q^k = 0$, then we proceed in one of the following *two* ways: either we

- *Adjoin* them to Equations 7.4.8a and b via the *multiplier rule*,* thus obtaining the *equations of Routh–Voss* (1877 to 1885):

$$\Lambda_k = \lambda_D\, A^D{}_k\,, \quad E_k(T) = Q_k + \Lambda_k, \tag{7.4.10a}$$

or, in extenso,

$$E_k(T) \equiv d/dt(\partial T/\partial V^k) - \partial T/\partial q^k = Q_k + \lambda_D\, A^D{}_k, \tag{7.4.10b}$$

where

$$\lambda_D = \lambda_D(t)\text{: Lagrangean multipliers;} \tag{7.4.10c**}$$

and which, along with the constraints, (6.6.6a): $A^D{}_k\, V^k + A^D{}_{n+1} = 0$, constitute a determinate but *coupled* system of $n + m$ equations for the $n + m$ unknown functions $q^k(t)$, $\lambda_D(t)$; or we

---

* We are reminded that this well-known theorem states that (with $M_k \equiv I_k - Q_k$): The variational equation $M_k\, \delta q^k = 0$, under the m ($< n$) independent Pfaffian constraints $A^D{}_k\, \delta q^k = 0$ (rank($A^D{}_k$) = m), is completely equivalent to the unconstrained variational equation: $(M_k - \lambda_D A^D{}_k)\delta q^k = 0$; which, then, leads immediately to the n equations: $M_k = \lambda_D\, A^D{}_k$, λ_D: Lagrangean multipliers. (The theorem holds whether the q^k are holonomic coordinates or not.) For a proof, see books on calculus (constrained extrema of functions) or mechanics; e.g., Gantmacher (1970, pp. 20–23), Hamel (1949, pp. 85–91), Rosenberg (1977, pp. 132, 212–214).

** Routh was first (1877), but Voss' study (1885) is far more extensive and deeper; in fact, it is the first systematic study of nonholonomic systems.

- *Embed* them into Equations 7.4.8a and b via the method of *quasi-coordinates*, i.e., substitute into them, à la Maggi (recalling Equations 6.6.9c; 6.7.5a, b) δq^k as a linear and homogeneous combination of *n*-*m independent* (i.e., *unconstrained*) *parameters* ε^I: $\delta q^k = P^k{}_I\, \varepsilon^I$, or à la Hamel: $\delta q^k = A^k{}_I\, \delta\theta^I$:

$$\delta' W_{(r)} = 0: \quad \Lambda_k\, \delta q^k = (\lambda_D\, A^D{}_k)(A^k{}_I\, \delta\theta^I) = (\lambda_D\, \delta^D{}_I)\delta\theta^I$$

$$= (0)\delta\theta^I = \Lambda_I\, \delta\theta^I = 0, \qquad (7.4.11a)$$

$$\delta I = \delta' W: \qquad [E_k(T) - Q_k]\,(A^k{}_I\, \delta\theta^I) = 0, \qquad (7.4.11b)$$

from which, since the n–m $\delta\theta^I$ are independent, the following equations result:

$$\Lambda_I = 0, \quad A^k{}_I\, E_k(T) = A^k{}_I\, Q_k\, (= Q_I), \qquad (7.4.12a)$$

or, *in extenso,*

$$[d/dt(\partial T/\partial V^k) - \partial T/\partial q^k]A^k{}_I = A^k{}_I\, Q_k. \qquad (7.4.12b)$$

The n–m kinetic equations (7.4.12b) are called *equations of Maggi* (1896, 1901, 1903); and along with the m constraints (Equations 6.6.6a): $A^D{}_k\, V^k + A^D{}_{n+1} = 0$, constitute a determinate system for the n unknown functions $q^k(t)$.
- To get the kinetostatic counterpart of the above, we abandon embedding (i.e., $\delta\theta^D = 0$) and, instead, employ *adjoining but in quasi-coordinates*: first, we represent δq^k as

$$\delta q^k = A^k{}_I\, \delta\theta^I + A^k{}_D\, \delta\theta^D,$$

but under the constraints

$$\delta\theta^D = 0 \text{ or } (1)\delta\theta^D = 0, \qquad (7.4.13a)$$

then substitute Equation 7.4.13a$_1$ into Equations 7.4.8a and b:

$$\delta' W_{(r)} = 0: \quad \Lambda_k\, \delta q^k = (\lambda_{D'}\, A^{D'}{}_k)(A^k{}_D\, \delta\theta^D + A^k{}_I\, \delta\theta^I)$$

$$= (\lambda_{D'}\, \delta^{D'}{}_D)\delta\theta^D + (\lambda_{D'}\, \delta^{D'}{}_I)\delta\theta^I = \lambda_D\, \delta\theta^D + (0)\delta\theta^I$$

$$= 0 \qquad (7.4.13b)$$

(= $\Lambda_D\, \delta\theta^D + \Lambda_I\, \delta\theta^I$, by invariance, as in Equation 7.4.11a)

$$\delta I = \delta' W: \qquad [E_k(T) - Q_k]\,(A^k{}_D\,\delta\theta^D + A^k{}_I\,\delta\theta^I)$$

$$= \{[E_k(T) - Q_k]\,A^k{}_D\}\delta\theta^D$$

$$+ \{[E_k(T) - Q_k]\,A^k{}_I\}\delta\theta^I = 0, \tag{7.4.13c}$$

and finally adjoin Equation 7.4.13a$_2$ to Equations 7.4.13b and c via multipliers μ_k, thus obtaining

$$\lambda_D + \mu_D = 0 \Rightarrow \lambda_D = \Lambda_D = -\mu_D, \quad 0 + \mu_I = 0 \Rightarrow \lambda_I = \Lambda_I = -\mu_I = 0;$$

and the two *uncoupled* sets of *equations of Maggi*:

$$[E_k(T) - Q_k]\,A^k{}_D + \mu_D = 0 \Rightarrow [E_k(T) - Q_k]\,A^k{}_D = \Lambda_D, \tag{7.4.13d}$$

$$[E_k(T) - Q_k]\,A^k{}_I + \mu_I = 0 \Rightarrow [E_k(T) - Q_k]\,A^k{}_I = 0; \tag{7.4.13e}$$

or, *in extenso*,

$$\text{Kinetostatic:} \quad [d/dt(\partial T/\partial V^k) - \partial T/\partial q^k]A^k{}_D = A^k{}_D\,Q_k + \Lambda_D, \tag{7.4.14a}$$

$$\text{Kinetic:} \quad [d/dt(\partial T/\partial V^k) - \partial T/\partial q^k]A^k{}_I = A^k{}_I\,Q_k. \tag{7.4.14b}^*$$

The uncoupling exhibited in Equations 7.4.13d and e and 7.4.14a and b, absent from Equation 7.4.10b, is the main advantage of our special choice of quasi-variables ["equilibrium form" (Equations 6.6.6a and b): $\delta\theta^D \equiv A^D{}_k\,\delta q^k = 0$, etc.]; other choices would not, in general, have resulted in such a neat separation of the equations of motion. In view of it, *first* we solve the $n - m$ second-order (generally coupled) differential equations (7.4.12b, 7.4.14b), plus constraints, and thus obtain the motion; and *then*, we calculate the *constraint reactions* from the remaining *m algebraic* equation (7.4.14a):

$$\Lambda_D = [d/dt(\partial T/\partial V^k) - \partial T/\partial q^k]A^k{}_D - A^k{}_D\,Q_k = \ldots = \Lambda_D(t). \tag{7.4.14c}$$

Clearly, the first task (finding the motion) constitutes the lion's share of the solution of the problem.

7.4.3.2 Nonholonomic Variables

We begin with the basic variational Equations 7.4.3 and 7.4.4, expressed in terms of the nonholonomic *system* variables (7.4.4e and g) and (7.4.11a):

---

* Actually, Maggi supplied only the kinetic Equations (7.4.12b, 7.4.13e, 7.4.14b). The kinetostatic Equations (7.4.13d, 7.4.14a), and associated unified derivation, seem to have been indicated by this author (late 1980s).

$$\delta' W_{(r)} = 0: \quad \Lambda_\Sigma\, \delta\theta^\Sigma = \Lambda_K\, \delta\theta^K = 0 \quad (\text{since } \delta\theta^{N+1} \equiv \delta t = 0), \tag{7.4.15a}$$

$$\delta I = \delta' W: \quad I_K\, \delta\theta^K \equiv [E_K(T^*) - \Gamma_K]\delta\theta^K = Q_K\, \delta\theta^K; \tag{7.4.15b}$$

$$[-\Gamma_K \equiv \gamma^L{}_{K\Sigma}(\partial T^*/\partial\omega^L)\omega^\Sigma \equiv h^L{}_K(\partial T^*/\partial\omega^L)].$$

Here, too, *all possible equations of motion, in nonholonomic variables, flow out of Equations 7.4.15a and b.* Indeed,

i. If the n $\delta\theta$ are *unconstrained*, the above lead at once to the generalizations of Equations 7.4.9a and b to nonholonomic variables:

$$\text{Kinetostatic: } \Lambda_K = 0, \tag{7.4.16a}$$

$$\text{Kinetic:} \quad I_K \equiv E_K(T^*) - \Gamma_K$$

$$\equiv d/dt(\partial T^*/\partial\omega^K) - \partial T^*/\partial\theta^K - \Gamma_K = Q_K. \tag{7.4.16b}*$$

ii. If the (m constraints are arranged so that the) *first* m $\delta\theta$, $\delta\theta^D$, are *constrained* by

$$\delta\theta^D = 0 \quad \text{or} \quad (1)\delta\theta^D = 0 \quad [D = 1, \ldots, m\ (< n)], \tag{7.4.17a}$$

while the *remaining* $n - m$ $\delta\theta$, $\delta\theta^I$, although unconstrained (i.e., $\delta\theta^I \neq 0$), are expressed *as if* they were also constrained by:

$$(0)\delta\theta^I = 0, \tag{7.4.17b}$$

then *adjoining* them to Equations 7.4.15a and b via the multiplier rule again (the constraint embedding being expressed by Equations 7.4.17a), i.e., proceeding as in Equations 7.4.13a through e, yields the two *uncoupled* sets of equations:

$$\text{Kinetostatic:} \quad I_D = Q_D + \Lambda_D \quad [\delta\theta^D = 0 \Rightarrow \Lambda_D \neq 0 \quad (\text{Stückler, 1955})] \tag{7.4.18a}$$

$$\text{Kinetic:} \quad I_I = Q_I \quad [\delta\theta^I \neq 0 \Rightarrow \Lambda_I = 0 \quad (\text{Hamel, 1903–4})]; \tag{7.4.18b}$$

or, explicitly,

$$\text{Kinetostatic:} \quad E_D(T^*) - \Gamma_D \equiv d/dt(\partial T^*/\partial\omega^D) - \partial T^*/\partial\theta^D - \Gamma_D$$

$$= Q_D + \Lambda_D, \tag{7.4.19a}$$

---

* Equations 7.4.16b, but *without any constraint considerations,* seem to be due to Volterra (1898).

Kinetic: $$E_I(T^*) - \Gamma_I \equiv d/dt(\partial T^*/\partial \omega^I) - \partial T^*/\partial \theta^I - \Gamma_I = Q_I. \quad (7.4.19b)^*$$

We notice that, physically, *Equations 7.4.18a and b and 7.4.19a and b are none other than Equations 7.4.14a and b, respectively,* but, while the latter (Maggi) are in holonomic variables, the former (Hamel) are in nonholonomic variables: as there, the decoupling of the equations of motion into kinetic and kinetostatic results from our particular choice of quasi-coordinates, and, because of it, the solution of the problem proceeds as there: first, we solve the $n - m$ first-order (generally coupled) *differential* Equations 7.4.19b, plus appropriate initial / temporal boundary conditions, and obtain the *motion*:

$$\omega_I = \omega_I(t) \Rightarrow dq^k/dt = dq^k/dt[t, q, \omega^I(t)] \equiv f^k(t) \Rightarrow q^k = q^k(t); \quad (7.4.19c)$$

and, then, we calculate the *reactions* from the remaining m *algebraic* Equations 7.4.19a:

$$\Lambda_D = I_D - Q_D = \ldots = \Lambda_D(t). \quad (7.4.19d)$$

Remarks

i. From the invariant equations (7.4.3, 7.4.8a, and 7.4.11a):

$$\delta' W_{(r)} = \Lambda_k\, \delta q^k = \Lambda_K\, \delta\theta^K \quad [= \Lambda_D(0) + (0)\delta\theta^I = 0], \quad (7.4.20a)$$

we readily obtain the earlier *covariant* vector-like transformation equation (7.3.3b):

$$\Lambda_k (A^k{}_K\, \delta\theta^K) = \Lambda_K\, \delta\theta^K \Rightarrow \Lambda_K = A^k{}_K\, \Lambda_k$$

$$[= (\partial V^k/\partial\omega^K)\Lambda_k, \text{ generally}]; \quad (7.4.20b)$$

and further, by Equation 7.4.10a$_1$,

$$\Lambda_K = A^k{}_K (\lambda_D\, A^D{}_k) = (\delta^D{}_K)\lambda_D = \lambda_K, \quad (7.4.20c)$$

$$\Lambda_k\, \delta q^k = \Lambda_K (A^K{}_k\, \delta q^k)$$

$$\Rightarrow \Lambda_k = A^K{}_k\, \Lambda_K\ [= (\partial\omega^K/\partial V^k)\Lambda_K, \text{ generally}]; \quad (7.4.20d)$$

---

* (a) Hamel called *his* equations (7.4.18b and 7.4.19b), gracefully, the *Lagrange–Euler* equations, because: if the θ^Ks are *holonomic* coordinates, the Γ terms vanish and these equations reduce to the Lagrangean equations (7.4.9b); while if they are *nonholonomic,* they constitute a far-reaching generalization of the well-known Eulerian equations of rotational rigid-body kinetics. (b) A special case of Equations 7.4.19a is given in Schouten (1929, 1954b, pp. 196–197); but Stückler's treatment, based on Hamel (1949, p. 480ff), is far more relevant, readable, and complete.

and similarly for the transformation relations between Q_k, Q_K (Equations 7.3.3a) and I_k, I_K (Equations 7.2.29a and following).* Also, Equation 7.4.20c, and the earlier Equations 7.4.11a and 7.4.13b, show that

Our Lagrangean multipliers, Equations 7.4.10a, are nothing but the covariant nonholonomic components of the system constraint reaction: $\lambda_D = \Lambda_D$;

and this provides additional geometrical insight into the PRC.

ii. All the above results are interrelated. Indeed,

- Dotting Equation 7.4.10b with $A^k{}_{D'}$ and $A^k{}_I$ yields, respectively, Maggi's equations (7.4.13d and 7.4.14a), (7.4.13e and 7.4.14b):

$$A^k{}_{D'}\, E_k(T) = A^k{}_{D'}\, Q_k + \lambda_D(A^D{}_k\, A^k{}_{D'})$$

$$= A^k{}_{D'}\, Q_k + \lambda_D(\delta^D{}_{D'}) = A^k{}_{D'}\, Q_k + \lambda_{D'}, \tag{7.4.21a}$$

$$A^k{}_I\, E_k(T) = A^k{}_I\, Q_k + \lambda_D(A^D{}_k\, A^k{}_I)$$

$$= A^k{}_I\, Q_k + \lambda_D(\delta^D{}_I) = A^k{}_{D'}\, Q_k + 0; \tag{7.4.21b}$$

while multiplying Equations 7.4.10a with $A^k{}_{D'}$ and $A^k{}_I$ yields, respectively, Equations 7.4.20c and 7.4.12a$_1$:

$$A^k{}_{D'}\, \Lambda_k = \lambda_D(A^k{}_{D'}\, A^D{}_k) = \lambda_D(\delta^D{}_{D'}) = \lambda_{D'} = \Lambda_{D'}, \tag{7.4.22a}$$

$$A^k{}_I\, \Lambda_k = \lambda_D(A^k{}_I\, A^D{}_k) = \lambda_D(\delta^D{}_I) = 0 = \Lambda_I. \tag{7.4.22b}$$

iii. Finally, it should be noted that Maggi's *method*, i.e., expressing δq^k, *or* $\delta\mathbf{r}$, as a linear and homogeneous combination of $n - m$ independent parameters, and then substituting it into LP, etc., thus obtaining $n - m$ kinetic equations, is not restricted to holonomic variables. Thus, if the hitherto unconstrained $n' \equiv n - m$ $\delta\theta^I$ are *later constrained* by the m' ($< n'$) constraints: $\delta\theta^{D'} \equiv A^{D'}{}_I\, \delta\theta^I = 0$ ($D' = 1, \ldots, m'$; $I = m + 1, \ldots, n$), then setting à la Maggi: $\delta\theta^I = A^I{}_{I'}\, \delta\eta^{I'}$ ($I' = m' + 1, \ldots, n'$) into LP for the formelrly unconstrained $n - m$ nonholonomic variables $\delta\theta^I$, ω^I: $I_I\, \delta\theta^I = Q_I\, \delta\theta^I$, and applying adjoining (i.e., PRC $\rightarrow$ multipliers) we obtain the new *$n' - m'$ kinetic Hamel-like equations in Maggi's form*: $A^I{}_{I'}\,(I_I - Q_I) = 0$;

* Further, with the help of (7.3.5d), we easily obtain

$$\Lambda^I = M^{IK}\, \Lambda_K = M^{ID}\, \Lambda_D + M^{II'}\, \Lambda_{I'} = M^{ID}\, \Lambda_D + M^{II'}\,(0) = M^{ID}\, \Lambda_D\ (\neq 0), \tag{7.4.20e}$$

$$\Lambda^D = M^{DK}\, \Lambda_K = M^{DD'}\, \Lambda_{D'} + M^{DI}\, \Lambda_I = M^{DD'}\, \Lambda_{D'} + M^{DI}\,(0) = M^{DD'}\, \Lambda_{D'}; \tag{7.4.20f}$$

i.e., $\Lambda^I \neq 0$, even though $\Lambda_I = 0$.

and similarly for the remaining m' kinetostatic equations: $A^I{}_{D'}(I_I - Q_I) = \Lambda_{D'}$ ($= \lambda_{D'}$). Further, applying embedding, we can express the above in *new quasi-variables*: $\omega^{I'} \equiv d\theta^{I'}/dt = A^{I'}{}_I\, \omega^I + A^{I'}{}_{N+1} \neq 0$, etc., since $\omega^{D'} = 0$, and thus express the above Hamel–Maggi kinetic equations in terms of them: $I_{I'} = Q_{I'}$; and similarly for the associated kinetostatic equations: $I_{D'} = Q_{D'} + \Lambda_{D'}$; and so on, for still additional constraints on the $\omega^{I'}$. Later we give a geometrical interpretation of Maggi's method as *projections* of $I_k = Q_k + \Lambda_k$ on the earlier virtual space (kinetic equations) and constraint space (kinetostatic equations) (Section 6.8, Figure 6.2).

7.4.4 Appellian Forms of the Equations of Motion

With the definition:

$$2S \equiv S\, dm\ \mathbf{a}\cdot\mathbf{a} = S\, dm\ a^2 = S\, dm\ \mathbf{a}^*\cdot\mathbf{a}^* \equiv 2S^*: \qquad (7.4.23a)$$

(inertial) *Gibbs–Appell* function of system; or, simply, *Appellian,* and recalling Equations 6.3.5a and 6.5.7a, we obtain the following useful kinematico-inertial identities:

$$\partial S/\partial(dV^k/dt) = S\, dm\ \mathbf{a}\cdot[\partial\mathbf{a}/\partial(dV^k/dt)]$$

$$= S\, dm\ \mathbf{a}\cdot\mathbf{e}_k = I_k, \qquad (7.4.23b)$$

$$\partial S^*/\partial(d\omega^K/dt) = S\, dm\ \mathbf{a}^*\cdot[\partial\mathbf{a}^*/\partial(d\omega^K/dt)]$$

$$= S\, dm\ \mathbf{a}\cdot\mathbf{e}_K = I_K; \qquad (7.4.23c)$$

i.e., the Appellian expression for the inertia force has the same *form* in both holonomic and nonholonomic variables. As a result of the above, in all preceding equations of motion, we can replace I_k with $\partial S/\partial(dV^k/dt)$ and I_K with $\partial S^*/\partial(d\omega^K/dt)$. The so-resulting equations from the kinetic (7.4.16b), i.e., $\partial S^*/\partial(d\omega^K/dt) = Q_K$, are due to *Gibbs* (1879; no constraints, but general quasi-velocities).*

7.4.5 Summary

Let us record all these basic types and forms of the equations of motion, in system variables:

---

* Other forms of I_k, I_K, due to Nielsen (1935), Tsenov (1950s), Mangeron–Deleanu (1950s) et al., are available, but they will not be examined here. See, e.g., Papastavridis, *Analytical Mechanics,* in press.

Motion and Reactions Coupled:

$$I_k = Q_k + \Lambda_k, \quad \text{(Routh-Voss)} \tag{7.4.24}$$

$$I_k \equiv S\, dm\, \mathbf{a} \cdot \mathbf{e}_k = E_k(T)$$

$$\equiv d/dt(\partial T/\partial V^k) - \partial T/\partial q^k = \partial S/\partial(dV^k/dt), \tag{7.4.24a}$$

$$\Lambda_k = \lambda_D\, A^D{}_k. \tag{7.4.24b}$$

Motion and Reactions Uncoupled:

Kinetic: $$I_I = Q_I; \tag{7.4.25a}$$

Kinetostatic: $$I_D = Q_D + \Lambda_D; \tag{7.4.25b}$$

In *Holonomic* Variables: $$I_K = A^k{}_K\, I_k, \quad Q_K = A^k{}_K\, Q_k \qquad \text{(Maggi)}; \tag{7.4.26a}$$

In *Nonholonomic* Variables: $$I_K \equiv S\, dm\, \mathbf{a}^* \cdot \mathbf{e}_K = E_K(T^*) - \Gamma_K \tag{7.4.26b}$$

$$= d/dt(\partial T^*/\partial \omega^K) - \partial T^*/\partial \theta^K$$

$$+ \gamma^L{}_{K\Lambda}(\partial T^*/\partial \omega^L)\omega^\Lambda \qquad \text{(Hamel)} \tag{7.4.26c}$$

$$= \partial S^*/\partial(d\omega^K/dt) \qquad \text{(Appell).} \tag{7.4.26d}$$

In conclusion, we have the following *three* basic types of equations of motion:

- **Lagrange/Routh–Voss** (holonomic variables; motion and reactions coupled)
- **Maggi** (holonomic variables; motion and reactions uncoupled)
- **Hamel** (nonholonomic variables; motion and reactions uncoupled);

and, *for each one of them, there exists an Appellian version,* due to the identities (7.4.23b and c).*

* The differences in *variables* and *form*, among these (and other) types of equations of motion, result not only in computational differences among them, but also in theoretical differences: *these equations are form invariant under different groups of transformations.* Specifically, the equations of: **I. Newton–Euler, II. Lagrange/Routh–Voss, III. Hamel,** and **IV. Hamilton** (not examined here) are form invariant under the following groups, respectively, **I. Galilean** (inertial frame and all other frames in accelerationless motion relative to it), **II. Point transformations in configuration space** $(t, q \to t' = t, q' = q'(t, q), T \to T' = T$. It is not hard to show that: $E_{k'}(T') = (\partial q^k/\partial q^{k'})E_k(T)$; also, recall Problem 7.2.1), **III. Linear velocity transformations on local affine tangent plane** $(\omega^K \to \omega^{K'} = A^{K'}{}_K(t, q)\, \omega^K + A^{K'}{}_{N+1}(t, q))$, **IV. Canonical transformations** (see texts on Hamiltonian mechanics). Such invariance-based classifications (initiated by F. Klein) are among the basic *ordering principles* of modern physics.

Remarks

Let $T^*(t, q, \omega_D = 0, \omega_I) = T^*_o(t, q, \omega_I) = T^*_o$: *constrained* T^*, and, generally: $(\ldots)_o$ or $\ldots|_o \equiv (\ldots)_{\text{constraints enforced}}$. Then, by Taylor's theorem, we have

$$T^* = T^*(t, q, \omega_D, \omega_I) = T^*_o + (\partial T^*/\partial \omega^D)_o\, \omega^D$$

$$+ \text{ higher order terms in the } \omega^D, \qquad (7.4.27a)$$

and, therefore,

$$(\partial T^*/\partial \omega^I)_o = \partial T^*_o/\partial \omega^I; \qquad (7.4.27b)$$

similarly,

$$(\partial T^*/\partial \theta^I)_o \equiv (\partial T^*/\partial q^k)_o\,(\partial V^k/\partial \omega^I)_o = A^k{}_I(\partial T^*/\partial q^k)_o = \partial T^*_o/\partial \theta^I, \quad (7.4.27c)$$

and so we can replace in the *first* and *second* terms of the kinetic Equations 7.4.25a, with (7.4.26b, c), T^* with T^*_o [$\Rightarrow E_I(T^*) = E_I(T^*_o)$]; but not in the *third* term $-\Gamma_I$. Indeed, the latter becomes

$$(-\Gamma_I)_o \equiv + \left[\gamma^L{}_{I\Lambda}(\partial T^*/\partial \omega^L)\omega^\Lambda\right]_o = (\gamma^L{}_{II'}\,\omega^{I'} + \gamma^L{}_I)(\partial T^*/\partial \omega^L)_o; \qquad (7.4.27d)$$

i.e., in there, L runs from 1 to n, and so we must set $\omega^D = 0$ in it *after* all $\partial(\ldots)/\partial\omega^I$ differentiations have been carried out. In conclusion, in the Hamel formalism, even for kinetic problems, T^* must be expressed as function of *all* the ω terms; and this is a (minor) drawback of that methodology.

As a result of the above, the kinetic Hamel equations, i.e., (7.4.25a) with (7.4.26b, c), become

$$d/dt(\partial T^*_o/\partial \omega^I) - \partial T^*_o/\partial \theta^I + (\gamma^L{}_{II'}\,\omega^{I'} + \gamma^L{}_I)(\partial T^*/\partial \omega^L)_o = (Q_I)_o. \quad (7.4.27e)$$

On the other hand, in the Appellian formalism, since (again expanding S^* à la Taylor, etc.):

$$\left[\partial S^*/\partial(d\omega^I/dt)\right]_o = \partial S^*_o/\partial(d\omega^I/dt), \qquad (7.4.28a)$$

where

$$S^*(t, q, \omega, d\omega/dt)|_o = S^*_o(t, q, \omega^I, d\omega^I/dt) = S^*_o, \qquad (7.4.28b)$$

i.e., *we may enforce the constraints* $\omega^D = 0$, $d\omega^D/dt = 0$ *into* S^* *right from the start,* and, therefore, we can rewrite the kinetic Appell equations, i.e., 7.4.25a with 7.4.26d, as

$$\partial S^*_o/\partial(d\omega^I/dt) = (Q_I)_o. \tag{7.4.28c)*}$$

Similarly, we find that the kinetostatic Hamel and Appell counterparts of Equations 7.4.27e and 7.4.28c are

$$d/dt(\partial T^*/\partial\omega^D)_o - (\partial T^*/\partial\theta^D)_o + (\gamma^L{}_{DI}\,\omega^I + \gamma^L{}_D)\,(\partial T^*/\partial\omega^L)_o$$

$$= \left[\partial S^*/\partial(d\omega^D/dt)\right]_o = (Q_D)_o + \Lambda_D. \tag{7.4.29}$$

Example 7.4.1. Geometrical Interpretation/Vectorial Forms of the Equations of Motion

The Routh–Voss equations (7.4.10a and b) are, in vector form:

$$I_k\,\mathbf{E}^k = Q_k\,\mathbf{E}^k + \Lambda_D\,A^D{}_k\,\mathbf{E}^k, \quad \text{or} \quad \mathbf{I} = \mathbf{Q} + \Lambda, \tag{a}$$

where (Figures 6.2 and 7.1):

$$\Lambda = \Lambda_k\,\mathbf{E}^k = \Lambda_D(A^D{}_k\,\mathbf{E}^k) = \Lambda_D\,\mathbf{E}^D, \tag{a1}$$

i.e., Λ *lies on the local constraint/range space* $\boldsymbol{C_m}$ (Figure 6.2). To eliminate Λ from Equation a, we simply *project the latter on the local virtual/null space* $\boldsymbol{T_f}$, that is, we dot it with its spanning vectors $\mathbf{E}_I$:

$$\mathbf{I}\cdot\mathbf{E}_I = \mathbf{Q}\cdot\mathbf{E}_I + \Lambda\cdot\mathbf{E}_I$$

$$\Rightarrow (I_k\,\mathbf{E}^k)\cdot\mathbf{E}_I = (Q_k\,\mathbf{E}^k)\cdot\mathbf{E}_I + (\Lambda_D\,\mathbf{E}^D)\cdot\mathbf{E}_I$$

$$\Rightarrow A^k{}_I\,I_k = A^k{}_I\,Q_k + \Lambda_D\,\delta^D{}_I \quad (\text{but } \Lambda_I = 0)$$

---

* (a) Equation 7.4.28c looks simpler than Equation 7.4.27e, but calculating S^*_o is usually harder than finding T^*! (b) In view of the kinematic identities (6.5.7a): $\partial\mathbf{r}^*/\partial\theta^K = \partial\mathbf{v}^*/\partial\omega^K = \partial\mathbf{a}^*/\partial(d\omega^K/dt) = \ldots = \mathbf{e}_K$ the so-called "Kane's equations" (1966) are nothing but a raw form of the kinetic Appellian equations (7.4.28c), rewritten as:

$$S\,d\mathbf{F}\cdot(\partial\mathbf{v}^*/\partial\omega^I) + S(-dm\,\mathbf{a}^*)\cdot(\partial\mathbf{v}^*/\partial\omega^I) = 0.$$

(Actually, these equations were formulated in 1951 by the German mechaniker H. Schaefer, for the far more general case of *nonlinear* velocity constraints, in a manner similar to our Equation 7.2.22ff.) The associated arcane terminology, such as "generalized speeds" (for the contravariant nonholonomic components of the system velocity ω^K) and "partial velocities" (for the particle and system basis vectors $\mathbf{e}_k$ (holonomic) and $\mathbf{e}_K$ (nonholonomic), reflects a primitive and incomplete understanding of the general principles and mathematical structure of the equations of dynamics, and is at odds with all reputable expositions on the subject; e.g., (alphabetically): Dobronravov, Funk, Gantmacher, Hamel, Lur'e, Neimark–Fufaev, Novoselov, Prange, Schaefer, Schouten, Synge, Vranceanu, Whittaker, to name the best. Also, from a pedagogical point of view, that is a conceptually impoverished, isolated, and unmotivated scheme that soon leads its student to a dynamical dead end.

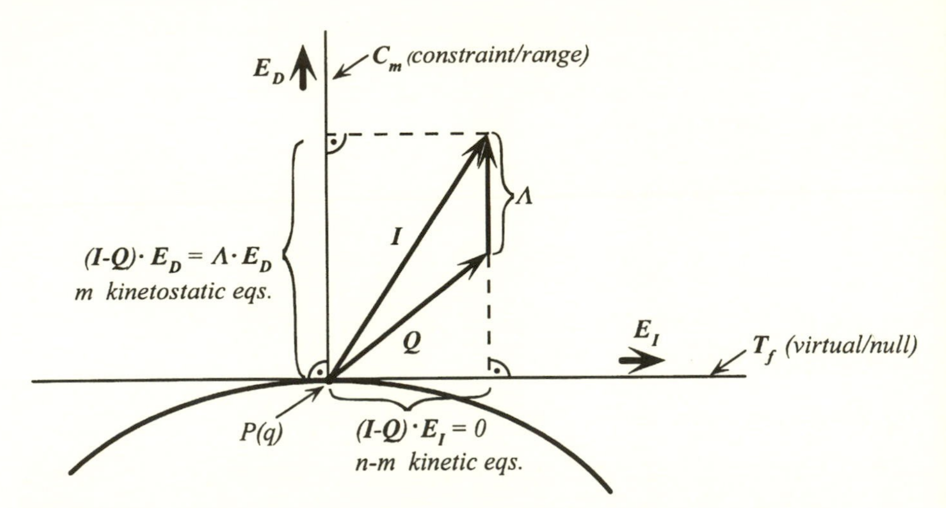

FIGURE 7.1 Geometrical interpretation of the kinetic and kinetostatic Maggi equations as projections of the Routh–Voss equations on the local virtual and constraint spaces, respectively:

n Routh–Voss Equations: $I_k\,\mathbf{E}^k = Q_k\,\mathbf{E}^k + \Lambda_D\,A^D{}_k\,\mathbf{E}^k$ or $\mathbf{I} = \mathbf{Q} + \Lambda$
$n - m$ Kinetic Maggi Equations: $\mathbf{I}\cdot\mathbf{E}_I = \mathbf{Q}\cdot\mathbf{E}_I + \Lambda\cdot\mathbf{E}_I$ or $A^k{}_I\,I_k = A^k{}_I\,Q_k$ (i.e., $\Lambda_I = 0$)
m Kinetostatic Maggi Equations: $\mathbf{I}\cdot\mathbf{E}_D = \mathbf{Q}\cdot\mathbf{E}_D + \Lambda\cdot\mathbf{E}_D$ or $A^k{}_D\,I_k = A^k{}_D\,Q_k + \Lambda_D$.

i.e.,

$$A^k{}_I\,I_k = A^k{}_I\,Q_k \quad [\textit{kinetic}\text{ Maggi equations (7.4.14b)}]. \tag{b}$$

Similarly, projecting Equation a on $\mathbf{C}_\mathrm{m}$, which is spanned by the constraint vectors $\mathbf{E}_D$, we get

$$\mathbf{I}\cdot\mathbf{E}_D = \mathbf{Q}\cdot\mathbf{E}_D + \Lambda\cdot\mathbf{E}_D$$

$$\Rightarrow (I_k\,\mathbf{E}^k)\cdot\mathbf{E}_D = (Q_k\,\mathbf{E}^k)\cdot\mathbf{E}_D + (\Lambda_{D'}\,\mathbf{E}^{D'})\cdot\mathbf{E}_D$$

$$\Rightarrow A^k{}_D\,I_k = A^k{}_D\,Q_k + \Lambda_{D'}\,\delta^{D'}{}_D \quad (D, D' = 1, \ldots, m),$$

i.e.,

$$A^k{}_D\,I_k = A^k{}_D\,Q_k + \Lambda_D \quad [\textit{kinetostatic}\text{ Maggi equations (7.4.14a)}] \tag{c}$$

[We hope that the above show the usefulness of the general tensorial notation; e.g., to get (nonholonomic ⇒ projected) covariant components from Equations a, which is also in holonomic covariant components, we dot them with $\mathbf{E}_I$, $\mathbf{E}_D$; i.e., inner multiply $I_k = Q_k + \Lambda_D\,A^D{}_k$ with $A^k{}_I$, $A^k{}_D$, respectively, not with $A^I{}_k$, $A^D{}_k$. Such important details, self-evident in the light of our indicial notation, are very difficult to catch and comprehend in the matrix/direct notation.]

Example 7.4.2. Special Cases/Forms of the Equations of Motion: Equilibrium Form of the Routh–Voss Equations

Let the additional m Pfaffian constraints be holonomic, and have the uncoupled "equilibrium form":

$$q^d = \text{constant} \equiv c^d, \quad \text{or } q^d = 0 \tag{a}$$

(i.e., holonomic counterpart of Equations 6.6.6a and 6.6.7) i.e., $f^D = \delta^D{}_d\, q^d = 0 \Rightarrow \partial f^D/\partial q^k = A^D{}_k \rightarrow \delta^D{}_k$: $A^D{}_d = \delta^D{}_d$, $A^D{}_i = 0$. Then we easily deduce that the Routh–Voss equations (7.4.10a and b) *decouple* into the following two groups:

$$\text{Kinetic:} \quad E_i(T_o) \equiv d/dt(\partial T_o/\partial V^i) - \partial T_o/\partial q^i$$

$$= (Q_i)_o \quad \left[\text{since } E_i(T)|_o = E_i(T_o)\right] \tag{b1}$$

$$\text{Kinetostatic:} \quad E_d(T)|_o \equiv d/dt(\partial T/\partial V^d)_o - (\partial T/\partial q^d)_o = (Q_d)_o + \lambda_d, \tag{b2}$$

where $d = 1, \ldots, m$; $i = m + 1, \ldots, n$. Equations b1 shows that if we are only interested in the motion, we can enforce the constraints (a) into T right from the start: $T|_o = T(t, q^i, dq^i/dt \equiv V^i) \equiv T_o$.* The explicit, covariant, form of Equations b1 and b2 is (recalling Equations 7.2.16c and noting that $dq^d/dt \equiv V^d = 0$):

$$\text{Kinetic:} \quad dV^i/dt + \Gamma^i{}_{i'i''}\, V^{i'}\, V^{i''} + 2\Gamma^i{}_{i'0}\, V^{i'} + \Gamma^i{}_{00} = Q^i, \tag{c1}$$

$$\text{Kinetostatic:} \quad \Gamma^d{}_{i'i''}\, V^{i'}\, V^{i''} + 2\Gamma^d{}_{i'0}\, V^{i'} + \Gamma^d{}_{00} = Q^d + \lambda^d$$

$$\Rightarrow \lambda^d = (\Gamma^d{}_{i'i''}\, V^{i'}\, V^{i''} + 2\Gamma^d{}_{i'0}\, V^{i'} + \Gamma^d{}_{00}) - Q^d; \tag{c2}$$

$$[d = 1, \ldots, m;\ i,\ i',\ i'' = m{+}1, \ldots, n]. \tag{c3}$$

The $n{-}m$ Equations b1 and c1 yield the motion $q^i(t)$; and, then, the m Equations b2 and c2 give the reactions $\lambda^d(t)$, if needed. For further computational details, see, e.g., Lur'e (1968, pp. 323–340).

Problem 7.4.1. Special Cases of the Routh–Voss and Maggi Equations: Equations of Hadamard (1895)–Chaplygin (1895, publ. 1897)

Show that if the Pfaffian constraints have the special form (c1 ff) of Example 6.6.1:

$$V^d = B^d{}_i(t, q)\, V^i + B^d{}_{n+1}(t, q) \equiv B^d{}_i(t, q)\, V^i + B^d(t, q), \tag{a}$$

---

* In fact, since practically all systems are brought to us somehow constrained, and then we assign to them qs, find T, etc., this is what is actually involved, although it is not acknowledged in most expositions, and our T_o, q^i are denoted as T, q, respectively.

i.e.,

$$\omega^D \equiv \delta^D{}_d(V^d - B^d{}_i\, V^i - B^d) = 0, \quad \omega^I \equiv \delta^I{}_i\, V^i \neq 0 \quad (\text{i.e., } \omega^I = V^i) \tag{b1}$$

$$V^d = \delta^d{}_D(\omega^D + B^D{}_I\, \omega^I + B^D), \quad V^i = \delta^i{}_I\, \omega^I, \tag{b2}$$

i. The Routh–Voss equations specialize to

$$E_d(T) = Q_d + \lambda_d, \quad E_i(T) = Q_i - \lambda_d\, B^d{}_i. \tag{c1, 2}$$

ii. Eliminating the λ_d among Equations c1 and c2 obtain the kinetic *Hadamard–Chaplygin* equations:

$$E_i(T) + B^d{}_i\, E_d(T) = Q_i + B^d{}_i\, Q_d. \tag{d}$$

iii. Confirm that Equations d and c1 are, respectively, the kinetic and kinetostatic specializations of the Maggi equations (7.4.14a and b) for Equations a, b1, and b2.

iv. Verify that Equation d also result if we substitute into LP: $[E_k(T) - Q_k]\,\delta q^k \equiv M_k\,\delta q^k = 0$, the virtual form of Equation a: $\delta q^d = B^d{}_i\,\delta q^i$. **Hint**: $M_k\,\delta q_k = M_d\,\delta q^d + M_i\,\delta q^i$.

Problem 7.4.2

Show that the *explicit form* of Equations d of the preceding Problem 7.4.1 is

$$M'_{ik}(dV^k/dt) + \Gamma'_{i,\alpha\beta}\, V^\alpha\, V^\beta = Q'_i, \tag{a}$$

where

$$M'_{ik} \equiv M_{ik} + B^d{}_i\, M_{dk}, \quad \Gamma'_{i,\alpha\beta} \equiv \Gamma_{i,\alpha\beta} + B^d{}_i\, \Gamma_{d,\alpha\beta}, \quad Q'_i \equiv Q_i + B^d{}_i\, Q_d \tag{b}$$

$$[d = 1, \ldots, m;\ i = m{+}1, \ldots, n;\ k = 1, \ldots, n;\ \alpha, \beta = 1, \ldots, n{+}1].$$

Remark: Recalling Equations 7.4.9b, and 7.2.16a, we may say that Equations a and b describe the *motion of an unconstrained system, but in a modified, or constrained, space with metric M'_{ik}, affinities/Christoffels $\Gamma'_{i,\alpha\beta}$ and forces Q'_i*; i.e., under such a "relativistic" interpretation, the constraints may be viewed as modifying the forces and geometry of the original (unconstrained) space. For further details, see Poliahov (1972, pp. 129–130; 1974, p. 111) and Poliahov et al. (1985, pp. 193–194).

Problem 7.4.3. Special Cases of the Hamel Equations: Equations of Voronets (1901) and Chaplygin (1895, publ. 1897)

Let the Pfaffian constraints have the special form (Equation c1 ff of Example 6.6.1):

$$V^d = B^d{}_i(t, q)V^i + B^d{}_{n+1}(t, q) \equiv B^d{}_i(t, q)V^i + B^d(t, q), \tag{a}$$

i.e.,

$$\omega^D \equiv \delta^D{}_d(V^d - B^d{}_i\, V^i - B^d) = 0, \quad \omega^I \equiv \delta^I{}_i\, V^i \neq 0 \quad (\text{i.e., } \omega^I = V^i) \tag{b1}$$

$$V^d = \delta^d{}_D(\omega^D + B^D{}_I\, \omega^I + B^D), \quad V^i = \delta^i{}_I\, \omega^I. \tag{b2}$$

i. Using the kinematical results of Example 6.6.1, verify that in this case the kinetic Hamel equations specialize to the $n - m$ kinetic equations of *Voronets*:

$$d/dt(\partial T_o/\partial V^i) - \partial T_o/\partial q^i - B^d{}_i(\partial T_o/\partial q^d)$$

$$- W^d{}_{ii'}(\partial T/\partial V^d)_o\, V^{i'} - W^d{}_i(\partial T/\partial V^d)_o = Q_i + B^d{}_i\, Q_d, \tag{c}$$

where (recalling Equations c8 ff of Example 6.6.1)

$$T = T(t, q, V^d, V^i) = T[t, q, V^d(t, q, V^i), V^i]$$

$$= T_o(t, q, V^i) = T_{o(\text{constrained kin.energy})},$$

$$\gamma^D{}_{II'} \Rightarrow -W^d{}_{ii'} = W^d{}_{i'i} \equiv \partial B^d{}_{i'}/\partial q^{(i)} - \partial B^d{}_i/\partial q^{(i')}$$

$$= [\partial B^d{}_{i'}/\partial q^i + B^{d'}{}_i(\partial B^d{}_{i'}/\partial q^{d'})] - [\partial B^d{}_i/\partial q^{i'} + B^{d'}{}_{i'}(\partial B^d{}_i/\partial q^{d'})]$$

$$= (\partial B^d{}_{i'}/\partial q^i - \partial B^d{}_i/\partial q^{i'}) + [B^{d'}{}_i(\partial B^d{}_{i'}/\partial q^{d'}) - B^{d'}{}_{i'}(\partial B^d{}_i/\partial q^{d'})]; \tag{c1}$$

$$\gamma^D{}_{I,N+1} \equiv \gamma^D{}_I \Rightarrow -W^d{}_{i,n+1} \equiv -W^d{}_i$$

$$\equiv \partial B^d{}_{n+1}/\partial q^{(i)} - \partial B^d{}_i/\partial q^{(n+1)} \equiv \partial B^d{}_{n+1}/\partial q^{(i)} - \partial B^d{}_i/\partial(t)$$

$$= [\partial B^d{}_{n+1}/\partial q^i + B^{d'}{}_i(\partial B^d{}_{n+1}/\partial q^{d'})] - [\partial B^d{}_i/\partial q^{n+1} + B^{d'}{}_{n+1}(\partial B^d{}_i/\partial q^{d'})]$$

$$= (\partial B^d{}_{n+1}/\partial q^i - \partial B^d{}_i/\partial q^{n+1}) + [B^{d'}{}_i(\partial B^d{}_{n+1}/\partial q^{d'}) - B^{d'}{}_{n+1}(\partial B^d{}_i/\partial q^{d'})]$$

$$\equiv [\partial B^d/\partial q^i + B^{d'}{}_i(\partial B^d/\partial q^{d'})] - [\partial B^d{}_i/\partial t + B^{d'}(\partial B^d{}_i/\partial q^{d'})]$$

$$= (\partial B^d/\partial q^i - \partial B^d{}_i/\partial t) + [B^{d'}{}_i(\partial B^d/\partial q^{d'}) - B^{d'}(\partial B^d{}_i/\partial q^{d'})]. \tag{c2}$$

ii. If $B^d{}_i = B^d{}_i\,(q^{i'})$, $B^d = 0$, and $T = T_o(t, q^i, V^i) = T_o$, in which case the only surviving transitivity coefficients are

$$-W^d{}_{ii'} \Rightarrow -T^d{}_{ii'} \equiv \partial B^d{}_{i'}/\partial q^i - \partial B^d{}_i/\partial q^{i'}, \tag{d1}$$

show that, Equations c reduces to the *Chaplygin* equations:

$$d/dt(\partial T_o/\partial V^i) - \partial T_o/\partial q^i + (\partial B^d{}_{i'}/\partial q^i - \partial B^d{}_i/\partial q^{i'})(\partial T/\partial V^d)_o\, V^{i'}$$

$$\equiv E_i(T_o) - T^d{}_{ii'}(\partial T/\partial V^d)_o\, V^{i'} = Q_i + B^d{}_i\, Q_d. \tag{d2}$$

Hints: Apply the chain rule to $T(t, q, V) = T^*(t, q, \omega)$, and use Equations c6 and c7 ff of Example 6.6.1:

$$(\partial T^*/\partial \omega^I)_o = \partial T^*{}_o/\partial \omega^I \to \partial T_o/\partial V^i, \tag{e1}$$

$$(\partial T^*/\partial \theta^I)_o = \partial T^*{}_o/\partial \theta^I = \partial T^*{}_o/\partial q^i + B^d{}_i(\partial T^*{}_o/\partial q^d)$$

$$\to \partial T_o/\partial q^i + B^d{}_i(\partial T_o/\partial q^d). \tag{e2}$$

Remarks

i. The Chaplygin equations (d2) are historically important because they showed conclusively that, in general, $E_i(T_o) \neq Q_i$, or even $Q_i + B^d{}_i\, Q_d$. (See Remark iv below.)

ii. In increasing order of generality, the nonholonomic variable equations are

Chaplygin (1895, publ. 1897) → Voronets (1901) → Hamel (1903–4)

iii. The Voronets equations are to the kinetic Hamel equations what the Hadamard–Chaplygin equations (Equations d of Problem 7.4.1) are to the kinetic Maggi equations. For additional explicit forms of these equations, see, e.g., Maißer (1981; 1983–4; 1991).

iv. Let S_o be the Appellian counterpart of T_o, i.e.,

$$S = S(t, q, V^d, V^i, dV^d/dt, dV^i/dt)$$

$$= S\left[t, q, V^d(t, q, V^i), V^i, dV^d(t, q, V^i, dV^i/dt)/dt, dV^i/dt\right]$$

$$\equiv S_o(t, q, V^i, dV^i/dt) = S_{o(\text{constrained Appellian})}. \tag{f1}$$

Then, by chain rule, and with the convenient notation $dV^k/dt \equiv a^k$, we find

$$\partial S_o/\partial a^i = \partial S/\partial a^i + (\partial S/\partial a^d)(\partial a^d/\partial a^i) = \partial S/\partial a^i + B^d{}_i(\partial S/\partial a^d)$$

$$= E_i(T) + B^d{}_i\, E_d(T), \tag{f2}$$

[$d/dt(\ldots)$–differentiating Equation a: $a^d = B^d{}_i\, a^i$ + function of t, q, V $\Rightarrow \partial a^d/\partial a^i = \partial V^d/\partial V^i = B^d{}_i$] and so the *Appellian counterpart* of the Voronets and Chaplygin equations is the following simple-looking equations:

$$\partial S_o/\partial a^i = Q_i + B^d{}_i\, Q_d \quad [\neq E_i(T_o)]. \tag{f3}$$

Equations d2 and f3, show clearly that, in general (e.g., for nonholonomic Pfaffian constraints), *Lagrange's "unconstrained equations" do not hold for the constrained kinetic energy; but Appell's "unconstrained equations" do, for the constrained Appellian.*

Problem 7.4.4

Verify that the explicit forms of the contravariant unconstrained Hamel equations (7.4.16b) are

$$d\omega^K/dt + \Gamma^K{}_{(\Lambda\Psi)}\, \omega^\Lambda\, \omega^\Psi$$

$$= d\omega^K/dt + \Gamma^K{}_{(LR)}\, \omega^L\, \omega^R + 2\Gamma^K{}_{(L,N+1)}\, \omega^L + \Gamma^K{}_{N+1,N+1} = Q^K. \tag{a}$$

For further details, see, e.g., Dobronravov (1976, pp. 195–201).
Remark: As already pointed out in Example 7.2.1, Remarks, these forms are more useful for theoretical arguments (e.g., linearization of the Hamel equations, for small oscillations in nonholonomic variables) rather than for application to concrete problems. For the latter we recommend calculating I_K from the Lagrange–Hamel form: $E_K(T^*) - \Gamma_K = \ldots$, and then I^K from

$$\delta I = I_K\, \delta\theta^K = I^K\, \delta\theta_K,$$

$$\delta\theta_K = M_{K\Lambda}\, \delta\theta^\Lambda = M_{KL}\, \delta\theta^L + M_{K,N+1}\, \delta\theta^{N+1} = M_{KL}\, \delta\theta^L \quad (\Rightarrow I^K = M^{KL}\, I_L). \tag{b}$$

Problem 7.4.5

Verify that the explicit forms of the contravariant constrained Hamel equations (7.4.27e and 7.4.29) are, respectively,

Kinetic: $$d\omega^I/dt + \Gamma^I{}_{(I'+1,I''+1)}\, \omega^{I'+1}\, \omega^{I''+1} = Q^I, \tag{a}$$

Kinetostatic: $$\Gamma^D{}_{(I'+1,I''+1)}\, \omega^{I'+1}\, \omega^{I''+1} = Q^D + \Lambda^D \quad (\text{since } \omega^D = 0) \tag{b}$$

where $I, I', I'' = m + 1, \ldots, n$; $I' + 1, I'' + 1 = m+1, \ldots, n + 1$.
Remark: The above constitute the nonholonomic variable counterpart of the explicit forms of Example 7.4.2. As there, the $n - m$ *differential* Equations a yield the motion: $\omega^I(t)$. Then, the m *algebraic* Equations b yield the reactions $\Lambda^D(t)$, if needed.

Example 7.4.3. Equations of Jacobi-Synge

Below, we combine the contravariant form of the Routh–Voss Equations [recalling (7.4.10b) and (7.2.16c)]:

$$I^k \equiv dV^k/dt + \Gamma^k{}_{bh}\, V^b\, V^h + 2\Gamma^k{}_{b0}\, V^b + \Gamma^k{}_{00}$$

$$= Q^k + \lambda_D\, A^{Dk} = Q^k + \lambda^D\, A^k{}_D, \tag{a}$$

with the constraints (6.6.6a):

$$\omega^D \equiv A^D{}_\beta\, V^\beta = A^D{}_k\, V^k + A^D{}_{n+1}\, V^{n+1} \equiv A^D{}_k\, V^k + A^D = 0, \tag{b}$$

and derive n *reactionless* equations of motion, and explicit expressions for the reactions λ_D. Indeed, $d/dt(\ldots)$–differentiating Equation b we find, successively,

$$0 = (dA^D{}_\beta/dt)V^\beta + A^D{}_\beta(dV^\beta/dt)$$

$$= [(\partial A^D{}_\beta/\partial q^\lambda)V^\lambda]V^\beta + A^D{}_\beta(dV^\beta/dt) \quad \text{(recalling Equation 3.4.1)}$$

$$= [(A^D{}_\beta|_\lambda + \Gamma^\sigma{}_{\beta\lambda}\, A^D{}_\sigma)V^\lambda]V^\beta + A^D{}_\beta(dV^\beta/dt)$$

$$= A^D{}_\beta|_\lambda\, V^\beta\, V^\lambda + A^D{}_\beta(dV^\beta/dt + \Gamma^\beta{}_{\sigma\lambda}\, V^\sigma\, V^\lambda)$$

$$= A^D{}_\beta|_\lambda\, V^\beta\, V^\lambda + A^D{}_k(dV^k/dt + \Gamma^k{}_{\sigma\lambda}\, V^\sigma\, V^\lambda) \quad (\text{since } \Gamma^{n+1}{}_{\sigma\lambda} = 0)$$

$$= A^D{}_\beta|_\lambda\, V^\beta\, V^\lambda + A^D{}_k\, I^k$$

(by Equations a)

$$= A^D{}_\beta|_\lambda\, V^\beta\, V^\lambda + A^D{}_k(Q^k + \lambda^{D'}\, A^k{}_{D'}),$$

(by Equation 6.4.4c: $A^D{}_k\, A^k{}_{D'} = \mathbf{E}^D \cdot \mathbf{E}_{D'} = \delta^D{}_{D'}$)

$$= A^D{}_\beta|_\lambda\, V^\beta\, V^\lambda + A^D{}_k\, Q^k + \lambda^{D'}\, \delta^D{}_{D'}$$

$$= A^D{}_\beta|_\lambda\, V^\beta\, V^\lambda + A^D{}_k\, Q^k + \lambda^D,$$

or, solving for the λ^D,

$$\lambda^D = -A^D{}_\beta|_\lambda\, V^\beta\, V^\lambda - A^D{}_k\, Q^k \quad (= -A^D{}_\beta|_\lambda\, V^\beta\, V^\lambda - Q^D); \tag{c}$$

or, since (recalling Equation 7.4.20f):

$$\lambda^D = \Lambda^D = M^{DK}\,\Lambda_K = M^{DD'}\,\Lambda_{D'} + M^{DI}\,\Lambda_I = M^{DD'}\,\Lambda_{D'} \quad [D, D' = 1, \ldots, m]$$

$$\Rightarrow \Lambda_{D'} = M_{D'D}\,\Lambda^D = \lambda_{D'}, \quad \text{where } M_{DD'}\text{: inverse of } M^{DD'}, \tag{d}$$

ultimately,

$$\lambda_{D'} = -M_{D'D}(A^D{}_\beta|_\lambda\, V^\beta\, V^\lambda + A^D{}_k\, Q^k) \tag{e}$$

$$[= -M_{D'D}(A^D{}_b|_h\, V^b\, V^h + A^D{}_k\, Q^k), \text{ for } \textit{stationary} \text{ constraints}]. \tag{e1}$$

Finally, substituting Equation e into the (covariant) Routh–Voss equations (7.4.10b): $E_k(T) = Q_k + \lambda_D\, A^D{}_k$, we obtain the n kinetic *equations of Jacobi–Synge:*

$$E_k(T) = Q_k - [M_{D'D}(A^D{}_\beta|_\lambda\, V^\beta\, V^\lambda + A^D{}_k\, Q^k)]A^{D'}{}_k. \tag{f}^*$$

Equations f give the motion: $q^k(t) \Rightarrow V^k(t)$. Then, Equations c or e yield the constraint reactions: $\lambda_D(t)$ in terms of the given constraints ($A^D{}_\beta$: $A^D{}_k$, $A^D{}_{n+1}$) and the motion.

Problem 7.4.6

Verify that the Jacobi–Synge equations of the preceding Problem can be written in the following two contravariant forms:

$$DV^k/Dt + \Gamma^k{}_{\beta\lambda}\, V^\beta\, V^\lambda = Q^k + \lambda^D\, A^k{}_D = Q^k - (A^D{}_\beta|_\lambda\, V^\beta\, V^\lambda + A^D{}_l\, Q^l)A^k{}_D, \tag{a}$$

and

$$DV^k/Dt + G^k{}_{\beta\lambda}\, V^\beta\, V^\lambda = F^k \tag{b}$$

(unconstrained form, but modified space and forces) where

$$F^k \equiv Q^k - A^k{}_D(A^D{}_l\, Q^l), \quad G^k{}_{\beta\lambda} \equiv \Gamma^k{}_{\beta\lambda} + (1/2)A^k{}_D(A^D{}_\beta|_\lambda + A^D{}_\lambda|_\beta). \tag{b1}$$

7.5 THE CENTRAL EQUATION

Lagrange's Principle (LP), Equations 7.4.4 ff, contains accelerations explicitly, i.e., *second-order* time derivatives. However, sometimes it is more expedient to express it in terms of *first-order* time derivatives, i.e., velocities. The resulting variational

---

* C. G. J. Jacobi described how to get these equations in his famous *Vorlesungen über Dynamik* (early 1840s), but it was J. L. Synge who obtained them explicitly, via the above steps (for scleronomic systems and orthonormal constraint vectors), in his pioneering memoir on tensorial dynamics (1927; pp. 53–55), hence, the term "Jacobi–Synge equations," proposed here for the first time. (We are grateful to Dr. F. Pfister for the Jacobi reference). For additional related expressions of the reactions, see, e.g., Maißer (1983–4, pp. 73–74); also Lur'e (1968, pp. 323–340).

equational of motion is called the *central equation of mechanics* (Heun, Hamel, early 1900s). Let us obtain it: we have, successively,

$$\delta I \equiv S\,dm\ \mathbf{a}\cdot\delta\mathbf{r} = d/dt\left(S\,dm\ \mathbf{v}\cdot\delta\mathbf{r}\right) - S\,dm\ \mathbf{v}\cdot[(\delta\mathbf{r})^{\cdot} - \delta\mathbf{v}]$$

$$= d/dt\left(S\,dm\ \mathbf{v}\cdot\delta\mathbf{r}\right) - \delta T - S\,dm\ \mathbf{v}\cdot[(\delta\mathbf{r})^{\cdot} - \delta\mathbf{v}], \tag{7.5.1}$$

where

$$2T \equiv S\,dm\ \mathbf{v}\cdot\mathbf{v} \Rightarrow \delta T = S\,dm\ \mathbf{v}\cdot\delta\mathbf{v}, \quad \text{and} \quad (\ldots)^{\cdot} \equiv d/dt(\ldots); \tag{7.5.2)*}$$

and inserting this δI-expression into LP yields the *generalized* central equation:

$$d/dt\left(S\,dm\ \mathbf{v}\cdot\delta\mathbf{r}\right) - \delta T - S\,dm\ \mathbf{v}\cdot[(\delta\mathbf{r})^{\cdot} - \delta\mathbf{v}] = \delta' W, \tag{7.5.3}$$

or, rearranged,

$$\delta T + \delta'\Gamma + \delta' W = d(\delta P)/dt, \tag{7.5.4}$$

where

$$\delta'\Gamma \equiv S\,dm\ \mathbf{v}\cdot[(\delta\mathbf{r})^{\cdot} - \delta\mathbf{v}]\text{: } \textit{virtual work of nonholonomic deviation,} \tag{7.5.4a}$$

$$\delta P \equiv S\,dm\ \mathbf{v}\cdot\delta\mathbf{r}\text{: } \textit{virtual work of (linear) momenta.} \tag{7.5.4b)**}$$

The variational equations (7.5.3 and 7.5.4) constitute a *second* route toward the derivation of Lagrange-type equations of motion, completely equivalent to LP; also, upon multiplying the form (7.5.4) with dt and integrating between two arbitrary time limits, produces a time–integral variational equation which is the basis for the various variational "principles" of mechanics. ***

To make Equation 7.5.3 more useful to Lagrangean mechanics, we must transform it to *system quasi-variables* (although special problems of rigid-body and continuum mechanics can be best treated directly with Equation 7.5.3). Indeed, we have, successively,

---

* We notice the difference between δT and $dT = S\,dm\ \mathbf{a}\cdot d\mathbf{v}$.

** Equations 7.5.3 or 7.5.4, but under the assumption $(\delta\mathbf{r})^{\cdot} = \delta\mathbf{v} \Rightarrow \delta'\Gamma = 0$, is called simply Central Equation. On the necessity, or not, of this assumption, see remarks / footnote near section's end.

*** On such "principles" for nonholonomic systems, see Papastavridis (1997).

a. $T = T(t, q, dq/dt) = T[t, q, dq/dt(t, q, \omega)] = T^*(t, q, \omega) = T^*$

δT and $dT = S\, dm\; a \cdot dv.$

$\Rightarrow \delta T = \delta T^* = (\partial T^*/\partial q^k)\delta q^k + (\partial T^*/\partial \omega^K)\delta\omega^K$

$= (\partial T^*/\partial q^k)(\partial V^k/\partial \omega^K)\delta\theta^K + (\partial T^*/\partial \omega^K)\delta\omega^K$ (recalling Equation 3.13.1a)

$$= (\partial T^*/\partial \theta^K)\delta\theta^K + (\partial T^*/\partial \omega^K)\delta\omega^K \equiv (\partial T^*/\partial \theta^K)\delta\theta^K + P_K\, \delta\omega^K; \quad (7.5.5a)$$

b. $\delta' W \equiv Q_k\, \delta q^k = Q_K\, \delta\theta^K;$ (7.5.5b)

c. $(\delta P)^{\cdot} = d/dt\left(S\, dm\, \mathbf{v} \cdot \delta\mathbf{r}\right) = d/dt\left[S\, dm\, \mathbf{v} \cdot (\mathbf{e}_K \delta\theta^K)\right]$

$$= d/dt(P_K\, \delta\theta^K) = (dP_K/dt)\delta\theta^K + P_K(\delta\theta^K)^{\cdot} \quad [= d/dt(P_k\, \delta q^k)]; \quad (7.5.5c)$$

$P_K \equiv S\, dm\; \mathbf{v} \cdot \mathbf{e}_K$: *nonholonomic covariant system momenta,* (7.5.5d)

$P_k \equiv S\, dm\; \mathbf{v} \cdot \mathbf{e}_k$: *holonomic covariant system momenta.* (7.5.5e)

Applying the results of Example 3.15.1, with $d_1(\ldots) \to d(\ldots)/dt \equiv (\ldots)^{\cdot}$ and $d_2(\ldots) \to \delta(\ldots)$, while recalling that $\delta\theta^{N+1} \equiv \delta t = 0$, we obtain the *particle transitivity equation:*

$$(\delta\mathbf{r})^{\cdot} - \delta\mathbf{v} = [(\delta\theta^K)^{\cdot} - \delta\omega^K]\mathbf{e}_K - (\gamma^L{}_{K\Sigma}\, \omega^\Sigma\, \delta\theta^K)\mathbf{e}_L, \quad (7.5.5f)$$

and substituting this expression into Equation 7.5.4a we find, successively,

$\delta'\Gamma \equiv S\, dm\; \mathbf{v} \cdot [(\delta\mathbf{r})^{\cdot} - \delta\mathbf{v}]$

$\equiv \left(S\, dm\, \mathbf{v} \cdot \mathbf{e}_K\right)\left[(\delta\theta^K)^{\cdot} - \delta\omega^K\right] - \left(S\, dm\, \mathbf{v} \cdot \mathbf{e}_L\right)\left(\gamma^L{}_{K\Sigma}\, \omega^\Sigma\, \delta\theta^K\right)$

$$= P_K[(\delta\theta^K)^{\cdot} - \delta\omega^K] - (\gamma^L{}_{K\Sigma}\, P_L\, \omega^\Sigma)\delta\theta^K. \quad (7.5.5g)$$

Finally, inserting Expressions 7.5.5a through 5c and 5g into Equation 7.5.3, and simplifying, we get the *central equation in quasi-variables*:

$$(dP_K/dt)\delta\theta^K - (\partial T^*/\partial \theta^K)\delta\theta^K + (\gamma^L{}_{K\Sigma}\, P_L\, \omega^\Sigma)\delta\theta^K = Q_K\, \delta\theta^K, \quad (7.5.6a)$$

or

$$I_K\, \delta\theta^K = (dP_K/dt - \partial T^*/\partial \theta^K + \gamma^L{}_{K\Sigma}\, P_L\, \omega^\Sigma)\delta\theta^K = Q_K\, \delta\theta^K; \quad (7.5.6b)$$

which is just one step away from the Hamel equations (7.4.19a and b).

It should be noted that, for the derivation of these equations, we did *not* have to make any special assumptions about $(\delta q^k)^{\cdot} - \delta q^k$, or about $(\delta \mathbf{r})^{\cdot} - \delta \mathbf{v}$; i.e., *whether we assume that they vanish or not, we still obtain the same equations of motion!**

Therefore, we can rewrite Equation 7.5.6a as

$$(dP_K/dt)\delta\theta^K - (\partial T^*/\partial\theta^K)\delta\theta^K + P_K[(\delta\theta^K)^{\cdot} - \delta\omega^K] = Q_K\, \delta\theta^K, \quad (7.5.6c)$$

under the (possible but *nonnecessary*) assumptions:

$$(\delta q^k)^{\cdot} - \delta V^k = 0 \Rightarrow (\delta\theta^K)^{\cdot} - \delta\omega^K = \gamma^K{}_{L\Sigma}\, \omega^\Sigma\, \delta\theta^L. \quad (7.5.6d)$$

The form (7.5.6c) is, *probably, the single most useful variational equation of mechanics for the direct derivation of equations of motion;* while Equation 7.5.6d allows us to calculate the γ terms *directly*, with no recourse to their other definitions, such as Equations 3.13.5 and 3.13.7a ff. On the other hand, Equation 7.5.6b is better suited for the understanding of the principle of relaxation of the constraints/multipliers, etc. (Section 7.4).

7.6 THE POWER, OR ENERGY RATE, EQUATIONS

In both Newtonian–Eulerian and Lagrangean mechanics, such equations are obtained by *multiplying each equation of motion with the corresponding velocity, adding all these equations together, and then transforming each term of the so-resulting scalar equation into meaningful rate of kinetic energy/work (i.e., power) expressions. Let us begin with power equations in holonomic variables.*

7.6.1 HOLONOMIC VARIABLES

Dotting the (covariant) Routh–Voss equations $I_k \equiv E_k(T) = Q_k + \lambda_D\, A^D{}_k$, with $dq^k/dt \equiv V^k$, and invoking the constraints $A^D{}_k\, V^k + A^D = 0$, we obtain

$$E_k(T)\, V^k = Q_k\, V^k + \lambda_D(A^D{}_k\, V^k) = Q_k\, V^k - \lambda_D\, A^D; \quad (7.6.1a)$$

or, due to the mathematical *identity* holding for any well-behaved function $f = f(t, q, V)$:

---

* If we assume that $(\delta \mathbf{r})^{\cdot} - \delta \mathbf{v} = \mathbf{0}$, then it follows that (recalling Equations 6.3.1ff):

$$\mathbf{0} = (\delta \mathbf{r})^{\cdot} - \delta \mathbf{v} = (\mathbf{e}_k\, \delta q^k)^{\cdot} - \delta(\mathbf{e}_k\, V^k + \mathbf{e}_{n+1}) = \ldots = [(\delta q^k)^{\cdot} - \delta(dq^k/dt)]\mathbf{e}_k$$

$$\Rightarrow (\delta q^k)^{\cdot} = \delta(dq^k/dt) \equiv \delta V^k \quad \text{or} \quad d(\delta q^k) = \delta(dq^k);$$

and vice versa; and then, by Equation 7.5.5f, $(\delta\theta^K)^{\cdot} - \delta\omega^K = \gamma^K{}_{L\Sigma}\, \omega^\Sigma\, \delta\theta^L$. In sum, *either* we assume that $d(\delta q^k) = \delta(dq^k)$, *or* that $(\delta\theta^K)^{\cdot} - \delta\omega^K = 0$, *but not both,* because, if we did, then we would also have $\gamma^K{}_{L\Sigma} = 0$, which, in general, does not hold.

$$E_k(f)\, V^k \equiv [d/dt(\partial f/\partial V^k) - \partial f/\partial q^k]V^k$$

$$= \ldots = d/dt[(\partial f/\partial V^k)V^k - f] + \partial f/\partial t \equiv dh(f)/dt + \partial f/\partial t, \quad (7.6.1b)^*$$

applied for $f \to T$ and $h(f) \to h(T)$, we finally get the *holonomic power equation* (HPE):

$$d/dt[(\partial T/\partial V^k)V^k - T] \equiv dh(T)/dt = -\partial T/\partial t + Q_k\, V^k - \lambda_D\, A^D. \quad (7.6.2)$$

If part of each of the Q_k (or even the entire Q_k) equals $-\partial\Pi/\partial q^k$, where $\Pi = \Pi(t, q^k)$: *potential* function (as in Chapter 5), then (7.6.2) transforms to

$$d/dt[(\partial L/\partial V^k)V^k - L] \equiv dh/dt = -\partial L/\partial t + Q_k\, V^k - \lambda_D\, A^D, \quad (7.6.3)$$

where $L \equiv T - \Pi$: *Lagrangean* function of the system, and from now on:

$$h \equiv h(L) = (\partial L/\partial V^k)V^k - L = h(t, q, V):$$

Generalized Energy function; (7.6.3a)**

and, of course, the Q_k comprise only the *nonpotential* part of the corresponding impressed forces. Next, application of the well-known *Eulerian homogeneous function theorem* to h yields (recalling notation in Equations 7.2.2ff.):

$$h \equiv L_{(2)} - L_{(0)} = T_{(2)} + (\Pi - T_{(0)}), \quad (7.6.3b)$$

where

$$L \equiv T(t, q, V) - \Pi(t, q) = T_{(2)} + T_{(1)} + T_{(0)} - \Pi$$

$$= T_{(2)} + T_{(1)} + (T_{(0)} - \Pi) = L_{(2)} + L_{(1)} + L_{(0)}. \quad (7.6.3c)$$

Now,

i. If $A^D{}_{n+1} \equiv A^D = 0$, i.e., if the constraints are *catastatic* (but not necessarily scleronomic, Section 6.6), and
ii. If all nonpotential forces vanish (or if $Q_k\, V^k = 0$, in which case the Q_k are called *gyroscopic*), and
iii. If $\partial L/\partial t = 0$ (even if neither $\partial T/\partial t$ nor $\partial\Pi/\partial t$ vanish separately), then Equation 7.6.3 reduces to the following *holonomic generalized energy*, or *Jacobi–Painlevé, integral*:

---

* Readers are encouraged to verify Equation 7.6.1b on their own.

** When h is expressed as a function of t, q, and $P_k \equiv p_k \equiv \partial T/\partial V^k$: holonomic covariant components of system momentum, (7.5.5e), i.e., $p_k = p_k(t, q, V) \Rightarrow V^k = V^k(t, q, p) \Rightarrow h\,(t, q, V(t, q, p)) \equiv H(t, q, p) = H$, it is called the *Hamiltonian* function of the system.

$$h = T_{(2)} + (\Pi - T_{(0)}) = \text{constant of motion}. \tag{7.6.4}$$

Of course, other combinations or physical circumstances than the above i to iii may nullify the right side of Equation 7.6.3 and thus re-produce Equation 7.6.4.

7.6.2 Nonholonomic Variables

Equation 7.6.3 has two drawbacks: (i) it contains (generally unknown) reactions λ_D, and (ii) it cannot distinguish between genuinely nonholonomic Pfaffian constraints and holonomic constraints in Pfaffian form. This should not surprise us: that equation is based on the Routh–Voss equations, which possess both these features. Hence, the starting point for power equations free of these limitations should be the *kinetic Hamel* equations:

$$E_I(L^*) - \Gamma_I \equiv d/dt(\partial L^*/\partial\omega^I) - \partial L^*/\partial\theta^I$$
$$+ (\gamma^K{}_{II'}\, \omega^{I'} + \gamma^L{}_I)(\partial T^*/\partial\omega^K) = Q_I. \tag{7.6.5}$$

As in the holonomic variable case, we dot the above equation with ω^I, and then transform the various terms:

i. For the first *two* terms of Equation 7.6.5 this process yields, successively,

$$E_I(L^*)\omega^I = d/dt[(\partial L^*/\partial\omega^I)\omega^I] - (\partial L^*/\partial\omega^I)(d\omega^I/dt) - (\partial L^*/\partial\theta^I)\omega^I,$$

or, since

$$-(\partial L^*/\partial\theta^I)\omega^I = -(\partial L^*/\partial q^k)(\partial V^k/\partial\omega^I)\omega^I \quad \text{(then recalling Equations 6.6.6c)}$$
$$= -(\partial L^*/\partial q^k)(A^k{}_I\, \omega^I) = -(\partial L^*/\partial q^k)(V^k - A^k), \quad A^k{}_{N+1} \equiv A^k,$$

finally,

$$E_I(L^*)\, \omega^I = d/dt[(\partial L^*/\partial\omega^I)\omega^I] - [(\partial L^*/\partial q^k)(dq^k/dt)$$
$$+ (\partial L^*/\partial\omega^I)(d\omega^I/dt)] + (\partial L^*/\partial q^k)A^k$$
$$= d/dt[(\partial L^*/\partial\omega^I)\omega^I] - (dL^*/dt - \partial L^*/\partial t) + (\partial L^*/\partial q^k)A^k$$
$$= dh^*/dt + \partial L^*/\partial t + (\partial L^*/\partial q^k)A^k, \tag{7.6.5a}$$

where

$$h^* \equiv (\partial L^*/\partial\omega^I)\omega^I - L^* = h^*(t,\, q,\, \omega): \tag{7.6.5b}$$

Generalized Energy function, in nonholonomic variables $[\neq h(L^*)]$

$$= T^*_{(2)} + (\Pi^* - T^*_{(0)}) \quad (\Pi = \Pi^*). \tag{7.6.5c}$$

Equation 7.6.5a is a mathematical *identity*, like Equation 7.6.1b; and for $A^k = 0$ coincides with it *in form.*

ii. For the *third* term of Equation 7.6.5 we readily find

$$\gamma^K{}_{II'}\ \omega^{I'}(\partial T^*/\partial\omega^K)\omega^I = (\partial T^*/\partial\omega^K)(\gamma^K{}_{II'}\ \omega^I\ \omega^{I'})$$

$$= 0, \quad \text{since} \quad \gamma^K{}_{II'} = -\gamma^K{}_{I'I}; \tag{7.6.5d}$$

i.e., that term is gyroscopic (a property showing the close connection between nonholonomic and gyroscopic systems).

Collecting all these results, we get the following general kinetic (i.e., reactionless) *nonholonomic power equation* (NPE):

$$dh^*/dt = -\partial L^*/\partial t + Q_I\ \omega^I + R, \tag{7.6.6}$$

where

$$R \equiv R' + R'': \textit{Rheonomic nonholonomic power,}$$

$$R' \equiv -(\partial L^*/\partial q^k)A^k, \quad R'' \equiv -\gamma^K{}_I\ \omega^I(\partial T^*/\partial\omega^K). \tag{7.6.6a}$$

All terms of Equation 7.6.6, except R, have counterparts in the HPE (Equation 7.6.3). Let us, therefore, examine the right side of this equation more closely:

i. Recalling Equation 6.4.6c we have

$$-\partial L^*/\partial t - (\partial L^*/\partial q^k)A^k \equiv -\partial L^*/\partial\theta^{N+1} \quad (\neq -\partial L^*/\partial t), \tag{7.6.7a}$$

despite the fact that $\theta^{N+1} = \delta^{N+1}{}_{n+1}\ q^{n+1} \equiv \delta^{N+1}{}_{n+1}\ t$.* Here,

- $-\partial L^*/\partial t$ results from the nonstationarity (or rheonomicity) of the original finite (holonomic) constraints; and possibly, that of the additional Pfaffian (holonomic or nonholonomic) constraints, i.e., $L[t, q, V(t, q, \omega)] = L^*(t, q, \omega)$; while
- $-(\partial L^*/\partial q^k)A^k$ results exclusively from the nonstationarity of the Pfaffian constraints; specifically, from their acatastaticity.

Hence, $-\partial L^*/\partial\theta^{N+1}$ represents, roughly, the sum of nonstationary (or rheonomic) contributions of the original and additional constraints.**

In view of the above, the NPE reads:

$$dh^*/dt = -\partial L^*/\partial\theta^{N+1} + Q_I\ \omega^I + R''. \tag{7.6.8}$$

ii. Finally, the term R'. Since (recalling Equation 6.5.3b):

---

* Had we defined $\partial(\ldots)/\partial\theta^{N+1} \equiv [\partial(\ldots)/\partial q^k](\partial q^k/\partial\theta^{N+1}) = A^k\,[\partial(\ldots)/\partial q^k]$, as some authors do, then we would have: $-\partial L^*/\partial t - (\partial L^*/\partial q^k)A^k = -\partial L^*/\partial t - \partial L^*/\partial\theta^{N+1}$.

** (See following page for footnote.)

$$\gamma^K{}_I \equiv \gamma^K{}_{I,N+1} = (A^K{}_{k,s} - A^K{}_{s,k})A^k{}_I\, A^s{}_{N+1} + (A^K{}_{k,n+1} - A^K{}_{n+1,k})A^k{}_I\, A^{n+1}{}_{N+1}$$

$$\equiv (\partial A^K{}_k/\partial q^s - \partial A^K{}_s/\partial q^k)A^k{}_I\, A^s + (\partial A^K{}_k/\partial t - \partial A^K/\partial q^k)A^k{}_I, \qquad (7.6.9a)$$

we have the following picture:

- If $A^K = 0 \Rightarrow A^k = 0$ (i.e., catastatic but possibly nonstationary Pfaffian constraints), then $-\partial L^*/\partial\theta^{N+1} \to -\partial L^*/\partial t$, $R' = 0$, and

$$R'' \equiv -\gamma^K{}_I\, \omega^I(\partial T^*/\partial\omega^K) = -[(\partial A^K{}_k/\partial t)A^k{}_I]\omega^I\, (\partial T^*/\partial\omega^K)$$

$$= -(\partial A^K{}_k/\partial t)(A^k{}_I\, \omega^I)(\partial T^*/\partial\omega^K) = -(\partial A^K{}_k/\partial t)(\partial T^*/\partial\omega^K)V^k. \quad (7.6.9b)$$

- If, *further*, $A^K{}_k = A^K{}_k(q)$ (i.e., stationary, or scleronomic, Pfaffian constraints), then $R', R'' \to 0$, and the NPE reduces to

$$dh^*/dt = -\partial L^*/\partial t + Q_I\, \omega^I. \qquad (7.6.9c)$$

- If, *in addition*, $\partial L^*/\partial t = 0$ and $Q_I\, \omega^I = 0$ (e.g., gyroscopic Q_I terms), then the above lead to the following *nonholonomic generalized energy, or Jacobi–Painlevé, integral:*

$$h^* = T^*_{(2)} + (\Pi^* - T^*_{(0)}) = \text{constant of motion}. \qquad (7.6.9d)$$

Problem 7.6.1

Obtain the power equation corresponding to the Voronets equations of motion; and then that corresponding to Chaplygin's equations (recall results of Problem 7.4.3).

** Had we used the definition:

$$d^*L^*/dt \equiv (\partial L^*/\partial\omega^I)(d\omega^I/dt) + (\partial L^*/\partial\theta^I)\omega^I + \partial L^*/\partial t, \qquad (7.6.7b)$$

instead of the more orthodox one made here:

$$dL^*/dt \equiv (\partial L^*/\partial\omega^I)(d\omega^I/dt) + (\partial L^*/\partial q^k)(dq^k/dt) + \partial L^*/\partial t, \qquad (7.6.7c)$$

then,

$$dL^*/dt - d^*L^*/dt = \ldots = (\partial L^*/\partial q^k)(A^k{}_I\, \omega^I + A^k) - (\partial L^*/\partial\theta^I)\omega^I$$

$$= (\partial L^*/\partial q^k)\, A^k \quad [= 0, \text{ for catastatic } dq \leftrightarrow d\theta \text{ relation}]; \qquad (7.6.7d)$$

and for a general function $f^*(t, q, \omega)$:

$$df^*/dt - d^*f^*/dt = (\partial f^*/\partial q^k)A^k. \qquad (7.6.7e)$$

Example 7.6.1. The (n + 1)th Equation of Motion

Dotting the fundamental Newton–Euler equation $dm\,\mathbf{a} = d\mathbf{F} + d\mathbf{R}$ with $\mathbf{e}_{n+1}$ and $\mathbf{e}_{N+1}$, and summing over the entire system yields, respectively, the "partial power equations":

$$I_{n+1} \equiv Q_{n+1} + \Lambda_{n+1}, \tag{a1}$$

$$I_{N+1} \equiv Q_{N+1} + \Lambda_{N+1}, \tag{a2}$$

where

$$I_{n+1} \equiv S\, dm\ \mathbf{a} \cdot \mathbf{e}_{n+1}, \quad I_{N+1} \equiv S\, dm\ \mathbf{a} \cdot \mathbf{e}_{N+1}, \tag{b1}$$

$$Q_{n+1} \equiv S\, d\mathbf{F} \cdot \mathbf{e}_{n+1}, \quad Q_{N+1} \equiv S\, d\mathbf{F} \cdot \mathbf{e}_{N+1}, \tag{b2}$$

$$\Lambda_{n+1} \equiv S\, d\mathbf{R} \cdot \mathbf{e}_{n+1}, \quad \Lambda_{N+1} \equiv S\, d\mathbf{R} \cdot \mathbf{e}_{N+1}. \tag{b3}$$

Let us transform the above to system variables.

Holonomic Partial Power Equation (or "temporal equation of motion")

We have

$$I_{n+1} \equiv S\, dm\ \mathbf{a} \cdot \mathbf{e}_{n+1} = S\, dm[d/dt(V^{\beta}\, \mathbf{e}_{\beta})] \cdot \mathbf{e}_{n+1}$$

$$= S\, dm[(dV^{\beta}/dt)\mathbf{e}_{\beta} + V^{\beta}(d\mathbf{e}_{\beta}/dt)] \cdot \mathbf{e}_{n+1}$$

$$= \left(S\, dm\, \mathbf{e}_{\beta} \cdot \mathbf{e}_{n+1}\right)(dV^{\beta}/dt) + \left(S\, dm\, \mathbf{e}_{\beta,\sigma} \cdot \mathbf{e}_{n+1}\right)(dV^{\beta}/dt)(dV^{\sigma}/dt)$$

$$= M_{\beta,n+1}(dV^{\beta}/dt) + \Gamma_{n+1,\beta\sigma}(dV^{\beta}/dt)(dV^{\sigma}/dt), \tag{c1}$$

or, since $2T = M_{kl}\, V^k\, V^l + 2M_{k,n+1}\, V^k\, V^{n+1} + M_{n+1,n+1}\, V^{n+1}\, V^{n+1}$, and $V^{n+1} = dq^{n+1}/dt = 1 \Rightarrow dV^{n+1}/dt = 0$, we finally obtain the partial power equation in the Lagrangean form:

$$I_{n+1} = E_{n+1}(T) \equiv d/dt(\partial T/\partial V^{n+1}) - \partial T/\partial q^{n+1}$$

$$= M_{n+1,k}(dV^k/dt) + \Gamma_{n+1,\beta\sigma}(dV^{\beta}/dt)(dV^{\sigma}/dt) = Q_{n+1} + \Lambda_{n+1}. \tag{c2}$$

(In view of Equations 7.2.11a through d and 7.3.5a ff, the contravariant form of the above is simply: $0 = 0$.*)

Relation of above to the "Elementary" Power Forms
(i.e., dT/dt = Power of all Acting Forces)

Applying Euler's homogeneous function theorem to $2T = M_{\alpha\beta} V^\alpha V^\beta$, we find

$$(\partial T/\partial V^\alpha)V^\alpha = (\partial T/\partial V^k)V^k + (\partial T/\partial V^{n+1})V^{n+1} = 2T. \quad \text{(d1)}$$

But also,

$$(\partial T/\partial q^k)V^k + (\partial T/\partial V^k)(dV^k/dt) = dT/dt - \partial T/\partial q^{n+1}. \quad \text{(d2)}$$

Hence, combining the above we obtain the following *purely analytical* result:

$$d/dt(\partial T/\partial V^{n+1}) - \partial T/\partial q^{n+1} = dT/dt - [d/dt(\partial T/\partial V^k) - \partial T/\partial q^k]V^k, \quad \text{(d3)}$$

i.e.,

$$I_{n+1} = dT/dt - I_k V^k \Rightarrow dT/dt = I_\beta V^\beta; \quad \text{(d4)}$$

and, therefore,

$$dT/dt = (Q_\beta + \Lambda_\beta)V^\beta\text{: Power of } all \text{ acting forces.} \quad \text{(d5)}$$

Next, with the help of the Routh–Voss equations, $I_k = Q_k + \Lambda_k$, where $\Lambda_k = \lambda_D A^D{}_k$, Equations d4 and d5 become

$$I_{n+1} = dT/dt - (Q_k + \Lambda_k)V^k$$

$$= dT/dt - Q_k V^k - \lambda_D(A^D{}_k V^k) = dT/dt - Q_k V^k + \lambda_D A^D; \quad \text{(e1)}$$

and when this is combined with Equation c2 it yields the elementary forms:

$$dT/dt = Q_\beta V^\beta + \Lambda_{n+1} + \Lambda_k V^k = Q_\beta V^\beta + \Lambda_{n+1} - \lambda_D A^D. \quad \text{(e2)}$$

Of course, Equations e1 and 2 can also be derived by dotting the Newton–Euler equation of each particle with its velocity $\mathbf{v} = V^k \mathbf{e}_k + V^{n+1} \mathbf{e}_{n+1}$, then summing over all particles, etc.

If $T = T(q^k, V^k)$ (i.e., *stationary* initial, holonomic, constraints), then $I_{n+1} = 0 \Rightarrow Q_{n+1} + \Lambda_{n+1} = 0$, and the above specialize to

$$dT/dt = Q_k V^k + \Lambda_k V^k = Q_k V^k - \lambda_D A^D; \quad \text{(e3)}$$

* For additional details on the $(n + 1)$th Lagrangean equation, see also Mattioli (1931–1932).

and if, *in addition*, the Pfaffian constraints are *catastatic*, then Equation e3 reduces to the familiar form:

$$dT/dt = Q_k\, V^k. \tag{e4}$$

Finally, in view of the above, the HPE Equation 7.6.3 can be rewritten as

$$dh/dt = -\partial L/\partial t + (dT/dt - I_{n+1}) = -\partial L/\partial t + [dT/dt - (Q_{n+1} + \Lambda_{n+1})]. \tag{e5}$$

The details of the corresponding transformation for I_{N+1} are left to the reader. For instance, it is not hard to show that:

$$I_{N+1} = A^{\beta}{}_{N+1}\, I_{\beta},\ Q_{N+1} = A^{\beta}{}_{N+1}\, Q_{\beta},\ \Lambda_{N+1} = A^{\beta}{}_{N+1}\, \Lambda_{\beta} \Rightarrow I_{N+1} = Q_{N+1} + \Lambda_{N+1};$$

$$dT/dt = I_k\, V^k + I_{n+1}\, V^{n+1} = I_k(A^k{}_I\, \omega^I + A^k) + I_{n+1}(A^{n+1}{}_I\, \omega^I + A^{n+1})$$

$$= \ldots = (A^{\beta}{}_I\, I_{\beta})\omega^I + (A^{\beta}\, I_{\beta})\omega^{N+1} = (Q_I + \Lambda_I)\omega^I + (Q_{N+1} + \Lambda_{N+1}). \tag{e6}$$

7.7 COMPREHENSIVE EXAMPLES AND PROBLEMS ON LAGRANGEAN DYNAMICS (SLED, SPHERE, RING)*

Example 7.7.1. The Knife (or Sled, or Skate, or Scissors, or Pizza Cutter, or Racing Boat)

Let us consider a knife K, modeled as a rigid thin and flat blade/bar AB bounded by a smooth and convex closed curve, moving on a fixed, say, horizontal, plane O–xy so that it remains perpendicular to it and in contact to it at point C (Figure 7.2a).

Let (x, y) be the coordinates of C, and e the (horizontal projection of the) distance, along K, between C and its center of mass G. Finally, let the angular orientation of K be given by the angle between, say GC and $+Ox$.

Geometry, Kinematics

Here the "natural" choice of Lagrangean (i.e., global positional system) coordinates is $q^1 = x$, $q^2 = y$, $q^3 = \phi$. The fact that, during such a motion, the resistance from the plane to the knife, at C, along the blade is much smaller than the resistance along the normal to it, is expressed by the constraint: *velocity of C,* $\mathbf{v} = (dx/dt, dy/dt)$, *must be along K,* or:

$$\frac{dy/dt}{dx/dt} = \tan\phi, \quad \text{or, simply,} \quad dy/dx = \tan\phi, \tag{a1}$$

or in Pfaffian form,

---

* Here we detail the *derivation/formulation* of the equations of motion etc. The systematic *solution* of the latter, analytically and/or computationally, must be sought elsewhere.

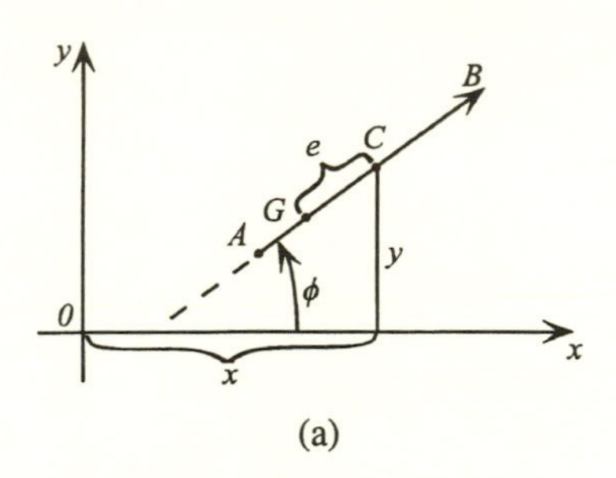

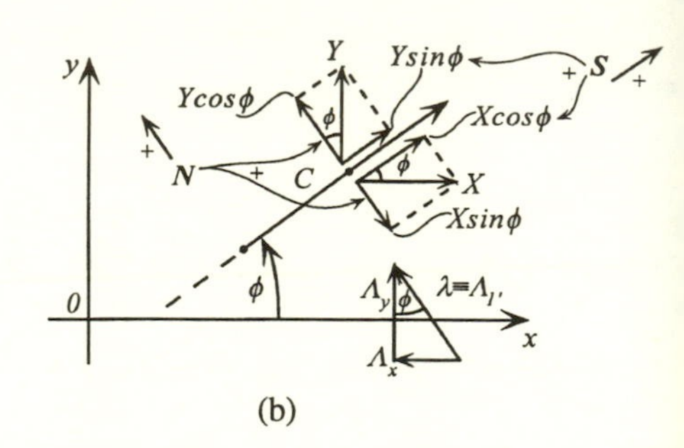

FIGURE 7.2 Geometry (a) and forces (b) of knife problem: K: knife (bar AB of mass m); $C(x, y)$: point of contact of K with fixed horizontal plane $O\text{–}xy$, $G(x' = x + e\cos\phi,\ y' = y + e\sin\phi)$: center of mass of K, e: (horizontal projection of) distance between C and G. $S = X\cos\phi + Y\sin\phi$: impressed force along knife, $N = -X\sin\phi + Y\cos\phi$: impressed force normal to knife, $\lambda = -\Lambda_x\sin\phi + \Lambda_y\cos\phi$: constraint reaction normal to knife.

$$(-\sin\phi)dx + (\cos\phi)dy + (0)d\phi = 0. \tag{a2}$$

Further, because this constraint is stationary (i.e., scleronomic system), its *virtual* form will be

$$(-\sin\phi)\delta x + (\cos\phi)\delta y + (0)\delta\phi = 0. \tag{a3}$$

Clearly, $n = 3$ and $m = 1$, and since (as shown below) Equations a1 or a2 are nonholonomic, this is a system with *three* global and *two* local freedoms, or $f \equiv n - m = 2$ DOF; and, as Frobenius' theorem teaches us, *it is the simplest nonholonomic problem possible.*

In view of Equations a1 through a3, we introduce the following three quasi-velocities (nonsingular transformation):

$$\omega^1 \equiv (-\sin\phi)(dx/dt) + (\cos\phi)(dy/dt) + (0)(d\phi/dt)$$

$$(= \text{component of } \mathbf{v} \text{ normal to } K) = 0, \tag{b1}$$

$$\omega^2 \equiv (\cos\phi)(dx/dt) + (\sin\phi)(dy/dt) + (0)(d\phi/dt)$$

$$= v \text{ (component of } \mathbf{v} \text{ along } K) \neq 0, \tag{b2}$$

$$\omega^3 \equiv (0)(dx/dt) + (0)(dy/dt) + (1)(d\phi/dt) \neq 0; \tag{b3}$$

which invert easily to (no constraint $\omega^1 = 0$ enforced):

$$dx/dt \equiv (-\sin\phi)\omega^1 + (\cos\phi)\omega^2 + (0)\omega^3, \tag{c1}$$

$$dy/dt \equiv (\cos\phi)\omega^1 + (\sin\phi)\omega^2 + (0)\omega^3, \tag{c2}$$

$$dz/dt \equiv (0)\omega^1 + (0)\omega^2 + (1)\omega^3. \tag{c3}$$

It is not hard to see that these transformations are nonsingular. Equations b1 through b3 and c1 through c3 also hold with all dq^k/dt replaced with dq^k or δq^k, and ω^K replaced with $d\theta^K$ or $\delta\theta^K$. Let us calculate the corresponding γ coefficients: $d(\ldots)$ and $\delta(\ldots)$–differentiating Equation b1 and its $d(\ldots)/\delta(\ldots)$ versions, we get

$$d(\delta\theta^1) - \delta(d\theta^1) = d(-\sin\phi\ \delta x + \cos\phi\ \delta y) - \delta(-\sin\phi\ dx + \cos\phi\ dy)$$

$$= (\sin\phi)(\delta dx - d\delta x) + (\cos\phi)(\delta dy - d\delta y)$$

$$+ \sin\phi(\delta\phi\ dy - d\phi\ \delta y) + \cos\phi(\delta\phi\ dx - d\phi\ \delta x)$$

(expressing dx, δx, ... in terms of $d\theta^K$, $\delta\theta^K$, ... from the d/δ versions of Equations c1 through 3)

$$= \delta\theta^3\ d\theta^2 - d\theta^3\ \delta\theta^2, \tag{d1}$$

and dividing by dt we finally get (no constraint $\omega^1 = 0$, $\delta\theta^1 = 0$ enforced) the transitivity equation:

$$(\delta\theta^1)^{\cdot} - \delta\omega^1 = (0)\delta\theta^1 + (-\omega^3)\delta\theta^2 + (\omega^2)\delta\theta^3; \tag{d2}$$

and, similarly,

$$(\delta\theta^2)^{\cdot} - \delta\omega^2 = (\omega^3)\delta\theta^1 + (0)\delta\theta^2 + (-\omega^1)\delta\theta^3, \tag{d3}$$

$$(\delta\theta^3)^{\cdot} - \delta\omega^3 = (0)\delta\theta^1 + (0)\delta\theta^2 + (0)\delta\theta^3. \tag{d4}$$

From Equations d2 through d4, we easily read off the *nonvanishing* γ coefficients:

$$\gamma^1{}_{23} = -\gamma^1{}_{32} = -1, \quad \gamma^2{}_{13} = -\gamma^2{}_{31} = 1. \tag{e)*}$$

Since $\gamma^D{}_{II'} \rightarrow \gamma^1{}_{23}, \gamma^1{}_{32} \neq 0$ (here $D = 1$; $I, I' = 2, 3$), by Frobenius' theorem (Section 6.7) the 1×3 constraint system: $\omega^1 \equiv (-\sin\phi)(dx/dt) + (\cos\phi)(dy/dt) + (0)(d\phi/dt) = 0$ is nonholonomic; i.e., it cannot be reduced to a finite relation, say, $f(x, y, \phi) = 0$ — the x, y, ϕ are independent.

Kinetics. The Routh–Voss equations.

The (inertial) kinetic energy of the knife, modeled as a bar AB of mass m and moment of inertia about an axis through C perpendicular to O–xy: $I_C \equiv I$, and of angular velocity ω is (see texts on dynamics):

* Since θ^3 is a holonomic coordinate, all $\gamma^3{}_{..}$ vanish; and since $\gamma^2{}_{13} = \gamma^2{}_{31} \neq 0$, $\omega^2 \equiv v$ is a quasi-velocity; and these results have nothing to do with constraints (recall Equations 3.12.5 and 6.4.1f).

$$2T = m\ v^2 + I(d\phi/dt)^2 + 2\ m\ \mathbf{v}_C \cdot (\omega \times \mathbf{r}_{G/C})$$

(we notice that $\omega \times \mathbf{r}_{G/C} = \mathbf{v}_{G/C}$)

$$= m[(dx/dt)^2 + (dy/dt)^2] + I(d\phi/dt)^2$$

$$+ 2m(d\phi/dt)^2\ [(dy/dt)(x' - x) - (dx/dt)(y' - y)]$$

$$= m[(dx/dt)^2 + (dy/dt)^2] + I(d\phi/dt)^2$$

$$+ 2m\ e(d\phi/dt)\ [(dy/dt)\cos\phi - (dx/dt)\sin\phi], \qquad \text{(f)}$$

(we do not enforce the constraint $\omega^1 \equiv \cos\phi\ (dy/dt) - \sin\phi\ (dx/dt) = 0$ in T, yet!).

Then, with $\delta'W \equiv X\,\delta x + Y\,\delta y + M\,\delta\phi$ (i.e., $k = 1, 2, 3 \Rightarrow x, y, z$), the Routh–Voss equations $I_k \equiv E_k(T) = Q_k + \lambda_D A^D{}_k$ become (with accents used to denote specific nonholonomic components, *wherever ambiguous*, i.e., not in ω^1, $\delta\theta^1$, $\gamma^1{}_{23}$, etc.):

$$I_x \equiv E_x(T) = d/dt[m(dx/dt) - m\ e\ \sin\phi\ (d\phi/dt)] = X + \lambda_{1'}(-\sin\phi), \quad \text{(g1)}$$

$$I_y \equiv E_y(T) = d/dt[m(dy/dt) + m\ e\ \cos\phi\ (d\phi/dt)] = Y + \lambda_{1'}(\cos\phi), \quad \text{(g2)}$$

$$I_\phi \equiv E_\phi(T) = d/dt\{m\ e[\cos\phi\ (dy/dt) - \sin\phi\ (dx/dt)]\} + I(d^2\phi/dt^2)$$

$$+ m\ e(d\phi/dt)\ [\sin\phi\ (dy/dt) + \cos\phi\ (dx/dt)] = M + \lambda_{1'}\ (0),$$

or

$$I(d^2\phi/dt^2) + m\ e(d\phi/dt)\ [\sin\phi\ (dy/dt) + \cos\phi\ (dx/dt)] = M. \qquad \text{(g3)}$$

Equations g1 through g3 plus the constraint (b1) constitute a determinate system of four equations for the four unknown functions $x(t)$, $y(t)$, $\phi(t)$, $\lambda_{1'}(t)$.*

For further details, see, e.g., Hamel (1949, pp. 465–470), Rosenberg, (1977, pp. 334–338).

---

* Had we enforced the constraint (b1) in T, i.e., $2T \to 2T_o = m\ v^2 + I(d\phi/dt)^2$ (that is, even without expressing T as a function of *two* velocities), and *then* applied the Routh–Voss equations, i.e., $E_k(T_o) = Q_k + \lambda_D\ A^D{}_k$, we would have obtained the following *incorrect* equations:

$$E_x(T_o) = d/dt[m(dx/dt)] = X + \lambda_{1'}\ (-\sin\phi), \qquad \text{(g4)}$$

$$E_y(T_o) = d/dt[m(dy/dt)] = Y + \lambda_{1'}\ (\cos\phi), \qquad \text{(g5)}$$

$$E_\phi(T_o) = d/dt[I(d\phi/dt)] = M + \lambda_{1'}\ (0). \qquad \text{(g6)}$$

Problem 7.7.1. Knife: Elimination of the Constraint Reactions, The Maggi Equations.

i. Continuing from the preceding example, show that elimination of $\lambda_{1'} \equiv \lambda$ among the Routh-Voss equations of that example, i.e., between its Equations g1 and g2, yields the following reactionless equation:

$$m \cos\phi\,(d^2x/dt^2) + m \sin\phi\,(d^2y/dt^2) - m\,e\,(d\phi/dt)^2$$

$$= m[dv/dt - e(d\phi/dt)^2] = m\ (\textit{acceleration of G along knife}) = S, \tag{a}$$

where

$$S \equiv X \cos\phi + Y \sin\phi\text{: } \textit{impressed force along knife;} \tag{a1}$$

while the already reactionless Equation g3 is rewritten as

$$I(d^2\phi/dt^2) + m\,e(d\phi/dt)\,[\sin\phi\,(dy/dt) + \cos\phi\,(dx/dt)] = M. \tag{b}$$

Hint: Multiply Equation g1 with $\cos\phi$, Equation g2 with $\sin\phi$, and add together.

ii. Equations a and b are the $n - m = 2$ reactionless Maggi equations of our problem; i.e., kinetic and in holonomic variables. The *third*, kinetostatic, Maggi equation is obtained by solving Equations g1 and g2 for λ. Show that after carrying this out, one obtains

$$m \cos\phi(d^2y/dt^2) - m \sin\phi(d^2x/dt^2) + m\,e(d^2\phi/dt^2) = m[v(d\phi/dt) + e(d^2\phi/dt^2)]$$

$$= m\ (\textit{acceleration of G normal to knife}) = N + \lambda, \tag{c}$$

where

$$N \equiv -X \sin\phi + Y \cos\phi\text{: } \textit{impressed force normal to knife} \tag{c1}$$

Hint: Multiply Equation g1 with $-\sin\phi$, Equation g2 with $\cos\phi$, and add together.

Once $x(t)$, $y(t)$, $\phi(t)$ have been found from the differential Equations a and b and the constraint $\cos\phi\,(dy/dt) - \sin\phi\,(dx/dt) = 0$, then $\lambda(t)$ can be immediately obtained from Equation c.

iii. As Equations g1 through g3 show, the holonomic covariant components of the system reaction are

$$\Lambda_x = -\lambda \sin\phi, \quad \Lambda_y = \lambda \cos\phi, \quad \Lambda_\phi = 0. \tag{d}$$

Verify that

$$\lambda = -\Lambda_x \sin\phi + \Lambda_y \cos\phi, \tag{e}$$

i.e., Equation c is the equation of linear momentum for the knife along the normal to it (Figure 7.2.b).

Hints: We have the following kinematical relations:

$$dx/dt = v \cos\phi,\ dy/dt = v \sin\phi \Rightarrow d^2x/dt^2 = (dv/dt) \cos\phi - v(d\phi/dt) \sin\phi,$$

$$d^2y/dt^2 = (dv/dt) \sin\phi + v(d\phi/dt) \cos\phi$$

$$\Rightarrow (d^2x/dt^2) \cos\phi + (d^2y/dt^2) \cos\phi = dv/dt\text{: } \textit{C-acceleration along knife,}$$

$$(d^2y/dt^2) \cos\phi - (d^2x/dt^2) \cos\phi = v\ (d\phi/dt)\text{: } \textit{C-acceleration normal to knife}$$

$$\Rightarrow dv/dt - e(d\phi/dt)^2\text{: } \textit{G-acceleration along knife,}$$

$$v(d\phi/dt) + e(d^2\phi/dt^2)\text{: } \textit{G-acceleration normal to knife.}$$

Example 7.7.2. The Knife (continued). The Maggi Equations

In the preceding Problem, the Maggi equations were derived in an *ad hoc* fashion. Here we derive them from the general theory and the results of the last Example. We have, recalling Equations c1 through c3 of Example 7.7.1 (and, again, using accents for specific nonholonomic components):

$$I_D = Q_D + \Lambda_D, \quad \text{or} \quad A^k{}_D\, I_k = A^k{}_D(Q_k + \Lambda_k) \Rightarrow A^k{}_D(I_k - Q_k - \Lambda_k) = 0:$$

$$0 = A^x{}_{1'}(I_x - Q_x - \Lambda_x) + A^y{}_{1'}(I_y - Q_y - \Lambda_y) + A^\phi{}_{1'}(I_\phi - Q_\phi - \Lambda_\phi)$$

$$= (-\sin\phi)\ [E_x(T) - X - (-\lambda \sin\phi)]$$

$$+ (\cos\phi)\ [E_y(T) - Y - (\lambda \cos\phi)] + (0)\ [E_\phi(T) - M - (0)]; \tag{a1}$$

$$I_I = Q_I, \quad \text{or} \quad A^k{}_I\, I_k = A^k{}_I\, Q_k \quad [\text{since } \Lambda_I = A^k{}_I\, \Lambda_k = 0] \Rightarrow A^k{}_I(I_k - Q_k) = 0:$$

$$0 = A^x{}_{2'}(I_x - Q_x) + A^y{}_{2'}(I_y - Q_y) + A^\phi{}_{2'}(I_\phi - Q_\phi)$$

$$= (\cos\phi)\ [E_x(T) - X] + (\sin\phi)\ [E_y(T) - Y] + (0)\ [E_\phi(T) - M], \tag{a2}$$

$$0 = A^x{}_{3'}(I_x - Q_x) + A^y{}_{3'}(I_y - Q_y) + A^\phi{}_{3'}(I_\phi - Q_\phi)$$

$$= (0)[E_x(T) - X] + (0)[E_y(T) - Y] + (1)[E_\phi(T) - M]; \tag{a3}$$

Equations a1 through a3 are the projections/componets of the Routh–Voss equations (system vector) respectively, *along the normal to the knife, along the knife, and along* ϕ (taken as the z-direction to O–xy).

It is not hard to verify that Equations a1 through a3 coincide, respectively, with Equations c, a, and b of Problem 7.7.1; i.e.,

$$Q_{1'} \to N, \quad Q_{2'} \to S, \quad Q_{3'} \to M;$$

$$\Lambda_{1'} \to \lambda, \quad \Lambda_{2'} \to 0, \quad \Lambda_{3'} \to 0. \tag{b}$$

Example 7.7.3. The Knife (continued). The Hamel Equations

To obtain them, we shall use the fundamental central equation in quasi-variables (Equation 7.5.6c):

$$(dP_K/dt)\delta\theta^K - [(\partial T^*/\partial q^k)A^k{}_K]\delta\theta^K + P_K[(\delta\theta^K)^{\cdot} - \delta\omega^K] = Q_K\,\delta\theta^K. \tag{a}$$

Recalling the results of Example 7.7.1, we get $T \to T^*$ [no enforcement of constraint $\omega^1 = 0$ yet; but we can safely neglect the *quadratic* term $m\,(\omega^1)^2$, below, at this stage (why?)]:

$$2T \to 2T^* = m[(\omega^1)^2 + (\omega^2)^2] + 2\,m\,e\,\omega^1\,\omega^3 + I(\omega^3)^2; \tag{b}$$

from which it follows that here: $\partial T^*/\partial q^k = 0$; and that the nonholonomic momenta are

$$P_1 = \partial T^*/\partial\omega^1 = m\,e\,\omega^1, \quad P_2 = \partial T^*/\partial\omega^2 = m\,\omega^2,$$

$$P_3 = \partial T^*/\partial\omega^3 = m\,e\,\omega^1 + I\,\omega^3. \tag{c}$$

Hence, utilizing in Equation a the transitivity equations (d2 through d4 of Example 7.7.1), and then collecting terms appropriately, we obtain the *constrained* variational equation:

$$[dP_1/dt + P_2\,\omega^3 - Q_1]\delta\theta^1 + [dP_2/dt - P_1\,\omega^3 - Q_2]\delta\theta^2$$

$$+ [dP_3/dt + P_1\,\omega^2 - Q_3]\delta\theta^1 = 0, \tag{d}$$

where the nonholonomic impressed forces Q_K: $Q_{1,2,3}$ (no need for accents here) are defined by

$$\delta' W \equiv X\,\delta x + Y\,\delta y + M\,\delta\phi \quad \text{(Recalling Equations a1 and c1 of Problem 7.7.1)}$$

$$= Q_1\,\delta\theta^1 + Q_2\,\delta\theta^2 + Q_3\,\delta\theta^3 = N\,\delta\theta^1 + S\,\delta\theta^2 + M\,\delta\theta^3. \tag{e}$$

By adjoining the constraint in virtual form, i.e., $\delta\theta^1 = 0$, to Equation d, via the method of multipliers (thus transforming it to an *unconstrained* variational equation) and then setting each of its $\delta\theta^K$-coefficients equal to zero, we get the three Hamel equations (and, in each, we enforce $\omega^1 = 0$):

$$\delta\theta^1: \quad dP_1/dt + P_2\,\omega^3 = Q_1 + \Lambda_1$$

(Recalling Equations b1 through b3 of Example 7.7.1, Equations b of Example 7.7.2, and Equations c)

$$\text{or} \quad me(d\omega^3/dt) + m\,\omega^2\,\omega^3 = m\,e(d^2\phi/dt^2) + m\,v(d\phi/dt) = N + \lambda, \tag{f1}$$

$$\delta\theta^2: \quad dP_2/dt - P_1\,\omega^3 = Q_2,$$

$$\text{or} \quad m(d\omega^2/dt) - m\,e(\omega^3)^2 = m(dv/dt) - m\,e(d\phi/dt)^2 = S, \tag{f2}$$

$$\delta\theta^3: \quad dP_3/dt + P_1\,\omega^2 = Q_3,$$

$$\text{or} \quad I(d\omega^3/dt) + m\,e\,\omega^2\,\omega^3 = I(d^2\phi/dt^2) + m\,e\,v(d\phi/dt) = M. \tag{f3}$$

In the *force-free* case, i.e., when N, S, $M = 0$, Equations f2 and f3 reduce to

$$dv/dt - e(d\phi/dt)^2 = 0 \quad \text{and} \quad I(d^2\phi/dt^2) + m\,e\,v(d\phi/dt) = 0, \tag{g1}$$

respectively; and, as can be verified easily, lead to the *energy integral:*

$$T = \text{constant:} \quad m\,v^2 + I(d\phi/dt)^2 = \text{constant}. \tag{g2}$$

For further details, see, e.g., Hamel (1949, pp. 465–470, 483), Rosenberg (1977, pp. 334–338).

Example 7.7.4. The Sphere

Let us consider a homogeneous sphere B, of mass m, center of mass G, and radius r, rolling (without slipping) on a rough horizontal plane P. The latter spins, relative to inertial space, about a fixed axis OZ perpendicular to it with *constant* angular velocity Ω.

Kinematics.

A generic configuration of B is specified by the following *five* independent Lagrangean coordinates $q^{1,\ldots,5}$: (i) the inertial coordinates of G ($x_G \equiv x$, $y_G \equiv y$) ($z_G = r$, always and independently of any further considerations) and (iii) the three Eulerian angles, $\phi \to \theta \to \psi$, defining the angular orientation of B-fixed axes G–xyz relative to translating but nonrotating axes G–XYZ [or relative to the inertial axes O–XYZ (with $+OZ$ upwards) (Example 6.5.1)]. The rolling constraint expresses the fact that *the (instantaneously) contacting points of the sphere and the plane, C, have equal inertial velocities* (with all components along O–XYZ):

$$(\mathbf{v}_C)_{\text{sphere}} = (\mathbf{v}_C)_{\text{plane}}, \tag{a}$$

$$(\mathbf{v}_C)_{\text{sphere}} = \mathbf{v}_G + \omega \times \mathbf{r}_{C/G}$$

$$= (dx/dt,\ dy/dt,\ 0) + (\omega_X,\ \omega_Y,\ \omega_Z) \times (0,\ 0,\ -r), \tag{a1}$$

$$(\mathbf{v}_C)_{\text{plane}} = \Omega \times \mathbf{r}_{C/O} = (0,\ 0,\ \Omega) \times (x,\ y,\ 0); \tag{a2}$$

here ω inertial angualr velocity of B; from which, calculating components and equating them (and with $dx/dt \equiv v^x$, $dy/dt \equiv v^y$), we get

$$v^x - r\ \omega_Y = -\Omega\ y, \quad v^y + r\ \omega_X = \Omega\ x, \quad 0 = 0; \tag{a3}$$

or, further expressing $\omega_{X,Y}$ in terms of the Eulerian rates $d\phi/dt \equiv v^\phi$, etc. (Equations b1 of Example 6.5.1), we obtain the following two Pfaffian *nonholonomic* (as shown below) and *acatastatic* constraints ($\sim\Omega$ terms):

$$v^x - r(\sin\phi\ v^\theta - \sin\theta\ \cos\phi\ v^\psi) + \Omega\ y = 0, \tag{b1}$$

$$v^y + r(\cos\phi\ v^\theta + \sin\theta\ \sin\phi\ v^\psi) - \Omega\ x = 0. \tag{b2}$$

In view of the above, we choose the following (constraint-satisfying) five quasivelocities:

$$\omega^1 \equiv v^x - r(\sin\phi\ v^\theta - \sin\theta\ \cos\phi\ v^\psi) + \Omega y = 0, \tag{c1}$$

$$\omega^2 \equiv v_y + r(\cos\phi\ v^\theta + \sin\theta\ \sin\phi\ v^\psi) - \Omega x = 0; \tag{c2}$$

$$\omega^3 \equiv \omega_X = (0)v^\phi + (\cos\phi)v^\theta + (\sin\phi\ \sin\theta)v^\psi \neq 0, \tag{c3}$$

$$\omega^4 \equiv \omega_Y = (0)v^\phi + (\sin\phi)v^\theta + (-\cos\phi\ \sin\theta)v^\psi \neq 0, \tag{c4}$$

$$\omega^5 \equiv \omega_Z = (1)v^\phi + (0)v^\theta + (\cos\theta)v^\psi \neq 0; \tag{c5}$$

$$[\text{and } \omega^6 \equiv V^6 = dt/dt = 1 \text{ (isochrony)}]; \tag{c6}$$

which invert easily to (no constraint enforcement yet!):

$$V^1 \equiv v^x = \omega^1 + r\ \omega^4 - \Omega\ y, \tag{d1}$$

$$V^2 \equiv v^y = \omega^2 - r\ \omega^3 + \Omega\ x, \tag{d2}$$

$$V^3 \equiv v^\phi = (-\text{cotan}\,\theta\ \sin\phi)\omega^3 + (-\text{cotan}\,\theta\ \cos\phi)\omega^4 + (1)\omega^5, \tag{d3}$$

$$V^4 \equiv v^\theta = (\cos\phi)\omega^3 + (\sin\phi)\omega^4 + (0)\omega^5, \tag{d4}$$

$$V^5 \equiv v^\psi = (\sin\phi/\sin\theta)\omega^3 + (-\cos\phi/\sin\theta)\omega^4 + (0)\omega^5; \tag{d5}$$

$$[\text{and } V^6 \equiv \omega^6 = dt/dt = 1 \text{ (isochrony)}]. \tag{d6}$$

The coefficients of Equations c1 through c6 and d1 through d6 are, respectively, the coefficients $A^\Lambda{}_\lambda$ and $A^\lambda{}_\Lambda$ of the $V^\lambda \leftrightarrow \omega^\Lambda$ transformation. Equations c1 through c6 also hold with all $d(\ldots)/dt$ replaced by $d(\ldots)$ (kinematically admissible/possible forms). However, the *virtual* forms of these equations are

$$\delta\theta^1 \equiv \delta x - r(\sin\phi\ \delta\theta - \sin\theta\ \cos\phi\ \delta\psi) = 0, \tag{e1}$$

$$\delta\theta^2 \equiv \delta y + r(\cos\phi\ \delta\theta + \sin\theta\ \sin\phi\ \delta\psi) = 0; \tag{e2}$$

$$\delta\theta^3 \equiv \delta\theta_X = (0)\delta\phi + (\cos\phi)\delta\theta + (\sin\phi\ \sin\theta)\delta\psi \neq 0, \tag{e3}$$

$$\delta\theta^4 \equiv \delta\theta_Y = (0)\delta\phi + (\sin\phi)\delta\theta + (-\cos\phi\ \sin\ \theta)\delta\psi \neq 0, \tag{e4}$$

$$\delta\theta^5 \equiv \delta\theta_Z = (1)\delta\phi + (0)\delta\theta + (\cos\theta)\delta\psi \neq 0; \tag{e5}$$

$$\delta q^1 \equiv \delta x = \delta\theta^1 + r\ \delta\theta^4, \tag{e6}$$

$$\delta q^2 \equiv \delta y = \delta\theta^2 - r\ \delta\theta^3, \tag{e7}$$

$$\delta\theta^3 \equiv \delta\phi = (-\text{cotan}\,\theta\ \sin\phi)\delta\theta^3 + (-\text{cotan}\,\theta\ \cos\phi)\delta\theta^4 + (1)\delta\theta^5, \tag{e8}$$

$$\delta q^4 \equiv \delta\theta = (\cos\phi)\delta\theta^3 + (\sin\phi)\delta\theta^4 + (0)\delta\theta^5, \tag{e9}$$

$$\delta q^5 \equiv \delta\psi = (\sin\phi/\sin\theta)\delta\theta^3 + (-\cos\phi/\sin\theta)\delta\theta^4 + (0)\delta\theta^5; \tag{e10}$$

$$[\text{and } \delta\theta^6 \equiv \delta q^6 = \delta t = 0 \text{ (virtualness)}]; \tag{e11}$$

i.e., *no nonhomogeneous terms* ($\sim\Omega$) *in the* $\delta q^\lambda \leftrightarrow \delta\theta^\Lambda$ *transformation.*

The Hamel coefficients are found from the following transitivity equations:

$$(\delta\theta^K)^\cdot - \delta\omega^K = \gamma^K{}_{R\Sigma}\,\omega^\Sigma\,\delta\theta^R \quad (K, R\text{: } 1, \ldots, 5;\ \Sigma = 1, \ldots, 6). \tag{f1}$$

If we only need the *kinetic* equations, then (since $\delta\theta^1, \delta\theta^2, \delta\theta^6 = 0$ and $\omega^1, \omega^2 = 0$), we use in the above: $K = 1, \ldots, 5$; $R = 3, 4, 5$; $\Sigma = 3, \ldots, 6$. Thus, we find successively:*

$$(\delta\theta^1)^\cdot - \delta\omega^1 = (\delta x - r\ \delta\theta_Y)^\cdot - \delta(v^x - r\ \omega_Y + \Omega\ y)$$

$$= [(\delta x)^\cdot - \delta(dx/dt)] - r[(\delta\theta_Y)^\cdot - \delta\omega_Y] - \Omega\ \delta y$$

* We note that, even if $\Omega = \Omega(t)$: given function of t, still $\delta\Omega = 0$.

(recalling Equations d1 through d5 of Example 6.5.1, with $\chi \to \theta$)

$$= (0) - r(\omega_Z\ \delta\theta_X - \omega_X\ \delta\theta_Z) - \Omega(\delta\theta^2 - r\ \delta\theta^3)$$

$$= -r(\omega^5\ \delta\theta^3 - \omega^3\ \delta\theta^5) - \Omega(\delta\theta^2 - r\ \delta\theta^3),$$

finally,

$$(\delta\theta^1)^{\cdot} - \delta\omega^1 = (-\Omega)\delta\theta^2 + (r\ \Omega)\delta\theta^3 + (-r\ \omega^5)\delta\theta^3 + (r)\omega^3\ \delta\theta^5; \qquad \text{(f2)}$$

and, similarly,

$$(\delta\theta^2)^{\cdot} - \delta\omega^2 = (\delta y + r\ \delta\theta_X)^{\cdot} - \delta(v^y + r\ \omega_X - \Omega\ x)$$

$$= [(\delta y)^{\cdot} - \delta(dy/dt)] + r[(\delta\theta_X)^{\cdot} - \delta\omega_X] + \Omega\ \delta x$$

$$= (0) + r(\omega_Y\ \delta\theta_Z - \omega_Z\ \delta\theta_Y) + \Omega(\delta\theta^1 + r\ \delta\theta^4)$$

$$= r(\omega^4\ \delta\theta^5 - \omega^5\ \delta\theta^4) + \Omega(\delta\theta^1 + r\ \delta\theta^4),$$

finally,

$$(\delta\theta^2)^{\cdot} - \delta\omega^2 = (\Omega)\delta\theta^1 + (r\ \Omega)\delta\theta^4 + (-r)\omega^5\ \delta\theta^4 + (r)\omega^4\ \delta\theta^5; \qquad \text{(f3)}$$

and the earlier found (Equations d1 ff of Example 6.5.1):

$$(\delta\theta^3)^{\cdot} - \delta\omega^3 = \omega^4\ \delta\theta^5 - \omega^5\ \delta\theta^4, \qquad \text{(f4)}$$

$$(\delta\theta^4)^{\cdot} - \delta\omega^4 = \omega^5\ \delta\theta^3 - \omega^3\ \delta\theta^5, \qquad \text{(f5)}$$

$$(\delta\theta^5)^{\cdot} - \delta\omega^5 = \omega^3\ \delta\theta^4 - \omega^4\ \delta\theta^3. \qquad \text{(f6)}$$

From Equations f1 through f6, we immediately read off the following nonvanishing Hamel coefficients:

$$\gamma^1{}_{35} = -\gamma^1{}_{53} = -r; \quad \gamma^1{}_{26} = -\gamma^1{}_{62} = -\Omega, \quad \gamma^1{}_{36} = -\gamma^1{}_{63} = r\ \Omega; \qquad \text{(f7)}$$

$$\gamma^2{}_{45} = -\gamma^2{}_{54} = -r; \quad \gamma^2{}_{46} = -\gamma^2{}_{64} = r\ \Omega, \quad \gamma^2{}_{16} = -\gamma^2{}_{61} = \Omega; \qquad \text{(f8)}$$

$$\gamma^3{}_{\dots},\ \gamma^4{}_{\dots},\ \gamma^5{}_{\dots}: -\ (\text{permutation symbol})\ [\gamma^K{}_{LR} = -\varepsilon_{KLR}\ (K, L, R: 3, 4, 5)]; \qquad \text{(f9)}$$

i.e., since $\gamma^D{}_{II'}$: $\gamma^1{}_{35}$, $\gamma^2{}_{45} \neq 0$, the constraints $\omega^1, \omega^2 = 0$ are nonholonomic (even if $\Omega = 0$).

Kinetics

The (inertial) kinetic energy of the sphere is (see texts on dynamics*), with $I_{G;X,Y,Z} \equiv I = 2mr^2/5$, $\omega_{X,Y,Z}$ given by Equations b1 of Example 6.5.1, and *no constraint enforcement yet*:

$$2T = m[(v^x)^2 + (v^y)^2 + (v^z)^2] + (I_X\, \omega_X^2 + I_Y\, \omega_Y^2 + I_Z\, \omega_Z^2), \qquad \text{(g1)}$$

and, therefore, in *holonomic variables:*

$$2T = m[(v^x)^2 + (v^y)^2] + I[(v^\phi)^2 + (v^\theta)^2 + (v^\psi)^2 + 2\, v^\phi\, v^\psi \cos\theta], \qquad \text{(g2)}$$

and in *nonholonomic* ones (recalling Equations d1 and d2):

$$2T \to 2T^* = m[(\omega^1 + r\,\omega_Y - \Omega\, y)^2 + (\omega^2 - r\,\omega_X + \Omega\, x)^2]$$

$$+ (I_X\, \omega_X^2 + I_Y\, \omega_Y^2 + I_Z\, \omega_Z^2)$$

$$= m[(\omega^1 + r\,\omega^4 - \Omega\, y)^2 + (\omega^2 - r\,\omega^3 + \Omega\, x)^2]$$

$$+ I[(\omega^3)^2 + (\omega^4)^2 + (\omega^5)^2]. \qquad \text{(g3)}$$

The *Routh–Voss Equations* (*R–V*): using T from Equation g2, the constraint coefficients by Equations c1 and c2 (with accents to denote specific nonholonomic indices):

$$A^{1'}{}_1 = 1,\quad A^{1'}{}_2 = 0,\quad A^{1'}{}_3 = 0,\quad A^{1'}{}_4 = -r\sin\phi,\quad A^{1'}{}_5 = r\sin\theta\cos\phi, \qquad \text{(h1)}$$

$$A^{2'}{}_1 = 0,\quad A^{2'}{}_2 = 1,\quad A^{2'}{}_3 = 0,\quad A^{2'}{}_4 = r\cos\phi,\quad A^{2'}{}_5 = r\sin\theta\sin\phi, \qquad \text{(h2)}$$

and the impressed forces calculated from $Q_k\, \delta q^k = Q_x\, \delta x + Q_y\, \delta y + Q_\phi\, \delta\phi + Q_\theta\, \delta\theta + Q_\psi\, \delta\psi$ (as if no constraints existed) we obtain the following *five* *R–V* equations: $E_k(T) = Q_k + \lambda_D\, A^D{}_k$ ($k = x, y, \phi, \theta, \psi$; $D = 1', 2'$, or $\lambda_{1'} \to \lambda$, $\lambda_{2'} \to \mu$):

$$m(d^2x/dt^2) = Q_x + \lambda, \qquad \text{(h3)}$$

$$m(d^2y/dt^2) = Q_y + \mu, \qquad \text{(h4)}$$

$$I[(d^2\phi/dt^2) + (d^2\psi/dt^2)\cos\theta - (d\theta/dt)(d\psi/dt)\sin\theta] = (I\,\omega_Z)^{\cdot} = Q_\phi, \qquad \text{(h5)}$$

$$I[(d^2\theta/dt^2) + (d\phi/dt)(d\psi/dt)\sin\theta] = Q_\theta + \lambda\,(-r\sin\phi) + \mu\,(r\cos\phi), \qquad \text{(h6)}$$

$$I[(d^2\psi/dt^2) + (d^2\phi/dt^2)\cos\theta - (d\phi/dt)(d\theta/dt)\sin\theta] =$$

$$Q_\psi + \lambda(r\sin\theta\cos\phi) + \mu(r\sin\theta\sin\phi). \qquad \text{(h7)}$$

* See, e.g., Synge and Griffith (1959, pp. 294ff). This is commonly called *König's theorem.*

Equation h5 readily shows that if $Q_\phi = 0$, then $\omega_Z = d\phi/dt + (d\psi/dt)\cos\theta = \text{constant}$. Frequently, in order to avoid errors in the enumeration of the $A^K{}_k$, $A^k{}_K$, we use LP *directly*:

$$\delta I = \delta' W, \quad \delta I \equiv E_k(T)\,\delta q^k, \quad \delta' W \equiv Q_k\,\delta q^k \quad (k = 1, 2, 3, 4, 5), \qquad \text{(h8)}$$

and then adjoin to it the constraints in virtual form. Thus, we find the *constrained* variational equation:

$$m[(d^2x/dt^2)\delta x + (d^2y/dt^2)\delta y] + I[(d^2\theta/dt^2) + (d\phi/dt)(d\psi/dt)\sin\theta]\delta\theta$$

$$+ I[(d^2\phi/dt^2) + (d^2\psi/dt^2)\cos\theta - (d\theta/dt)(d\psi/dt)\sin\theta]\delta\phi$$

$$+ I[(d^2\psi/dt^2) + (d^2\phi/dt^2)\cos\theta - (d\phi/dt)(d\theta/dt)\sin\theta]\delta\psi$$

$$= Q_x\,\delta x + Q_y\,\delta y + Q_\phi\,\delta\phi + Q_\theta\,\delta\theta + Q_\psi\,\delta\psi; \qquad \text{(h9)}$$

and then add to it:

$$(-\lambda)[\delta x - r(\sin\phi\,\delta\theta - \sin\theta\cos\phi\,\delta\psi)]$$

$$+ (-\mu)[\delta y + r(\cos\phi\,\delta\theta + \sin\theta\sin\phi\,\delta\psi)] = 0.$$

Finally, setting equal to zero the five coefficients of the so resulting *unconstrained* variational equation, $(\ldots)_k\,\delta q^k = 0$, we obtain Equations h3 through h7.

The *Maggi Equations* (*M*): With the help of Equations d1 through d5, which give the coefficients $A^k{}_K$, and the notation $M_k \equiv E_k(T) - Q_k$, we can immediately write down the five Maggi equations:

$K \to D$ (kinetostatic): 1: $M_x = \lambda$, 2: $M_y = \mu$; (i1,2)

$K \to I$ (kinetic):
3: $(-r)M_y + (-\cot\theta\sin\phi)M_\phi + (\cos\phi)M_\theta + (\sin\phi/\sin\theta)M_\psi = 0$,
or explicitly:

$$(-r)[m(d^2y/dt^2) - Q_y] + (-\cot\theta\sin\phi)\{I[(d^2\phi/dt^2) + (d^2\psi/dt^2)\cos\theta$$

$$- (d\theta/dt)(d\psi/dt)\sin\theta] - Q_\phi\} + (\cos\phi)\{I[(d^2\theta/dt^2)$$

$$+ (d\phi/dt)(d\psi/dt)\cos\theta] - Q_\theta\} + (\sin\phi/\sin\theta)\{I[(d^2\psi/dt^2)$$

$$+ (d^2\phi/dt^2)\cos\theta - (d\phi/dt)(d\theta/dt)\sin\theta] - Q_\psi\} = 0; \qquad \text{(i3)}$$

4: $(r)M_x + (-\cot\theta \cos\phi)M_\phi + (\sin\phi)M_\theta + (-\cos\phi/\sin\theta)M_\psi = 0,$

or explicitly:

$$(r)[m(d^2x/dt^2) - Q_x] + (-\cot\theta \cos\phi)\{I[(d^2\phi/dt^2) + (d^2\psi/dt^2)\cos\theta$$

$$- (d\theta/dt)(d\psi/dt)\sin\theta] - Q_\phi\} + (\sin\phi)\{I[(d^2\theta/dt^2)$$

$$+ (d\phi/dt)(d\psi/dt)\cos\theta] - Q_\theta\} + (-\cos\phi/\sin\theta)\{I[(d^2\psi/dt^2)$$

$$+ (d^2\phi/dt^2)\cos\theta - (d\phi/dt)(d\theta/dt)\sin\theta] - Q_\psi\} = 0. \qquad \text{(i4)}$$

5: $M_\phi = 0.$ (i5)

We notice that Equations i1, i2, and i5 coincide, respectively, with the earlier *R–V* Equations (h3, h4, and h5). [For additional, equivalent, forms of the *M* equations (i3 through i5), see Problem 7.7.2 below.] Also, we point out that the equations resulting by *eliminating* d^2x/dt^2 and d^2y/dt^2 from Equations i3 and i4 with the help of the $d(\ldots)/dt$–derivatives of the constraints (b1 and b2), are none other than the corresponding *two* Chaplygin–Vornets Equations (*C–V*); the *third* such equation being Equation i5 (see Problem 7.7.3 below).

The *Hamel Equations* (*H*): Using Equations d1 through d5 for the constraint coefficients $A^k{}_K$ (with accents to denote specific nonholonomic indices; only wherever needed to eliminate ambiguity): $A^1{}_{3'} = 0$, $A^1{}_{4'} = r$, $A^1{}_{5'} = 0$, $A^2{}_{3'} = -r$, $A^2{}_{4'} = 0$, $A^2{}_{5'} = 0$, (f7 through f9) for the γ coefficients, Equation g3 for $T^* \Rightarrow P_K$, $\partial T^*/\partial\theta^K \equiv A^k{}_K(\partial T^*/\partial q^k)$, and enforcing the constraints $\omega^1, \omega^2 = 0$ *after* all differentiations, we obtain, after some careful algebra, the following three *kinetic H* equations:

$$dP_3/dt - \partial T^*/\partial\theta^3 + P_4\,\omega^5 - P_5\,\omega^4 + P_1\,r(\Omega - \omega^5) = Q_{3'}:$$

$$(7mr^2/5)d\omega_X/dt - m\,r\,\Omega(r\,\omega_Y - \Omega\,y) = Q_X, \qquad \text{(j1)}$$

(after using the constraints b1, b2 to eliminate dx/dt and dy/dt)

$$dP_4/dt - \partial T^*/\partial\theta^4 + P_5\,\omega^3 - P_3\,\omega^5 + P_2\,r(\Omega - \omega^5) = Q_{4'}:$$

$$(7mr^2/5)d\omega_Y/dt + m\,r\,\Omega(r\,\omega_X - \Omega\,x) = Q_Y, \qquad \text{(j2)}$$

$$dP_5/dt - \partial T^*/\partial\theta^5 + P_3\,\omega^4 - P_4\,\omega^3 + r(P_1\,\omega^3 + P_2\,\omega^4) = Q_{5'}:$$

$$(2mr^2/5)d\omega_Z/dt = Q_Z. \qquad \text{(j3)}$$

The Q_K can be found from the invariant relation (as if the constraints did not exist):

$$\delta'W = Q_x\,\delta x + Q_y\,\delta y + Q_\phi\,\delta\phi + Q_\theta\,\delta\theta + Q_\psi\,\delta\psi$$

(recalling Equations e6 through e10)

$$= Q_x(\delta\theta^1 + r\,\delta\theta^4) + Q_y(\delta\theta^2 - r\,\delta\theta^3) + Q_\phi[(-\text{cotan}\,\theta\,\sin\phi)\delta\theta^3$$

$$+ (-\text{cotan}\,\theta\,\cos\phi)\delta\theta^4 + (1)\delta\theta^5] + Q_\theta[(\cos\phi)\delta\theta^3 + (\sin\phi)\delta\theta^4]$$

$$+ Q_\psi[(\sin\phi/\sin\theta)\delta\theta^3 + (-\cos\phi/\sin\theta)\delta\theta^4]$$

$$= \ldots = Q_{1'}\,\delta\theta^1 + \ldots + Q_{5'}\,\delta\theta^5,$$

$$\Rightarrow \quad Q_{1'} = Q_x, \quad Q_{2'} = Q_y,$$

$$Q_{3'} = -r\,Q_y - \text{cotan}\,\theta\,\sin\phi\,Q_\phi + \cos\phi\,Q_\theta + (\sin\phi/\sin\theta)Q_\psi,$$

$$Q_{4'} = r\,Q_x - \text{cotan}\,\theta\,\cos\phi\,Q_\phi + \sin\phi\,Q_\theta - (\cos\phi/\sin\theta)Q_\psi,$$

$$Q_{5'} = Q_\phi. \tag{j4}$$

Solving the *first-order* system (Equations j1 through j3) (plus initial conditions) we find $\omega_{X,Y,Z}(t)$; and then integrating Equations c3 through c5, we obtain $\phi(t)$, $\theta(t)$, $\psi(t)$.*

The above, *plus the kinetostatic equations*, can also be found by direct application of the central equation (7.5.6c): $\delta I \equiv (dP_K/dt)\,\delta\theta^K + P^K\,[(\delta\theta^K)^\cdot - \delta\omega^K] - A^k{}_K\,(\partial T^*/\partial q^k)\,\delta\theta^K = Q_K\,\delta\theta^K$, $K = 1, \ldots, 5$ (or $1', \ldots, 5'$). Indeed, proceeding in this fashion we find:

$$\delta I = (dP_1/dt)\delta\theta^1 + \ldots + (dP_5/dt)\delta\theta^5$$

$$+ P_1[(-\Omega)\delta\theta^2 + r(\Omega - \omega^5)\delta\theta^3 + (r\,\omega^3)\delta\theta^5]$$

$$+ P_2\,[(\Omega)\delta\theta^1 + r(\Omega - \omega^5)\delta\theta^4 + (r\,\omega^4)\delta\theta^5] + P_3(\omega^4\,\delta\theta^5 - \omega^5\,\delta\theta^4)$$

$$+ P^4(\omega^5\,\delta\theta^3 - \omega^3\,\delta\theta^5) + P_5(\omega^3\,\delta\theta^4 - \omega^4\,\delta\theta^3) - [A^1{}_{1'}(\partial T^*/\partial q^1)\delta\theta^1$$

$$+ A^2{}_{1'}(\partial T^*/\partial q^2)\delta\theta^1 + A^1{}_{2'}(\partial T^*/\partial q^1)\delta\theta^2 + A^2{}_{2'}(\partial T^*/\partial q^2)\delta\theta^2$$

$$+ A^1{}_{3'}\,(\partial T^*/\partial q^1)\delta\theta^3 + A^2{}_{3'}(\partial T^*/\partial q^2)\delta\theta^3 + A^1{}_{4'}(\partial T^*/\partial q^1)\delta\theta^4$$

$$+ A^2{}_{4'}(\partial T^*/\partial q^2)\delta\theta^4 + A^1{}_{5'}(\partial T^*/\partial q^1)\delta\theta^5 + A^2{}_{5'}(\partial T^*/\partial q^2)\delta\theta^5]$$

$$= Q_{1'}\,\delta\theta^1 + \ldots + Q_{5'}\,\delta\theta^5, \tag{j5}$$

---

* In Equations j1 and j2, $-m\,r^2\,\Omega\,\omega_Y$ and $+m\,r^2\,\Omega\,\omega_X$ are called *gyroscopic* or Coriolis terms; i.e., their combined power vanishes: $(-m\,r^2\,\Omega\,\omega_Y)\,\omega_X + (m\,r^2\,\Omega\,\omega_X)\,\omega_Y = 0$ (recall mention in Section 7.6); and this is a general result: such terms result from the gradients of $T^*_{(1)}$, $T_{(1)}$, and that part of the kinetic energy does not appear in h^*, h, although it may appear in the generalized power equations (7.6.3 and 7.6.6) through the nonstationary term: $-\partial L^*/\partial t$, $-\partial L/\partial t$. Finally, $+m\,r\,\Omega^2\,y$ and $-m\,r\,\Omega^2\,y$ are the well-known *centripetal* terms.

and collecting $\delta\theta^K$ terms, we obtain the following constrained variational equation:

$$0 = \{dP_1/dt + P_2\,\Omega - [A^1{}_{1'}(\partial T^*/\partial q^1) + A^2{}_{1'}(\partial T^*/\partial q^2)] - Q_{1'}\}\delta\theta^1$$

$$+ \{dP_2/dt - P_1\,\Omega - [A^1{}_{2'}(\partial T^*/\partial q^1) + A^2{}_{2'}(\partial T^*/\partial q^2)] - Q_{2'}\}\delta\theta^2$$

$$+ \{dP_3/dt + P_1\, r(\Omega - \omega^5) + (P_4\,\omega^5 - P_5\,\omega^4) - [A^1{}_{3'}(\partial T^*/\partial q^1)$$

$$+ A^2{}_{3'}(\partial T^*/\partial q^2)] - Q_{3'}\}\delta\theta^3 + \{dP_4/dt + P_2\, r(\Omega - \omega^5)$$

$$+ (P_5\,\omega^3 - P_3\,\omega^5) - [A^1{}_{4'}(\partial T^*/\partial q^1) + A^2{}_{4'}(\partial T^*/\partial q^2)] - Q_{4'}\}\delta\theta^4$$

$$+ \{dP_5/dt + r(P_1\,\omega^3 + P_2\,\omega^4) + (P_3\,\omega^4 - P_4\,\omega^3) - [A^1{}_{5'}(\partial T^*/\partial q^1)$$

$$+ A^2{}_{5'}(\partial T^*/\partial q^2)] - Q_{5'}\}\delta\theta^5,$$

then, adjoining to it the constraints $\delta\theta^1 = 0$ and $\delta\theta^2 = 0$ via the method of multipliers, and setting the coefficients of the so-resulting unconstrained variational equation equal to zero, we obtain the five H equations:

$$dP_1/dt + P_2\,\Omega - (1)(\partial T^*/\partial x) + (0)(\partial T^*/\partial y) = Q_{1'} + \Lambda_{1'} \quad (\Lambda_{1'} = \lambda), \tag{j6}$$

$$dP_2/dt - P_1\,\Omega - (0)(\partial T^*/\partial x) + (1)(\partial T^*/\partial y) = Q_{2'} + \Lambda_{2'} \quad (\Lambda_{2'} = \mu); \tag{j7}$$

$$dP_3/dt + P_1\, r(\Omega - \omega^5) + (P_4\,\omega^5 - P_5\,\omega^4) - (-r)(\partial T^*/\partial y) = Q_{3'}, \tag{j8}$$

$$dP_4/dt + P_2\, r(\Omega - \omega^5) + (P_5\,\omega^3 - P_3\,\omega^5) - (r)(\partial T^*/\partial x) = Q_{4'}, \tag{j9}$$

$$dP_5/dt + r(P_1\,\omega^3 + P_2\,\omega^4) + (P_3\,\omega^4 - P_4\,\omega^3) = Q_{5'}\,. \tag{j10}$$

Problem 7.7.2

Show that direct elimination of λ, μ among the five Routh–Voss sphere equations results in *three* kinetic Maggi equations:

$$I[(d^2\phi/dt^2) + (d^2\psi/dt^2)\cos\theta - (d\theta/dt)(d\psi/dt)\sin\theta] = Q_\phi, \tag{a}$$

$$I[(d^2\theta/dt^2) + (d\phi/dt)(d\psi/dt)\sin\theta] + m\, r[(d^2x/dt^2)\sin\phi - (d^2y/dt^2)\cos\phi]$$

$$= Q_\theta + r(Q_x \sin\phi - Q_y\cos\phi), \tag{b}$$

$$I[(d^2\psi/dt^2) + (d^2\phi/dt^2)\cos\theta - (d\phi/dt)(d\theta/dt)\sin\theta]$$

$$- m\, r[(d^2x/dt^2)\cos\phi + (d^2y/dt^2)\sin\phi]$$

$$= Q_\psi - r\sin\theta(Q_x\cos\phi + Q_y\sin\phi); \tag{c}$$

which, along with the *two* constraint equations (b1 and b2 of Example 7.7.4) constitute a determinate system of five coupled ordinary differential equations for the five functions $x(t)$, $y(t)$, $\phi(t)$, $\theta(t)$, $\psi(t)$.

Problem 7.7.3

Show that the (three) *Chaplygin–Voronets* equations of the sphere, in the "independent velocities/accelerations" of $\phi(t)$, $\theta(t)$, $\psi(t)$ [obtained by elimination of (d^2x/dt^2) and (d^2y/dt^2) from Equations a, b, and c of Problem 7.7.2 via the constraints and their $d/dt(\ldots)$–derivatives], are

$$(2mr^2/5)[(d^2\phi/dt^2) + (d^2\psi/dt^2)\cos\theta - (d\theta/dt)(d\psi/dt)\sin\theta] = Q_\phi, \tag{a}$$

$$(7mr^2/5)[(d^2\theta/dt^2) + (d\phi/dt)(d\psi/dt)\sin\theta]$$

$$= Q_\theta + r(Q_x \sin\phi - Q_y \cos\phi), \tag{b}$$

$$(mr^2)[(d^2\psi/dt^2)\sin^2\theta + (2/5)(d^2\psi/dt^2) + (d^2\phi/dt^2)(2\cos\theta/5)$$

$$+ (d\theta/dt)(d\psi/dt)(\sin\theta\cos\theta) - (d\phi/dt)(d\theta/dt)(7\sin\theta/5)]$$

$$= Q_\psi - r\sin\theta\,(Q_x\cos\phi + Q_y\sin\phi). \tag{c}$$

Remark: In concrete problems, as this one, it is much easier to, first, formulate the kinetic Maggi equations, and then eliminate the dependent velocities/accelerations, rather than apply the general Chaplygin–Voronets equations (Problem 7.4.3).

Example 7.7.5. The Sphere: Power Equations

Holonomic variables

Here: $T = T_{(2)}$, $\Pi_{(\text{potential energy})} = 0$, $Q_k = 0$, $A^{1'}{}_{n+1} = \Omega\, y$, $A^{2'}{}_{n+1} = \Omega\, y$, and, therefore,

$$dh/dt = dT_{(2)}/dt = -A^{1'}{}_{n+1}\,\lambda_{1'} - A^{2'}{}_{n+1}\,\lambda_{2'} = (-\Omega\, y)(\lambda) - (-\Omega\, x)(\mu)$$

$$= \Omega(x\,\mu - y\,\lambda) = \Omega\, M_O \tag{a}$$

[M_O: Moment of rolling contact reaction $(\lambda, \mu, 0)$ about O (i.e., OZ)], as expected by the Newton–Euler power equations; i.e., $d/dt(T + \Pi)$ = power of all external (non-potential) forces and couples. Further, by the principle of *linear* momentum,

$$dh/dt = \Omega[x\, m(d^2y/dt^2) - y\, m(d^2x/dt^2)] = \Omega(dH_O/dt), \tag{b}$$

where $H_O \equiv m\,[x(dy/dt) - y\,(dx/dt)]$: angular momentum of fictitious particle of mass m, at G. Then, integrating, and since Ω = constant, we get

$$d/dt(T + \Pi) = dT/dt = \Omega(dH_O/dt),$$

$$\Rightarrow T = \Omega\ H_O + \text{constant} \Rightarrow T - \Omega\ H_O = \text{constant}. \qquad \text{(c)}$$

Nonholonomic variables

We find, successively,

i. $\Pi \to \Pi^* = 0 \Rightarrow \partial\Pi^*/\partial q^k = 0 \Rightarrow \partial\Pi^*/\partial\theta^K \equiv A^k{}_K(\partial\Pi/\partial q^k) = 0;$

$$\partial\Pi^*/\partial\theta^{N+1} = \partial\Pi^*/\partial t + A^k{}_{N+1}(\partial\Pi^*/\partial q^k)$$

$$= 0 + A^k{}_{N+1}(0) = 0; \text{ also } Q_I = 0; \qquad \text{(d1)}$$

ii. $\partial T^*/\partial t = \ldots = m[\Omega(x^2 + y^2) - r(r\ \omega_X + y\ \omega_Y)]\ (d\Omega/dt) = 0, \qquad \text{(d2)}$

$$R' \equiv - A^k{}_{N+1}(\partial L^*/\partial q^k) = -A^k{}_{N+1}(\partial T^*/\partial q^k)$$

$$= A^1{}_{N+1}(\partial T^*/\partial x) + A^2{}_{N+1}(\partial T^*/\partial x)$$

$$= -\{(-\Omega\ y)\ [m\ \Omega\ (\Omega\ x - r\ \omega_3)] + (\Omega\ x)\ [m\ \Omega\ (\Omega\ y - r\ \omega_4)]\}$$

$$= -[m\ r\ \Omega^2(y\ \omega_X - x\ \omega_Y)] = m\ r\ \Omega^2(x\ \omega_Y - y\ \omega_X); \qquad \text{(d3)}$$

i.e.,

$$-\ \partial T^*/\partial\theta^{N+1} \equiv -[\partial T^*/\partial t + A^k{}_{N+1}(\partial T^*/\partial q^k)] = m\ r\ \Omega^2(x\ \omega_Y - y\ \omega_X); \qquad \text{(d4)}$$

iii. $R'' \equiv -\gamma^K{}_{I,N+1}\ \omega^I(\partial T^*/\partial\omega^K)$

$$= -[\gamma^1{}_{36}\ \omega^3\ P_1 + \gamma^2{}_{46}\ \omega^4\ P_2 + \gamma^1{}_{46}\ \omega^4\ P_1 + \gamma^2{}_{36}\ \omega^3\ P_2]$$

$$= -\{(r\ \Omega\ \omega_X)\ [m(r\ \omega_Y - \Omega\ y)] + (r\ \Omega\ \omega_Y)\ [m(-r\ \omega_X + \Omega\ x)] + 0 + 0\}$$

$$= -[m\ r\ \Omega^2(x\ \omega_Y - y\ \omega_X)]; \qquad \text{(d5)}$$

$$\Rightarrow R \equiv R' + R'' = 0; \quad \text{or} \quad -\partial L^*/\partial\theta^{N+1} + R'' = 0. \qquad \text{(d6)}$$

In view of the above, the power equation (7.6.6 or 7.6.8) integrates at once to Equation 7.6.9d:

$$h^* = T^*_{(2)} - T^*_{(0)} = \text{constant} \equiv C, \qquad \text{(e1)}$$

which, since here (recalling that $I = 2mr^2/5$):

$$2T^*_{(2)} = m\ r^2(\omega_3{}^2 + \omega_4{}^2) + I(\omega_3{}^2 + \omega_4{}^2 + \omega_5{}^2)$$

$$= \ldots = (mr^2/10)\ [7(\omega_X{}^2 + \omega_Y{}^2) + 2\omega_Z{}^2],$$

$$2T^*_{(0)} = m\ \Omega^2(x + y^2) \tag{e2}$$

$$[\text{also: } T^*_{(1)} = -m\ r\ \Omega(y\ \omega_4 + x\ \omega_3) = -m\ r\ \Omega(x\ \omega_X + y\ \omega_Y)] \tag{e3}$$

results in the following integral (*found without recourse to the equations of motion*):

$$7(\omega_X{}^2 + \omega_Y{}^2) + 2\omega_Z{}^2 = 5(\Omega/r)^2\,(x^2 + y^2) + C; \tag{e4}$$

or, since $d/dt(I\omega_Z) = Q_Z = 0 \Rightarrow \omega_Z = \text{constant}$, finally,

$$7(\omega_X{}^2 + \omega_Y{}^2) = 5(\Omega/r)^2\,(x^2 + y^2) + \text{constant}. \tag{e5}$$

[We notice that, as a result of Equation e1, we have, successively,

$$E^* \equiv T^* + \Pi^* = T^*_{(2)} + T^*_{(1)} + T^*_{(0)} + \Pi^*$$

$$= T^*_{(2)} + T^*_{(1)} + T^*_{(0)} + \text{constant}$$

$$= 2T^*_{(2)} + T^*_{(1)} + \text{constant}$$

$$= T^*_{(1)} + 2T^*_{(0)} + \text{constant} \neq \text{constant}, \tag{e6}$$

i.e., the *classical energy* E^* is not conserved, even though the generalized one h^* is.]*

---

* An additional, independent derivation of $dT^*_{(2)}/dt = dT^*_{(0)}/dt$ results by $d/dt(\ldots)$–differentiation of $T^*_{(2)}$, $T^*_{(0)}$, Equations e2, and subsequent utilization of the angular equations (Equations j1 through j3 of Example 7.7.4):

$$(7mr^2/5)d\omega_X/dt - m\ r\ \Omega(r\ \omega_Y - \Omega\ y) = Q_X = 0 \Rightarrow d\omega_X/dt = \ldots, \tag{e7}$$

$$(7mr^2/5)d\omega_Y/dt + m\ r\ \Omega(r\ \omega_X - \Omega\ x) = Q_Y = 0 \Rightarrow d\omega_Y/dt = \ldots, \tag{e8}$$

$$(2mr^2/5)d\omega_Z/dt = Q_Z = 0 \Rightarrow d\omega_Z/dt = 0, \tag{e9}$$

and the constraints: $dx/dt = r\ \omega_Y - \Omega\ y$, $dy/dt = -r\ \omega_X + \Omega\ x$. In this way, we find

$$dT^*_{(2)}/dt = (mr^2/5)\{7[\omega_X(d\omega_X/dt) + \omega_Y(d\omega_Y/dt)] + 2\omega_Z(d\omega_Z/dt)\}$$

$$= \ldots = m\ r\ \Omega^2(x\ \omega_Y - y\ \omega_X), \tag{e10}$$

$$dT^*_{(0)}/dt = m\ \Omega(x^2 + y^2)(d\Omega/dt) + m\ \Omega^2[x(dx/dt) + y(dy/dt)]$$

$$= \ldots = m\ r\ \Omega^2(x\ \omega_Y - y\ \omega_X), \quad \text{q.e.d.} \tag{e11}$$

Direct Transition from Holonomic to Nonholonomic Variables

We have, successively,

$$H_O\,\Omega \equiv m\,\Omega[x(dy/dt) - y(dx/dt)]$$

[using the constraints (a3 of Example 7.7.4)]

$$= m\,\Omega[x(-r\,\omega_X + \Omega\,x) - y(r\,\omega_Y - \Omega\,y)]$$

$$= m\,\Omega^2(x^2 + y^2) - m\,r(x\,\omega_X + y\,\omega_Y) = 2T^*_{(0)} + T^*_{(1)}; \qquad \text{(f1)}$$

and combining this with the earlier Equation c we recover Equation e1:

$$T - \Omega\,H_O = \text{constant}$$

$$\Rightarrow (T^*_{(2)} + T^*_{(1)} + T^*_{(0)}) - (2T^*_{(0)} + T^*_{(1)}) = T^*_{(2)} - T^*_{(0)} = \text{constant}. \qquad \text{(f2)}$$

Example 7.7.6. The Ring/Hoop

Let us consider a thin homogeneous ring, or hoop (or coin, or wheel, or disk) R, of mass m, center of mass G, and (average) radius r, rolling on a rough and fixed horizontal plane P, instantaneously in contact with it at point C (Figure 7.3).

Kinematics

A generic configuration of R is specified by the following *five* independent Lagrangean coordinates: (i) the inertial coordinates of G: $q^1 = x_G \equiv x$, $q^2 = y_G \equiv y$, $q^3 = z_G \equiv z$ and (ii) the Eulerian angles: $q^4 = \phi \rightarrow q^5 = \theta \rightarrow q^6 = \psi$, defining the angular orientation of R–fixed axes G–xyz relative to *translating but nonrotating* axes G–XYZ [or relative to the inertial axes O–XYZ (with + OZ upwards) (Example 6.5.1)].* However, it is easily seen that z and θ are related by the *holonomic* constraint: $z = r\sin\theta$; i.e., only *five* of these qs are independent; say, x, y, ϕ, θ, ψ. The rolling constraint is $\mathbf{v}_C = \mathbf{0}$. Let us express this in convenient components. We have successively [with G–$\mathbf{u}_{x',y',z'}/x'y'z'$: "semi-mobile" orthonomal basis/axes (notation slightly different from that of Example 6.5.1)]:

$$\mathbf{v}_C = \mathbf{v}_G + \omega \times \mathbf{r}_{C/G} = \mathbf{v}_G - r\,\omega \times \mathbf{u}_{y'}$$

(ω: inertial angular velocity of R)

$$= \mathbf{v}_G - r(\omega_{x'}\,\mathbf{u}_{x'} + \omega_{y'}\,\mathbf{u}_{y'} + \omega_{z'}\,\mathbf{u}_{z'}) \times \mathbf{u}_{y'}$$

$$= \mathbf{v}_G - r\,\omega_{x'}(\mathbf{u}_{x'} \times \mathbf{u}_{y'}) - r\,\omega_{z'}(\mathbf{u}_{z'} \times \mathbf{u}_{y'})$$

$$= \mathbf{v}_G - r\,\omega_{x'}\,\mathbf{u}_{z'} - r\,\omega_{z'}(-\mathbf{u}_{x'}), \qquad \text{(a1)}$$

* Other authors use as Lagrangean coordinates: x_C, y_C, z_C; ϕ, θ, ψ ($z_C = 0$).

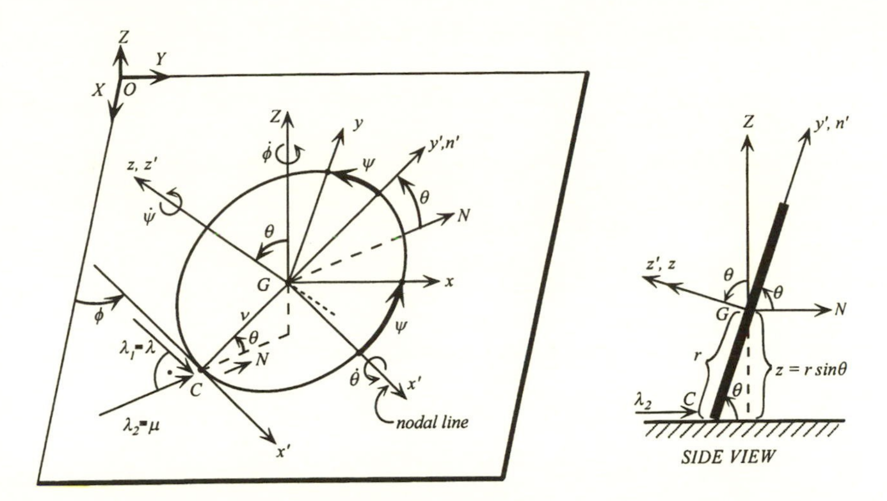

FIGURE 7.3 Rolling ring: geometry/kinematics [ϕ (precession) $\rightarrow$ θ (nutation) $\rightarrow$ ψ (proper spin)] and reactions: axes/bases used: O–XYZ: fixed (inertial) axes, G–xyz: body-fixed axes, G–$x'y'z'/\mathbf{u}_{x',y',z'}$: semi-mobile (intermediate) axes/basis, G–$x'NZ/\mathbf{u}_{x',N,Z}$: semi-fixed (intermediate) axes/basis.

where, clearly,

$$\omega_{x'} = d\theta/dt \equiv \nu^{\theta}, \qquad \omega_{y'} = (d\phi/dt)\sin\theta \equiv \nu^{\phi}\sin\theta,$$

$$\omega_{z'} = d\psi/dt + (d\phi/dt)\cos\theta \equiv \nu^{\psi} + \nu^{\phi}\cos\theta, \tag{a2}$$

are the semi-mobile components of the (inertial) angular velocity of R.

Therefore, the components of the constraint $\mathbf{v}_C = 0$ along $\mathbf{u}_{x'}$ and $\mathbf{u}_N$, of the also convenient (and kinetostatically natural, for this problem) "semi-fixed" orthonormal basis C–$\mathbf{u}_{x'NZ}$, are

- $0 = v_{C,x'} \equiv \mathbf{v}_C \cdot \mathbf{u}_{x'} = \mathbf{v}_G \cdot \mathbf{u}_{x'} - r\,\omega_{x'}(\mathbf{u}_{z'} \cdot \mathbf{u}_{x'}) + r\,\omega_{z'}(\mathbf{u}_{x'} \cdot \mathbf{u}_{x'})$

$$= v_{G,x'} - r\,\omega_{x'}(0) + r\,\omega_{z'}(1) = v_{G,x'} + r\,\omega_{z'}, \tag{b1}$$

- $0 = v_{C,N} \equiv \mathbf{v}_C \cdot \mathbf{u}_N = \mathbf{v}_G \cdot \mathbf{u}_N - r\,\omega_{x'}(\mathbf{u}_{z'} \cdot \mathbf{u}_N) + r\,\omega_{z'}(\mathbf{u}_{x'} \cdot \mathbf{u}_N)$

$$= v_{G,N} - r\,\omega_{x'}[\cos(\pi/2 + \theta)] + r\,\omega_{z'}(0) = v_{G,N} + r\,\omega_{x'}\sin\theta; \qquad \text{(b2)*}$$

or, further, since

---

* The $\mathbf{u}_Z$-component of $\mathbf{v}_C$ gives the earlier integrable constraint:

$$0 = v_{C,Z} \equiv \mathbf{v}_C \cdot \mathbf{u}_Z = \mathbf{v}_G \cdot \mathbf{u}_Z - r\,\omega_{x'}(\mathbf{u}_{z'} \cdot \mathbf{u}_Z) + r\,\omega_{z'}(\mathbf{u}_{x'} \cdot \mathbf{u}_Z)$$

$$= v_{G,Z} - r\,\omega_{x'}(\cos\theta) + r\,\omega_{z'}(0) = dz/dt - r(d\theta/dt)\cos\theta \Rightarrow z = r\sin\theta.$$

$$v_{G,x'} = (v^x \equiv dx/dt,\ v^y \equiv dy/dt) \cdot (\cos\phi,\ \sin\phi) = v^x \cos\phi + v^y \sin\phi, \quad \text{(b3)}$$

$$v_{G,N} = (v^x \equiv dx/dt,\ v^y \equiv dy/dt) \cdot (-\sin\phi,\ \cos\phi) = -v^x \sin\phi + v^y \cos\phi, \quad \text{(b4)}$$

and invoking Equation a2, the constraints b1 and b2 become, respectively,

$$v_{C,x'} = v_{G,x'} + r\ \omega_{z'} = v^x \cos\phi + v^y \sin\phi + r(v^\psi + v^\phi \cos\theta) = 0, \quad \text{(b5)}$$

$$v_{C,N} = v_{G,N} + r\ \omega_{x'} \sin\theta = -v^x \sin\phi + v^y \cos\phi + r\ v^\theta \sin\theta = 0. \quad \text{(b6)}$$

In view of Equations b5 and b6, we choose the following five quasi-velocities:

$$\omega^1 \equiv v_{C,x'} = (\cos\phi)v^x + (\sin\phi)v^y + (r\cos\theta)v^\phi + (r)v^\psi = 0, \quad \text{(c1)}$$

$$\omega^2 \equiv v_{C,N} = (-\sin\phi)v^x + (\cos\phi)v^y + (r\sin\theta)v^\theta = 0; \quad \text{(c2)}$$

$$\omega^3 \equiv \omega_{x'} = v^\theta \neq 0, \quad \text{(c3)}$$

$$\omega^4 \equiv \omega_{y'} = (\sin\theta)v^\phi \neq 0, \quad \text{(c4)}$$

$$\omega^5 \equiv \omega_{z'} = (\cos\theta)v^\phi + v^\psi \neq 0 \quad (\omega^6 = V^6 = dt/dt = 1,\ \text{always}) \quad \text{(c5)}$$

(i.e., $f \equiv n - m = 5 - 2 = 3$); which invert easily to (no constraint enforcement yet):

$$V^1 \equiv v^x = (\cos\phi)\omega^1 + (-\sin\phi)\omega^2 + (r\sin\theta\ \sin\phi)\omega^3 + (-r\cos\phi)\omega^5, \quad \text{(d1)}$$

$$V^2 \equiv v^y = (\sin\phi)\omega^1 + (\cos\phi)\omega^2 + (-r\sin\theta\ \cos\phi)\omega^3 + (-r\sin\phi)\omega^5, \quad \text{(d2)}$$

$$V^3 \equiv v^\phi = (1/\sin\theta)\omega^4, \quad \text{(d3)}$$

$$V^4 \equiv v^\theta = \omega^3, \quad \text{(d4)}$$

$$V^5 \equiv v^\psi = \omega^5 - (\cot\theta)\omega^4. \quad \text{(d5)}$$

Next, to the transivity equations and their Hamel coefficients: By direct $d(\ldots)$ and $\delta(\ldots)$ operations on the kinematically admissible/possible and virtual forms of Equations c1 through c5, and use of the corresponding forms of Equations d1 through d5 in them, we find, after some long but straightforward algebra,

$$(\delta\theta^1)^{\cdot} - \delta\omega^1 = (1/\sin\theta)(\omega^4\ \delta\theta^2 - \omega^2\ \delta\theta^4), \quad \text{(e1)}$$

$$(\delta\theta^2)^{\cdot} - \delta\omega^2 = (1/\sin\theta)(\omega^1\ \delta\theta^4 - \omega^4\ \delta\theta^1) + (r/\sin\theta)(\omega^4\ \delta\theta^5 - \omega^5\ \delta\theta^4), \quad \text{(e2)}$$

$$(\delta\theta^3)^{\cdot} - \delta\omega^3 = 0 \quad (\theta^3 = \theta\text{: holonomic coordinate}), \quad \text{(e3)}$$

$$(\delta\theta^4)^{\cdot} - \delta\omega^4 = (\cot\theta)(\omega^3\ \delta\theta^4 - \omega^4\ \delta\theta^3), \quad \text{(e4)}$$

$$(\delta\theta^5)^{\cdot} - \delta\omega^5 = \omega^4\ \delta\theta^3 - \omega^3\ \delta\theta^4; \quad \text{(e5)}$$

from which the nonvanishing γ coefficients are easily read off (no constraint enforcement in Equations e1 through e5 yet):

$$\gamma^1{}_{24} = -\gamma^1{}_{42} = 1/\sin\theta, \quad \gamma^2{}_{41} = -\gamma^2{}_{14} = 1/\sin\theta,$$

$$\gamma^2{}_{54} = -\gamma^2{}_{45} = r/\sin\theta; \tag{f1}$$

$$\gamma^4{}_{43} = -\gamma^4{}_{34} = \cot\text{an}\,\theta, \quad \gamma^5{}_{34} = -\gamma^5{}_{43} = 1. \tag{f2}$$

Since here $\gamma^D{}_{II'}$: $\gamma^2{}_{45} = -\gamma^2{}_{54} \equiv \gamma^N{}_{y'z'} \neq 0$, the constraints $\omega^1, \omega^2 = 0$ are nonholonomic.

Kinetics

By König's theorem, the (inertial) kinetic energy of R is (with the well-known values: $I_{G,z'} \equiv I_{z'} = m\,r^2$, $I_{G,x'} = I_{G,y'} \equiv I_{x'} = I_{y'} = m\,r^2/2$; and no constraint enforcement yet):

$$2T = m[(v^x)^2 + (v^y)^2 + (v^z)^2] + (I_{x'}\,\omega_{x'}{}^2 + I_{y'}\,\omega_{y'}{}^2 + I_{z'}\,\omega_{z'}{}^2), \tag{g1}$$

and, therefore, (by Equations a2, etc.), in *holonomic variables*:

$$2T = m[(v^x)^2 + (v^y)^2 + r^2(v^\theta)^2\cos^2\theta] + [(m\,r^2/2)(v^\theta)^2$$

$$+ (m\,r^2/2)(v^\phi\sin\theta)^2 + (m\,r^2)\,(v^\psi + v^\phi\cos\theta)^2], \tag{g2}$$

and in *nonholonomic* ones [recalling Equations d1 through d5, and noting that $(v^x)^2 + (v^y)^2 + (v^z)^2 = v_{G,x}{}^2 + v_{G,N}{}^2 + v_{G,Z}{}^2$]:

$$2T \to 2T^* = m[(\omega^1 - r\,\omega^5)^2 + (\omega^2 - r\,\omega^3\sin\theta)^2 + r^2(\omega^3)^2\cos^2\theta]$$

$$+ (m\,r^2)[(\omega^3)^2/2 + (\omega^4)^2/2 + (\omega^5)^2]$$

$$= m\{(\omega^1)^2 + (\omega^2)^2 - 2r\,\omega^1\,\omega^5 - 2r\,\omega^2\,\omega^3\sin\theta$$

$$+ r^2[2(\omega^5)^2 + 3(\omega^3)^2/2 + (\omega^4)^2/2]\}. \tag{g3}$$

The *Routh–Voss Equations* (*R–V*): With Equation g2, $\lambda_{1'} \equiv \lambda$, $\lambda_{2'} \equiv \mu$, and since here: $Q_{x,y,\phi,\psi} = 0$, $Q_\theta = -\partial\Pi/\partial\theta = -\partial(mgr\sin\theta)/\partial\theta = -mgr\cos\theta$, the *R–V* equations yield

$$m(d^2x/dt^2) = \lambda\cos\phi - \mu\sin\phi, \tag{h1}$$

$$m(d^2y/dt^2) = \lambda\sin\phi + \mu\cos\phi, \tag{h2}$$

$$d/dt[(mr^2/2)(v^\phi\sin^2\theta) + (mr^2\cos\theta)(v^\psi + v^\phi\cos\theta)] = \lambda\,r\cos\theta, \tag{h3}$$

$$d/dt[(mr^2)(v^\theta\cos^2\theta) + (mr^2/2)v^\theta] + (mr^2)(v^\theta)^2\sin\theta\cos\theta$$

$$-(mr^2/2)(v^\phi)^2\sin\theta\cos\theta + (mr^2)v^\phi\,(v^\psi + v^\phi\cos\theta)\sin\theta$$

$$= -mgr\cos\theta + \lambda(0) + \mu r\sin\theta, \tag{h4}$$

$$d/dt[(mr^2)(v^\psi + v^\phi \cos\theta)] = \lambda\, r + \mu(0). \tag{h5}$$

The *Maggi Equations* (*M*): Next, $(\ldots)^\cdot$–differentiating the constraints (b5 and b6) once, solving them for d^2x/dt^2, d^2y/dt^2 and substituting the so-resulting expressions into Equations h1 and h2, and then solving them for λ, μ, we obtain:

$$\lambda = mr[v^\phi\, v^\theta \sin\theta - (v^\psi + v^\phi \cos\theta)^\cdot], \tag{i1}$$

$$\mu = -mr[v^\phi(v^\psi + v^\phi \cos\theta) + (v^\theta \sin\theta)^\cdot]; \tag{i2}$$

or, eliminating λ, μ among Equations h1 through h5 we find three reactionless equations, which are none other than the kinetic *M* equations of this problem; the kinetostatic ones being Equations i1 and i2. Let us derive all five of these equations more systematically. Reading off the coefficients $A^k{}_K$ from Equations d1 through d5, and with the notation $M_k \equiv E_k(T) - Q_k \equiv I_k - Q_k$ ($k = x, y, \phi, \theta, \psi$), we obtain

Kinetostatic Maggi:

$$A^k{}_{1'}\, M_k = \Lambda_{1'}\ (= \lambda_{1'} \equiv \lambda):$$

$$(\cos\phi)M_x + (\sin\phi)M_y = \lambda \Rightarrow \lambda = I_x \cos\phi + I_y \sin\phi, \tag{j1}$$

$$A^k{}_{2'}\, M_k = \Lambda_{2'}\ (= \lambda_{2'} \equiv \mu):$$

$$(-\sin\phi)M_x + (\cos\phi)M_y = \mu \Rightarrow \mu = -I_x \sin\phi + I_y \cos\phi, \tag{j2}$$

Kinetic Maggi:

$$A^k{}_{3'}\, M_k = 0: \quad (r \sin\theta \sin\phi)M_x + (-r \sin\theta \cos\phi)M_y + (1)M_\theta = 0,$$

$$\text{or:} \quad r \sin\theta\,(I_x \sin\phi - I_y \cos\phi) + I_\theta = Q_\theta, \tag{j3}$$

$$A^k{}_{4'}\, M_k = 0: \quad (1/\sin\theta)M_\phi + (-\text{cotan}\,\theta)M_\psi = 0,$$

$$\text{or:} \quad I_\phi - \cos\theta\, I_\psi = 0, \tag{j4}$$

$$A^k{}_{5'}\, M_k = 0: \quad (-r \cos\phi)M_x + (-r \sin\phi)M_y + (1)M_\psi = 0,$$

$$\text{or:} \quad I_x \cos\phi + I_y \sin\phi - (1/r)I_\psi = 0. \tag{j5}$$

Naturally, Equations j3 through j5 coincide with those found by elimination of λ, μ among the *R–V* equations; and, the equations resulting by eliminating the (chosen here as) *dependent* velocities v^x, v^y, and their $(\ldots)^\cdot$–derivatives, via Equations b5 and b6, from Equations j3 through j5 are the three Chaplygin–Voronets equations of our problem, in terms of the *independent* velocities v^ϕ, v^θ, v^ψ and their $(\ldots)^\cdot$–derivatives.

The *Hamel Equations* (*H*): With the notation: $(\ldots)_o \equiv (\ldots, \omega^1 = 0, \omega^2 = 0)$, wherever needed for extra clarity, and Equation g3, we obtain

i. $P_1 \equiv (\partial T^*/\partial\omega^1)_o = (m\ \omega^1 - m\ r\ \omega^5)_o = -m\ r\ \omega^5 \Rightarrow dP_1/dt$

$$= -m\ r(d\omega^5/dt), \tag{k1}$$

$$P_2 \equiv (\partial T^*/\partial\omega^2)_o = (m\ \omega^2 - m\ r\ \omega^3 \sin\theta)_o = -m\ r\ \omega^3 \sin\theta$$

$$\Rightarrow dP_2/dt = -m\ r[(d\omega^3/dt)\sin\theta + (\omega^3)^2 \cos\theta], \tag{k2}$$

$$P_3 \equiv (\partial T^*/\partial\omega^3)_o = [(3/2)m\ r^2\ \omega^3 - m\ r\ \omega^2 \sin\theta]_o = (3/2)m\ r^2\ \omega^3$$

$$\Rightarrow dP_3/dt = (3/2)m\ r^2\ (d\omega^3/dt), \tag{k3}$$

$$P_4 \equiv (\partial T^*/\partial\omega^4)_o = (m\ r^2/2)\omega^4 \Rightarrow dP_4/dt = (m\ r^2/2)(d\omega^4/dt), \tag{k4}$$

$$P_5 \equiv (\partial T^*/\partial\omega^5)_o = 2m\ r^2\ \omega^5 \Rightarrow dP_5/dt = (2\ m\ r^2)\ (d\omega^5/dt); \tag{k5}$$

ii. $$\partial T^*/\partial\theta^K = A^k{}_K(\partial T^*/\partial q^k) = A^4{}_K(\partial T^*/\partial\theta) = \ldots = [(\ldots)\ \omega^2]_o = 0; \tag{k6}$$

iii. $P_K\ [(\delta\theta^K)^{\cdot} - \delta\omega^K]$ (with the constraints $\omega^1, \omega^2 = 0$ enforced, but not $\delta\theta^1$, $\delta\theta^2 = 0$ yet!)

$$= P_1[(\omega^4/\sin\theta)\delta\theta^2]$$

$$+ P_2[(-\omega^4/\sin\theta)\delta\theta^1 + (-r\ \omega^5/\sin\theta)\delta\theta^4 + (r\ \omega^4/\sin\theta)\delta\theta^5]$$

$$+ P_3(0) + P_4[(\cot\theta)\omega^3\ \delta\theta^4 - (\cot\theta)\omega^4\ \delta\theta^3]$$

$$+ P_5(\omega^4\ \delta\theta^3 - \omega^3\ \delta\theta^4)$$

$$= (-P_2\ \omega^4/\sin\theta)\delta\theta^1 + (P_1\ \omega^4/\sin\theta)\delta\theta^2 + [(-\cot\theta)P_4\ \omega^4 + P_5\ \omega^4]\delta\theta^3$$

$$+ [(\cot\theta)P_4\ \omega^3 - P_5\ \omega^3 - (r/\sin\theta)P_2\ \omega^5]\delta\theta^4$$

$$+ [(r/\sin\theta)P_2\ \omega^4]\delta\theta^5. \tag{k7}$$

Inserting the results of i through iii (k1 through k7) into the fundamental central equation (7.5.6c), collecting $\delta\theta^K$ terms and recalling that $\delta\theta^1$, $\delta\theta^2$ are constrained, while $\delta\theta^3$, $\delta\theta^4$, $\delta\theta^5$ are free, we obtain the following *H* equations:

Kinetostatic Hamel:

$$dP_1/dt - (\omega^4/\sin\theta)P_2 = m\ r[-(d\omega^5/dt) + \omega^3\ \omega^4]$$

$$= Q_{1'} + \Lambda_{1'} \quad (\Lambda_{1'} = \lambda), \tag{l1}$$

$$dP_2/dt + (\omega^4/\sin\theta)P_1 = -m\, r[\sin\theta\, (d\omega^3/dt) + \cos\theta(\omega^3)^2 + \omega^4\, \omega^5/\sin\theta]$$

$$= Q_{2'} + \Lambda_{2'} \quad (\Lambda_{2'} = \mu); \tag{12}$$

Kinetic Hamel:

$$dP_3/dt + \omega^4\, P_5 - (\text{cotan}\,\theta\, \omega^4)P_4$$

$$= m\, r^2[(3/2)(d\omega^3/dt) + 2\omega^4\, \omega^5 - (\text{cotan}\,\theta/2)(\omega^4)^2] = Q_{3'}, \tag{13}$$

$$dP_4/dt - \omega^3\, P_5 + (\text{cotan}\,\theta\, \omega^3)\, P_4 - (r\, \omega^5/\sin\theta)\, P_2$$

$$= m\, r^2[(1/2)(d\omega^4/dt) + (\text{cotan}\,\theta/2)\omega^3\, \omega^4 - \omega^3\, \omega^5] = Q_{4'}, \tag{14}$$

$$dP_5/dt + (r\, \omega^4/\sin\theta)P_2 = m\, r^2[2(d\omega^5/dt) - \omega^3\, \omega^4] = Q_{5'}. \tag{15}$$

The reader may verify that Equations 13 through 15 are indeed the earlier kinetic Maggi equations (j3 through j5), but while the latter are in holonomic variables, the above are in nonholonomic variables.

Next, if the sole impressed force is *gravity* (as assumed here), then $Q_{1',2',4',5'} = 0$ and $\Pi = mgz = mgr\sin\theta = mgr\sin\theta^{3'} \Rightarrow Q_{3'} = -\,\partial\Pi/\partial\theta^{3'} = -\,mgr\cos\theta\ (= Q_\theta)$, and Equations 13 through 15 reduce to the following coupled nonlinear first-order system for ω^3, ω^4, ω^5:

$$(3/2)(d\omega^3/dt) - (\text{cotan}\,\theta/2)(\omega^4)^2 + 2\omega^4\, \omega^5 = -(g/r)\cos\theta, \tag{m1}$$

$$(1/2)(d\omega^4/dt) + (\text{cotan}\,\theta/2)\omega^3\, \omega^4 - \omega^3\, \omega^5 = 0, \tag{m2}$$

$$2(d\omega^5/dt) - \omega^3\, \omega^4 = 0; \tag{m3}$$

or, in terms of the clearer *semi-mobile* components of the angular velocity (recalling Equations a2 and c3 through c5):

$$\omega^3 \equiv \omega_{x'} = v^\theta, \quad \omega^4 \equiv \omega_{y'} = v^\phi \sin\theta, \quad \omega^5 \equiv \omega_{z'} = v^\psi + v^\phi\cos\theta,$$

$$3(d\omega_{x'}/dt) + (4\omega_{z'} - \omega_{y'}\,\text{cotan}\,\theta)\omega_{y'} = -(2g/r)\cos\theta, \tag{n1}$$

$$d\omega_{y'}/dt + (\omega_{y'}\,\text{cotan}\,\theta - 2\omega_{z'})\omega_{x'} = 0, \tag{n2}$$

$$2(d\omega_{z'}/dt) - \omega_{x'}\, \omega_{y'} = 0. \tag{n3}$$

The solution of the above is facilitated by the introduction of the new independent variable θ, instead of t: with the notation $d(\ldots)/d\theta \equiv (\ldots)'$, and since

$$d\theta/dt = \omega^3 \quad \text{and} \quad d\omega^K/dt = (\omega^K)'(d\theta/dt) = (\omega^K)'\omega^3, \tag{o1}$$

Equations n2 and n3 transform, respectively, to (assuming that $\omega^3 \equiv \omega_{x'} = v^\theta \neq 0$):

$$(\omega^4)' + \cot\!\text{an}\,\theta\; \omega^4 - 2\omega^5 = 0, \quad 2(\omega^5)' - \omega^4 = 0; \tag{o2}$$

and eliminating ω^4 between them, we, finally, get the single θ-equation:

$$(\omega^5)'' + \text{cotan}\,\theta\,(\omega^5)' - \omega^5 = 0, \tag{o3}$$

or, in terms of the $\omega_{z'}$:

$$(\omega_{z'})'' + \text{cotan}\,\theta\,(\omega_{z'})' - \omega_{z'} = 0. \tag{o4}$$

Then, since $d/d\theta[(\omega^3)^2] = 2\omega^3(d\omega^3/d\theta) = 2\omega^3(d\omega^3/dt)\,(dt/d\theta) = 2(d\omega^3/dt)\,[\omega^3/(d\theta/dt)] = 2(d\omega^3/dt)$, the first kinetic equation, n1, reduces to

$$d/d\theta[(3/4)(\omega^3)^2] =$$

$$-(g/r)\cos\theta + (1/2)(\omega^4)^2\,\text{cotan}\,\theta - 2\omega^4\omega^5\text{: known function of } \theta, \tag{o5}$$

from which $\omega^3(\theta)$ can be found by a quadrature; and, a final integration of $d\theta/dt = \omega^3(\theta)$ yields $\theta(t)$. Finally, the kinetostatic equations (l1 and l2): with the help of Equations a2 and c3 through c5, they become

$$mr[-(d\omega_{z'}/dt) + \omega_{x'}\,\omega_{y'}] = Q_{x'} + \Lambda_{x'} \equiv Q_{x'} + \lambda, \tag{p1}$$

$$-mr[\sin\theta(d\omega_{x'}/dt) + \cos\theta(\omega_{x'})^2 + \omega_{y'}\,\omega_{z'}/\sin\theta]$$

$$= Q_N + \Lambda_N \equiv Q_N + \mu; \tag{p2}$$

and, naturally, coincide with the earlier found kinetostatic Maggi equations (j1 and j2).

Substituting in them $d\omega_{z'}/dt$ and $d\omega_{x'}/dt$ from the third and first kinetic ones, respectively, and assuming that $Q_{x'}, Q_N = 0$, we obtain

$$\lambda = (mr/2)\omega_{x'}\,\omega_{y'}, \tag{q1}$$

$$\mu = -mr\{\cos\theta\,(\omega_{x'}{}^2 + \omega_{y'}{}^2/3) + [(1/\sin\theta) - (4\sin\theta/3)]\omega_{y'}\,\omega_{z'}$$

$$- (2g/3r)\sin\theta\cos\theta\}. \tag{q2}$$

Once the motion has been found, i.e., $\omega_{x',y',z'}$, θ: functions of time, Equations q1 and q2 yield $\lambda(t)$, $\mu(t)$.*

* (See following page for footnote.)

Problem 7.7.7

By eliminating the "dependent velocities and accelerations" v^x, v^y, dv^x/dt, dv^y/dt, from Equations j3 through j5 of the preceding Example 7.7.6, via the constraints (b5 and b6) and their $d/dt(\ldots)$–derivatives, obtain the (three) *Chaplygin–Voronets* equations of the ring/hoop in the "independent velocities/accelerations" v^ϕ, v^θ, v^ψ; dv^ϕ/dt, dv^θ/dt, dv^ψ/dt.

Problem 7.7.8

Formulate the ring/hoop problem, for a ring rolling on a horizontal and rough plane, which spins uniformly about a fixed vertical axis OZ.

Example 7.7.7. Plane Motion of a Particle in Quasi-Variables; Appell's Equations

Let us consider the free (unconstrained) motion of a particle P, of mass m, moving on a fixed and smooth horizontal plane O–xy; i.e., $n = 2$, $m = 0$.

Kinematics

To specify the configurations of P let us choose [instead of the commonly used rectangular Cartesian coordinates (x, y), or polar coordinates (r, ϕ)]: (i) its first polar coordinate $r = |\mathbf{OP}|$, i.e., $\theta^1 = r$, and (ii) $\theta^2 = \sigma$, such that $d\sigma/dt$ = *twice the area of the sector swept by the radius vector* **OP** *between the time instants* $t \rightarrow (r, \phi)$ *and* $t + dt \rightarrow (r + dr, \phi + d\phi)$, i.e.,

$$d\sigma/dt = 2[r^2(d\phi/dt)/2] = r^2(d\phi/dt) \tag{a1}$$

$$\Rightarrow d\sigma = r^2\, d\phi \Rightarrow \delta\sigma = r^2\, \delta\phi; \tag{a2}$$

i.e., the $\omega^K \Leftrightarrow dq^k/dt \equiv v^k$ $(K, k = 1, 2)$ transformation equations are

---

* It has been shown (by Appell, Korteweg, et al., 1890s) that under the initial conditions: $\theta(0) = \theta_o$, $\omega^3(0) \equiv \omega^3{}_o$, $\omega^4(0) \equiv \omega^4{}_o = 2(d\omega^5/d\theta)_o$, $\omega^5(0) \equiv \omega^5{}_o$, the *general* solution of Equations o3, o4, integrated via a "hypergeometric series" will have the form: $\omega^5 = \omega^5(\theta;\, \theta_o, \omega^4{}_o, \omega^5{}_o)$. The additional variable change: $\theta \rightarrow \eta \equiv \cos^2\theta$ transforms Equation o4 to the *hypergeometric equation*:

$$\eta(1 - \eta)(d^2\omega_{z'}/d\eta^2) + (1/2 - 3\eta/2)(d\omega_{z'}/d\eta) - (1/4)\omega_{z'} = 0, \tag{o6}$$

which has the following *particular* series solution: $\omega_{z'} = a_0 + a_1\,\eta^1 + a_2\,\eta^2 + \ldots$

$$a_k/a_{k-1} = (4k^2 - 6k + 3)\,/\,[2k(2k - 1)] \quad [k = 1, 2, 3, \ldots]$$

for $k = 1$: $a_1/a_0 = 1/2 \Rightarrow a_1 = a_0/2$; for $k = 2$: $a_2/a_1 = \ldots = 7/12 \Rightarrow a_2 = 7a_1/12 = 7a_0/24$, …. On the integration of these (historically famous) equations, there exists an extensive literature; e.g., (alphabetically): Hamel (1949, pp. 470–471, 478–479, 489–492, 778–781; includes stability of motion of ring), Neimark and Fufaev (1972, pp. 55–60, 155–156), Lur'e (1968, pp. 409–410; Newton–Euler derivation of the above, i.e., via momentum principles, and use of the "semi-fixed" axes G–$x'NZ$), Rosenberg (1977, pp. 265–268, 338–340).

$$\omega^1 \equiv dr/dt = (1)dr/dt + (0)d\phi/dt \equiv (1)v^1 + (0)v^2, \tag{b1}$$

$$\omega^2 \equiv d\sigma/dt = (0)dr/dt + (r^2)d\phi/dt \equiv (0)v^1 + (r^2)v^2; \tag{b2}$$

$$v^1 \equiv v^r \equiv dr/dt = (1)\omega^1 + (0)\omega^2, \tag{b3}$$

$$v^2 \equiv v^\phi \equiv d\phi/dt = (0)\omega^1 + (1/r^2)\omega^2. \tag{b4}$$

Kinetics

i. As a result of the above, the kinetic energy of P equals

$$2T = m[(dx/dt)^2 + (dy/dt)^2] = m[(dr/dt)^2 + r^2(d\phi/dt)^2]$$

$$= m[(dr/dt)^2 + r^{-2}(d\sigma/dt)^2] = m[(\omega^1)^2 + r^{-2}(\omega^2)^2] = 2T^*; \tag{c1}$$

and its Appellian function is [recalling Equations 7.4.23a through c; and with $\ldots \equiv$ *up to first d/dt(...)-derivative terms;* or "Appell-nonimportant" terms, i.e., S to within "Appell-important" terms]:

$$2S = m[(d^2x/dt^2)^2 + (d^2y/dt^2)^2]$$

$$= m([(d^2r/dt^2) - r(d\phi/dt)^2]^2 + \{r^{-1}\, d/dt[r^2(d\phi/dt)]\}^2)$$

$$\Rightarrow 2S^* = m\{[(d^2r/dt^2) - r^{-3}(d\sigma/dt)^2]^2 + [r^{-1}(d^2\sigma/dt^2)]^2\}$$

$$= m[(d^2r/dt^2)^2 - 2r^{-3}(d^2r/dt^2)(d\sigma/dt)^2 + r^{-2}(d^2\sigma/dt^2)^2] + \ldots \tag{c2}$$

$$= m[(d\omega^1/dt)^2 - 2r^{-3}(\omega^2)^2\,(d\omega^1/dt) + r^{-2}(d\omega^2/dt)^2] + \ldots. \tag{c3}$$

ii. The transitivity equations corresponding to Equations b1 through b4 are

$$d(\delta\theta^1) - \delta(d\theta^1) = d(\delta r) - \delta(dr) = 0$$

$$[r\text{: } \textit{holonomic coordinate}\text{, i.e., } \gamma^1{}_{\ldots} = 0] \tag{d1}$$

$$d(\delta\theta^2) - \delta(d\theta^2) = d(\delta\sigma) - \delta(d\sigma) = d(r^2\delta\phi) - \delta(r^2 d\phi)$$

$$= 2r\, dr\, \delta\phi + r^2\, d(\delta\phi) - 2\, r\, \delta r\, d\phi - r^2\, \delta(d\phi) = 2\, r(dr\, \delta\phi - \delta r\, d\phi)$$

$$= 2r[dr\, (\delta\sigma/r^2) - \delta r(d\sigma/r^2)] = (2/r)(d\theta^1\, \delta\theta^2 - d\theta^2\, \delta\theta^1), \tag{d2}$$

$$\text{i.e., } \gamma^2{}_{12} = -\gamma^2{}_{21} = -2/r \neq 0 \Rightarrow \theta^2 = \sigma\text{: nonholonomic coordinate!} \tag{d3}$$

iii. The *holonomic* and *nonholonomic* components of the impressed force are calculated from

$$\delta' W = X\,\delta x + Y\,\delta y = Q_r\,\delta r + Q_\phi\,\delta\phi \quad (Q_1 \equiv Q_r,\ Q_2 \equiv Q_\phi) \qquad \text{(e1)}$$

$$= Q_r\,\delta r + Q_\phi(\delta\sigma/r^2) = (Q_r)\delta r + (Q_\phi/r^2)\delta\sigma$$

$$= Q_r\,\delta r + Q_\sigma\,\delta\sigma \equiv Q_{1'}\,\delta\theta^1 + Q_{2'}\,\delta\theta^2$$

$$(Q_{1'} \equiv Q_r,\ Q_{2'} \equiv Q_\sigma = Q_\phi/r^2). \qquad \text{(e2)}$$

With the help of these results, we readily find that:

i. The *Routh–Voss* equations, and the Appell equations in holonomic variables are (with $d^2q^k/dt^2 \equiv dv^k/dt \equiv a^k$; $k = r, \phi$):

$$I_r \equiv E_r(T) \equiv (\partial T/\partial v^r)^{\cdot} - \partial T/\partial r = \partial S/\partial a^r$$

$$= [m(dr/dt)]^{\cdot} - m\,r(d\phi/dt)^2 = m[(d^2r/dt^2) - r\,(d\phi/dt)^2] = Q_r, \qquad \text{(f1)}$$

$$I_\phi \equiv E_\phi(T) \equiv (\partial T/\partial v^\phi)^{\cdot} - \partial T/\partial\phi = \partial S/\partial a^\phi$$

$$= [m\,r^2(d\phi/dt)]^{\cdot} = m\,r[2(dr/dt)(d\phi/dt) + r(d^2\phi/dt^2)] = Q_\phi; \qquad \text{(f2)}$$

ii. The *Maggi* equations in this unconstrained case [with $M_k \equiv E_k(T) - Q_k$], are

$$A^r{}_{1'}\,M_r + A^\phi{}_{1'}\,M_\phi = (1)M_r + (0)M_\phi = 0 \Rightarrow M_r = 0, \qquad \text{(g1)}$$

$$A^r{}_{2'}\,M_r + A^\phi{}_{2'}\,M_\phi = (0)M_r + (1/r^2)M_\phi = 0 \Rightarrow M_\phi = 0; \qquad \text{(g2)}$$

i.e., nothing new.

iii. The *Hamel* equations. By Equation c1 we have

$$P_1 \equiv \partial T^*/\partial\omega^1 = m\,\omega^1 = m(dr/dt) \Rightarrow dP_1/dt = m(d\omega^1/dt) = m(d^2r/dt^2), \ \text{(h1)}$$

$$P_2 \equiv \partial T^*/\partial\omega^2 = (m/r^2)\omega^2 = (m/r)(d\sigma/dt) \Rightarrow dP_2/dt = m(r^{-2}\omega^2)^{\cdot}; \qquad \text{(h2)}$$

$$\partial T^*/\partial\theta^1 = A^r{}_{1'}(\partial T^*/\partial r) + A^\phi{}_{1'}(\partial T^*/\partial\phi) = (1)(\partial T^*/\partial r) + (0)(0)$$

$$= \partial T^*/\partial r = (m/2)(-2)(1/r^3)(\omega^2)^2 = -(m/r^3)(\omega^2)^2, \qquad \text{(h3)}$$

$$\partial T^*/\partial\theta^2 = A^r{}_{2'}(\partial T^*/\partial r) + A^\phi{}_{2'}(\partial T^*/\partial\phi) = (0)(\partial T^*/\partial r) + (1)(0) = 0; \ \text{(h4)}$$

and so, and by Equations d1 through d3 and e2, the central equation (7.5.6c), after collecting $\delta\theta^K$ terms, yields

$$(I_{1'} - Q_{1'})\delta\theta^1 + (I_{2'} - Q_{2'})\delta\theta^2 = 0, \qquad \text{(i1)}$$

where

$$I_{1'} \equiv dP_1/dt - \partial T^*/\partial\theta^1 - P_2(2\omega^2/r)$$

$$= m(d\omega^1/dt) + (m/r^3)(\omega^2)^2 - (m\omega^2/r^2)(2\omega^2/r)$$

$$= m[(d^2r/dt^2) - r^{-3}(d\sigma/dt)^2], \tag{i2}$$

$$I_{2'} \equiv dP_2/dt + P_2(2\omega^1/r)$$

$$= [(m/r^2)\ \omega^2]^{\cdot} + (m\omega^2/r^2)(2\omega^1/r) = \ldots = (m/r^2)(d^2\sigma/dt^2); \tag{i3}$$

($\Rightarrow I_{1'} = I_r$, $I_{2'} = I_\phi/r^2$) from which, since $\delta\theta^1$, $\delta\theta^2$ are unconstrained, we obtain the equations: $I_{1'} = Q_{1'}$, $I_{2'} = Q_{2'}$; i.e., Equations f1 and f2. If $Q_{2'} = Q_\sigma = 0 \Rightarrow d\sigma/dt = r^2(d\phi/dt) = 0$ (Kepler's second "law").

iv. The *Appell* equations in nonholonomic variables are (with $d^2\theta^K/dt^2 \equiv d\omega^K/dt \equiv \alpha^K$):

$$I_{1'} = \partial S^*/\partial\alpha^1 \equiv \partial S^*/\partial(d^2r/dt^2) = Q_{1'}, \tag{j1}$$

$$I_{2'} = \partial S^*/\partial\alpha^2 \equiv \partial S^*/\partial(d^2\sigma/dt^2) = Q_{2'}, \tag{j2}$$

coincide with the just found Hamel equations, from Equation i1; and confirm the earlier kinematico-inertial result:

Measure of inertial nonholonomicity $\equiv E_K(T^*) - \partial S^*/\partial\alpha^K \neq 0$, in general.

Here,

$$[(\partial T^*/\partial\omega^1)^{\cdot} - \partial T^*/\partial\theta^1] - \partial S^*/\partial\alpha^1$$

$$= (2/r)\omega^2\ P_2 = (2m/r^2)(d\sigma/dt)^2 \neq 0, \tag{j3}$$

$$[(\partial T^*/\partial\omega^2)^{\cdot} - \partial T^*/\partial\theta^2] - \partial S^*/\partial\alpha^2$$

$$= (-2/r)\omega^1\ P_2 = (-2m/r^2)(dr/dt)\ (d\sigma/dt) \neq 0. \tag{j4}$$

Example 7.7.8. Moving Axes (or Relative Motion) via the Lagrangean Method (Rheonomic Systems)

We recall that (Equations 7.2.2ff):

$$2T = M_{\alpha\beta}\ V^\alpha\ V^\beta = M_{kl}\ V^k\ V^l + 2M_k\ V^k + M_0 = 2(T_{(2)} + T_{(1)} + T_{(0)}), \tag{a}$$

where

$$2T_{(2)} = M_{kl}\ V^k\ V^l, \quad T_{(1)} = M_{k,n+1}\ V^k\ V^{n+1} \equiv M_{k0}\ V^k \equiv M_k\ V^k,$$

$$2T_{(0)} = M_{n+1,n+1} V^{n+1}\ V^{n+1} \equiv M_{00} V^0\ V^0 \equiv M_0; \quad M_{\alpha\beta} = M_{\alpha\beta}(t,\ q). \quad \text{(a1)}$$

Hence, recalling Equations 7.2.16a through 7.2.17i, Lagrange's *covariant* equations (7.4.9b), $E_k(T) = Q_k$, assume the following explicit forms:

$$E_k(T) = E_k(T_{(2)}) + E_k(T_{(1)}) + E_k(T_{(0)})$$

$$= [M_{kl}\ (dV^l/dt) + \Gamma_{k,bh}\ V^b\ V^h + (\partial M_{kb}/\partial t) V^b]$$

$$+ [(\partial M_k/\partial q^b - \partial M_b/\partial q^k) V^b + \partial M_k/\partial t] + [0 - (1/2)(\partial M_0/\partial q^k)]$$

$$= M_{kl}(dV^l/dt) + \Gamma_{k,bh}\ V^b\ V^h + 2\Gamma_{k,b,n+1}\ V^b + \Gamma_{k;n+1,n+1}$$

$$\equiv M_{kl}(dV^l/dt) + \Gamma_{k,bh}\ V^b\ V^h + 2\Gamma_{k,b0}\ V^b + \Gamma_{k,00} = Q_k, \quad \text{(b)}$$

where

$$2\Gamma_{k,bh} = (\partial M_{kb}/\partial q_h + \partial M_{kh}/\partial q_b - \partial M_{bh}/\partial q_k), \quad \text{(b1)}$$

$$2\Gamma_{k,b0}\ V^b = [(\partial M_k/\partial q^b - \partial M_b/\partial q^k) + \partial M_{kb}/\partial t] V^b$$

$$= E_k(T_{(1)}) + (\partial M_{kb}/\partial t) V^b - \partial M_k/\partial t$$

$$[= E_k(T_{(1)}),\ \text{for scleronomic systems}] \quad \text{(b2)}$$

$$\Gamma_{k,00} = \partial M_k/\partial t - (1/2)(\partial M_0/\partial q^k) = \partial M_k/\partial t + E_k(T_{(0)}); \quad \text{(b3)}$$

or, rearranging, we can, finally, put them in the "scleronomic" or "inertial" form:

$$M_{kb}(dV^b/dt) + \Gamma_{k,bh}\ V^b\ V^h = Q_k - (\partial M_{kb}/\partial t) V^b - E_k(T_{(1)}) - E_k(T_{(0)})$$

$$= Q_k - 2\Gamma_{k,b0}\ V^b - \Gamma_{k,00}; \quad \text{(c)}$$

$$= Q_k + T_k + G_k, \quad \text{(d)}$$

where

$T_k \equiv -(\partial M_{kb}/\partial t) V^b - \partial M_k/\partial t + C_k$: *transport* part of "apparent force," (d1)

$C_k \equiv +\ (1/2)\partial M_0/\partial q^k = -E_k(T_{(0)})$: *centrifugal* part of transport "force," (d2)

$G_k \equiv (\partial M_b/\partial q^k - \partial M_k/\partial q^b) V^b \equiv G_{kb}\ V^b = -E_k(T_{(1)}) + \partial M_k/\partial t$: (d3)

gyroscopic (or Coriolis) part of apparent "force" [$\Rightarrow$ Power $\equiv (G_{kb}\ V^b)\ V^k = 0$]

$G_{kb} = -G_{bk} \equiv \partial_k M_b - \partial_b M_k$: *gyroscopic coefficients* (curl of "vector" M_k). (d4)*

Now, these results may be given the following geometrical intepretation: *The motion of the rheonomic system is still represented by the motion of a figurative particle in a generally moving and deforming Riemannian space with metric* $ds^2 \equiv 2T_{(2)}\, dt^2 = M_{kl}\, dq^k\, dq^l$, *but also subject to the (nonimpressed and nonconstraint) "apparent, or fictitious, forces"* $-2\Gamma_{k,b0} V^b$ *and* $-\Gamma_{k,00}$, *i.e., those arising from the "temporal" part of the Christoffels,* $\Gamma_{k,b0}$, $\Gamma_{k,00}$.**

In nonrelativistic mechanics, both impressed and constraint forces are real forces, i.e., frame-independent, or objective; and, therefore, their components can change only as a result of coordinate transformations (local axes orientation), not because of the kinematical state of the frame. That explains why only the spatial part of the metric, i.e., M_{kl} (and its inverse M^{kl}), is needed to raise/lower their indices (recall Equation 7.2.11a and following); i.e., strictly speaking, the $Q^k = M^{kl}\, Q_l$ (instead of $Q^k = M^{kl}\, Q_l + M^{k,n+1}\, Q_{n+1}$) are not the contravariant components of the Q_k — the event manifold R_{n+1} is non-Riemannian! And similarly for the momentum components:

$$p_k \equiv \partial T/\partial V^k = M_{kb}\, V^b + M_{k,n+1}\, V^{n+1} \equiv M_{kb}\, V^b + M_k \Rightarrow V^b = M^{bk}(p_k - M_k);$$

the V^b are not really the covariant components of the velocity vector $V^k \equiv dq^k/dt$. On the other hand, the apparent forces are frame-dependent; i.e., they result solely from the description of motion in the "wrong" (noninertial) frame.

This asymmetry between the transformation properties of real and inertial forces can be resolved satisfactorily only by the general theory of relativity; one of the basic axioms of which is the *principle of equivalence,* i.e., indistinguishability, between inertial (fictitious) and gravitational (real) forces; see, e.g., Adler et al. (1975), Misner, et al. (1973).

An Application

Let us obtain the equations of motion of a particle P, of mass m, moving in a uniformly rotating frame of reference (represented by the rectangular Cartesian axes $O\text{–}xyz$), relative to inertial space (represented by the rectangular Cartesian axes $O\text{–}XYZ$; such that, always, $OZ = Oz$). With $\phi = \omega t$, ω: constant angular velocity of $O\text{–}xyz$ relative to $O\text{–}XYZ$, we have

$$X = x\cos\phi - y\sin\phi, \quad Y = x\sin\phi + y\cos\phi, \quad Z = z = 0, \tag{e1}$$

$$\Rightarrow x = X\cos\phi + Y\sin\phi, \quad y = -X\sin\phi + Y\cos\phi, \quad Z = z = 0; \tag{e2}$$

---

* It is not hard to show that the *contravariant* counterparts of Equation b can be put in the *geodesic* form:

$$DV^k/Dt \equiv dV^k/dt + G^{\alpha}{}_{\beta\gamma}\, V^{\beta}\, V^{\gamma} = 0, \tag{d4}$$

where (recalling Equations 7.2.11a through d): M_{kh} is the inverse of M_{kh}, $G^{n+1}{}_{\beta\gamma} = 0$, and $2G^k{}_{\beta\gamma} \equiv 2\Gamma^k{}_{\beta\gamma} - 2\delta^{n+1}{}_{\beta}\, \delta^{n+1}{}_{\gamma}\, Q^k = M^{kh}(M_{h\beta,\gamma} + M_{h\gamma,\beta} - M_{\beta\gamma,h}) - 2\delta^{n+1}{}_{\beta}\, \delta^{n+1}{}_{\gamma}\, Q^k$

** For a related treatment of relative motion, see Ishlinskii (1987, pp. 264–274).

and so, the inertial kinetic energy of P equals

$$2T = m[(dX/dt)^2 + (dY/dt)^2] = \ldots = 2(T_{(2)} + T_{(1)} + T_{(0)}), \quad \text{(e3)}$$

where

$$2T_{(2)} \equiv m[(dx/dt)^2 + (dy/dt)^2], \quad \text{i.e., } M_{11} = M_{22} = m, \quad M_{12} = 0, \quad \text{(e4)}$$

$$T_{(1)} \equiv -m\omega[y(dx/dt) - x(dy/dt)], \quad \text{i.e., } M_{10} \equiv M_1 = -m\omega y,$$

$$M_{20} \equiv M_2 = +m\omega x, \quad \text{(e5)}$$

$$2T_{(0)} \equiv m\omega^2(x^2 + y^2), \quad \text{i.e., } M_{00} \equiv M_0 = m\omega^2(x^2 + y^2). \quad \text{(e6)}$$

Therefore,

i. The *centrifugal* "force," C_k (the only surviving part of the *transport* "force," T_k) equals

$$C_k \equiv \partial T_{(0)}/\partial q^k = (1/2)(\partial M_0/\partial q^k); \quad \text{i.e., } C_1 = m\omega^2 x, \quad C_2 = m\omega^2 y; \quad \text{(f1)}$$

ii. The *gyroscopic/Coriolis* "force" $G_k = G_{kb}\, V^b = (\partial M_b/\partial q^k - \partial M_k/\partial q^b)V^b$ is

$$k = 1 \rightarrow x: \quad (M_{1,1} - M_{1,1})V^1 + (M_{2,1} - M_{1,2})V^2 = [m\omega - (-m\omega)]v^y$$

$$= 2m\omega v^y, \quad \text{(f2)}$$

$$k = 2 \rightarrow y: \quad (M_{1,2} - M_{2,1})V^1 + (M_{2,2} - M_{2,2})V^2 = (-m\omega - m\omega)v^x$$

$$= -2m\omega v^x; \quad \text{(f3)}$$

i.e.,

$$G_1 = 2m\omega v^y, \quad G_2 = -2m\omega v^x. \quad \text{(f4)}$$

Hence, with covariant holonomic components of the impressed force on P equal to $Q_{x,y}$, the covariant equations of relative motion, $E_k(T_{(2)}) = Q_k + G_k + C_k$, become

$$k = 1 \rightarrow x: \quad (mv^x)^{\cdot} = Q_x + 2m\omega v^y + m\omega^2 x$$

$$\Rightarrow m(dv^x/dt - 2\omega v^y - \omega^2 x) = Q_x, \quad \text{(g1)}$$

$$k = 2 \rightarrow y: \quad (mv^y)^{\cdot} = Q_y - 2m\omega v^x + m\omega^2 y$$

$$\Rightarrow m(dv^y/dt + 2\omega v^x - \omega^2 y) = Q_y. \quad \text{(g2)}$$

Finally, if $Q_{x,y} = 0$, the holonomic power equation,

$$dh/dt \to d/dt[(\partial T/\partial v^k)v^k - T] = -\partial T/\partial t \to 0,$$

reduces to the generalized energy integral:

$$h = (\partial T/\partial v^k)v^k - T = T_{(2)} - T_{(0)}$$

$$= \text{constant} \Rightarrow (v^x)^2 + (v^y)^2 = \omega^2(x^2 + y^2) + \text{constant}. \qquad \text{(g3)}$$

(Of course, Equation g3 also results by multiplying Equation g1 with v^x and Equation g2 with v^y, adding them, etc.)

BIBLIOGRAPHY AND REFERENCES

This is a *cumulative* and *alphabetical* listing of all books, papers, etc. referenced in the text,* which complements those found in such excellent expositions as (alphabetically): Moon and Spencer (1986), Reich (1994), Schouten (1954a, b),** Synge (1936), Vranceanu (1936); also Ericksen (1960) and Papastavridis (1998). Specific instructions and/or suggestions for its use are given at the beginning of each chapter, and wherever else is deemed appropriate.

Abram, J., *Tensor Calculus*, Through Differential Geometry, Butterworths, London, 1965.

Adler, R., Bazin, M., and Schiffer, M., *Introduction to General Relativity*, 2nd ed. McGraw-Hill, New York, 1975.

Altmann, S. L., *Rotations, Quaternions, and Double Groups*, Clarendon Press, Oxford, 1986.

Appell, P., *Eleménts de Calcul Tensoriel*, Applications Geometriques et Mécaniques (Vol. 5 of *Traité de Mécanique Rationelle*), 2nd ed., Gauthier-Villars, Paris, 1933.

Aris, R., *Vectors, Tensors, and the Basic Equations of Fluid Mechanics*, Prentice-Hall, Englewood Cliffs, NJ, 1962 (reprinted by Dover, New York, 1990).

Arnold, V. I., *Mathematical Methods in Classical Mechanics*, 2nd ed., Springer, New York , 1988 (original in Russian, 1974; 1st Engl. ed., 1978).

Bahar, L. Y., The Theory of Rigid-Body Rotations, private communication, 1987.

Barbotte, J., *Le Calcul Tensoriel*, Bordas, Paris, 1948.

Bergmann, P. G., *Introduction to the Theory of Relativity*, Prentice-Hall, Englewood Cliffs, NJ, 1942 (reprinted by Dover, New York).

Bergmann, P. G., "The Special Theory of Relativity," and "The General Theory of Relativity," pp. 109–272, vol. 4 of *Handbuch der Physik*, ed. by S. Flügge. Springer, Berlin, 1962.

Berwald, L., "Differentialinvarianten in der Geometrie. Riemannsche Mannigfaltigkeiten und ihre Verallgemeinerungen," pp. 73–181, vol. 3, part 3, article D 11 of *Encyklopädie der Mathematischen Wissenschaften*, Teubner, Leipzig, 1927 (article completed in October 1923).

Betten, J., *Tensorrrechnung für Ingenieure*, Teubner, Stuttgart, 1987.

Bewley, L. V., *Tensor Analysis of Electric Circuits and Machines*, Ronald Press, New York, 1961.

Blaschke, W. and Leichtweiss, K., *Elementare Dofferentialgeometrie*, 5th ed., Springer, Berlin, 1973.

Block, H. D., *Introduction to Tensor Analysis*, C. E. Merrill Books, Columbus, OH, 1962.

Boer, R. de, *Vektor- und Tensorrechnung für Ingenieure*, Springer, Berlin, 1982.

Borisenko, A. I. and Tarapov, I. E., *Vector and Tensor Analysis with Applications*, revised Engl. ed., Prentice-Hall, Englewood Cliffs, NJ, 1968 (reprinted by Dover, New York, 1979).

Brand, L., *Vector and Tensor Analysis*, Wiley, New York, 1947.

Bricas, M. A., *Lectures on Tensor Calculus* (mimeographed notes in Greek), University of Athens, Athens, 1958 (reprinted in 1968).

* In cases of multiple authorship, the alphabetical rule applies to the last name of the *first* author.

** The various publications of an author *within a certain year* are denoted with lower case Latin characters, e.g., Schouten (1954a), Schouten (1954b).

Brillouin, L., *Tensors in Mechanics and Elasticity,* Academic Press, New York, 1964 (original in French, 1938).

Bronshtein, I. N. and Semendyayev, K. A. "Tensor Calculus," pp. 808–826, Ch. 8.3 of *Handbook of Mathematics*, 3rd ed., Van Nostrand Reinhold, New York, 1985 (original in Russian, also transl. in German; several editions).

Budiansky, B., "Tensors," pp. 179-225 of *Handbook of Applied Mathematics*, Selected Results and Methods, ed. by C. Pearson, Van Nostrand-Reinhold, New York, 1974.

Bureau, F., *Calcul Vectoriel et Calcul Tensoriel,* University of Liége Press, Liége, Belgium, and Masson, Paris, 1945.

Burke, W. L., *Applied Differential Geometry,* Cambridge University Press, Cambridge, 1985 (reprinted: 1987, 1989, 1992, 1994, 1996).

Carmeli, M., *Classical Fields: General Relativity and Gauge Theory,* McGraw-Hill, New York, 1975.

Cartan, E., *Leçons sur les Invariants Intégraux,* Hermann, Paris, 1922.

Cartan, E., "Sur la Représentation Géométrique des Systèmes Matériels Non-holonomes," *Atti Congr. Int. Mat.*, 4, 253–261, 1928.

Cartan, E., *La Théorie des Groupes Finis et Continus, et la Géométrie Différentielle, Traitées par la Méthode du Repère Mobile,* Gauthier-Villars, Paris, 1937.

Cartan, E., *Leçons sur la Géométrie des Espaces de Riemann*, 2nd ed., Gauthier-Villars, Paris, 1946.

Cartan, E., *La Géométrie des Espaces de Riemann*, 2nd ed., Gauthier-Villars, Paris, 1951 (1st ed., 1925. Engl. transl. of 2nd ed., by J. Glazebrook, with notes and appendices by R. Hermann, Math Science Press, Brookline, MA, 1983).

Chetaev, N. G., *Theoretical Mechanics,* Mir-Springer, Moscow, 1989 (original in Russian, 1987).

Cisotti, U., *Lezioni di Calcolo Tensoriale,* C. Tamburini fu Camillo, Milano, 1928.

Clauser, E., "Geometrizzazione Della Dinamica dei Sistemi a Vincoli Mobili," *Atti Accad. Naz. Lincei, Rendiconti, Cl. Sci. Fis., Mat. Nat.*, 8th Ser., 29, 33–39, 1955.

Coburn, N., *Vector and Tensor Analysis,* Macmillan, New York, 1955 (reprinted by Dover, New York, 1970).

Craig, H. V., *Vector and Tensor Analysis,* McGraw-Hill, New York, 1943.

D'Abro, A., *The Evolution of Scientific Thought, From Newton to Einstein,* 2nd ed., Dover, New York, 1950.

Danielson, D. A., *Vectors and Tensors in Engineering and Physics,* Addison-Wesley, Redwood City, CA, 1992.

Delachet, A., *Le Calcul Tensoriel*, 2nd ed., Presses Universitaires de France, Paris, 1974 (1st ed., 1969).

Denis-Papin, M. and Kaufmann, A., *Cours de Calcul Tensoriel Appliqué (Geométrie Différentielle Absolue),* 3rd ed., Albin Michel, Paris, 1966.

Dirac, P. A. M., *General Theory of Relativity,* Wiley, New York, 1975.

Dobronravov, V. V., "Analytical Dynamics in Nonholonomic (or Anholonomic) Coordinates," (in Russian) *Uch. Zap. Mosk. Gos. Univ.*, 2 (122), 77–182, 1948.

Dobronravov, V. V., *Fundamentals of Nonholonomic System Mechanics* (in Russian). Vishaya Shkola, Moscow, 1970.

Dobronravov, V. V., *Foundations of Analytical Mechanics* (in Russian). Vishaya Shkola, Moscow, 1976.

Dörrie, H., *Vektoren,* R. Oldenbourg, München, 1941.

Dubrovin, B. A., Fomenko, A. T., and Novikov, S. P., *Modern Geometry,* Methods and Applications, in three parts, 2nd ed., Springer, New York, 1992 (1st Engl. ed., 1984; original in Russian, 1979).

Duschek, A. and Hochrainer, A., *Tensorrechnung in Analytischer Darstellung*, in three parts (*Tensoralgebra, Tensoranalysis, Anwendungen in Physik und Technik*), 4th ed., Springer, Wien, 1960 (1st ed., 1946–1955).

Duschek, A. and Mayer, W., *Lehrbuch der Differentialgeometrie*, 2 Vol., Teubner, Leipzig, 1930.

Eddington, A. S., *The Mathematical Theory of Relativity*, 2nd ed., Cambridge University Press, Cambridge, 1924 (1st ed., 1923; reprinted by Chelsea, New York).

Eisenhart, L. P., *Riemannian Geometry,* Princeton University Press, Princeton, NJ, 1926 (2nd corrected printing, 1949).

Eisenhart, L. P., *Non-Riemannian Geometry,* American Mathematical Society, New York, 1927.

Eisenhart, L. P., *Continuous Groups of Transformations,* Princeton University Press, Princeton, NJ, 1933 (reprinted, with corrections, by Dover, New York, 1961).

Eisenhart, L. P., *An Introduction to Differential Geometry, with Use of the Tensor Calculus,* Princeton University Press, Princeton, NJ, 1947.

Eisenreich, G., *Vorlesungen über Vektor- und Tensorrechnung*, Teubner, Leipzig, 1971.

Ericksen, J. L., "Tensor Fields," pp. 794–858, Vol. III/1 of *Handbuch der Physik*, ed. by S. Flügge, Springer, Berlin, 1960.

Eringen, A. C., "Tensor Analysis," pp. 1–155, Vol. 1 of *Continuum Physics*, ed. by A. C. Eringen, Academic Press, New York, 1971.

Essén, H., "On the Geometry of Nonholonomic Dynamics," *J. Appl. Mech.* (ASME), 61, 689–694, 1994.

Farrashkhalvat, M. and Miles, J. P., *Tensor Methods for Engineers,* Ellis Horwood, New York, 1990.

Favard, J., *Cours de Géometrie Differentielle Locale,* Gauthier-Villars, Paris, 1957.

Ferrarese, G., "Sulle Equazioni di Moto di un Sistema Soggetto a un Vincolo Anolonomo Mobile," *Rend. Mat., Inst. Naz. Alta Mat.*, 22, 351–370, 1963.

Ferrarese, G., *Lezioni di Meccanica Razionale*, 2 Vol., Pitagora Editrice, Bologna, 1980.

Finkbeiner, D. T., *Introduction to Matrices and Linear Transformations,* Freeman, San Francisco, 1978.

Finzi, B. and Pastori, M., *Calcolo Tensoriale e Applicazioni*, 2nd ed., Zanichelli, Bologna, 1961.

Fock, V., *The Theory of Space, Time and Gravitation*, 2nd ed., Pergamon-Macmillan, New York, 1964 (original in Russian, 1961).

Forsyth, A. R., *Theory of Differential Equations*, Vol. I: *Exact Equations and Pfaff's Problem,* Cambridge University Press, 1890 (reprinted by Dover, 1959).

Foster, J. and Nightingale, J. D., *A Short Course in General Relativity,* Longman, London, 1979.

Funk, P., *Variationsrechnung und ihre Anwendung in Physik und Technik,* Springer, Berlin, 1962.

Gallissot, F., "Les Formes Exterieures en Mécanique," *Ann. Inst. Fourier Univ. Grenoble*, 4, 145–297, 1954 (for the year 1952).

Gantmacher, F., *Lectures in Analytical Mechanics,* Mir, Moscow, 1970 (original in Russian, 1966).

Gelfand, I. M., *Lectures on Linear Algebra,* Interscience, New York, 1961 (reprinted by Dover, New York, 1989).

Gerretsen, J. C. H., *Lectures on Tensor Calculus and Differential Geometry,* Noordhoff, Groningen, 1962.

Gibbs, W. J., *Tensors in Electrical Machine Theory,* Chapman and Hall, London, 1952.

Golab, S., *Tensor Calculus,* Elsevier, Amsterdam, 1974 (original in Polish, 1970).

Golomb, M., *Lectures on Theoretical Mechanics*, prepared by I. Marx, Purdue University, West Lafayette, IN, 1961.

Gontier, G., *Mécanique des Milieux Déformables*, Principes et Théorèmes Généraux, Dunod, Paris, 1969.

Green, A. E. and Zerna, W., *Theoretical Elasticity,* 2nd ed., Clarendon Press, Oxford, 1968 (reprinted by Dover, New York).

Guldberg, A., "Partielle und Totale Differentialgleichungen," pp. 561–578, Vol. 1, 2nd ed., of *Repertorium der Höheren Mathematik,* ed. by E. Pascal, E. Salkowski, and H. E. Timerding. Teubner, Leipzig, 1927.

Hamel, G., *Elementare Mechanik,* 2nd ed., Teubner, Leipzig, 1922 (1st ed., 1912).

Hamel, G., *Theoretische Mechanik,* Springer, Berlin, 1949.

Hawkins, G. A., *Multilinear Analysis for Students in Engineering and Science,* Wiley, New York, 1963.

Heil, M. and Kitzka, F., *Grundkurs Theoretische Mechanik,* Teubner, Stuttgart, 1984.

Hessenberg, G., "Vektorielle Begründung der Differentialgeometrie," *Math. Ann.*, 78, 187–217, 1918.

Hladik, J., *Le Calcul Tensoriel en Physique*, avec Exercices Corrigés, 2nd ed., Masson, Paris, 1995.

Horák, Z., "Sur les Systèmes Non Holonomes," *Bull. Int. Acad. Sci. Bohême (Prague)*, 29, 83–100, 1928.

Horák, Z., "Theorie Générale du Choc dans les Systèmes Materiels," *J. Ec. Polytech.*, 28 (2), 15–64, 1931.

Horák, Z., "Sur la Dynamique Absolue des Systèmes Rhéonomes," *Prac. Mat. Fiz.*, 41, 25–37, 1933.

Horák, Z., "Mécanique Absolue et sa Représentation dans l'Espace-Temps des Configurations," *Prac. Mat. Fiz.*, 42, 59–108, 1935 (extensive reference list on pp. 106–107).

Horák, Z., "Sur le Calcul Absolu des Variations," *Prac. Mat. Fiz.*, 43, 119–149, 1936.

Ishlinskii, A. Yu., *Classical Mechanics and Inertial Forces* (in Russian), Nauka, Moscow, 1987.

Jaunzemis, W., *Continuum Mechanics,* Macmillan, New York, 1967.

Jeffreys, H., *Cartesian Tensors,* Cambridge University Press, Cambridge, 1931 (several reprintings).

Johnsen, L., "Dynamique Générale des Systèmes Non-holonomes," *Skr. Utigitt Nor. Vidensk. Akad. Oslo, I. Mat. Naturv. Klass.*, 1–75, 1941.

Juvet, G., *Introduction au Calcul Tensoriel et au Calcul Différentiel Absolu.,* Blanchard, Paris, 1922.

Kähler, E., *Einführung in die Theorie der Systeme von Differentialgleichungen.,* Springer, Berlin, 1934 (reprinted by Chelsea, 1949).

Kästner, S., *Vektoren, Tensoren, Spinoren,* Akademie-Verlag, Berlin, 1960.

Kay, D. C., *Tensor Calculus,* McGraw-Hill (Schaum's Outline Series), New York, 1988.

Kil'chevskii, N. A., *Elements of Tensor Calculus and Its Applications to Mechanics*, 2nd ed. (in Russian), Naukova Dumka, Kiev, 1972.

Kil'chevskii, N. A., *Course of Theoretical Mechanics*, 2 Vols. (in Russian), Nauka, Moscow, 1977.

Kilmister, C. W., *Hamiltonian Dynamics,* Wiley, New York, 1964.

Kilmister, C. W. and Reeve, J. E., *Rational Mechanics,* Longman, London, 1966.

Klein, F., *Elementary Mathematics from an Advanced Viewpoint*, Vol. 1, *Arithmetic, Algebra, Analysis;* Vol. 2, *Geometry,* Dover, New York, 1939 (transl. from 3rd German ed., of 1924–1925).

Kline, M., *Mathematical Thought from Ancient to Modern Times,* Oxford University Press, New York, 1972.

Klingbeil, E., *Tensorrechnung für Ingenieure,* Bibliographisches Institut, Mannheim, 1966.

Klingbeil, E., *Variationsrechnung,* Bibliographisches Institut/Wissenschaftsverlag, Mannheim, 1977.

Korenev, G. V., *The Mechanics of Guided Bodies,* Iliffe Books, London, 1967 (transl. from Russian).

Korenev, G. V., *Mechanics of Controlled Manipulators* (in Russian), Nauka, Moscow, 1979.

Korn, A. and Korn, A., "Representation of Mathematical Models: Tensor Algebra and Analysis," pp. 469-496 of *Mathematical Handbook for Scientists and Engineers,* McGraw-Hill, New York, 1962.

Kramer, E. E., *The Nature and Growth of Modern Mathematics,* 2 Vol., Fawcett, Greenwich, CT, 1974 (originally published as a single volume by Princeton University Press, 1970).

Kreyszig, E., *Differential Geometry,* University of Toronto Press, Toronto, 1959 (reprinted by Dover, New York, 1991).

Lagally, M., *Vorlesungen über Vektorrechnung,* 7th ed., Geest & Portig K.-G., Leipzig, 1964 (1st ed., 1928).

Lagrange, R., *Calcul Différentiel Absolu,* Gauthier-Villars, Paris, 1926.

Lanczos, C., *Space Through the Ages,* The Evolution of Geometrical Ideas from Pythagoras to Hilbert and Einstein, Academic Press, London and New York, 1970.

Landau, L. and Lifshitz, E., *The Classical Theory of Fields,* 3rd rev. Engl. ed., (Vol. 2 of *Course of Theoretical Physics*), Addison-Wesley, Reading, MA; and Pergamon Press, London, 1971 (originally in Russian, early 1960s).

Langhaar, H. L., *Energy Methods in Applied Mechanics,* Wiley, New York, 1962.

Lass, H., *Vector and Tensor Analysis,* McGraw-Hill, New York, 1950.

Laugwitz, D., *Differentialgeometrie,* 2nd ed., Teubner, Stuttgart, 1968 (English transl. of 1st ed. of 1960 is also available).

Lawden, D. F., *Analytical Mechanics,* Allen and Unwin, London, 1972.

Lawrie, I. D., *A Unified Grand Tour of Theoretical Physics,* A. Hilger, Bristol, 1990.

Levi-Civita, T., "Sugli Integrali Algebrici delle Equazioni Dinamiche," *Atti R. Accad. Sci. Torino,* 31, 816–823, 1895 (for years 1895–1896).

Levi-Civita, T., *The Absolute Differential Calculus (Calculus of Tensors),* Blackie, London, 1926 (reprinted by Dover, New York, 1977; original in Italian, 1925).

Lianis, G., *Theory of Continuous Media, Appendix on Tensors* (mimeographed lecture notes), Purdue University, West Lafayette, IN; and National Technical University of Athens; 1970–1975.

Lichnerowicz, A., *Elements of Tensor Calculus,* Methuen, London, 1962 (transl. of 2nd French ed., 1958; 1st ed., 1950).

Lindsay, R. B. and Margenau, H., *Foundations of Physics,* Dover, New York, 1957 (reprint of original version of 1936).

Lippmann, H., *Angewandte Tensorrechnung,* Für Ingenieure, Physiker und Mathematiker, Springer, Berlin, 1993.

Lipschutz, M. M., *Differential Geometry,* Schaum's/McGraw-Hill, New York, 1969.

Lodge, A. S., *Body Tensor Fields in Continuum Mechanics,* With Applications to Polymer Rheology, Academic Press, New York, 1974.

Lotze, A., *Vektor- und Affinor-Analysis,* R. Oldenbourg, München, 1950.

Lovelock, D. and Rund, H., *Tensors, Differential Forms and Variational Principles,* Wiley, New York, 1975 (reprinted with corrections by Dover, New York, 1988).

Lur'e, A. I., *Mécanique Analytique*, 2 Vol., Librairie Universitaire, Louvain, 1968 (original in Russian, 1961).

Lur'e, A. I., *Nonlinear Theory of Elasticity,* North Holland, Amsterdam, 1990 (original in Russian, 1980).

Lynn, J. W., *Tensors in Electrical Engineering,* Arnold, London, 1963.

Maißer, P. and Steigenberger, J., "Der Lagrange-Formalismus für Discrete Elektromechanische Systeme," *Z. Angew. Math. Mech.*, 59, 717–730, 1979.

Maißer, P., "Der Lagrange-Formalismus für Diskrete Elektromechanische Systeme in Anholonomen Koordinaten und Seine Anwendung in der Theorie Elektrischer Maschinen," *Wiss. Z. Tech. Hochsch. Ilmenau*, 27 (2), 131–145, 1981.

Maißer, P., "Modellgleichungen für Manipulatoren," *Tech. Mech.*, 3 (2), 64–77, 1982.

Maißer, P., "Ein Beitrag zur Theorie diskreter elektromechanische Systeme mit Anwendungen in der Manipulator-/Robotertechnik," Dr. Sc. Nat. Dissertation, Faculty of Mathematics and Natural Sciences, Technische Hochschule Ilmenau, Germany, 1983–4.

Maißer, P., "A Differential-Geometric Approach to the Multi Body System Dynamics," *Z. Angew. Math. Mech.*, 71, T116-T119, 1991.

Malvern, L. E., *Introduction to the Mechanics of a Continuous Medium,* Prentice-Hall, Englewood Cliffs, NJ, 1969.

Marsden, J. E. and Ratiu, T. S., *Introduction to Mechanics and Symmetry,* A Basic Exposition of Classical Mechanical Systems, Springer, New York, 1994.

Martin, D., *Manifold Theory, An Introduction for Mathematical Physicists,* Ellis Horwood, New York, 1991.

Mattioli, G. D., "Su una Forma delle Equazioni di Lagrange Nelle Quali il Tempo è Introdotto Come *n + 1* — esima Coordinata," *Atti R. Inst. Veneto Sci., Lett. Arti*, 91 (Pt. 2), 79–91, 1931–1932.

Maxwell, E., A., *Coordinate Geometry, with Vectors and Tensors,* Clarendon Press, Oxford, 1958.

McCauley, J. L., *Classical Mechanics*, Transformations, Flows, Integrable and Chaotic Dynamics, Cambridge University Press, Cambridge, 1997.

Mc Cleary, J., *Geometry from a Differentiable Viewpoint,* Cambridge University Press, Cambridge, 1994.

Mc Connell, A. J., *Applications of the Absolute Differential Calculus,* Blackie, London, 1931 (reprinted by Dover, New York, as *Applications of Tensor Analysis*, 1957).

Michal, A., *Matrix and Tensor Calculus, with Applications to Mechanics, Elasticity, and Aeronautics,* Wiley, New York, 1947.

Mishchenko, A. and Fomenko, A., *A Course of Differential Geometry and Topology,* Mir, Moscow, 1988 (original in Russian, 1980).

Misner, C. W., Thorne, K. S., and Wheeler, J. A., *Gravitation,* Freeman, San Francisco, 1973.

Mittelstaedt, P., *Klassische Mechanik,* Bibliographisches Institut/Hochschultaschenbücherverlag, Mannheim, 1970.

Moon, P. and Spencer, D. E., *Theory of Holors,* A Generalization of Tensors, Cambridge University Press, Cambridge, 1986.

Neimark, J. I. and Fufaev, N. A., *Dynamics of Nonholonomic Systems,* American Mathematical Society, Providence, RI, 1972 (original in Russian, 1967).

Norden, A. P., *Differentialgeometrie*, Parts I, II, VEB, Deutscher Verlag der Wissenschaften, Berlin, 1956.

Novoselov, V. S., "On Classes of Equivalent Nonholonomic Coordinates," (in Russian), *Vestn. Leningr. Univ.*, Nr. 7, 82–85, 1981.

Ogden, R. W., *Non-linear Elastic Deformations,* Ellis Horwood, New York, 1984.

Panagiotounakos, D. E., *Vector and Tensor Analysis,* With Applications to the Mechanics of the Deformable Body and to Theoretical Mechanics (in Greek), Foudas/Frame, Athens, 1989.

Papastavridis, J. G., "Time-Integral Principles for Nonlinear Nonholonomic Systems," *J. Appl. Mech.* (ASME), 64, 985–991, 1997.

Papastavridis, J. G., "A Panoramic Overview of the Principles and Equations of Motion of Advanced Engineering Dynamics," *Appl. Mech. Rev.*, 51, 239–265, 1998.

Papastavridis, J. G., *Analytical Mechanics,* Oxford University Press, New York (in press, to appear in 2000).

Pars, L. A., *A Treatise on Analytical Dynamics,* Wiley/Heine; London, 1965 (reprinted by Ox Bow Press, Woodbridge, CT, 1979).

Pascal, E., "Totale Differentialgleichungen und Differentialformen," pp. 579–600, Vol. 1, 2nd ed., of *Repertorium der Höheren Mathematik,* ed. by E. Pascal, E. Salkowski, and H. E. Timerding, Teubner, Leipzig, 1927.

Pastori, M., "Vincoli e Riferimenti Mobili in Meccanica Analitica," *Ann. Mat.*, Ser. IV, 50, 476–484, 1960.

Pastori, M., "Apparent Forces in Analytical Mechanics," *Meccanica,* 2, 75–81, 1967.

Peschl, E., *Differentialgeometrie,* Bibliographisches Institut, Mannheim, 1973.

Pipes, L. A., *Matrix Methods for Engineering,* Prentice-Hall, Englewood Cliffs, NJ, 1963.

Poliahov, N. N., "Equations of Motion of Mechanical Systems under Nonlinear, Nonholonomic Constraints," (in Russian), *Vestn. Leningr. Univ.*, Nr. 1, 124–132, 1972.

Poliahov, N. N., "On the Differential Principles of Mechanics Obtained from the Equations of Motion of Nonholonomic Systems," (in Russian), *Vestn. Leningr. Univ.*, Nr. 13, 106–114, 1974.

Poliahov, N. N., Zegzhda, S. A., and Yushkov, M. P., "Generalization of Gauss' Principle to Nonholonomic Systems of Higher Order," *Dokl. Akad. Nauk. SSSR*, 269, 1328–1330, 1983 (*Sov. Phys., Dokl.* 28(4), 330–332, 1983).

Poliahov, N. N., Zegzhda, S. A., and Yushkov, M. P., *Theoretical Mechanics* (in Russian), Leningrad University Press, Leningrad (St. Petersbourg), 1985 (English version under preparation).

Pomey, J. B., *Notions de Calcul Tensoriel,* Gauthier-Villars, Paris, 1934.

Pozniak, E. G. and Shikin, E. V., *Differential Geometry* (in Russian), Moscow University Press, Moscow, 1990.

Prange, G., "Die allgemeinen Integrationsmethoden der analytischen Mechanik," pp. 505–804, vol. 4, Pt. 2, articles 12 and 13 of *Encyklopädie der Mathematischen Wissenschaften,* Teubner, Leipzig, 1935 (article completed in December 1934).

RAAG *Memoirs of the Unifying Study of the Basic Problems in Engineering Sciences by Means of Geometry,* 4 Vols., ed. K. Kondo. Gakujutsu Bunken Fukyu-Kai, Tokyo, 1955–1968.

Radakovic, Th. and Lense, J., "Vector und Tensorrechnung, Riemannsche Geometrie," pp. 182-214, Vol. 3 of (old) *Handbuch der Physik,* ed. by H. Geiger and K. Scheel, Springer, Berlin, 1928.

Rainich, G. Y., *Mathematics of Relativity,* Wiley, New York, 1950.

Raschewski, P. K., *Riemannsche Geometrie und Tensoranalysis,* VEB Deutscher Verlag der Wissenschaften, Berlin, 1959 (original in Russian, 1953). First chapter also published separately as *Elementare Einführung in die Tensorrechnung* (same publisher and year).

Reich, K., *Die Entwicklung des Tensorkalküls*, Vom Absoluten Differentialkalkül zur Relativitätstheorie, Birkhäuser, Basel, 1994.

Reichardt, H., *Vorlesungen über Vector und Tensorrechnung*, 2nd ed., VEB Deutscher Verlag der Wissenschaften, Berlin, 1968.

Richtmyer, R. D., *Principles of Advanced Mathematical Physics*, Vol. 2, Springer, New York, 1981.

Rosenberg, R. M., *Analytical Dynamics of Discrete Systems,* Plenum, New York, 1977.

Schild, A., "Tensor Analysis," pp. 7-1–7-20 of *Handbook of Engineering Mechanics*, ed. by W. Flügge, Mc Graw-Hill, New York, 1962.

Schmutzer, E., *Relativistische Physik* (Klassische Theorie), Akademische Verlagsgesellschaft, Geest und Portig K.-G., Leipzig, 1968.

Schouten, J. A., "Ueber Nichtholonome Uebertragungen in Einer L_n," *Math. Z.*, 30, 149–172, 1929.

Schouten, J. A., *Ricci-Calculus*, 2nd ed., Springer, Berlin, 1954a (1st ed. 1924).

Schouten, J. A., *Tensor Analysis for Physicists*, 2nd ed., Clarendon Press, Oxford, 1954b (reprinted by Dover, New York, 1989).

Schouten, J. A. and Struik, D. J., *Einführung in die Neueren Methoden der Differentialgeometrie*, Vol. 1: *Algebra und Uebertragungslehre* (1935), Vol. 2: *Geometrie* (1938), 2nd ed., Noordhoff, Groningen.

Schultz-Piszachich, W., *Tensor algebra und -analysis*, 4th ed., Teubner, Leipzig, 1988.

Schutz, B. F., *Geometrical Methods of Mathematical Physics,* Cambridge University Press, Cambridge, 1980.

Shilov, G. E., *An Introduction to the Theory of Linear Spaces,* Prentice-Hall, Englewood Cliffs, NJ, 1961 (reprinted by Dover, 1974).

Simmonds, J. G., *A Brief on Tensor Analysis*, 2nd ed., Springer, Berlin, 1994 (1st ed. 1982).

Sokolnikoff, I. S., *Tensor Analysis, Theory and Applications to Geometry and Mechanics of Continua,* 2nd ed., Wiley, New York, 1964 (1st ed., 1951).

Somoff, J., *Theoretische Mechanik,* Theil 1: *Kinematik* (1879), Theil 2: *Einleitung in die Statik und Dynamik,* Statik (1879). Teubner, Leipzig (original in Russian, written between 1870 and 1876).

Spain, B., *Tensor Calculus*, 3rd ed., Oliver and Boyd, Edinburgh, 1960 (1st ed., 1953; 2nd ed., 1957).

Spiegel, M., *Theoretical Mechanics,* McGraw-Hill–Schaum, New York, 1967.

Spielrein, J., *Lehrbuch der Vektorrechnung*, Nach der Bedürfnissen in der Technischen Mechanik und Elektrizitätslehre, 2nd ed., K. Wittwer, Stuttgart, 1926 (1st ed., 1916).

Stephani, H., *General Relativity, An Introduction to the Theory of the Gravitational Field,* Cambridge University Press, Cambridge, 1982 (original in German, 1980).

Synge, J. L., "On the Geometry of Dynamics," *Philos. Trans. R. Soc. London, Ser. A*, 226, 31–106, 1927.

Synge, J. L., "Geodesics in Non-holonomic Geometry," *Math. Ann.*, 99, 738–751, 1928.

Synge, J. L., *Tensorial Methods in Dynamics,* University of Toronto Press, Toronto, 1936.

Synge, J. L., "Classical Dynamics," pp. 1–225, Vol. 3/1 of *Handbuch der Physik,* ed. by S. Flügge, Springer, Berlin, 1960.

Synge, J. L. and Griffith, A. B., *Principles of Mechanics*, 3rd ed., McGraw-Hill, New York, 1959.

Synge, J. L. and Schild, A., *Tensor Calculus,* University of Toronto Press, Toronto, 1949 (reprinted by Dover, New York, 1978).

Taylor, J. H., *Vector Analysis,* With an Introduction to Tensor Analysis, Prentice-Hall, Englewood Cliffs, NJ, 1939.

Teichmann, H., *Physikalische Anwendungen der Vector- und Tensorrrechnung*, 2nd ed., Bibliographisches Institut/Hochschultaschenbücher, Mannheim, 1964 (1st ed., 1963; Engl. transl. of 2nd ed. available from G. G. Harrap & Co. Ltd, London, 1969).

Thomas, T. Y., *The Elementary Theory of Tensors,* With Applications to Geometry and Mechanics, McGraw-Hill, New York, 1931.

Thomas, T. Y., *Concepts from Tensor Analysis and Differential Geometry,* 2nd ed., Academic Press, New York, 1965.

Tietz, H., "Geometrie," pp. 117-197, Vol. 2 of *Handbuch der Physik,* ed. by S. Flügge, Springer, Berlin, 1955.

Tonnelat, M. A., *The Principles of Electromagnetic Theory and of Relativity,* Reidel, Dordrecht (Holland), 1966 (original in French, 1959).

Truesdell, C. A., "The Physical Components of Vectors and Tensors," *Z. Angew. Math. Mech.*, 33, 345–356, 1953; 34, 69–70, 1954.

Truesdell, C. A. and Toupin, R., "The Classical Field Theories," pp. 226–793, Vol. III/1 of *Handbuch der Physik,* ed. by S. Flügge, Springer, Berlin, 1960.

Udeschini Brinis, E., "Sul Significato delle Caratteristiche Cinetiche e Delle Equazioni del Maggi per Sistemi Anolonomi," *Atti Accad. Naz. Lincei, Rendiconti, Cl. Sci. Fisiche, Matematiche e Naturali,* 8th Ser., 35, 301–311, 1963.

Vagner, V., "Geometrical Interpretation of the Motion of Nonholonomic Dynamical Systems (in Russian)," *Trudi Semin. Vektornomu Tensornomu Anal.*, 5, 301–327, 1941.

Veblen, O., *Invariants of Quadratic Differential Forms,* Cambridge University Press, Cambridge, 1927.

Veblen, O. and Whitehead, J. H. C., *The Foundations of Differential Geometry,* Cambridge University Press, Cambridge, 1932.

Visconti, A., *Introductory Differential Geometry for Physicists,* World Scientific, River Edge, NJ, London, 1992.

Voigt, W., *Lehrbuch der Kristallphysik,* Teubner, Leipzig, 1910.

Vold, T. G., "Review of Book *New Foundations for Classical Mechanics*," *Am. J. Phys.*, 58, 703–704, 1990.

Vranceanu, G., "Studio Geometrico dei Sistemi Anolonomi, "*Ann. Mat., Pura Appli.,* 4th ser., 6, 9–43, 1929.

Vranceanu, G., *Les Espaces Non Holonomes et Leurs Applications Mécaniques,* Gauthier-Villars, Paris, 1936 (extensive reference list on pp. 66–68).

Vranceanu, G., *Vorlesungen über Differentialgeometrie*, 2 Vols., Akademie-Verlag, Berlin, 1961 [originally in French (part I: 1947, part II: 1951), then in Romanian (1952), new ed. in French (1957)].

Vujanovic, B. D. and Jones, S. E., *Variational Methods in Nonconservative Phenomena,* Academic Press, Boston, 1989.

Vujicic, V., *Kovarijantna Dinamika,* Mathematical Institute, Special Editions, Belgrade, 1981.

Vujicic, V., *Dynamics of Rheonomic Systems,* Mathematical Institute, Belgrade, 1990.

Wang, C.-C., *Mathematical Principles of Mechanics and Electromagnetism, Part A: Analytical and Continuum Mechanics*, pp. 1–86. Plenum, New York, 1979.

Warsi, Z. U. A., "Operations on the Physical Components of Tensors," *Z. Angew. Math. Mech. (ZAMM)*, 76, 361–363, 1996.

Weatherburn, C. E., *An Introduction to Riemannian Geometry and the Tensor Calculus,* Cambridge University Press, Cambridge, 1938 (last reprinting 1963).

Weber, E. v., " Partielle Differentialgleichungen," pp. 294–399, Vol. 2, Pt. 1 (first half), article A 5 of *Encyklopädie der Mathematischen Wissenschaften*, ed. by H. Burkhardt, W. Wirtinger, and R. Fricke, Teubner, Leipzig, 1899–1916 (article completed in March 1900).

Weitzenböck, R., "Neuere Arbeiten der Algebraischen Invariantentheorie. Differentialinvarianten," pp. 1–71, Vol. 3, Pt. 3, article E 1 of *Encyklopädie der Mathematischen Wissenschaften*, ed. by W. F. Meyer and H. Mohrmann, Teubner, Leipzig, 1927 (article completed in March 1921).

Wempner, G., *Mechanics of Solids,* With Applications to Thin Bodies, McGraw-Hill, New York, 1973.

Westenholz, C. v., *Differential Forms in Mathematical Physics*, rev. ed., North Holland, Amsterdam, 1981.

Weyl, H., *Space, Time, Matter*, 4th ed., Methuen, London, 1922 (reprinted by Dover, New York, 1952; transl. from 4th German ed., of 1921).

Whittaker, E. T., *A Treatise on the Analytical Dynamics of Particles and Rigid Bodies,* With an Introduction to the Problem of Three Bodies, 4th ed., Cambridge University Press, Cambridge, U.K., 1937 (earlier editions: 1904, 1917, 1927; reprinted by Dover, New York, 1944, and by Cambridge University Press, 1944, 1959, 1960, 1961, 1970, 1988).

Willmore, T. J., *An Introduction to Differential Geometry,* Clarendon, Oxford, 1959.

Wills, A. P., *Vector Analysis, with an Introduction to Tensor Analysis,* Prentice-Hall, Englewood Cliffs, NJ, 1931 (reprinted by Dover, New York).

Winogradzki, J., *Les Méthodes Tensorielles de la Physique*, Vol. 1: *Calcul Tensoriel dans un Continuum Amorphe,* 1979, Vol. 2: *Calcul Tensoriel dans un Continuum Structuré,* (1987). Masson, Paris.

Wrede, R. C., *Introduction to Vector and Tensor Analysis,* Wiley, New York, 1963 (reprinted by Dover, New York, 1972).

Wright, J. E., *Invariants of Quadratic Differential Forms,* Cambridge University Press, Cambridge, 1908.

Wundheiler, A., "Ueber die Variationsgleichungen für Affine Geodätische Linien und Nicht Holonome, Nichtconservative Dynamische Systeme," *Pr. Matematyczno-fizyczne* (Warsaw), 38, 129–147, 1931.

Wundheiler, A., "Rheonome Geometrie. Absolute Mechanik," *Pr. Matematyczno-fizyczne* (Warsaw), 40, 97–142, 1932.

Zander, W., "Sätze und Formeln der Mechanik und Elektrotechnik, I. Mechanik," pp. 248–418, Vol. 4 of *Mathematische Hilfsmittel des Ingenieurs*, ed. by R. Sauer and I. Szabó, Springer, Berlin, 1970.

Index

(Principal Authors and Subjects)

D

E

V

W